Management for Professionals

More information about this series at
http://www.springer.com/series/10101

Marcus Schulmerich
Yves-Michel Leporcher
Ching-Hwa Eu

Applied Asset and Risk Management

A Guide to Modern Portfolio Management and Behavior-Driven Markets

Springer

Marcus Schulmerich
State Street Global Advisors (SSgA)
Munich
Germany

Yves-Michel Leporcher
Crédit Agricole
Paris
France

Ching-Hwa Eu
Deutsche Bank
London
U.K.

ISSN 2192-8096 ISSN 2192-810X (electronic)
ISBN 978-3-662-52575-3 ISBN 978-3-642-55444-5 (eBook)
DOI 10.1007/978-3-642-55444-5
Springer Heidelberg New York Dordrecht London

Printed on acid-free paper

Springer is part of Springer Science+Business Media (www.springer.com)

To my family
Marcus Schulmerich

To the only lady I have ever loved
Yves-Michel Leporcher

To my family who supported me throughout my life
Ching-Hwa Eu

Foreword

Since the financial crisis, risk and finance practitioners have been even more keenly aware of the need to combine successfully practical and theoretical knowledge. Finance is an area where there are competing theories of asset pricing and empirical observations that can undermine confidence in our theoretical beliefs. Knowledge is evolving, even though some of the most intractable challenges and market failures appear to repeat themselves with alarming regularity; for example, the problems of spotting bubbles and avoiding market panics.

This book is designed to meet the needs of students and practitioners in finance who wish to advance their knowledge of asset and risk management especially in equities, with a particular focus on bringing together theory and practical applications. It will bring experienced practitioners up to date with recent developments and advances in theory and puts into context behavioral finance, traditional asset pricing, and observations from financial markets such as the flash crash, the Greek crisis, and others. It will help those who wish to understand many different types of stock market anomalies, and why they might arise in a nearly efficient market.

The global financial crisis caused many to reexamine practices in the finance industry and to wonder whether there had been too strict an adherence to theoretical concepts at the expense of a more pragmatic or practical viewpoint. Thankfully, practitioners in the industry have not neglected theory, but have endeavored to build on it and blend it with practice, and by doing so learn from the events of 2007–2009. This book will help in that important job.

Global Chief Investment Officer — Richard Lacaille
State Street Global Advisors (SSgA)
August 2014

Preface

In March 2014, when this book was finished and handed over to the publishers, more than 5 years had passed since the collapse of the U.S. investment bank Lehman Brothers. Yet, the consequences of this bankruptcy, which was the culminating point of the subprime crisis, can still be felt today. And when we look at the current status of the euro crisis, which itself was triggered by the subprime crisis, we find few convincing signs that the financial issues in Europe have really been fundamentally solved.

The cepDefault-Index 2014, published in February 2014 by the cep (Center for European Policy, a think tank), shows that while the creditworthiness of countries like Ireland and Spain has improved, Greece is still far from having regained the trust of the investors.[1] The situation of France has remained unchanged, whereas the credit standing of Italy continues to decline. Surprisingly, for the first time, there are troubling signs of a deteriorating creditworthiness of Belgium and Finland, two core countries of the eurozone.

The euro crisis and the subprime crisis are only the latest crises in a long line of historical financial meltdowns and stock market crashes. However, the frequency of these crashes has become higher and their impact has become more severe in the recent past. Therefore, when we deal with asset and risk management today, these extreme market situations should be considered, as we have endeavored to do in this book.

This book takes a practical look at the rational and irrational aspects of investing. In financial research, these two sides of the coin are represented by modern portfolio theory and behavioral finance. The significance of both was recently highlighted by the decision of the Nobel Prize Committee to award the Nobel Prize in Economics to Eugene Fama, a proponent of a rational view of finance, as well as to Robert Shiller, who follows the approach of behavioral finance.

This book is intended to serve as a comprehensive introduction to asset and risk management for bachelor and master students in this field as well as for young professionals in the asset management industry. In addition, the account of the actual

[1] Gerken and Kullas (2014, p. 1).

investment behavior of investors given in this book may be appreciated also by more senior professionals.

There are two central questions that this book refers to, which were provoked by the worldwide events in the stock markets since the middle of the last decade:

- Why do crashes happen when in theory they should not?
- How do investors deal with such crises in terms of their risk measurement and management and, as a consequence, what are the implications for the chosen investment strategies?

While the first question concerns critical situations that we have all encountered over the last two decades, the second question is much more important to financial investors, be it big institutional investors, like pension funds or sovereign wealth funds, or retail investors, like you and me: how do investors respond to these crises? How do they manage tail risk and prevent drawdowns?

We start, in Chap. 1, with presenting the necessary basic concepts used to measure the risk and the return of a portfolio or a single investment. The mathematical definitions will be introduced using simple math and many illustrations. This is supported by many detailed examples and a step-by-step business case. All calculations are done in Microsoft® Excel®. The Excel® file with the calculations and solutions for all 17 examples as well as all business case calculations can be downloaded at http://www.pecundus.com/publications/springer-solutions.

In Chap. 2 we utilize these concepts to introduce modern portfolio theory. We will present the capital asset pricing model (CAPM) and investigate its validity. Our result is that while CAPM is an elegant theoretical concept, it rarely reflects reality. Its most famous extension, the Fama–French three-factor model, is also not supported by empirical results, as we will show in the final part of Chap. 2.

In Chaps. 3–5 we shift our attention to events and developments which cannot be sufficiently explained by MPT and contradict the concept of the investor as a rational *Homo economicus*. Chapter 3 is devoted to various forms of stock market anomalies. Some are persistent over a very long time and in various asset classes like the turn-of-the-month effect while others are not. Chapter 4 focuses on stock market crashes and looks at crashes with a regional impact and crashes with a global effect. The chapter ends with the analysis of the possibility of a crash in China.

Chapter 5 introduces behavioral finance, a field of research that evolved in the 1970s and includes the psychology of investing. We present the key behavioral biases that seem to have the most significant explanatory power for stock market crashes. At the end of this chapter we look especially into the October 1987 crash and try to find a potential answer to the question why these crashes happen.

After having introduced a lot of concepts and empirical tests concerning the theory and practice of stock markets, the final Chap. 6 answers the second key question of this book: How do investors deal with such crises in terms of their risk measurement and management? Doing so, we distinguish between different EMEA regions and investor types, i.e., institutional investors, retail investors, sovereign wealth funds, and central banks. We describe how investors managed risk in the time before the Lehman demise, i.e., up to the middle of 2008, which risk measures were applied during the crisis and which changes occurred since mid-2009 until early 2014. We take a detailed look at a study of the Economist Intelligence Unit from summer 2012 that shows that drawdown protection was a priority of investors in the years 2011 and 2012 and we discuss the consequences for product development in light of the market situation.

This book would not have been possible without the assistance and support from various people. First of all I want to thank my co-authors Yves-Michel Leporcher and Ching-Hwa Eu, Ph.D., who were ready to join me in this publication project and to contribute their expertise and their share of work. Ching-Hwa Eu produced Chap. 2 and Yves-Michel Leporcher authored Chaps. 3, 4, and 5, while I wrote Chaps. 1 and 6 and was in charge of the general oversight of our work. We developed the book concept together, but each of us is responsible for the content that he has provided.

I am particularly grateful to Christian Theis who doublechecked the whole text and gave valuable input. Many thanks go to Alexander Schmid whose support when finalizing the book was a great aid.

Last but not least, we all wish to thank our wives, partners, and significant others. If it had not been for their ongoing encouragement, we might not have been able to achieve the publication of this book.

Munich, Germany — Marcus Schulmerich
March 2014

Reference

Gerken, L., & Kullas, M. (2014, February). *cepDefault-Index 2014*, cepStudie, cep - Centrum für europäische Politik, Freiburg, Germany. http://www.cep.eu/fileadmin/user_upload/cepDefault-Index/cepStudie_cepDefault-Index_2014.pdf. Accessed 1 Mar 2014.

Contents

Risk Measures in Asset Management 1

1.1 Introduction

Everyone has heard of stock market crashes in the time between 2008 and now, whether he or she followed the events on the financial markets or not. The *subprime crisis*, which unfolded in 2008, has affected every person in almost every country in the world. When we try to explore what happened, as we intend to do in this book, we are immediately drawn to key questions of today's discussions in finance and economics: Why does traditional finance theory fail to explain these crises and why seems behavioral finance, the psychology of investing, better suited to provide an explanatory model?

The first of the two focal questions of this book is, therefore: Why do crashes happen when in theory they should not? The second question is: How have investors reacted to the recent crisis in terms of their risk measurement and management and what are the implications for the chosen investment strategies?

Centered around these two key questions, this book provides a thorough introduction to applied asset and risk measurement in today's markets which are more and more driven by behavioral finance as could be observed in 2008 (a stock market crash) and 2009 (a stock market rally). The six chapters are designed to answer the two questions above by offering a structured introduction to modern portfolio management, behavioral finance and abnormal market behavior. Thereby, this book is ideally suited for bachelor and master students in the field of modern portfolio management as well as for young professionals in the asset and risk management industry.

Chapter 1 presents the mathematical prerequisites to understand the mathematical part of what follows thereafter. While the mathematical description is *not* the focus of this book, formulas cannot be completely avoided as finance, and here in particular asset management, is about performance and risk. Therefore, Chap. 1 presents the measurement of both with many applications. Chapter 2 then explains

M. Schulmerich et al., *Applied Asset and Risk Management*,
Management for Professionals, DOI 10.1007/978-3-642-55444-5_1

the theory of asset management, also known as *modern portfolio theory* (MPT) as introduced in the 1950s by Harry Markowitz. The focus is placed on the capital asset pricing model (CAPM) and its extension, the Fama-French three-factor model. In particular, the key historical tests undertaken in the past 50 years will be presented to see how the theory holds against reality.

The remaining four chapters of this book deal with the reality of investor behavior and stock markets. Chapter 3 provides an overview of stock market anomalies which should not exist according to MPT. However, they *do* exist and have been extensively researched since the early 1980s. Thereafter, Chap. 4 presents a historical survey of the most significant market crashes.

Chapter 5 then returns to the question how crashes can happen. To answer it, the psychology of investors has to be taken into consideration. The research field of behavioral finance and the behavioral biases associated with it will be introduced and used to explain crashes, in particular the crash of 1987.

Finally, Chap. 6 answers the second essential question of this book: How do investors respond to these crises in terms of their risk measurement and management and what are the implications for the chosen investment strategies? Broken down by country and investor type, a market intelligence report covering the period 2004–2014 will be presented.

A changed investor view of risk would have important implications in the way risk is handled by the asset managers and on how new strategies have to be developed. As clients are more and more concerned with the downside of their investment, which is a direct result of the catastrophic stock market performance in 2008, the bad equity returns in 2011 and the ongoing euro crisis, we need to take a closer look on downside risk measures.

Since symmetrical traditional risk measures like volatility or tracking error are not effective in this respect, investors are now increasingly leaning towards asymmetrical risk measures, for example, semi-volatility, drawdown or shortfall risk. Therefore, in the following sections, we present the most important risk and risk-adjusted return measures used in asset management practice.

We start in Sect. 1.2 with looking into measuring investment returns in general, introducing also the notation used throughout this chapter. This section concludes with the key formulas for calculating returns. In Sect. 1.3, we will present traditional measures. At first we look at absolute return portfolio management, i.e., when the portfolio is not managed against any benchmark. Then, we look at relative portfolio management against a certain index as the portfolio's benchmark. The overview of the presented traditional risk and risk-adjusted return measures is listed in Table 1.1.

In Sect. 1.4 we will describe non-traditional measures, especially asymmetrical and downside risk measures. This will be done again for absolute and relative portfolio management. An overview of these non-traditional return, risk and risk-adjusted return measures is given in Table 1.2.

Table 1.1 Examples of traditional risk and risk-adjusted return measures

Absolute portfolio management	Relative portfolio management
• Volatility (Sect. 1.3.1)	• Tracking error (Sect. 1.3.2)
• Sharpe ratio (Sect. 1.3.7)	• Covariance and correlation (Sect. 1.3.4) • Beta (Sect. 1.3.5) • Bull and bear beta (Sect. 1.3.6) • Information ratio (Sect. 1.3.8) • Treynor ratio (Sect. 1.3.9)

Source: Own

Table 1.2 Examples of non-traditional risk and risk-adjusted return measures

Absolute portfolio management	Relative portfolio management
• Maximum absolute drawdown (Sect. 1.4.1)	• Maximum relative drawdown (Sect. 1.4.2) • Semi-deviation and Semi-variance (Sect. 1.4.3) • Shortfall risk/probability (Sect. 1.4.4) • Sortino ratio (Sect. 1.4.5)

Source: Own

Downside risk measures are not new, but until now they were primarily used for portfolios with nonsymmetrical return distributions. Such a pattern is typical for hedge funds, see, for example, Lhabitant (2004). However, after the erratic stock market behavior since early 2008 this kind of return pattern could also be observed in the more traditional part of institutional asset management which of course scared investors.

Therefore, Sect. 1.2 is a necessary first step for the beginner and a brief reminder to the experienced reader. After being familiar with all of these risk and return measures we then turn our attention to modern portfolio theory in Chap. 2 and see, how assets should be managed in theory before we switch to reality in Chaps. 3–6.

1.2 Measuring Investment Returns

1.2.1 Notation

Before we can look at the relevant return and risk measures in detail, we need to introduce a suitable notation. We start with the historical time period, using a notation which has to be consistent for any investment analysis, be it short-term or long-term. Its mathematical expression is the interval $[0, T]$, where T is the end point of the time period, expressed in years. The initial time point (for example, today) is set as $t_0 = 0$. Time point t is always expressed in years or a fraction of years.

If, for example, we want to analyze the upcoming 3 years, we will look at the time period $[0, 3]$ with $T = 3$ years. Time points t and T can be any real positive number, for example, $t = \frac{5}{12}$ is the time point $\frac{5}{12}$ years or 5 months starting from 0.

For analysis purposes, we now divide time interval $[0, T]$ into N (where N can be any positive integer[1]) equidistant subintervals with length $\Delta t = \frac{T}{N}$. The time points are $t_k = k \cdot \Delta t = k \cdot \frac{T}{N}, i = 0, 1, \ldots, N$. For example, if we want to look at the months over the next 3 years we have to split our time period $[0, 3]$ into 36 subintervals with length $\Delta t = \frac{3}{36} = \frac{1}{12}$ years (i.e., 1 month).

Let us summarize the notation:

$$\begin{aligned}
t_0 &= 0 = \text{starting point (today)},\\
T &= \text{time point at the end of the whole period } [0, T],\\
N &= \text{number of time periods in } [0, T],\\
\Delta t &= T/N,\\
t_k &= k \cdot \Delta t \quad = \quad \text{time point } k, 1 \leq k \leq N,\\
[t_{k-1}, t_k] &= \text{subinterval } k, 1 \leq k \leq N.
\end{aligned}$$

Figure 1.1 displays this general framework. In the center of the figure is the time axis, starting on the left with time point $t_0 = 0$ and ending with time point $t_N = T$. Time point t_1 is $\Delta t = \frac{T}{N}$ after t_0, time point t_2 is Δt after t_1 and so on. The next to last time point t_{N-1} is Δt before the end time point t_N. In general, time point $t_k = k \cdot \Delta t$. The top of the figure shows the N equidistant subperiods, and the bottom of the figure illustrates their length, which is $\Delta t = \frac{T}{N}$.

Let us look at our example again with its 3-year time period using monthly data, i.e., $T = 3$ years and $N = 36$ months, yielding $\Delta t = 3/36 = 1/12$ years. Each subperiod is exactly 1 month. If we start our analysis on, say, January 1, 2008, and finish on December 31, 2010, then the first month is January 2008, the second month is February 2008 and the 36th month is December 2010.

This example is displayed in Fig. 1.2. The time axis starts from time point $t_0 = 0$ and ends with time point $T = t_{36} = 3$ years. Time point t_1 is $\frac{1}{12}$ years (or 1 month) after t_0, time point t_2 is $\frac{1}{12}$ years after t_1, etc. Time point t_k is $k \cdot \frac{1}{12}$. Above the time axis are the subperiods, i.e., the different months. Their length Δt which is $\frac{1}{12}$ years is indicated below.

[1]An integer is a number that can be written without a fractional or decimal component. For example, 1, 55, −11, 0, and 1,270 are integers, but 0.25 or 9.76 are not. The name derives from the Latin word *integer*, meaning literally *untouched*, hence *whole*. *Source*: http://www.wikipedia.org.

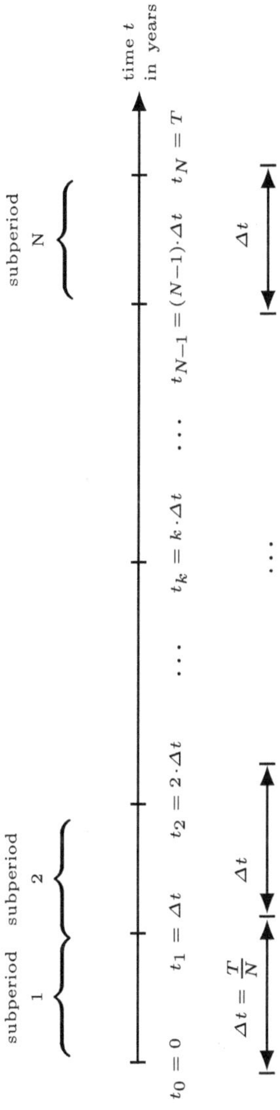

Fig. 1.1 Notation. *Source*: Own, for illustrative purposes only

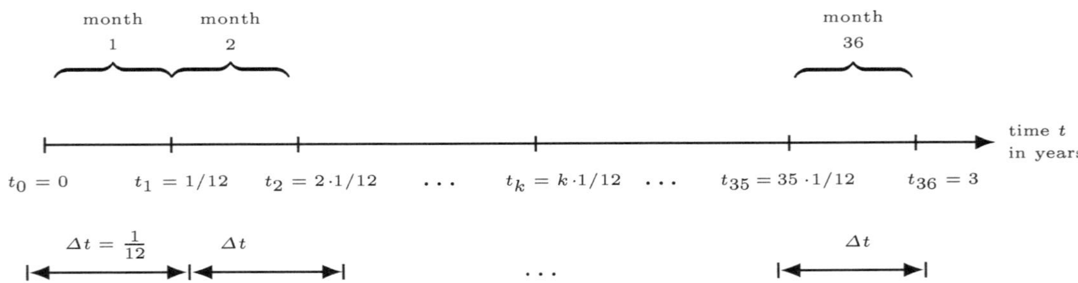

Fig. 1.2 Graphical illustration of our framework example. The 3-year time period [0, 3] is divided in 36 equidistant subintervals of $\frac{1}{12}$ years or 1 month length. *Source*: Own, for illustrative purposes only

1.2.2 Basic Performance Measures

In this section we will introduce the return as a measure to evaluate historic investment performances. At first we will look at the absolute return of an investment.[2] Thereafter, we define the relative return of an investment versus a benchmark.

Definition: Time Value and Absolute Return
For an asset i we make the following definitions:

$$\begin{aligned} V_i^{t_k} &= \text{value of the asset at time point } t_k, \\ V_i^{t_{k-1}} &= \text{value of the asset at time point } t_{k-1}, \\ r_i^k &= \text{percentage change of the asset value in subperiod } k, \\ C_i^{t_k} &= \text{cash amount that the asset pays out at the end} \\ &\quad \text{of the time period } k \text{ (for example, coupon, dividend).} \end{aligned}$$

If no cash flow happens in the subperiod, the asset's *absolute return* in subperiod k is

$$r_i^k = \frac{V_i^{t_k} - V_i^{t_{k-1}}}{V_i^{t_{k-1}}}. \tag{1.1}$$

If the asset pays out a cash amount at the end of subperiod k, the return of the investment is

$$r_i^k = \frac{V_i^{t_k} + C_i^{t_k} - V_i^{t_{k-1}}}{V_i^{t_{k-1}}}. \tag{1.2}$$

In the specific case of V_i^0 being the value of the asset i at time 0 and $V_i^{T,adj}$ being the value at time T, adjusted for stock splits and dividends (i.e., dividends are assumed to be reinvested in the stock), then the *cumulative absolute return* for the entire period $[0, T]$ is r_i^{cum} where:

$$r_i^{cum} = \frac{V_i^{T,adj} - V_i^0}{V_i^0}. \tag{1.3}$$

The cumulative absolute return for the period $[0, T]$ can also be calculated from the absolute returns of all the subperiods, i.e., we continue to look at the return of the portfolio on a stand-alone basis and not relative to a benchmark. We do not

[2] Lhabitant (2004, p. 27). Often, the term *holding period return* is used (Lhabitant (2004)).

distinguish between a single asset and a portfolio but rather we generally talk about the portfolio return. If the portfolio consists of a single asset, the portfolio return is the single asset return.

Definition: Cumulative and Annualized Absolute Return

Using the notation introduced above for time measurement we can now proceed to look at the calculation for a portfolio return (percentage return) over time using the subperiods' percentage returns. For each subinterval k we define:

$$
\begin{aligned}
r_{Pf}^{k} &= \text{return of the portfolio during k-th subinterval } [t_{k-1}, t_k],\\
r_{Bm}^{k} &= \text{return of the benchmark during k-th subinterval } [t_{k-1}, t_k],\\
r_{Pf} &= \text{(cumulative) return of the portfolio in time period } [0, T],\\
r_{Bm} &= \text{(cumulative) return of the benchmark in time period } [0, T],\\
r_{Pf}^{p.a.} &= \text{annualized return of the portfolio in time period } [0, T],\\
r_{Bm}^{p.a.} &= \text{annualized return of the benchmark in time period } [0, T].
\end{aligned}
$$

The cumulative return (percentage return) of the portfolio and the benchmark over time period $[0, T]$ can then be calculated as

$$
\left.
\begin{aligned}
r_{Pf} &= \left(\prod_{k=1}^{N}(1 + r_{Pf}^{k})\right) - 1\\
&\text{and}\\
r_{Bm} &= \left(\prod_{k=1}^{N}(1 + r_{Bm}^{k})\right) - 1.
\end{aligned}
\right\}
\tag{1.4}
$$

If T is at least 1 year we can calculate the annualized percentage return of the portfolio and the benchmark over the time period $[0, T]$ as[3]

$$
r_{Pf}^{p.a.} = (1 + r_{Pf})^{\frac{1}{T}} - 1 \qquad \text{and} \qquad r_{Bm}^{p.a.} = (1 + r_{Bm})^{\frac{1}{T}} - 1. \tag{1.5}
$$

In the specific case that the subintervals are months, fraction $\frac{1}{T}$ in Eq. (1.5) gets replaced by $\frac{12}{N}$ since then $12 \cdot T = N$.

As mentioned before, in the special case that the portfolio is a single asset (i.e., a single asset portfolio), Eqs. (1.4) and (1.5) can also be used to only calculate the

[3]Esch, Kieffer, and Lopez (2005, p. 36).

return of the single asset. Let us now look at the average returns for each of the N subintervals within $[0, T]$.

Definition: Arithmetic and Geometric Mean

Using the notation introduced above, the arithmetic means $\overline{r}_{Pf}$ and $\overline{r}_{Bm}$ of the N subinterval portfolio and benchmark returns are[4]

$$\overline{r}_{Pf} = \frac{1}{N} \sum_{k=1}^{N} r_{Pf}^{k} \quad \text{and} \quad \overline{r}_{Bm} = \frac{1}{N} \sum_{k=1}^{N} r_{Bm}^{k}. \tag{1.6}$$

The arithmetic mean will be used when looking at the standard deviation of returns (i.e., volatility) below. The geometric means for any one of the N subintervals over the time period $[0, T]$ are

$$\hat{r}_{Pf} = (1 + r_{Pf})^{\frac{1}{N}} - 1 \quad \text{and} \quad \hat{r}_{Bm} = (1 + r_{Bm})^{\frac{1}{N}} - 1. \tag{1.7}$$

The key part of this and the next chapter are exercises to further demonstrate the presented concepts. Two types of exercises will be used:

- **Examples:**
  Examples with hypothetical data are used throughout Chap. 1 for illustrative purposes. All calculations will be explained in detail and the Microsoft® Excel® formulas will be provided that were used to generate the result.
- **Business Case:**
  We will use a business case based on real data which will be extended as we progress through Chaps. 1 and 2. The business case starts with data displayed in Table 1.3.

An Excel® file with the calculations and solutions for all 17 examples as well as all business case calculations can be downloaded at http://www.pecundus.com/publications/springer-solutions (username: *solutions*; password: *springer-book-sle*).

Business Case

The business case serves as a large example using historical data on the Lufthansa stock and other assets. We will frequently return to it and expand it over the course of Chaps. 1 and 2.

(continued)

[4]Esch et al. (2005, p. 36).

Table 1.3 Calculation of absolute returns for Lufthansa common stock from time series of end-of-month stock values

i	Time point t_k	Month	Stock value $V_{LHA}^{t_k}$ (in EUR)	Dividend (in EUR)	Absolute return r_{LHA}^{k} (in %, based on EUR)
0	0	Dec 2008	11.19		
1	1/12	Jan 2009	9.50		−15.10
2	2/12	Feb 2009	8.70		−8.42
3	3/12	Mar 2009	8.17		−6.09
4	4/12	Apr 2009	9.66	0.70	26.81
5	5/12	May 2009	9.77		1.14
6	6/12	Jun 2009	8.93		−8.60
7	7/12	Jul 2009	9.48		6.16
8	8/12	Aug 2009	11.20		18.14
9	9/12	Sep 2009	12.11		8.13
10	10/12	Oct 2009	10.50		−13.29
11	11/12	Nov 2009	10.63		1.24
12	1	Dec 2009	11.75		10.54
13	13/12	Jan 2010	11.62		−1.11
14	14/12	Feb 2010	10.96		−5.68
15	15/12	Mar 2010	12.28		12.04

Dividend was paid on April 27, 2009, but for simplicity, we assume that the payment was made at the end of April 2009. The difference is only minor. *Ticker*: LHA.DE. *Source*: Yahoo! Finance

To start, let us look at the price of Lufthansa (LHA) stock over the 15 months in the time period January 2009–March 2010. First, we will calculate the stock's return. Table 1.3 shows the end-of-month values of one Lufthansa (LHA) stock, together with the absolute returns in the respective months, using Eqs. (1.1) and (1.2) in the case of dividends being paid.

Please note that deviations from the stated results can occur due to rounding errors and depending if the input data are truncated in decimal places 3 and higher or not. For example, the cumulative return using the original data without truncating decimals is 17.69 % while it is 17.71 % if only two digits after the decimal point are used.

For the month February 2009, the absolute return of the Lufthansa stock can be calculated as follows:

$$r_{LHA}^{2} = \frac{V_{LHA}^{2/12} - V_{LHA}^{1/12}}{V_{LHA}^{1/12}} = \frac{€8.70 - €9.50}{€9.50} = -8.42\,\%. \qquad (1.8)$$

(continued)

Since the Lufthansa stocks paid a dividend in April 2009, we have to take this into account for the absolute return calculation for that month, using Eq. (1.2):

$$\begin{aligned} r^4_{LHA} &= \frac{V^{4/12}_{LHA} + C^{4/12}_{LHA} - V^{3/12}_{LHA}}{V^{3/12}_{LHA}} \\ &= \frac{€9.66 + €0.70 - €8.17}{€8.17} \\ &= 26.81\,\%. \end{aligned} \tag{1.9}$$

Using Eq. (1.3), the cumulative absolute return for the period from January 2009 to March 2010 is calculated as

$$\begin{aligned} r^{cum}_{LHA} &= (1 + r^1_{LHA}) \cdot (1 + r^2_{LHA}) \cdot \ldots \cdot (1 + r^{15}_{LHA}) - 1 \\ &= (1 - 15.10\,\%) \cdot (1 - 8.42\,\%) \cdot \ldots \cdot (1 + 12.04\,\%) - 1 \\ &= 17.69\,\%. \end{aligned} \tag{1.10}$$

If Lufthansa had not paid any dividends in this period, we could simply calculate the cumulative return from Eq. (1.1) by using the end values and initial values of Lufthansa stock:

$$\begin{aligned} r^{cum}_{LHA} &= \frac{V^{15}_{LHA} - V^0_{LHA}}{V^0_{LHA}} = \frac{€12.28 - €11.19}{€11.19} \\ &= 9.74\,\%. \end{aligned} \tag{1.11}$$

But in the case of a dividend payment, this gives us a *wrong* number! For the cumulative return calculation, we have to account for the payment: We assume that at the end of April 2009, the €0.70 dividend is reinvested in €0.70/€9.66 = 0.07246 stocks, so that by this time, the investor holds 1.07246 shares of Lufthansa.

Hence, we have to adjust the end value of Lufthansa stock by the factor 1.07246:

$$\begin{aligned} V^{15/12,adj}_{LHA} &= 1.07246 \cdot V^{15/12}_{LHA} \\ &= 1.07246 \cdot €12.28 = €13.17. \end{aligned} \tag{1.12}$$

This time, we use Eq. (1.3) to calculate the cumulative return

(continued)

$$r_{LHA}^{cum} = \frac{V_{LHA}^{15/12,adj} - V_{LHA}^{0}}{V_{LHA}^{0}}$$

$$= \frac{€\,13.17 - €\,11.19}{€\,11.19} = 17.69\,\%, \qquad (1.13)$$

and get the same result as in Eq. (1.10). The annualized return is calculated from Eq. (1.5) (with $T = 15/12$):

$$r_{LHA}^{p.a.} = (1 + r_{LHA}^{cum})^{1/T} - 1$$

$$= (1 + 17.69\,\%)^{12/15} - 1 = 13.92\,\%. \qquad (1.14)$$

At the end of this section let us have a look on the return of an investment versus a benchmark.

Definition: Relative Return

When considering a portfolio which is managed against an index as a benchmark, the difference of the portfolio return and the benchmark return is crucial. This *excess return* of the portfolio versus its benchmark is often named *alpha*.[5] With α^k being the alpha in the k-th subperiod, α being the (cumulative) alpha over time period $[0, T], T \geq 1$, and $\alpha^{p.a.}$ being the annualized alpha over this time period, we define:

$$\alpha^k = r_{Pf}^k - r_{Bm}^k \quad \text{and} \quad \alpha^{p.a.} = r_{Pf}^{p.a.} - r_{Bm}^{p.a.}. \qquad (1.15)$$

Example 1

To illustrate the concept, let us use a hypothetical example of the monthly performance of an actively managed portfolio and its benchmark over a period of 18 months, as shown in Table 1.4. In our terminology, we have $T = 1.5$ years, $t_0 = 0$, $N = 18$ and $\Delta t = 1/12$ years.

Like in Microsoft® Excel®, the columns and rows are marked with letters and numbers, respectively. The first column is labeled A (month) and the second B (monthly portfolio performance). Applying Excel® *functions*, the portfolio return, the benchmark return and alpha are calculated as follows below. Please

[5] Chincarini and Kim (2006).

Table 1.4 Example 1: Return calculation

	A	B	C	D
1	Month	Monthly portfolio performance	Monthly benchmark performance	Monthly alpha
2	07/2012	6.10 %	6.01 %	0.09 %
3	08/2012	5.50 %	5.45 %	0.05 %
4	09/2012	4.70 %	4.63 %	0.07 %
5	10/2012	−5.00 %	−6.99 %	1.99 %
6	11/2012	−5.10 %	−4.16 %	−0.94 %
7	12/2012	6.70 %	7.07 %	−0.37 %
8	01/2013	6.03 %	5.97 %	0.06 %
9	02/2013	−3.23 %	−2.95 %	−0.28 %
10	03/2013	5.12 %	4.66 %	0.46 %
11	04/2013	5.21 %	4.91 %	0.30 %
12	05/2013	−4.10 %	−4.01 %	−0.09 %
13	06/2013	−4.50 %	−3.87 %	−0.63 %
14	07/2013	1.75 %	−2.95 %	4.70 %
15	08/2013	3.71 %	4.52 %	−0.81 %
16	09/2013	−4.20 %	−3.93 %	−0.27 %
17	10/2013	4.26 %	4.99 %	−0.73 %
18	11/2013	−4.00 %	−3.84 %	−0.16 %
19	12/2013	5.10 %	4.99 %	0.11 %
20		$r_{Pf} = 24.60\,\%$	$r_{Bm} = 20.15\,\%$	$\alpha = 4.45\,\%$
21		$r_{Pf}^{p.a.} = 15.79\,\%$	$r_{Bm}^{p.a.} = 13.02\,\%$	$\alpha^{p.a.} = 2.77\,\%$

Source: Own, for illustrative purposes only

note that the brackets {...} are generated by hitting the three keys CTRL + SHIFT + RETURN at the same time rather than hitting RETURN alone:

- Cumulative portfolio return r_{Pf} in cell $B20$:

  $$24.60\,\% = \{PRODUCT(1 + B2 : B19) - 1\}$$

- Cumulative benchmark return r_{Bm} in cell $C20$:

  $$20.15\,\% = \{PRODUCT(1 + C2 : C19) - 1\}$$

- Cumulative alpha α in cell $D20$:

  $$4.45\,\% = B20 - C20$$

- Annualized portfolio return $r_{Pf}^{p.a.}$ in cell $B21$:

  $$15.79\,\% = \{PRODUCT(1 + B2 : B19)\hat{}(12/18) - 1\}$$

- Annualized benchmark return $r_{Bm}^{p.a.}$ in cell $C21$:

  $$13.02\,\% = \{PRODUCT(1 + C2 : C19)\hat{}(12/18) - 1\}$$

- Annualized alpha $\alpha^{p.a.}$ in cell $D21$:

$$2.77\,\% \quad = \quad B21 - C21$$

End of Example 1

Until now, we have calculated the historical returns. However, every investor is even more interested in the returns of the future. But while the past return of an asset can always be expressed as a certain number, we do not know the future asset return with certainty and, therefore, treat it as a *random variable*. Using this notion, we introduce the *expected value* to describe

- the return that we can expect to earn despite the uncertainties, i.e., the *expected return* (see the following section)
- the uncertainty in our return forecast, i.e., the *volatility* of the investment's percentage return (see Sect. 1.3.1).

1.2.3 Random Variable and Expected Value

A *random variable*[6] is a quantity whose future outcomes (possible values) are uncertain, but have certain probabilities. In general, we denote a random variable with capital letters, for example, as random variable X.

The absolute return of a portfolio is a random variable. But although the exact return on an investment is unknown, you can make a forecast and estimate the average return. This is the purpose of the *expected value*.

The *expected value*[7] of a random variable X is the probability-weighted average of the possible outcomes of X. It is written as $\mathbb{E}[X]$.

The future asset return is also a random variable. We will use a notation for future returns in analogy to the notation for past returns (see Sect. 1.2.2) and apply the same time period framework introduced in Sect. 1.2.1. For distinction, we use R when describing future returns instead of r. Always remember that R is a random variable while r is a specific number. To use a small letter in Sect. 1.2.2 was appropriate since only past data were analyzed. For the future time period $[0, T]$ and an asset i, we write

- R_i^{cum} for the *cumulative absolute return*,
- $R_i^{p.a.}$ for the *annualized return*,

---

[6]See the non-formal definition of *random variable* in DeFusco, McLeavey, Pinto, and Runkle (2004, p. 232). The book mentioned here is the standard reference for the quantitative part of the CFA program. It offers a good summary of the basic mathematical/statistical methods in finance. This book is helpful as a starting point for those with a non-mathematical background who would like to get some basic understanding in this topic without too much mathematical formality. The material in this book is also useful for the preparation of the CFA exam if you need to review quantitative methods.

[7]DeFusco et al. (2004, p. 194).

both as random variables.

Since the future asset return is a random variable, it has an expected value. When talking about the expected value of returns, we simply use the term *expected return*:

- $\mathbb{E}[R_i^{cum}]$, the *expected cumulative absolute return*, is the cumulative absolute return we expect to earn on average.
- $\mathbb{E}[R_i^{p.a.}]$, the *expected annualized return*, is the annualized return we expect to earn on average.

Example 2

The following hypothetical example in Table 1.5 shows three scenarios (bear, normal, bull market[8]) for an asset i, together with their probabilities and the outcomes for the absolute return R_i^{cum} in a future time period:

The expected return R_i^{cum} of asset i is

$$\mathbb{E}[R_i^{cum}] = 0.20 \cdot (-10\,\%) + 0.50 \cdot 5\,\% + 0.30 \cdot 18\,\% = 5.9\,\%,$$

i.e., the investor should expect to earn 5.9 % on this asset.

End of Example 2

Having covered the return side, we can now look at risk as the other side of the coin. In Sect. 1.3 we will look at traditional risk measures and the so-called risk-adjusted return measures. Risk-adjusted return measures are return measures that also take into consideration the engaged risk. Obviously, of two investments that have the same return over the same time period but different risk, the investor prefers the investment with the least risk. Clearly, the exact meaning of risk needs to be specified.

Thereafter, in Sect. 1.4, we will present the most important risk and risk-adjusted return measures. In each section we have to distinguish if we analyze a portfolio which is managed against a benchmark (relative portfolio management which has an alpha target) or if the portfolio is managed on an absolute return basis (i.e., with an absolute return target).

Table 1.5 Example 2: Calculation of the expected return with three different scenarios for an asset

Scenario	Probability	Absolute return R_i^{cum}
Bear market	0.20	−10 %
Normal market	0.50	5 %
Bull market	0.30	18 %

Source: Own

[8]The use of *bull* and *bear* to describe markets comes from the way the animals attack their opponents. A bull thrusts its horns up into the air while a bear swipes its paws down. These actions are metaphors for the movement of a market. In a bull market, the trend is up, in a bear market, the trend is down. *Source*: http://www.investopedia.com.

1.3 Traditional Risk and Risk-Adjusted Return Measures

1.3.1 Volatility

All investments offer returns, but also carry risk. These are the two sides of the coin. But what is risk and how do we describe risk mathematically? The following list provides a brief overview of financial disasters that affected not only professional investors but also the man on the street. They reflect different kinds of risk:

- The fixed exchange rate system broke down in 1971, leading to flexible and volatile exchange rates.
- The oil-price shocks starting in 1973 were accompanied by high inflation and wild swings in interest rates.
- On Black Monday, October 19, 1987, U.S. stocks collapsed by 23 %.
- In the bond debacle of 1994, the FED, after having kept interest rates low for 3 years, started a series of six consecutive interest rate hikes that erased $1.5 tn. in global capital. A specific example is the Orange County case: Orange County went bankrupt in 1994 due to losses of $1.7 bn. as their treasurer Robert Citron speculated with reverse floaters on decreasing interest rates while rates actually increased dramatically.[9]
- The Russian default in August 1998 sparked a global financial crisis that culminated in a near failure of the big and prominent hedge fund Long Term Capital Management (LTCM).[10]

Here, we will *not* look at all kinds of risk but only at a specific type known as *market price risk*, i.e., the risk of the returns fluctuating around a mean value. In the world of finance, the standard measure of risk is called *volatility*.

The understanding of risk as the variation of investment returns (in percentage) goes back to the definition of risk used by Harry Markowitz in 1952 when he invented what is now known as *modern portfolio theory*.[11] To illustrate this, let us compare the price movements of Lufthansa stock with the gold price in the time period December 2008–March 2010. Figure 1.3 shows the end-of-month prices of Lufthansa (in EUR), Fig. 1.4 the end-of-month prices of one ounce of gold (in EUR).

While the Lufthansa stock price shows stronger fluctuations, the gold price remains very stable. Therefore, at that time, Lufthansa was the riskier investment. The stronger the up-and-down movements of the stock price, the greater the perceived risk, because if you sell your asset at the wrong time, you may end up with a large loss. This is the idea of *volatility*, the standard measure for absolute risk, which measures the strength of the fluctuations.

---

[9]Zagst, Goldbrunner, and Schlosser (2010), *Orange County*, Chap. 3.

[10]Lowenstein (2002).

[11]See Markowitz (1959) for the original English version or Markowitz (2008) for a German translation. A short summary of the key ideas of modern portfolio theory with focus on the concepts rather than on formulas can be found in Schulmerich (2013).

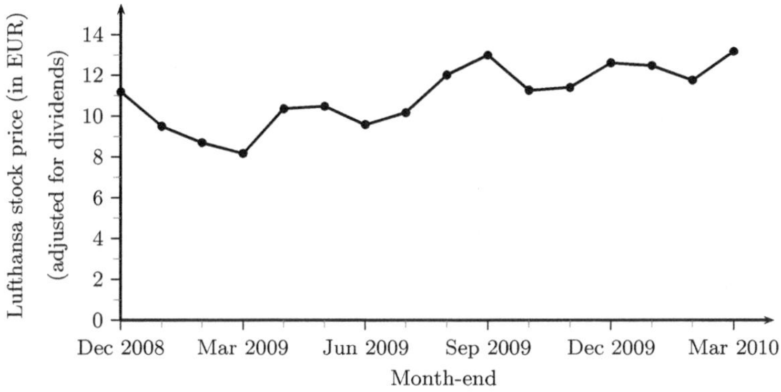

Fig. 1.3 End-of-month prices for Lufthansa stock in the time period December 2008–March 2010, adjusted for dividends, i.e., dividend is assumed to be reinvested in the stock. *Source*: Yahoo! Finance

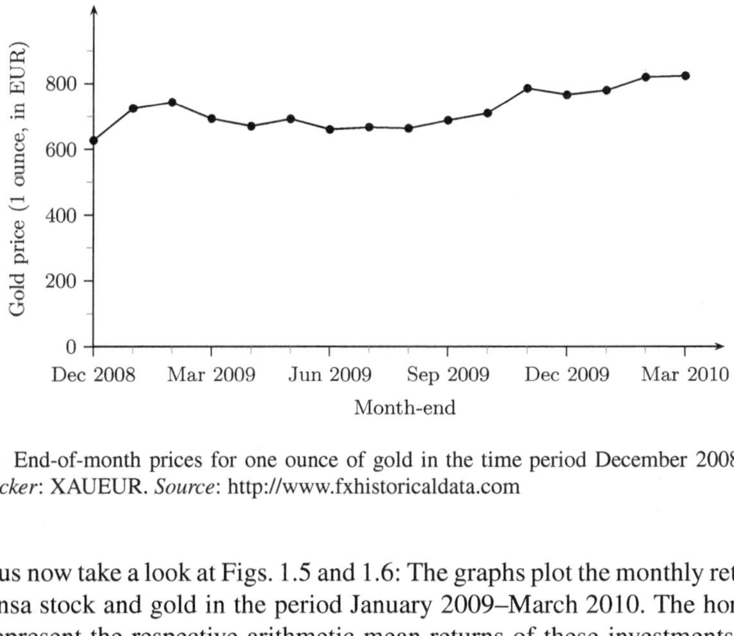

Fig. 1.4 End-of-month prices for one ounce of gold in the time period December 2008–March 2010. *Ticker*: XAUEUR. *Source*: http://www.fxhistoricaldata.com

Let us now take a look at Figs. 1.5 and 1.6: The graphs plot the monthly returns of Lufthansa stock and gold in the period January 2009–March 2010. The horizontal lines represent the respective arithmetic mean returns of these investments in this period. In this plot, the monthly returns of Lufthansa stock which are shown in Fig. 1.5 are much further from the line than the gold returns (see Fig. 1.6). The short-term fluctuations of Lufthansa stock prices are greater.

As a measure for risk, we could use the average distance of the plotted points to the line. But in practice, the expected squared distance is used because it puts a greater weight on individual big outliers besides measuring the overall fluctuations, and this is exactly what the variance and volatility of an investment describes.

Now, we need to measure volatility. This means we have to use the concept of random variables.

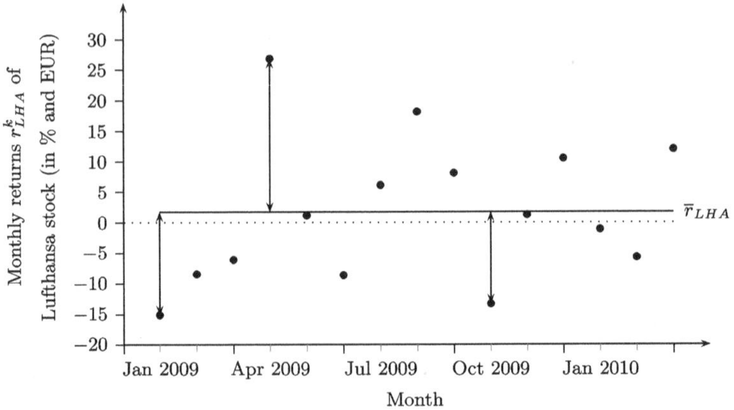

Fig. 1.5 The plot shows the monthly returns of Lufthansa stock in the time period January 2009–March 2010. The *line* represents the arithmetic mean return of Lufthansa stock in this period. *Source*: Yahoo! Finance

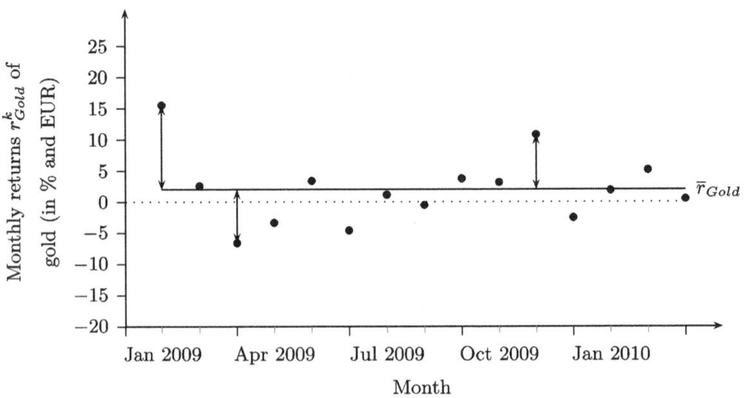

Fig. 1.6 The plot shows the monthly returns of gold (based on EUR) in the time period January 2009–March 2010. The *line* represents the arithmetic mean return of gold in this period. *Source*: http://www.fxhistoricaldata.com

The *variance*[12] $Var(X)$ of a random variable X is

$$Var(X) = \mathbb{E}[(X - \mathbb{E}[X])^2]. \tag{1.16}$$

The *standard deviation*[13] $\sigma(X)$ of a random variable X is defined as

$$\sigma(X) = \sqrt{Var(X)}. \tag{1.17}$$

[12] DeFusco et al. (2004, p. 195).

[13] DeFusco et al. (2004, p. 195).

Now, we apply this concept to the annualized return of an asset i, replacing X with $R_i^{p.a.}$. We call the standard deviation of the annualized return the *annualized volatility* σ_i:

$$\sigma_i = \sigma_i^{p.a.} = \sigma(R_i^{p.a.}) = \sqrt{\mathbb{E}[(R_i^{p.a.} - \mathbb{E}[R_i^{p.a.}])^2]}. \tag{1.18}$$

The variance of the asset returns,

$$Var(R_i) = \mathbb{E}[(R_i - \mathbb{E}[R_i])^2] = \sigma_i^2, \tag{1.19}$$

is also called the *variance* of the asset.

These definitions can also be applied to a portfolio, including single asset portfolios. As before, in the special case of a single asset portfolio we get the formulas above. However, we will now define the portfolio volatility without explicitly using the concept of random variables.

Definition: Volatility and Variance

The volatility σ_{Pf} of a portfolio is an absolute risk measure which evaluates the magnitude of fluctuations of the portfolio's percentage return around its arithmetic mean[14] return $\overline{r}_{Pf}$ within time period $[0, T]$.

When we split time interval $[0, T]$ into N equidistant subintervals, which are usually days or months, then, the volatility is the standard deviation of the subperiod percentage returns $r_{Pf}^1, r_{Pf}^2, \ldots, r_{Pf}^N$ of the portfolio. Depending on the chosen subperiods we get the respective volatility: Daily returns as input lead to a daily volatility, monthly returns lead to a monthly volatility. The monthly volatility can be calculated as[15]

$$\sigma_{Pf}^{monthly} = \sqrt{Var(r_{Pf}^1, r_{Pf}^2, \ldots, r_{Pf}^N)} = \text{Stdev}(r_{Pf}^1, r_{Pf}^2, \ldots, r_{Pf}^N), \tag{1.20}$$

with N representing the number of months in $[0, T]$. Formula (1.20) can easily be modified for using daily returns as input. Then, N is the number of days in $[0, T]$:

$$\sigma_{Pf}^{daily} = \sqrt{Var(r_{Pf}^1, r_{Pf}^2, \ldots, r_{Pf}^N)} = \text{Stdev}(r_{Pf}^1, r_{Pf}^2, \ldots, r_{Pf}^N). \tag{1.21}$$

In practice the volatility can easily be calculated using the percentage subinterval returns of the portfolio within time period $[0, T]$. By using monthly returns, this means mathematically:

(continued)

[14] Also simply called *mean* or *average*.

[15] Esch et al. (2005, p. 41).

$$\sigma_{Pf}^{monthly} = \sqrt{\frac{1}{N-1}\sum_{k=1}^{N}(r_{Pf}^{k} - \overline{r}_{Pf})^2}. \quad (1.22)$$

The variance of the portfolio is then simply the squared volatility, noted by $\sigma_{Pf}^{2,monthly}$ for the monthly and $\sigma_{Pf}^{2,daily}$ for the daily variance.

Since the volatility concept does not include any benchmark, volatility is an absolute risk measure. As the standard deviation from a mean return, it is a symmetrical risk measure which looks at deviations both above and below the average of the subperiod returns. This is not what the layman person would understand as risk. However, it makes sense in the financial world (at least in normal times) as it refers to the risk of withdrawing money at the wrong time, i.e., it is relevant for *market timing decisions*.

It is important to note though that Eq. (1.22) needs some clarification since in practice we calculate volatility based on sample data. Therefore, we first have to discuss the relationship of *sample* vs. *population*.

1.3.1.1 Sample vs. Population

Ideally, when doing a statistical analysis, one would like to have all data points available. For example, if the daily returns of the Dow Jones Industrial Average Index were known since its inception in 1884, then all important key figures, such as arithmetic mean, variance or standard deviation, could be exactly calculated. The totality of data points for a random variable is called *population*. However, in reality, we usually only have access to a *sample* of this population. Furthermore, it is sometimes not meaningful to take all data points into consideration, for example, the standard deviation of the Dow Jones Industrial Average Index since its creation might be of no use since the market activity may have changed over time. Therefore, data of a recent time period is preferred and inferences are made from a sample to the population.

Let us have a look at, for example, the standard deviation. The formula for the population standard deviation of a random variable X with N observations is[16]

$$\sigma = \sqrt{\frac{1}{N}\sum_{i=1}^{N}(X_i - \overline{X})^2} \cdot \quad (1.23)$$

with $\overline{X}$ being the arithmetic mean $\frac{1}{N}\sum_{i=1}^{N} X_i$. When we observe a population and calculate the arithmetic mean, then this is the expected value of X. If you

[16]Compare Esch et al. (2005, p. 41).

Table 1.6 The importance of sample vs. population

	A	B	C
1	Month	Monthly portfolio performance	Monthly benchmark performance
2	07/2012	6.10 %	6.01 %
3	08/2012	5.50 %	5.45 %
4	09/2012	4.70 %	4.63 %

Source: Own, for illustrative purposes only

compare Eq. (1.23) with the sample standard deviation in Eq. (1.22), then the only difference is the factor $\frac{1}{N}$ instead of $\frac{1}{N-1}$.

The reason for this distinction between the factors $\frac{1}{N}$ or $\frac{1}{N-1}$ is the so-called *unbiased estimator*. If the expected value of the sample standard deviation is equal to the population standard deviation, the estimator for the population standard deviation in Eq. (1.22) is unbiased. This is achieved using $\frac{1}{N-1}$ instead of $\frac{1}{N}$ in Eq. (1.22).

In the following, we will illustrate the difference in the formulas for population and sample estimators. Let us first look at Table 1.6 where you find the monthly performance of a hypothetical portfolio and its benchmark for a period of 3 months, i.e., $N = 3$.

Below you see the outcome of the different Excel® formulas either using $.P$ in case the three data points characterize a population and $.S$ if they represent only a sample of a population.

- Population monthly portfolio variance $\sigma_{Pf}^{2,monthly}$:

  $0.003\,\% \quad = \quad VAR.P(B2 : B4)$

- Sample monthly portfolio variance $\sigma_{Pf}^{2,monthly}$:

  $0.005\,\% \quad = \quad VAR.S(B2 : B4)$

- Population monthly portfolio volatility $\sigma_{Pf}^{monthly}$:

  $0.57\,\% \quad = \quad STDEV.P(B2 : B4)$

- Sample monthly portfolio volatility $\sigma_{Pf}^{monthly}$:

  $0.70\,\% \quad = \quad STDEV.S(B2 : B4)$

Obviously, there are differences that cannot be ignored. They are evident, because we look at a population or sample that consists only of three data points. It makes a big difference if one divides by 2 or by 3. Since, as a rule of thumb, the deviations vanish if the number of observations N is greater than 30, the size of available data is crucial. On the other hand, the factors $\frac{1}{N}$ or $\frac{1}{N-1}$ have no impact on the correlation as long as it is computed by using either the population formulas or the sample formulas. This is because in the formula for the correlation, see Eq. (1.47), the terms $\frac{1}{N}$ or $\frac{1}{N-1}$ simply vanish.

Let us look at a concrete example.

Example 3

An analyst follows the stock of company *XYC Ltd.* over a 3-month period. At the beginning of the first month, the stock has a value of € 100. It rises to € 110 at the end of the first month, and to € 122 at the end of the second month. At the end of the third month the stock is priced at € 136.

In order to calculate the volatility using the formula in Eq. (1.22), we first have to calculate the percentage stock returns every month. Then we calculate the arithmetic average of these three percentage returns and plug the results into Eq. (1.22). Knowing that, in practice, only three relative returns are insufficient to get a meaningful volatility value, this example is only for illustrative purposes.

Calculation for the percentage returns r_{XYC}^k for each subperiod $k = 1, 2$ and $3 = N$:

$$r_{XYC}^1 = \frac{€\,110 - €\,100}{€\,100} = 10\,\%$$

$$r_{XYC}^2 = \frac{€\,122 - €\,110}{€\,110} = 10.91\,\%$$

$$r_{XYC}^3 = \frac{€\,136 - €\,122}{€\,122} = 11.48\,\%.$$

Calculation of the arithmetic average of the percentage returns in the three subperiods:

$$\begin{aligned}\overline{r}_{XYC} &= \frac{r_{XYC}^1 + r_{XYC}^2 + r_{XYC}^3}{N} \\ &= \frac{10\,\% + 10.91\,\% + 11.48\,\%}{3} \\ &= 10.80\,\%.\end{aligned}$$

According to Eq. (1.22) the stock's monthly volatility $\sigma_{XYC}^{monthly}$ is then

$$\begin{aligned}\sigma_{XYC}^{monthly} &\overset{(1.22)}{=} \sqrt{\frac{1}{N-1}\sum_{k=1}^{3}(r_{XYC}^k - \overline{r}_{XYC})^2} \\ &= \sqrt{\begin{matrix}\frac{1}{3-1}[(10\,\% - 10.80\,\%)^2 + (10.91\,\% - 10.80\,\%)^2 + \\ +(11.48\,\% - 10.80\,\%)^2]\end{matrix}} \\ &= 0.75\,\%.\end{aligned}$$

End of Example 3

In practice, neither a monthly nor a daily volatility is used, rather an annual volatility. This would call for annual returns over non-overlapping consecutive years and the time series would need to be sufficiently long, for example, 20 years, in order to get a meaningful volatility figure. However, this is not practicable: Over such a long time, the management of a fund will most likely have changed or the fund will have existed only for a much shorter time period. But to calculate an annual volatility using only a few years of returns is meaningless. This brings us to the question of how to calculate an *annualized* volatility using less data.

In order to do this we use daily or monthly return data and *annualize* them. At least 1 year's worth of data are required, i.e., $T \geq 1$ year. For example, we look at a historical 3-year period using either 36 monthly percentage returns or (roughly) 750 daily percentage returns.

Let us assume we have $N \geq 12$ monthly percentage returns. Using this data in Eq. (1.22) yields the monthly volatility $\sigma_{Pf}^{monthly}$. We now can annualize this monthly volatility which yields the annualized volatility $\sigma_{Pf}^{p.a.}$ by scaling $\sigma_{Pf}^{monthly}$ with the square root of 12, i.e.:

$$\sigma_{Pf}^{p.a.} = \sqrt{12} \cdot \sigma_{Pf}^{monthly}. \tag{1.24}$$

Similarly, if daily percentage return data are used, the calculated daily volatility σ_{Pf}^{daily} can be annualized by scaling it with the square root of business days per year which is roughly 252, i.e.:

$$\sigma_{Pf}^{p.a.} = \sqrt{252} \cdot \sigma_{Pf}^{daily}. \tag{1.25}$$

For the interpretation of volatility we have to turn to a research field called *financial engineering*. Assuming that the portfolio value follows a so-called *Geometric Brownian Motion* (GBM), the percentage returns of the portfolio in the subperiods are normally distributed.[17] Accordingly, we assume that for a future 1-year time period $[0, 1]$ the annualized return $R_i^{p.a.}$, as a random variable, is normally distributed. The interpretation of the annualized volatility $\sigma_i^{p.a.}$ of an asset i, as shown in Fig. 1.7, is then as follows:

Let $\mu_i = \mathbb{E}[R_i^{p.a.}]$ be the expected annualized return and $\sigma_i^{p.a.}$ its annualized volatility. Then the probability distribution for the annualized returns $R_i^{p.a.}$ is described by a symmetrical bell-shaped curve which peaks at $R_i^{p.a.} = \mu_i$. In this figure, the probability of the return being between two return values is the stated area under the curve (the entire area under the graph is 1). The percentages in the diagram show the area under the graph in the respective intervals, for example, the probability for the return to lie between μ_i and $\mu_i + \sigma_i^{p.a.}$ is 34.13 %. Thus, the probability of the annualized return $R_i^{p.a.}$ to be

- at most $\sigma_i^{p.a.}$ off from the expected annualized return μ_i is 68.3 %.
- at most $2 \cdot \sigma_i^{p.a.}$ off from the expected annualized return μ_i is 95.4 %.
- at most $3 \cdot \sigma_i^{p.a.}$ off from the expected annualized return μ_i is 99.7 %.

[17] Please note that no further mathematical analysis on this will be provided here. For further details, please see the basic introductions in Schulmerich (2010a, Chap. 2), Schulmerich (2005, Chap. 2), Schulmerich (2008a, Chaps. 1 and 2) or Schulmerich (1997, Chaps. 1 and 2). For a rigorous mathematical introduction to this topic see, for example, Neftci (2000) or Øksendal (1995).

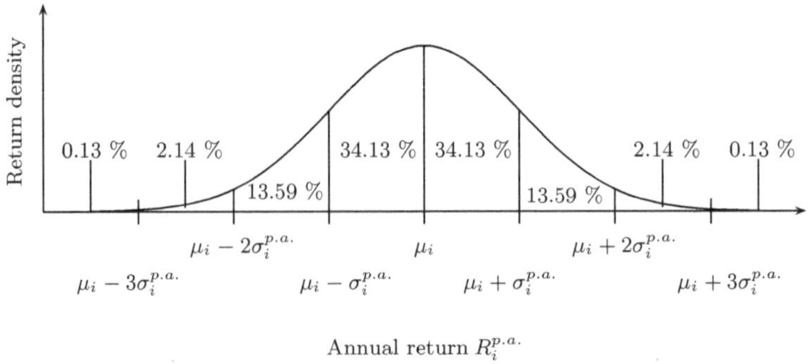

Fig. 1.7 Graphical illustration of annualized volatility for normally distributed returns. *Source*: Reilly and Brown (1997, Appendix D, p. 1047)

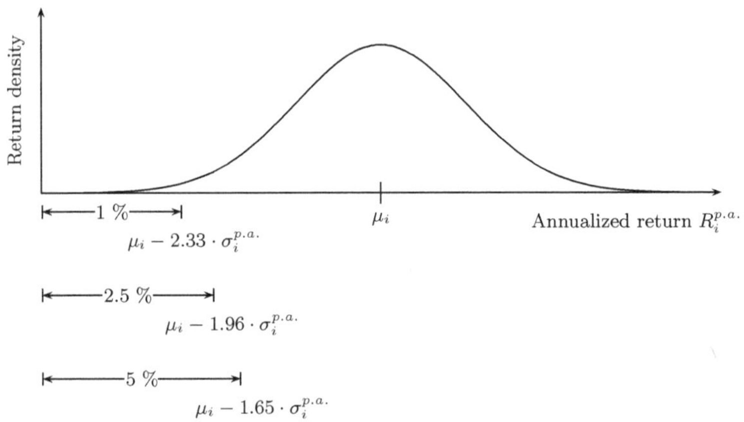

Fig. 1.8 Graphical illustration of the worst-case scenarios (1, 2.5, 5 %) for normally distributed returns. *Source*: Reilly and Brown (1997, Appendix D, p. 1047)

Volatility can also be used to illustrate worst-case scenarios, as shown in Fig. 1.8.

Typically investors look at the worst 5, 2.5 and 1 % outcomes when they want to assess risk. The lines below the return axis show the left tails which make up 5, 2.5 and 1 % of the total area under the graph, corresponding to our worst-case scenarios.

The probability that the annualized return $R_i^{p.a.}$ is less than

- the expected annualized return μ_i minus 1.65 times the annualized volatility $\sigma_i^{p.a.}$ is 5.0 %.
- the expected annualized return μ_i minus 1.96 times the annualized volatility $\sigma_i^{p.a.}$ is 2.5 %.
- the expected annualized return μ_i minus 2.33 times the annualized volatility $\sigma_i^{p.a.}$ is 1.0 %.

Let us now look at a practical example for calculating volatility.

Example 4

In Table 1.7 the monthly and annualized volatility is calculated based on the percentage returns of a portfolio and its benchmark for 18 consecutive months. The return data are the same as shown in Table 1.4 for Example 1. Using the column and row notation from Microsoft® Excel®, the first column is labeled *A* (month) and the second *B* (monthly portfolio performance). Applying Excel® *functions*, the volatility of the portfolio can easily be calculated:

- Monthly portfolio volatility $\sigma_{Pf}^{monthly}$ in cell *B*20:

  $4.76\,\% \;=\; STDEV.S(B2 : B19)$

- Annualized portfolio volatility $\sigma_{Pf}^{p.a.}$ in cell *B*21:

  $16.49\,\% \;=\; SQRT(12) * STDEV.S(B2 : B19)$

Table 1.7 Example 4: Volatility calculation

	A	B	C	D
1	Month	Monthly portfolio performance	Monthly benchmark performance	Monthly alpha
2	07/2012	6.10 %	6.01 %	0.09 %
3	08/2012	5.50 %	5.45 %	0.05 %
4	09/2012	4.70 %	4.63 %	0.07 %
5	10/2012	−5.00 %	−6.99 %	1.99 %
6	11/2012	−5.10 %	−4.16 %	−0.94 %
7	12/2012	6.70 %	7.07 %	−0.37 %
8	01/2013	6.03 %	5.97 %	0.06 %
9	02/2013	−3.23 %	−2.95 %	−0.28 %
10	03/2013	5.12 %	4.66 %	0.46 %
11	04/2013	5.21 %	4.91 %	0.30 %
12	05/2013	−4.10 %	−4.01 %	−0.09 %
13	06/2013	−4.50 %	−3.87 %	−0.63 %
14	07/2013	1.75 %	−2.95 %	4.70 %
15	08/2013	3.71 %	4.52 %	−0.81 %
16	09/2013	−4.20 %	−3.93 %	−0.27 %
17	10/2013	4.26 %	4.99 %	−0.73 %
18	11/2013	−4.00 %	−3.84 %	−0.16 %
19	12/2013	5.10 %	4.99 %	0.11 %
20		$\sigma_{Pf}^{monthly} = 4.76\,\%$	$\sigma_{Bm}^{monthly} = 4.91\,\%$	
21		$\sigma_{Pf}^{p.a.} = 16.49\,\%$	$\sigma_{Bm}^{p.a.} = 17.02\,\%$	

Source: Own, for illustrative purposes only

- Monthly benchmark volatility $\sigma_{Bm}^{monthly}$ in cell $C20$:
  $4.91\,\% \quad = \quad STDEV.S(C2:C19)$
- Annualized benchmark volatility $\sigma_{Bm}^{p.a.}$ in cell $C21$:
  $17.02\,\% \quad = \quad SQRT(12) * STDEV.S(C2:C19)$

Using the interpretation of a normal distribution in Fig. 1.7 this means: With a probability of 68 %, the 1-year percentage return of our portfolio lies between $\mu_{Pf} - 16.49\,\%$ and $\mu_{Pf} + 16.49\,\%$ with μ_{Pf} being the average of the historical yearly portfolio returns.

Using $\mu_{Pf} = 15.79\,\%$ as calculated in Example 1 (same data series as in Example 4), there is a probability of 68 % that the 1-year percentage return of our portfolio lies in the interval $[-0.70\,\%, 32.28\,\%]$.

End of Example 4

We will now return to our business case and expand it to include gold. For our calculations we will use Table 1.8 which shows the end-of-month gold prices in EUR together with the monthly absolute returns (in percentages) in the period January 2009–March 2010.

Table 1.8 Calculation of absolute returns for one ounce of gold from time series of end-of-month stock values

k	Time point t_k	Month	Gold value (one ounce) $V_{Gold}^{t_k}$ (in EUR)	Absolute return $r_{Gold}^{t_k}$ (in %)
0	0	Dec 2008	627.10	
1	1/12	Jan 2009	724.18	15.48
2	2/12	Feb 2009	742.54	2.54
3	3/12	Mar 2009	693.13	−6.65
4	4/12	Apr 2009	669.67	−3.38
5	5/12	May 2009	692.11	3.35
6	6/12	Jun 2009	659.94	−4.65
7	7/12	Jul 2009	667.30	1.12
8	8/12	Aug 2009	663.53	−0.56
9	9/12	Sep 2009	688.26	3.73
10	10/12	Oct 2009	709.76	3.12
11	11/12	Nov 2009	786.08	10.75
12	1	Dec 2009	765.89	−2.57
13	13/12	Jan 2010	780.06	1.85
14	14/12	Feb 2010	820.11	5.13
15	15/12	Mar 2010	824.08	0.48

Ticker: XAUEUR. *Source*: http://www.fxhistoricaldata.com

Business Case (cont.)
In our business case, the annualized volatility for Lufthansa stock is 41.26 %, whereas the annualized volatility for gold is 19.71 %, implying that the Lufthansa stock fluctuated roughly twice as much as gold. These quantities, which show the greater risk of Lufthansa stock, will be calculated below [Eqs. (1.28) and (1.31)]. Let us first calculate the volatility for Lufthansa based on the monthly returns in the period January 2009–March 2010 from Table 1.3. First, we start with the arithmetic mean monthly return:

$$\begin{aligned}\overline{r}_{LHA} &= \frac{1}{15}\sum_{k=1}^{15} r_{LHA}^{k} \\ &= \frac{1}{15}(-15.10\,\% + (-8.42\,\%) + \ldots + 12.04\,\%) \\ &= 1.73\,\%. \end{aligned} \tag{1.26}$$

Then, we use Eq. (1.22) to calculate the volatility ($\sigma_{LHA}^{monthly}$ in our example):

$$\begin{aligned}\sigma_{LHA}^{monthly} &= \sqrt{\frac{1}{14}\sum_{k=1}^{15}(r_{LHA}^{k} - \overline{r}_{LHA})^2} \\ &= \sqrt{\frac{1}{14}[(-15.10\,\% - 1.73\,\%)^2 + \ldots + (12.04\,\% - 1.73\,\%)^2]} \\ &= 11.91\,\%. \end{aligned} \tag{1.27}$$

To obtain the annualized volatility, multiply the monthly volatility by the factor $\sqrt{12}$, see Eq. (1.24):

$$\sigma_{LHA}^{p.a.} = \sqrt{12}\cdot\sigma_{LHA}^{monthly} = \sqrt{12}\cdot 11.91\,\% = 41.26\,\%. \tag{1.28}$$

Let us now calculate the volatility for one ounce of gold based on the monthly returns in the period January 2009–March 2010 from Table 1.8. First, we start with the arithmetic mean monthly return:

$$\begin{aligned}\overline{r}_{Gold} &= \frac{1}{15}\sum_{k=1}^{15} r_{Gold}^{k} = \frac{1}{15}(15.48\,\% + (2.54\,\%) + \ldots + 0.48\,\%) \\ &= 1.98\,\%. \end{aligned} \tag{1.29}$$

Then, we use Eq. (1.22) to calculate the volatility ($\sigma_{Gold}^{monthly}$ in our example):

(continued)

$$\begin{aligned}
\sigma_{Gold}^{monthly} &= \sqrt{\frac{1}{14}\sum_{k=1}^{15}(r_{Gold}^{k} - \bar{r}_{Gold})^2} \\
&= \sqrt{\frac{1}{14}[(15.48\,\% - 1.98\,\%)^2 + \ldots + (0.48\,\% - 1.98\,\%)^2]} \\
&= 5.69\,\%. \qquad (1.30)
\end{aligned}$$

To obtain the annualized volatility, multiply the monthly volatility by the factor $\sqrt{12}$, see Eq. (1.24):

$$\sigma_{Gold}^{p.a.} = \sqrt{12} \cdot \sigma_{Gold}^{monthly} = \sqrt{12} \cdot 5.69\,\% = 19.71\,\%. \qquad (1.31)$$

To illustrate the volatility, we need to calculate the annualized return of gold. The cumulative return is [using Eq. (1.3)]

$$\begin{aligned}
r_{Gold}^{cum} &= \frac{V_{Gold}^{15/12} - V_{Gold}^{0}}{V_{Gold}^{0}} \\
&= \frac{€824.08 - €627.10}{€627.10} \\
&= 31.41\,\%, \qquad (1.32)
\end{aligned}$$

and the annualized return is [using Eq. (1.5), with $T = 15/12$]

$$\begin{aligned}
r_{Gold}^{p.a.} &= (1 + r_{Gold}^{cum})^{1/T} - 1 \\
&= (1 + 31.41\,\%)^{12/15} - 1 \\
&= 24.42\,\%. \qquad (1.33)
\end{aligned}$$

Based on our results the

- annualized return for Lufthansa stocks is 13.9 % [see Eq. (1.14)] and the annualized volatility is 41.3 % [Eq. (1.28)].
- annualized return for gold is 24.4 % [Eq. (1.33)] and the annualized volatility is 19.7 % [Eq. (1.31)].

Assuming that the expected return equals the past annualized return, we can illustrate the volatilities in Figs. 1.9 and 1.10, based on the general pictures of Figs. 1.7 and 1.8.

(continued)

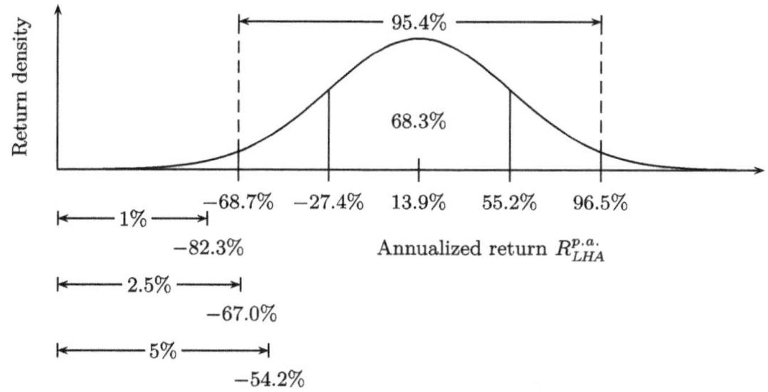

Fig. 1.9 Graphical illustration of annualized volatility for Lufthansa stock, assuming normally distributed returns. *Sources*: Reilly and Brown (1997, Appendix D, p. 1047), Yahoo! Finance

The graph shows that for Lufthansa stock, the probability that

- the annualized return lies between −27.4 and 55.2 % is 68.3 %.
- the annualized return lies between −68.6 and 96.4 % is 95.4 %.

For gold, the probability that

- the annualized return lies between 4.7 and 44.1 % is 68.3 %.
- the annualized return lies between −15.0 and 63.8 % is 95.4 %.

Comparing Figs. 1.9 and 1.10 we see that because of the smaller volatility the ranges of the annual returns are narrower for gold.

Let us compare the left tails: For Lufthansa stock, the expected return is

- less than −54.2 % with a probability of 5 %.
- less than −66.9 % with a probability of 2.5 %.
- less than −82.2 % with a probability of 1 %.

For gold, the expected return is

- less than −8.1 % with a probability of 5 %.
- less than −14.2 % with a probability of 2.5 %.
- less than −21.5 % with a probability of 1 %.

The lower volatility of gold has the effect that the left tail risks are very low compared to Lufthansa.

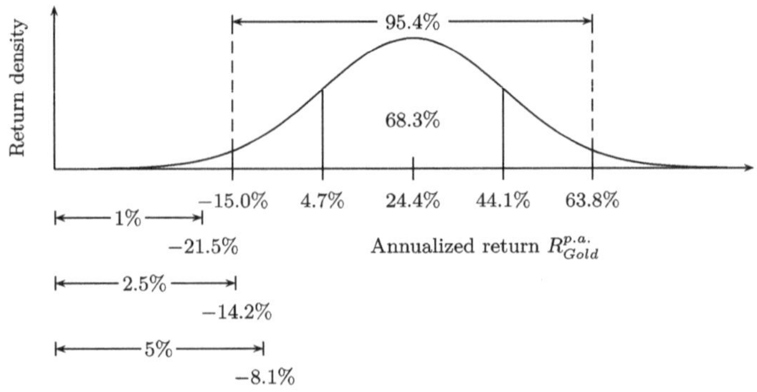

Fig. 1.10 Graphical illustration of annualized volatility for gold (in EUR), assuming normally distributed returns. *Sources*: Reilly and Brown (1997, Appendix D, p. 1047), http://www.fxhistoricaldata.com

1.3.1.2 Conclusion

The volatility, i.e., the standard deviation of the subperiods' portfolio percentage returns, is the traditional risk measure of a portfolio. Since there is no benchmark needed in the calculation, this risk measure can also be called an *absolute risk measure*. As it is a standard deviation and, therefore, considers positive and negative deviations from the mean in the same way, it is a typical *symmetrical* risk measure.

In active portfolio management both the portfolio and the benchmark volatility are usually calculated on a stand-alone basis and then compared with each other. Active portfolio management ideally achieves an excess return over the benchmark with the same level of volatility as the benchmark and not with a higher level. We will look more into the relationship of risk and return for active portfolio management once we have introduced an appropriate measure for risk versus a benchmark.

Up to now, we have only considered portfolios that are not measured against an index and dealt with *absolute return portfolio management* which is typical for hedge funds. In traditional institutional asset management, a portfolio is, however, in most of the cases managed against an index as a benchmark and the portfolio manager aims to beat this benchmark by generating a higher return. This calls for another risk measure.

Therefore, in the next step, we will introduce a risk measure that is benchmark-oriented: Managing a portfolio versus the benchmark with a relative return (i.e., alpha) target, the corresponding risk has to be relative, i.e., benchmark-oriented, as well. This relative risk is defined in line with volatility and is called *tracking error*.

1.3.2 Tracking Error

Tracking error (*TE*) is one of the most commonly used relative risk measures in active management.[18] Its definition is in line with the definition of volatility. However, *TE* does not use an absolute return time series but a relative return time series, i.e., a time series of alphas.

Tracking error is *the* typical risk measure used when managing a portfolio versus an index as a benchmark. Therefore, *TE* is a relative risk measure, similarly constructed like its absolute counterpart volatility. Like volatility, tracking error is a symmetrical risk measure since it looks at both deviations above and below the average of the subperiod alphas.

Definition: Tracking Error
Like in the previous sections, we again split time interval $[0, T]$ into N equidistant subintervals which are usually days or months. Let us denote the subperiod percentage returns of the portfolio as always with

$$r_{Pf}^1,\ r_{Pf}^2,\ \ldots\ ,\ r_{Pf}^N,$$

thc subpcriod percentage returns of the benchmark with

$$r_{Bm}^1,\ r_{Bm}^2,\ \ldots\ ,\ r_{Bm}^N,$$

and the alpha for the k-th subperiod with $\alpha^k, 1 \leq k \leq N$.

Then, the tracking error of the portfolio versus its benchmark is the standard deviation of the subperiod percentage alpha returns $\alpha^1, \alpha^2, \ldots, \alpha^N$.

Depending on the chosen subperiods, we get the respective tracking error: daily alphas as input lead to a daily tracking error. Monthly alphas lead to a monthly tracking error. Mathematically, the monthly tracking error can be calculated as[19]

$$TE^{monthly} = \sqrt{Var(\alpha^1, \alpha^2, \ldots, \alpha^N)} = \text{Stdev}(\alpha^1, \alpha^2, \ldots, \alpha^N) \qquad (1.34)$$

with N representing the number of months in $[0, T]$. Formula (1.34) can easily be modified for using daily returns as input. Then, N is the number of days in $[0, T]$:

$$TE^{daily} = \sqrt{Var(\alpha^1, \alpha^2, \ldots, \alpha^N)} = \text{Stdev}(\alpha^1, \alpha^2, \ldots, \alpha^N). \qquad (1.35)$$

(continued)

[18] Lhabitant (2004, p. 59).

[19] Lhabitant (2004, p. 59).

In practice the tracking error can easily be calculated using the subperiod alphas within time period $[0, T]$. By using monthly returns, this means mathematically:

$$\alpha^{monthly} = \sqrt{\frac{1}{N-1}\sum_{k=1}^{N}(\alpha^k - \overline{\alpha})^2}. \tag{1.36}$$

1.3.2.1 Note

The name *tracking error* stems from passive portfolio management where the goal is to achieve an alpha of zero. If, for example, for each month alpha is zero, then the alpha of the whole period is of course zero as well. If, however, one or more of the monthly alphas are different from zero it is most likely that over the whole time period alpha is different from zero as well. This deviation for the whole period can be measured with the standard deviation of each subperiod's alpha, i.e., tracking error.[20] More specifically:

- Tracking error quantifies the *error* made by the passive portfolio manager when tracking the benchmark.
- A perfect passive portfolio manager generates an alpha of zero in each subperiod, i.e., $\alpha^k = 0$ for $1 \leq k \leq N$. This automatically leads to $TE = 0$.
- The less perfect the passive portfolio manager, the higher TE and the more the portfolio manager will deviate from his overall alpha target of zero.

It is important for tracking error calculation that the type of input determines the interpretation of the output. If the alphas used as input to calculate tracking error are monthly alphas in % over the last 3 years, the calculated tracking error is a monthly tracking error and also in %.

As with volatility, a daily or monthly tracking error is not very meaningful. Only an annualized tracking error derived from a daily or monthly tracking error is used. This annualized tracking error can then be compared to an annualized alpha over the same time interval $[0, T]$. To annualize a tracking error, T has to be at least 1 year, i.e., $T \geq 1$.

In order to annualize a monthly tracking error $TE^{monthly}$, we have to scale it by the square root of 12, i.e.:

$$TE^{p.a.} = \sqrt{12} \cdot TE^{monthly}. \tag{1.37}$$

Similarly, if a daily tracking error is given, this has to be scaled with the square root of the number of business days per year which is roughly 252, i.e.:

[20] Natenberg (1994, pp. 60–61).

$$TE^{p.a.} = \sqrt{252} \cdot TE^{daily}. \tag{1.38}$$

The reasons for the scaling in Eqs. (1.37) and (1.38) are (like in the case of annualizing a volatility) rooted in financial engineering and will not be discussed further here. As a standard deviation, the tracking error can never be negative. And, if we assume a normal distribution for subperiod alphas, as we did when interpreting volatility, the standard deviation concept allows a neat interpretation of the tracking error.

1.3.2.2 Interpretation

For interpretation purposes, we assume that the future annualized alpha $\alpha^{p.a}$ is a normally distributed random variable. Let then $TE^{p.a.}$ be the corresponding annualized tracking error. Then the probability distribution for the annualized alpha $\alpha^{p.a}$ is described by a symmetrical bell-shaped curve which peaks at $\mathbb{E}[\alpha^{p.a.}]$. This expected value can be interpreted as the p.a. alpha target of the actively managed portfolio versus its benchmark, see Fig. 1.11:

In Fig. 1.11, the probability of the return being in the interval $[\alpha', \alpha'']$ is the area under the curve between $\alpha^{p.a.} = \alpha'$ and $\alpha^{p.a.} = \alpha''$ (the entire area under the graph is 1). The percentages in the diagram show the area under the graph in the respective intervals, for example, the probability for the return to lie between $\alpha' = \mathbb{E}[\alpha^{p.a.}]$ and $\alpha'' = \mathbb{E}[\alpha^{p.a.}] + TE^{p.a.}$ is 34.13 %.

Thus, the probability of the annualized return $\mathbb{E}[\alpha^{p.a.}]$ to be

- at most $TE^{p.a.}$ off from the expected annualized return $\mathbb{E}[\alpha^{p.a.}]$ is 68.3 %,
- at most $2 \cdot TE^{p.a.}$ off from the expected annualized return $\mathbb{E}[\alpha^{p.a.}]$ is 95.4 %,
- at most $3 \cdot TE^{p.a.}$ off from the expected annualized return $\mathbb{E}[\alpha^{p.a.}]$ is 99.7 %.

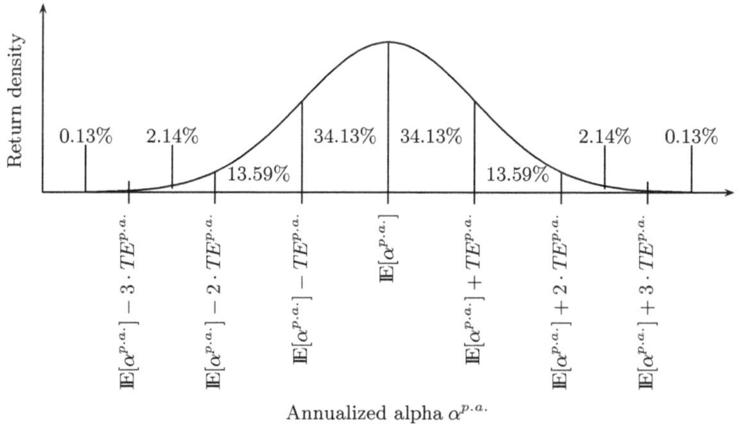

Fig. 1.11 Graphical interpretation of the tracking error: normal distribution of the annualized alpha for a portfolio managed against a benchmark with alpha target $\mathbb{E}[\alpha^{p.a.}]$ and tracking error $TE^{p.a.}$. *Source*: Own

Table 1.9 Example 5: Tracking error calculation

	A	B	C	D
1	Month	Monthly portfolio performance	Monthly benchmark performance	Monthly alpha
2	07/2012	6.10 %	6.01 %	0.09 %
3	08/2012	5.50 %	5.45 %	0.05 %
4	09/2012	4.70 %	4.63 %	0.07 %
5	10/2012	−5.00 %	−6.99 %	1.99 %
6	11/2012	−5.10 %	−4.16 %	−0.94 %
7	12/2012	6.70 %	7.07 %	−0.37 %
8	01/2013	6.03 %	5.97 %	0.06 %
9	02/2013	−3.23 %	−2.95 %	−0.28 %
10	03/2013	5.12 %	4.66 %	0.46 %
11	04/2013	5.21 %	4.91 %	0.30 %
12	05/2013	−4.10 %	−4.01 %	−0.09 %
13	06/2013	−4.50 %	−3.87 %	−0.63 %
14	07/2013	1.75 %	−2.95 %	4.70 %
15	08/2013	3.71 %	4.52 %	−0.81 %
16	09/2013	−4.20 %	−3.93 %	−0.27 %
17	10/2013	4.26 %	4.99 %	−0.73 %
18	11/2013	−4.00 %	−3.84 %	−0.16 %
19	12/2013	5.10 %	4.99 %	0.11 %
20			$TE^{monthly} =$	1.29 %
21			$TE^{p.a.} =$	4.48 %

Source: Own, for illustrative purposes only

Example 5

Let us now take a look at the example of a fund and its benchmark. Table 1.9 shows the monthly performance of both and the resulting alpha over a time period of 18 months, i.e., $T = 1.5$ years and $N = 18$ months.

Using Eqs. (1.34) and (1.37) to calculate the monthly and annualized tracking error over time period $[0, T]$, respectively, we need the following functions in cells $D20$ and $D21$:

- Monthly tracking error $TE^{monthly}$ in cell $D20$:

  $$1.29\,\% \quad = \quad STDEV.S(D2 : D19)$$

- Annualized tracking error $TE^{p.a.}$ in cell $D21$:

  $$4.48\,\% \quad = \quad SQRT(12) * STDEV.S(D2 : D19)$$

End of Example 5

1.3.2.3 Conclusion

Tracking error is *the* risk measure in benchmark-oriented portfolio management. As a relative risk measure, it corresponds in its definition and interpretation to volatility as an absolute risk measure in benchmark-agnostic portfolio management. Both risk measures are symmetrical, and their interpretation which is based on the mathematical concept of confidence intervals is only valid, if the underlying data (either an absolute return or a relative return time series) are normally distributed. This especially means that tracking error (as well as volatility) treats positive and negative deviations in the same way.

1.3.3 Relationship of Tracking Error and Alpha

As already mentioned above, the interpretation of *TE* depends on the underlying relative performance data: daily alphas result in a daily *TE* and monthly alphas result in a monthly *TE*. Irrespective of the underlying data, it can be concluded that the higher *TE*, the more volatile the daily or monthly alphas and the further the portfolio performance will deviate from the benchmark. While this is not wanted in passive portfolio management (hence, *TE* and alpha targets are zero), it is desirable (to a certain and controlled degree) in active portfolio management.

In fact, there exists an interesting relationship between tracking error and the possible alpha. This relationship is known as the *fundamental law of active management*. The fundamental law of active management states a basic relationship between relative risk and return that holds true for both fundamentally[21] and quantitatively[22] managed portfolios.[23] It was first presented[24] by Richard Grinold[25]

---

[21]A fundamentally managed portfolio is a portfolio that follows the fundamental approach to investing. This means that fundamental portfolio managers when constructing a portfolio use the evaluations of fundamental analysts who gather and analyze information on potential investments, for example, by examining balance sheets and income statements.

[22]A quantitatively managed portfolio is a portfolio that follows the quantitative approach to investing. Quantitative analysts develop computer algorithms that evaluate the potential return of an investment. Quantitative portfolio managers use this return estimation in order to achieve an optimal portfolio using portfolio construction software that includes trading costs and risk limits.

[23]A detailed comparison between the fundamental and quantitative approach to asset management can be found in Glavin and Reinganum (2013), Schulmerich, Hooker, McGoldrick, and Mallik (2008), Hooker and Schulmerich (2008) and Schulmerich and Hooker (2008). Interesting articles about quantitative equity investing in crisis times or shortly thereafter can be found in Schulmerich (2008b), Schulmerich et al. (2009), Schulmerich (2009) and Schulmerich (2010b).

[24]Grinold (1989).

[25]Richard C. Grinold, Ph.D., was until 2009 managing director of the Advanced Strategies and Research group at Barclays Global Investors. Dr. Grinold was for 20 years on the faculty at the School of Business Administration at the University of California, Berkeley. He has published extensively and is widely known in the industry for his pioneering work on risk models, portfolio optimization, and trading analysis; equity, fixed income, and international investing; and quantitative approaches to active management.

in 1989. This version was then improved in 2002 by Roger Clark,[26] Harindra de Silva[27] and Steven Thorley,[28] see Clarke, de Silva, and Thorley (2002).

The link between the engaged relative risk (measured by tracking error) and the relative return (measured by alpha) in benchmark-oriented portfolio management is based on the assumption that in the portfolio all desired positions can be implemented. No restrictions, for example, on short selling or portfolio weights are considered. This, of course, is unrealistic in active portfolio management practice as there are severe restrictions, like the long-only constraint which prohibits short positions in portfolios.[29] These restrictions were included into the fundamental law of active portfolio management by Clark, de Silva and Thorley in 2002. De Silva and Clark work for Analytic Investors, a quantitative asset management firm in Los Angeles.[30] Their version is what nowadays is understood under the *fundamental law of active management* and this is the version that will be presented here.

Simply speaking, the basic message is "no risk no fun". In active portfolio management, the fun part is alpha while the risk part is the engaged tracking error. Before we can mathematically state the fundamental law of active management we need to define a few variables:

In portfolio management, the *investment universe* is the set of all securities a portfolio manager chooses from when creating his portfolio. For a European equity portfolio, the *investment universe* comprises all European equities. Let IC be the *information coefficient*, TC the *transfer coefficient* and N the *breadth* of the *investment universe*:

1. **Breadth** N**:**
   N is the number of stocks in the investment universe adjusted by correlations between the stocks to achieve *independence*.[31] Each stock can be seen as a

---

[26]Roger G. Clarke, Ph.D., is the chairman of Analytic Investors. Recognized as an authority with more than 20 years experience in quantitative investment research, Roger Clarke has authored numerous articles and papers including two tutorials for the CFA Institute. He also served on the faculty of Brigham Young University for 8 years where he specialized in investment and options theory and continues to lecture as a guest professor.

[27]Harindra de Silva, Ph.D., CFA, is the president of Analytic Investors and a portfolio manager. De Silva has authored several articles and studies on finance-related topics including stock market anomalies, market volatility and asset valuation.

[28]Steven Thorley, Ph.D., CFA, is the H. Taylor Peery Professor of Finance at the Marriott School of Management at Brigham Young University in Provo, UT.

[29]This is most common in retail funds, i.e., funds available for public distribution.

[30]Analytic Investors, LLC was founded in 1970. The original firm was known for its expertise in derivatives strategies. Nowadays, Analytic is part of Old Mutual Asset Management (U.S.), a group of affiliate firms selected by Old Mutual that have complementary investment styles (non-overlapping) and are considered top-quality investment management firms.

[31]This is, for example, done by using risk models like BARRA. It is a highly mathematical task, so we do not go into further detail here.

"strategy",[32] i.e., as the outcome of a decision on how to weight a particular stock in the portfolio. In active portfolio management, its weight is mainly expressed as the relative security weight in the portfolio versus the index. This is also called over- or underweight. For example, if a security has 1.5 % weight in the index but only 1.0 % weight in the portfolio, the security has an underweight of 0.5 % versus the benchmark. This over- or underweight decision compared to the benchmark is a strategic decision of the portfolio manager. In this respect, N is the number of independent strategies the portfolio manager can invest in.

2. **Information Coefficient** IC**:**
   The IC is a coefficient that measures the correlation between the predicted and the finally realized alphas, i.e., the quality of the predictions of the portfolio manager about the future alpha of each stock in the investment universe.
3. **Transfer Coefficient** TC**:**
   The TC is a coefficient that measures the correlation between the portfolio manager's alpha predictions and the implemented strategies in the portfolio, i.e., the implementation quality of a portfolio manager.

Using these definitions the *ex-ante* alpha of the fundamental law of active management can be written as

$$\alpha_{ex-ante} \approx IC \cdot TC \cdot TE \cdot \sqrt{N}. \tag{1.39}$$

Equation (1.39) is only valid when the impact of fees like management fees, administration fees, depot bank cost, etc. is neglected. In this case alpha is called *gross-of-fee alpha.*[33] These costs only pertain to the portfolio but not to the benchmark.

It is important to repeat that the fundamental law of active management holds for all types of active asset management vs. a benchmark. The term *fundamental* does *not* refer to fundamental asset management (as opposed to quantitative asset management) but simply states that this is a general (i.e., fundamental) rule governing all types of benchmark-related active asset management, let it be fundamental or quantitative, equity or fixed income portfolio management.

Example 6

Let us consider an equity portfolio comprising European large-cap stocks managed against the MSCI Europe Index. While the index comprises roughly 450 equities, the number of stocks an active portfolio manager can usually invest in, i.e., the investment universe, is almost 1,600.[34] However, some of these securities are highly correlated. For example, stocks within the same industry and

[32] Sometimes also called *bet*.

[33] If these costs are already subtracted, alpha is called *net-of-fee alpha*.

[34] Assuming the portfolio manager is not restricted to invest only in benchmark securities, i.e., off-benchmark positions are allowed.

within the same country are likely to have a higher correlation than stocks from different industries and countries. Breadth N is corrected for this correlation using financial engineering software. As a result, breadth N of a European investment universe is roughly 900.

Assuming the allowed tracking error as given by the client is $3\,\% = \frac{3}{100}$ we can calculate the expected alpha of the portfolio. But while N (via the investment universe) and TE are both specified by the client in the portfolio's investment guidelines,[35] IC and TC depend solely on the quality of the portfolio manager.

For a very good portfolio manager, we roughly have the long-run $IC = \frac{1}{10}$ and $TC = \frac{1}{3}$. Putting these data in Eq. (1.39), we get as the expected (i.e., ex-ante) alpha:

$$\begin{aligned} \alpha_{ex-ante} &\approx IC \cdot TC \cdot TE \cdot \sqrt{N} \\ &= \frac{1}{10} \cdot \frac{1}{3} \cdot \frac{3}{100} \cdot \sqrt{900} \\ &= 3\,\%. \end{aligned} \tag{1.40}$$

This means, the fundamental law of active management predicts the expected (therefore, *ex-ante*) p.a. alpha based on the client's investment universe and tracking error specification in combination with the skills of the portfolio manager.

End of Example 6

1.3.4 Covariance and Correlation

When investing in more than one security, another risk measure besides the individual volatility of each stock has to be taken into consideration. Investors want to know: What is the relationship or interaction between the percentage return for two assets in the equidistant subperiods of, for example, 1 day or 1 month?

When you create a portfolio of single assets, you should not only consider the volatility of the single assets but also how the return streams interact. This interaction is the reason for what we have called *diversification* in Chap. 2. In order to measure it, we first have to calculate the covariance or, more standardized, the correlation between the subperiods' percentage returns for each pair of securities. This will then allow us to calculate the portfolio volatility by using the volatility of the single assets and the correlations of the assets amongst each other.

For illustration, let us take a look at the price movements of the airlines Delta and US Airways and oil: Fig. 1.12 plots the end-of-month values of US Airways (black),

[35] Also known as IMA, *Investment Management Agreement*.

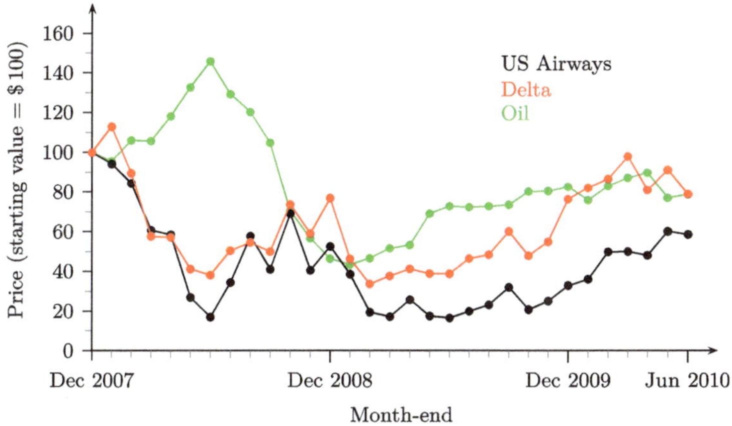

Fig. 1.12 End-of-month prices of Delta (NYSE:DAL), US Airways (NYSE:LCC) and oil (Crushing, OK Crude Oil Futures Contract, price per barrel) in the period December 2008–June 2010. Based on starting value of $100. *Sources*: Yahoo! Finance, U.S. Energy Information Administration

Delta (red) and oil (green) during the time period January 2007–June 2010, based on a starting value of $100. We can observe that the oil price and the airline stock prices tend to move in opposite directions, while the airline stock prices tend to move in tandem. The idea of *covariance* and *correlation* is to measure how asset prices move together. In our case, oil prices and Delta (or US Airways) have *negative* covariance and are *negatively* correlated while Delta and US Airways show *positive* covariance and are *positively* correlated.

Let us now take a look at Fig. 1.13 which plots the monthly returns of US Airways (black), Delta (red) and oil (green) during the time period January 2008–June 2010. US Airways tends to perform relatively well (relative to the average US Airways performance) when Delta performs relatively well (relative to the average Delta performance), while it is the opposite for US Airways and oil. US Airways and Delta tend to have above-average and below-average returns at the same time, and consequently, the product of US Airways relative returns (by *relative*, we mean relative to its average monthly return, i.e., we subtract its average monthly return from its monthly return) with Delta relative returns is typically positive, and tends to be bigger, when the return values move closely together. US Airways and oil show the opposite behavior, i.e., above-average returns of US Airways tend to coincide with below-average returns of oil, and below-average returns of US Airways tend to coincide with above-average returns of oil and vice versa. A a result, the product of US Airways relative returns and oil relative returns is typically negative. This is the idea of the covariance and correlation measures.

After having discussed the motivation for covariance and correlation, we will proceed with the relevant general definitions. As with the expected value and variance, we will first provide the correct mathematical definition which is then followed by the definition that should be used in practical application.

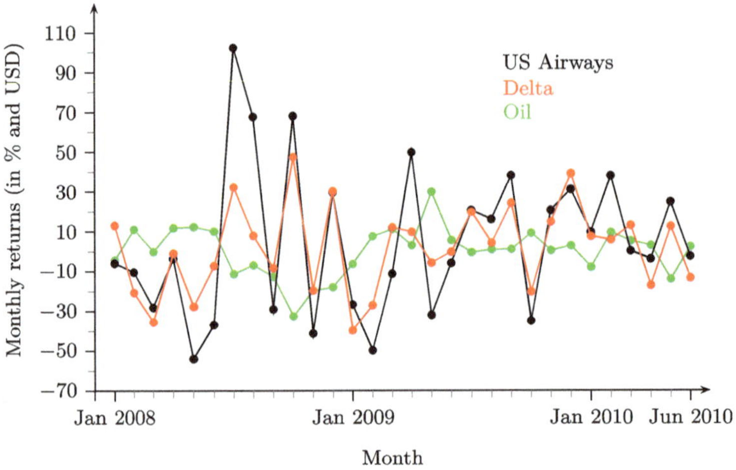

Fig. 1.13 Monthly returns of Delta (NYSE:DAL), US Airways (NYSE:LCC) and oil (Crushing, OK Crude Oil Futures Contract, price per barrel) in the period January 2008–June 2010. *Sources*: Yahoo! Finance, U.S. Energy Information Administration

The covariance[36] $\sigma_{X,Y}$ of two random variables X and Y is defined as

$$\sigma_{X,Y} = \mathbb{E}[(X - \mathbb{E}[X])(Y - \mathbb{E}(Y))]. \tag{1.41}$$

In particular, we can specify the two random variables X and Y to be the returns R_1 and R_2 of the two assets 1 and 2, respectively, over a certain future time period $[0, T]$. Then,

$$\sigma_{1,2} = \mathbb{E}[(R_1 - \mathbb{E}[R_1])(R_2 - \mathbb{E}[R_2])] \tag{1.42}$$

is the *covariance of subperiod returns in time period* $[0, T]$ *between the two assets 1 and 2* or just the *covariance between the assets 1 and 2*. Please note that as always the returns are percentage values.

The covariance $\sigma_{1,2}$ alone does not tell the strength of the relationship between the asset returns. Its absolute value lies between zero and the product of the volatilities σ_1 and σ_2. The covariance can therefore be *normed* by dividing it by the product $\sigma_1 \cdot \sigma_2$. The resulting measure is called *correlation*[37] ($\rho_{1,2}$):

$$\rho_{1,2} = \frac{\sigma_{1,2}}{\sigma_1 \cdot \sigma_2}. \tag{1.43}$$

Please note that $\rho_{1,2}$ is zero if σ_1 or σ_2 is zero. In general, for any pair of random variables X and Y, the correlation $\rho_{X,Y}$ is defined as

[36]DeFusco et al. (2004, p. 204).

[37]DeFusco et al. (2004, p. 207).

$$\rho_{X,Y} = \frac{\sigma_{X,Y}}{\sigma_X \cdot \sigma_Y}. \tag{1.44}$$

After having introduced the mathematical definitions of covariance and correlation using random variables, we now will present the definition of covariance that can be used in practice in order to calculate covariance using historical return data. Thereafter, we will look at the properties of covariance and correlation and provide several examples for illustration. To do this, we first have to look at time period $[0, T]$ and divide it, as usual, into subintervals.

Definition: Covariance and Correlation

We divide time period $[0, T]$ into N equidistant subintervals. The return of security 1 over the k-th subinterval $[t_{k-1}, t_k], 0 \leq k \leq N$, is denoted by r_1^k. Accordingly, let r_2^k be the return of security 2 over the k-th subinterval.

Then, the covariance $\sigma_{1,2}$ of the securities' subinterval returns measured by N subintervals of time period $[0, T]$ is defined as[38]:

$$\sigma_{1,2} = \frac{1}{N-1}\sum_{k=1}^{N}(r_1^k - \overline{r}_1)(r_2^k - \overline{r}_2) \tag{1.45}$$

where

$$\overline{r}_1 = \frac{1}{N}\sum_{k=1}^{N} r_1^k \quad \text{and} \quad \overline{r}_2 = \frac{1}{N}\sum_{k=1}^{N} r_2^k$$

are the arithmetic average returns of the two assets, respectively. If we have monthly subintervals, we also denote the covariance as

$$\sigma_{1,2} = \sigma_{1,2}^{monthly}.$$

For daily subintervals, we denote the covariance as

$$\sigma_{1,2} = \sigma_{1,2}^{daily}.$$

If the two assets are identical (1=2), the calculated covariance is the asset's variance, i.e.:

$$\sigma_{1,1}^{monthly} = \sigma_1^{2,monthly} \quad \text{and} \quad \sigma_{1,1}^{daily} = \sigma_1^{2,daily} \tag{1.46}$$

(continued)

[38]Based on Lhabitant (2004, p. 127).

Using Eq. (1.20) to calculate the volatilities σ_1, σ_2, respectively, allows us to calculate the correlation [see Eq. (1.43)] as[39]:

$$\rho_{1,2} = \frac{\sigma_{1,2}}{\sigma_1 \cdot \sigma_2} \tag{1.47}$$

with σ_1 and σ_2 being the volatility of the percentage return of security 1 and 2, respectively, over N subintervals.

1.3.4.1 Notes

When calculating covariance and correlation, the following points should be remembered:

- In order to calculate covariance and correlation, percentage return data for consecutive subintervals (for example, days or months) are needed without any data missing for any of the two relative return time series.
- Therefore, a monthly covariance is more common than a daily covariance since there may be days where one asset has a return and the other one has not. For example, if asset 1 is a stock listed in a Middle East country where Friday is a weekend day, then it does not have a return for Fridays. If you want to calculate the correlation of this security with another security listed in London where you do have a Friday return, this poses mathematical difficulties. This does not happen if monthly data are used.
- Although less common in practice, one can annualize a covariance if $T \geq 1$ (like for return and volatility) according to the following formulas depending on whether the subintervals are, for example, months or days[40]:

$$\sigma_{1,2}^{p.a.} = 12 \cdot \sigma_{1,2}^{monthly}, \tag{1.48}$$

$$\sigma_{1,2}^{p.a.} = 252 \cdot \sigma_{1,2}^{daily}. \tag{1.49}$$

1.3.4.2 Properties of Covariance and Correlation

A covariance has the following properties:

- If the covariance is positive, then one asset tends to have high (low) returns whenever the other one also has high (low) returns. Usually, this is the case for stocks of companies from the same industry.

[39]Esch et al. (2005, p. 42).

[40]Hull (2009, p. 284). We assume 252 trading days per year.

- If the covariance is negative, then one asset tends to have high (low) returns when the other one has low (high) returns. For example, high oil prices negatively affect airlines. Hence, the covariance between the returns of airline stocks and oil prices is negative.
- If the covariance is zero, then there is no linear relation between the asset returns: knowledge of the return of one asset will not lead to any knowledge about the return of the other asset.

A correlation has the following properties[41]:

- The correlation between two assets lies between −1 and 1:

$$-1 \leq \rho_{1,2} \leq 1. \tag{1.50}$$

- If the correlation is positive, then one asset tends to have high (low) returns whenever the other one also has high (low) returns. If the correlation is 1, then the relationship between the asset returns is positively linear.
- If the correlation is negative, then one asset tends to have high (low) returns whenever the other one has low (high) returns. If the correlation is −1, then the relationship between the asset returns is negatively linear.
- If the correlation is 0, then there is no linear relationship between the asset returns.
- The greater the absolute value of the correlation, the stronger the association between the asset returns.

To illustrate the last property of this list, let us first look at Table 1.10. It lists different absolute values for the correlation coefficient together with the corresponding strength of the association between the asset returns. The degree of association is very high for 0.80 and above, and very low for values below 0.20.

Next, we will visualize various correlations by using a *scatter plot*. A scatter plot[42] is a type of mathematical diagram using Cartesian coordinates to display

Table 1.10 Absolute value of correlation coefficient and strength of corresponding association

Absolute value of correlation	Strength of association between asset returns
0.80–1.00	Very strong association
0.60–0.79	Strong association
0.40–0.59	Moderate association
0.20–0.39	Weak association
0.00–0.19	Little if any association

Source: Lhabitant (2004, p. 129)

[41] DeFusco et al. (2004, pp. 207–208).

[42] Also called scatter chart, scattergram, scatter diagram or scatter graph.

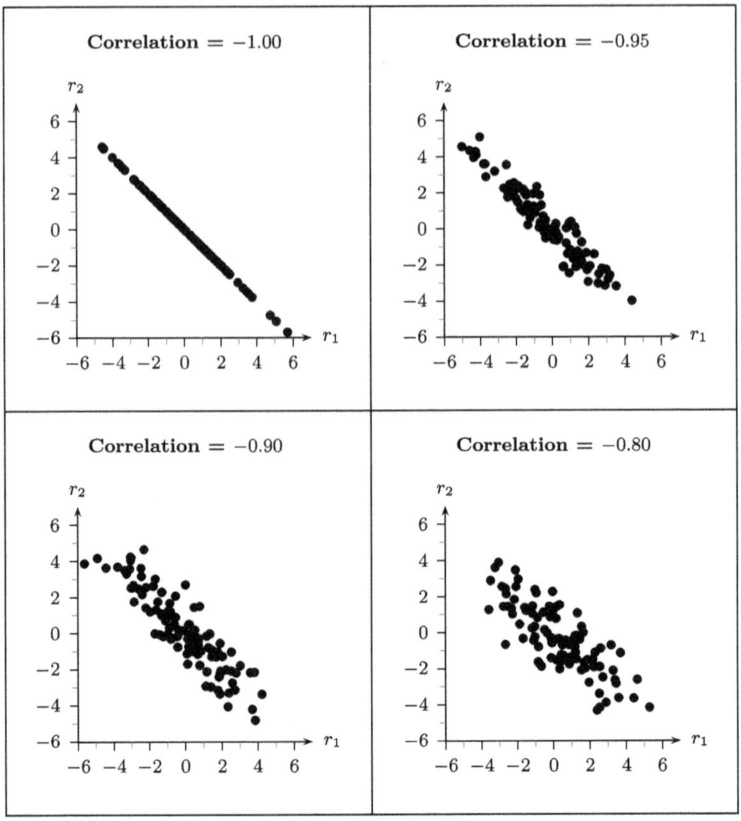

Fig. 1.14 Scatter plots for different correlations. Each graph illustrates returns of assets 1 and 2 during some time period $[0, T]$ with 100 subintervals in which the returns are measured. *Source*: Own, for illustrative purposes only

values for two variables for a set of data. The data are displayed as a collection of points, each having the value of one variable determining the position on the horizontal axis and the value of the other variable determining the position on the vertical axis.

The scatter plots in Figs. 1.14, 1.15, 1.16, and 1.17 illustrate correlation graphically. They show examples with two different assets 1 and 2 during some time period $[0, T]$ with 100 subintervals where returns

$$r_1^1, r_1^2, \ldots, r_1^{100}$$

and

$$r_2^1, r_2^2, \ldots, r_2^{100}$$

are measured. The points

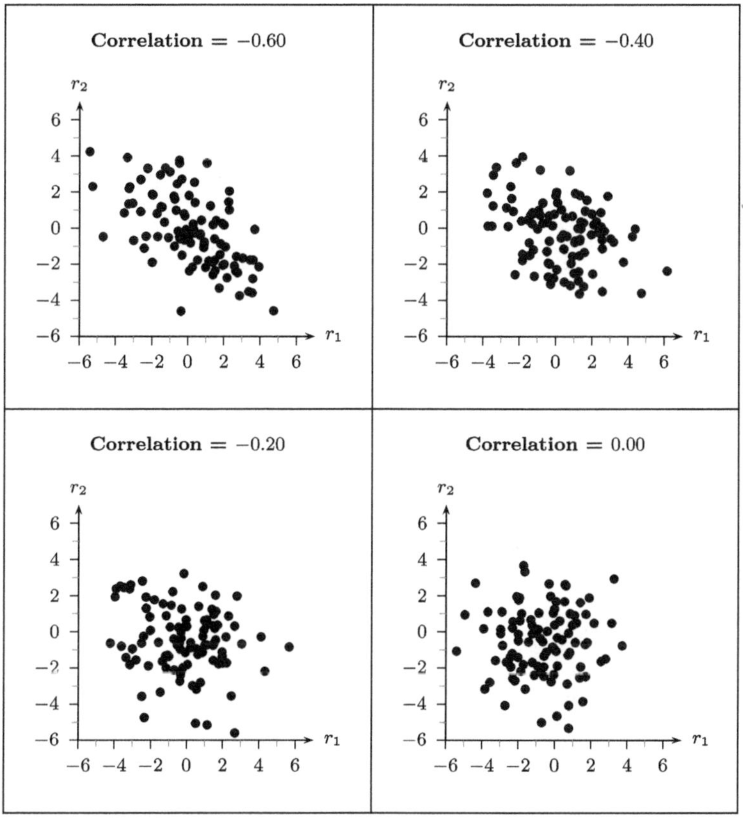

Fig. 1.15 Scatter plots for different correlations. Each graph illustrates the returns of assets 1 and 2 during some time period $[0, T]$ with 100 subintervals in which the returns are measured. *Source*: Own, for illustrative purposes only

$$(r_1^1, r_2^1),\ (r_1^2, r_2^2), \ldots,\ (r_1^{100}, r_2^{100})$$

are plotted on the graphs which represent the absolute returns of these assets in the respective subintervals, for example, (r_1^{50}, r_2^{50}) represents the returns of assets 1 and 2 in the 50th subinterval.

These scatter plots show the typical pictures for different correlations.

- For correlation 1 (-1), the points lie exactly on a line with positive (negative) slope.
- For ± 0.95, the points are still close to a line.
- For ± 0.90 and ± 0.80, the graphs show a strong relationship between the asset returns.
- The relationship is weaker for ± 0.60 and ± 0.40.
- For ± 0.20, you can hardly spot any relationship at first sight, and for 0.00, there is no relationship between the asset returns.

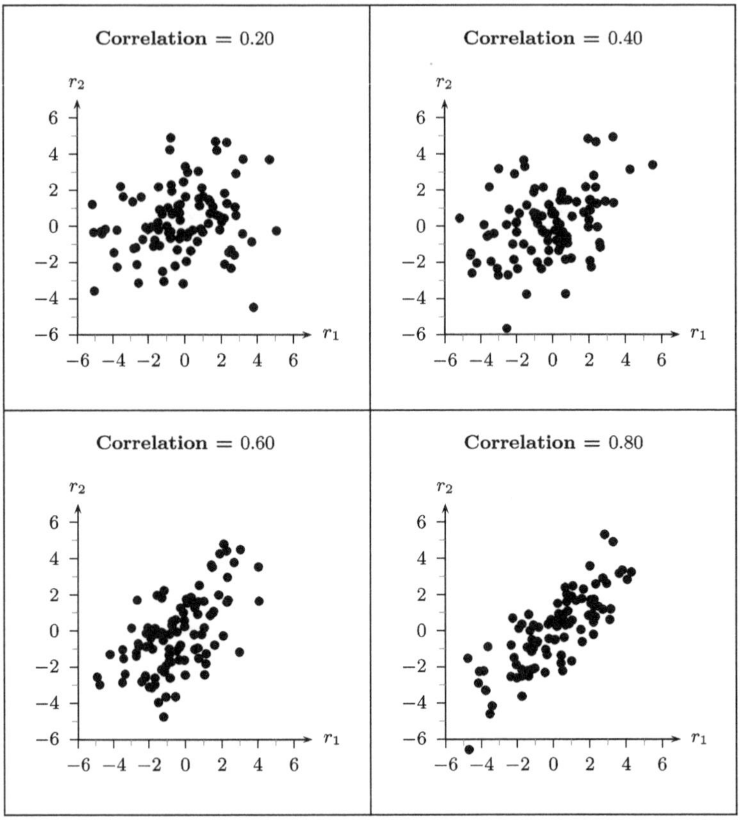

Fig. 1.16 Scatter plots for different correlations. Each graph illustrates the returns of assets 1 and 2 during some time period $[0, T]$ with 100 subintervals in which the returns are measured. *Source*: Own, for illustrative purposes only

In Fig. 1.18, we see the scatter plot of the monthly returns of US Airways (LCC) against Delta (DAL) for the period January 2008–June 2010. Given the monthly returns

$$r_{DAL}^1, r_{DAL}^2, \ldots, r_{DAL}^{30}$$

of Delta and

$$r_{LCC}^1, r_{LCC}^2, \ldots, r_{LCC}^{30}$$

of US Airways, the graph plots the points

$$(r_{DAL}^1, r_{LCC}^1), (r_{DAL}^2, r_{LCC}^2), \ldots, (r_{DAL}^{30}, r_{LCC}^{30}).$$

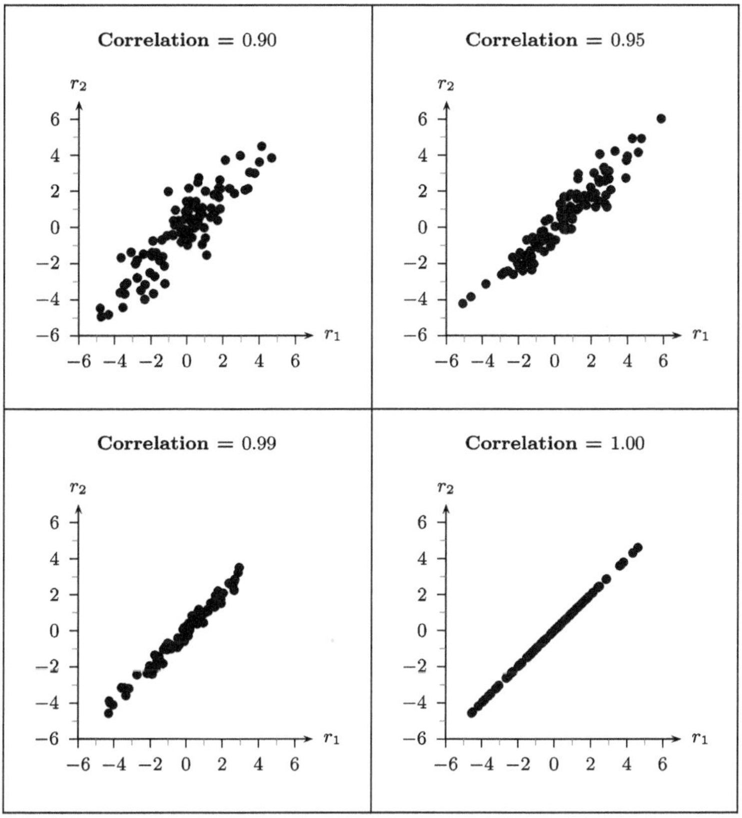

Fig. 1.17 Scatter plots for different correlations. Each graph illustrates the returns of assets 1 and 2 during some time period $[0, T]$ with 100 subintervals in which the returns are measured. *Source*: Own, for illustrative purposes only

The correlation, which is calculated below in Eq. (1.66), is 0.789, i.e., the relationship between the monthly returns of the respective airline stocks is strong. This is graphically supported by the plot in Fig. 1.18.

Figure 1.19 shows the scatter plot of monthly returns of US Airways (LCC) against oil for the period January 2008–June 2010. Given the monthly returns

$$r_{LCC}^1, r_{LCC}^2, \ldots, r_{LCC}^{30}$$

of US Airways and

$$r_{Oil}^1, r_{Oil}^2, \ldots, r_{Oil}^{30}$$

of oil, the graph plots the points

$$(r_{Oil}^1, r_{LCC}^1), (r_{Oil}^2, r_{LCC}^2), \ldots, (r_{Oil}^{30}, r_{LCC}^{30}).$$

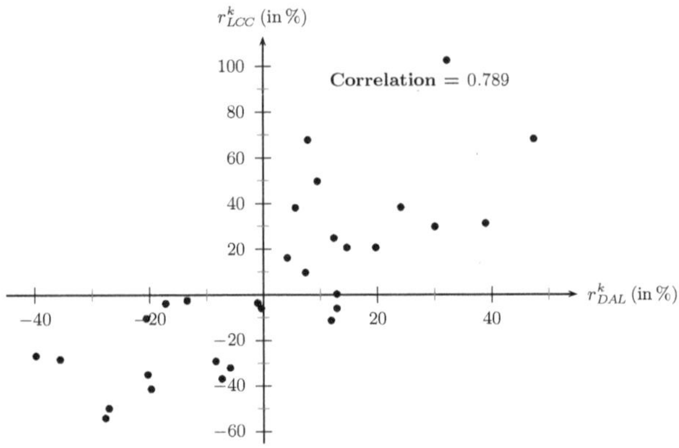

Fig. 1.18 Scatter plot of monthly returns of US Airways (LCC) against Delta (DAL) for the period January 2008–June 2010. *Source*: Yahoo! Finance

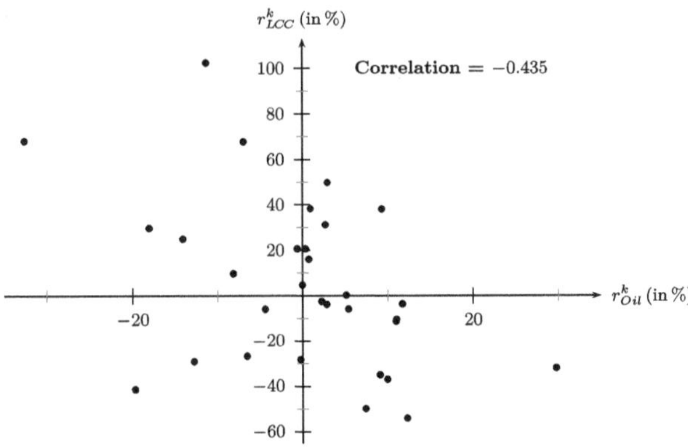

Fig. 1.19 Scatter plot of monthly returns of US Airways (LCC) against oil (Crushing, OK Crude Oil Futures Contract) for the period January 2008–June 2010. *Sources*: Yahoo! Finance and U.S. Energy Information Administration

The correlation, which is calculated below in Eq. (1.68), is −0.435. The plot shows that the monthly returns of US Airways tend to be negatively affected by oil.

We will now extend our business case with additional data displayed in Table 1.11. Please note, that we will refer to Table 1.11 again in Chap. 2 when we continue with this business case in the context of regression analysis.

Business Case (cont.)

Let us determine the covariances and correlations between the monthly returns of the airline stocks Delta (DAL) & US Airways (LCC), and the monthly returns on crude oil. The calculation is based on the monthly data from the period January 2008–June 2010 as shown in Table 1.11. To calculate the different correlations, we have to start with the covariances using Eq. (1.45). This requires to calculate the arithmetic mean monthly returns first:

(continued)

Table 1.11 Time series of end-of-month stock values of Delta (NYSE:DAL), US Airways (NYSE:LCC), both paying no dividends, and oil (Crushing, OK Crude Oil Futures Contract, price per barrel) in the period January 2008–June 2010, together with the respective monthly returns

k	Time point t_k	Month	$V_{DAL}^{t_k}$ (in USD)	r_{DAL}^{k} (in %)	$V_{LCC}^{t_k}$ (in USD)	r_{LCC}^{k} (in %)	$V_{Oil}^{t_k}$ (in USD)	r_{Oil}^{k} (in %)
0	0	Dec 2007	14.89		14.71		95.98	
1	1/12	Jan 2008	16.82	12.96	13.84	−5.91	91.75	−4.41
2	2/12	Feb 2008	13.35	−20.63	12.40	−10.40	101.84	11.00
3	3/12	Mar 2008	8.60	−35.58	8.91	−28.15	101.58	−0.26
4	4/12	Apr 2008	8.51	−1.05	8.59	−3.59	113.46	11.70
5	5/12	May 2008	6.15	−27.73	3.96	−53.90	127.35	12.24
6	6/12	Jun 2008	5.70	−7.32	2.50	−36.87	140.00	9.93
7	7/12	Jul 2008	7.54	32.28	5.06	102.40	124.08	−11.37
8	8/12	Aug 2008	8.13	7.82	8.49	67.79	115.46	−6.95
9	9/12	Sep 2008	7.45	−8.36	6.03	−28.98	100.64	−12.84
10	10/12	Oct 2008	10.98	47.38	10.14	68.16	67.81	−32.62
11	11/12	Nov 2008	8.81	−19.76	5.96	−41.22	54.43	−19.73
12	1	Dec 2008	11.46	30.08	7.73	29.70	44.60	−18.06
13	13/12	Jan 2009	6.90	−39.79	5.67	−26.65	41.68	−6.55
14	14/12	Feb 2009	5.03	−27.10	2.85	−49.74	44.76	7.39
15	15/12	Mar 2009	5.63	11.93	2.53	−11.23	49.66	10.95
16	16/12	Apr 2009	6.17	9.59	3.79	49.80	51.12	2.94
17	17/12	May 2009	5.81	−5.83	2.58	−31.93	66.31	29.71
18	18/12	Jun 2009	5.79	−0.34	2.43	−5.81	69.89	5.40
19	19/12	Jul 2009	6.93	19.69	2.93	20.58	69.45	−0.63
20	20/12	Aug 2009	7.22	4.18	3.40	16.04	69.96	0.73
21	21/12	Sep 2009	8.96	24.10	4.70	38.24	70.61	0.93
22	22/12	Oct 2009	7.14	−20.31	3.06	−34.89	77.00	9.05
23	23/12	Nov 2009	8.19	14.71	3.69	20.59	77.28	0.36
24	2	Dec 2009	11.38	38.95	4.84	31.17	79.36	2.69
25	25/12	Jan 2010	12.23	7.47	5.31	9.71	72.89	−8.15
26	26/12	Feb 2010	12.92	5.64	7.33	38.04	79.66	9.29
27	27/12	Mar 2010	14.59	12.93	7.35	0.27	83.76	5.15
28	28/12	Apr 2010	12.08	−17.20	7.07	−3.81	86.15	2.85
29	29/12	May 2010	13.58	12.42	8.83	24.89	73.97	−14.14
30	30/12	Jun 2010	11.75	−13.48	8.61	−2.49	75.63	2.24

Sources: Yahoo! Finance (for DAL and LCC) and U.S. Energy Information Administration (for oil)

$$\begin{aligned}\overline{r}_{DAL} &= \frac{1}{30}\sum_{k=1}^{30} r_{DAL}^{k} \\ &= \frac{1}{30}(12.96\,\% + (-20.63\,\%) + \ldots + (-13.48\,\%)) \\ &= 1.59\,\%. \end{aligned} \tag{1.51}$$

$$\begin{aligned}\overline{r}_{LCC} &= \frac{1}{30}\sum_{k=1}^{30} r_{LCC}^{k} \\ &= \frac{1}{30}(-5.91\,\% + (-10.40\,\%) + \ldots + (-2.49\,\%)) \\ &= 4.73\,\%. \end{aligned} \tag{1.52}$$

$$\begin{aligned}\overline{r}_{Oil} &= \frac{1}{30}\sum_{k=1}^{30} r_{Oil}^{k} \\ &= \frac{1}{30}(-4.41\,\% + 11.00\,\% + \ldots + 2.24\,\%) \\ &= -0.04\,\%. \end{aligned} \tag{1.53}$$

Using Eq. (1.45), we calculate the covariances between the monthly returns:

$$\begin{aligned}\sigma_{DAL,LCC}^{monthly} &= \frac{1}{29}\sum_{k=1}^{30}(r_{DAL}^{k} - \overline{r}_{DAL}) \cdot (r_{LCC}^{k} - \overline{r}_{LCC}) \\ &= \frac{1}{29}[(12.96\,\% - 1.59\,\%) \cdot (-5.91\,\% - 4.73\,\%) + \ldots \\ &\quad + (-13.48\,\% - 1.59\,\%) \cdot (-2.49\,\% - 4.73\,\%)] \\ &= 0.065368. \end{aligned} \tag{1.54}$$

$$\begin{aligned}\sigma_{DAL,Oil}^{monthly} &= \frac{1}{29}\sum_{k=1}^{30}(r_{DAL}^{k} - \overline{r}_{DAL}) \cdot (r_{Oil}^{k} - \overline{r}_{Oil}) \\ &= \frac{1}{29}[(12.96\,\% - 1.59\,\%) \cdot (-4.41\,\% - (-0.04\,\%)) + \ldots \\ &\quad + (-13.48\,\% - 1.59\,\%) \cdot (2.24\,\% - (-0.04\,\%)] \\ &= -0.010083. \end{aligned} \tag{1.55}$$

(continued)

$$\begin{aligned}
\sigma_{LCC,Oil}^{monthly} &= \frac{1}{29}\sum_{k=1}^{30}(r_{LCC}^{k} - \overline{r}_{LCC})\cdot(r_{Oil}^{k} - \overline{r}_{Oil}) \\
&= \frac{1}{29}[(-5.91\,\% - 4.73\,\%)(-4.41\,\% - (-0.04\,\%)) + \ldots \\
&\quad + (-2.49\,\% - 4.73\,\%)\cdot(2.24\,\% - (-0.04\,\%)] \\
&= -0.019952. \qquad (1.56)
\end{aligned}$$

Now, we annualize the covariances using Eq. (1.48):

$$\sigma_{DAL,LCC}^{p.a.} = 12\cdot\sigma_{DAL,LCC}^{monthly} = 12\cdot 0.065368 = 0.7844, \qquad (1.57)$$

$$\sigma_{DAL,Oil}^{p.a.} = 12\cdot\sigma_{DAL,Oil}^{monthly} = 12\cdot(-0.010083) = -0.1210, \qquad (1.58)$$

$$\sigma_{LCC,Oil}^{p.a.} = 12\cdot\sigma_{LCC,Oil}^{monthly} = 12\cdot(-0.019952) = -0.2394. \qquad (1.59)$$

The next step is to calculate the monthly volatilities using Eq. (1.22):

$$\begin{aligned}
\sigma_{DAL}^{monthly} &= \sqrt{\frac{1}{29}\sum_{k=1}^{30}(r_{DAL}^{k} - \overline{r}_{DAL})^2} \\
&= \sqrt{\frac{1}{29}[(12.96\,\% - 1.59\,\%)^2 + \ldots + (-13.48\,\% - 1.59\,\%)^2]} \\
&= 0.21914. \qquad (1.60)
\end{aligned}$$

$$\begin{aligned}
\sigma_{LCC}^{monthly} &= \sqrt{\frac{1}{29}\sum_{k=1}^{30}(r_{LCC}^{k} - \overline{r}_{LCC})^2} \\
&= \sqrt{\frac{1}{29}[(-5.91\,\% - 4.73\,\%)^2 + \ldots + (-2.49\,\% - 4.73\,\%)^2]} \\
&= 0.37816. \qquad (1.61)
\end{aligned}$$

(continued)

$$\sigma_{Oil}^{monthly} = \sqrt{\frac{1}{29}\sum_{k=1}^{30}(r_{LCC}^{k} - \overline{r}_{LCC})^2}$$

$$= \sqrt{\frac{1}{29}[(-4.41\,\% - (-0.04\,\%))^2 + \ldots + (2.24\,\% - (-0.04\,\%))^2]}$$

$$= 0.12133. \tag{1.62}$$

Using Eq. (1.24), the annualized monthly volatilities are:

$$\sigma_{DAL}^{p.a.} = \sqrt{12}\cdot\sigma_{DAL}^{monthly} = \sqrt{12}\cdot 0.21914 = 0.7591, \tag{1.63}$$

$$\sigma_{LCC}^{p.a.} = \sqrt{12}\cdot\sigma_{LCC}^{monthly} = \sqrt{12}\cdot 0.37816 = 1.3100, \tag{1.64}$$

$$\sigma_{Oil}^{p.a.} = \sqrt{12}\cdot\sigma_{Oil}^{monthly} = \sqrt{12}\cdot 0.12133 = 0.4203. \tag{1.65}$$

We use Eq. (1.43) to calculate the correlations:

$$\rho_{DAL,LCC}^{p.a.} = \frac{\sigma_{DAL,LCC}}{\sigma_{DAL}\cdot\sigma_{LCC}} = \frac{0.7844}{0.7591\cdot 1.3100} = 0.789, \tag{1.66}$$

$$\rho_{DAL,Oil}^{p.a.} = \frac{\sigma_{DAL,Oil}}{\sigma_{DAL}\cdot\sigma_{Oil}} = \frac{-0.1210}{0.7591\cdot 0.4203} = -0.379, \tag{1.67}$$

$$\rho_{LCC,Oil}^{p.a.} = \frac{\sigma_{LCC,Oil}}{\sigma_{LCC}\cdot\sigma_{Oil}} = \frac{-0.2394}{1.3100\cdot 0.4203} = -0.435. \tag{1.68}$$

The correlations indicate that the stock prices of both airlines are strongly correlated and have a tendency to move together, whereas their correlations with the oil price are negative, i.e., their stock prices and the oil price tend to move in opposite directions.

Having demonstrated several applications of our formulas of covariance and correlation in the business case, we now turn to our hypothetical examples in order to show how these values can be easily calculated using Excel® formulas.

Example 7

Table 1.12 shows the consecutive monthly relative returns of two assets A and B. In this example we assume that the two assets are two portfolios available to investors as retail funds. The goal is to calculate the covariance and correlation of these two funds over a given time period of 18 consecutive months.

Using again the column and row notation from Microsoft® Excel®, the first column is labeled A (month), the second B (monthly performance of fund A) and the third C (monthly portfolio of fund B). Applying Excel® *functions*, the

Table 1.12 Example 7: Covariance and correlation calculation

	A	B	C
1	Month	Monthly performance of fund A	Monthly performance of fund B
2	07/2012	6.10 %	7.71 %
3	08/2012	5.50 %	6.38 %
4	09/2012	4.70 %	4.83 %
5	10/2012	−5.00 %	−7.70 %
6	11/2012	−5.10 %	−4.90 %
7	12/2012	6.70 %	5.32 %
8	01/2013	6.03 %	6.48 %
9	02/2013	−3.23 %	−3.95 %
10	03/2013	5.12 %	3.80 %
11	04/2013	5.21 %	5.20 %
12	05/2013	−4.10 %	−3.00 %
13	06/2013	−4.50 %	−2.01 %
14	07/2013	1.75 %	−3.54 %
15	08/2013	3.71 %	4.90 %
16	09/2013	−4.20 %	−1.62 %
17	10/2013	4.26 %	6.05 %
18	11/2013	−4.00 %	−4.10 %
19	12/2013	5.10 %	6.20 %
20		$\sigma_{A,B}^{monthly} =$	0.00224
21		$\rho_{A,B} =$	0.9261

Source: Own, for illustrative purposes only

covariance and the correlation of funds A and B over this time period can be calculated as:

- Covariance of the monthly percentage fund returns $\sigma_{A,B}^{monthly}$ in cell $C20$:
  $0.00224 = COVARIANCE.S(B2 : B19, C2 : C19)$
- Correlation of the monthly percentage fund returns $\rho_{A,B}^{monthly}$ in cell $C21$:
  $0.9261 = CORREL(B2 : B19, C2 : C19)$

This concept will be needed in the next section where beta is introduced as a measure of the systematic risk of a portfolio (or single security) versus an index.

End of Example 7

1.3.5 Beta

A key principle of MPT is the idea of diversification, i.e., the reduction of the overall portfolio volatility through the combination of securities in a portfolio. As seen

Fig. 1.20 Systematic and unsystematic risk in a portfolio in relation to the number of portfolio constituents. *Source*: Own, for illustrative purposes only

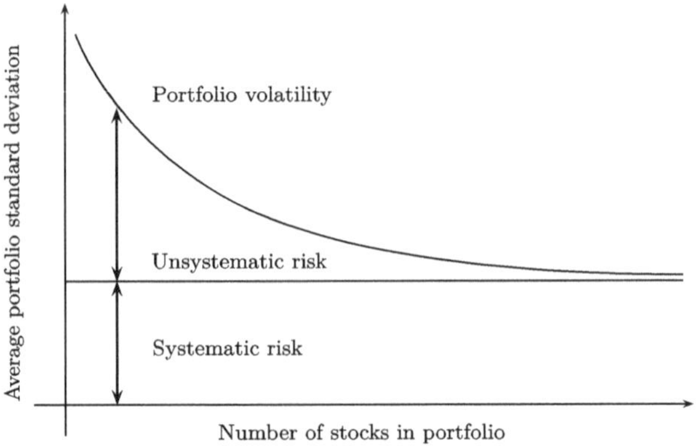

above, the critical component when constructing this portfolio is the covariance of all its securities. This idea is illustrated in Fig. 1.20.

Figure 1.20 distinguishes between *systematic* and *unsystematic risk*. Systematic risk is the inherent risk in the market which cannot be diversified away, hence it is *systematic*. The second risk component of the portfolio volatility is unsystematic risk which is the unique risk of the portfolio. As this figure shows, the effect of diversification reduces unsystematic risk. Systematic risk is called *beta* (β).

Figure 1.21 illustrates the idea of beta. In this diagram, the horizontal axis shows the 250 daily relative returns (not all data points are shown) of a benchmark for a portfolio. The corresponding 250 daily returns of the portfolio are displayed on the vertical axis. The daily returns of the benchmark and the portfolio are denoted with r_{Bm}^{daily} and r_{Pf}^{daily}, respectively. This diagram also applies if the portfolio only comprises one single security. In such a situation, beta is the security's beta versus the index. Now, a simple linear regression is run (for example, using Microsoft® Excel®) which yields a linear regression line shown in Fig. 1.21. The regression output shows two parameters which determine the regression line as a function of r_{Bm}^{daily}:

- The slope of the regression line which is the systematic risk measure β_{Pf}^{daily}.
- The value γ_{Pf}^{daily} where the regression line intersects with the vertical axis.

A detailed introduction to linear regression will be presented in Sect. 2.2 in order to explain the tests of the capital asset pricing model (CAPM). However, regression is not needed in the calculation of beta as we will see in the following definition.

Definition: Beta

β_{Pf} measures the interaction of a portfolio with an index as a benchmark, i.e., how the portfolio return changes depending on the returns of the benchmark. To measure this interaction mathematically, we again split time interval $[0, T]$

(continued)

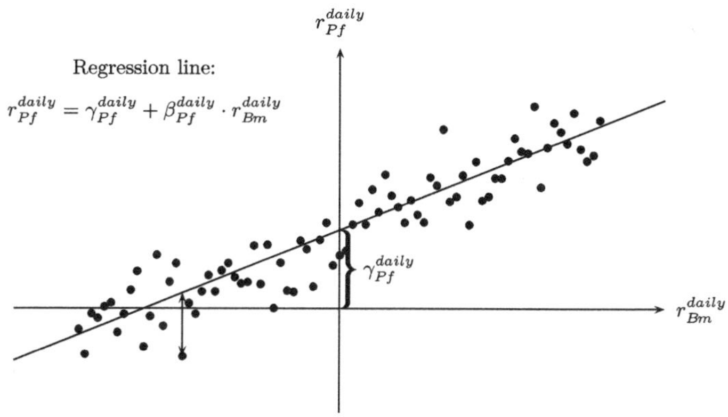

Fig. 1.21 Description of the regression's beta. See also Fig. 2.3 on page 106. Parameters γ_{Pf}^{daily} and β_{Pf}^{daily} are the regression parameters (real values). *Source*: Own, for illustrative purposes only

into N equidistant subintervals. These are usually days or months. Let further

$$r_{Pf}^{1}, r_{Pf}^{2}, \ldots, r_{Pf}^{N}$$

be the subperiod percentage returns of the portfolio and

$$r_{Bm}^{1}, r_{Bm}^{2}, \ldots, r_{Bm}^{N}$$

the subperiod percentage returns of the benchmark. Depending on the chosen subperiods we get the respective beta: daily returns as input lead to a daily beta, monthly returns lead to a monthly beta. Using the monthly covariance $\sigma_{Pf,Bm}^{2,monthly}$ of portfolio and benchmark as well as the monthly variance $\sigma_{Bm}^{2,monthly}$ of the benchmark, the monthly beta can be calculated as[43]

$$\beta_{Pf}^{monthly} = \frac{\sigma_{Pf,Bm}^{monthly}}{\sigma_{Bm}^{2,monthly}} \tag{1.69}$$

with N representing the number of months in $[0, T]$. Formula (1.69) can be easily modified for daily returns as input. Then, N is the number of days in $[0, T]$:

$$\beta_{Pf}^{daily} = \frac{\sigma_{Pf,Bm}^{daily}}{\sigma_{Bm}^{2,daily}} \tag{1.70}$$

Note that the use of daily data is more common in practice.

[43] Esch et al. (2005, p. 91).

1.3.5.1 Interpretation

By using Fig. 1.21 and our notation from above, it is easy to interpret beta. The regression equation for the regression line in Fig. 1.21 is

$$r_{Pf}^{daily} = \gamma_{Pf}^{daily} + \beta_{Pf}^{daily} \cdot r_{Bm}^{daily}. \tag{1.71}$$

To be precise: In statistics, the left term r_{Pf}^{daily} is called *estimator*. If we interpret this term as a scalar parameter we would have to include an error term on the right hand side of the equation.

Equation (1.71) leads to the interpretation of beta as a sensitivity measure. If the benchmark value changes by r_{Bm}^{daily} in 1 day, then the portfolio value on that day changes by

$$\beta_{Pf}^{daily} \cdot r_{Bm}^{daily}.$$

For example, if the portfolio beta for daily data is 0.8 and the benchmark return on a certain day is $-2.4\,\%$, then the expected percentage return of the portfolio on that day is $0.8 \cdot (-2.4) = -1.92$, i.e., the portfolio is less sensitive than the benchmark.

Obviously, a beta below one is advantageous in falling markets and a beta of above one is advantageous in rising markets. A beta of one indicates that the portfolio return is in line with the benchmark return. The choice of beta, based on the portfolio manager's expectations on the future development of the benchmark, is often referred to as *market timing*. However, smart market timing is a very difficult task.

Example 8

Table 1.13 shows the monthly percentage returns of fund A and its benchmark. The goal is to calculate the portfolio beta using Eq. (1.70).

A beta of 0.9345 as calculated in Table 1.13 means that the monthly percentage returns of the portfolio are highly correlated with the respective benchmark returns. Simply speaking, a 1 % return of the benchmark will lead (on average) to a 0.9345 % return of the portfolio.

As usual, the first column is column A (month). In column B and C the monthly performance of the portfolio and the benchmark are displayed. Using Excel®, the calculation of beta is easy if the covariance and the correlation are calculated up-front as done in the previous section:

- Covariance of the monthly portfolio and benchmark returns $\sigma_{Pf,Bm}^{monthly}$ in cell $C21$:

  $0.00226 = COVARIANCE.S(B2:B19, C2:C19)$

- Monthly benchmark volatility $\sigma_{Bm}^{monthly}$ in cell $C23$:

  $4.91\,\% = STDEV.S(C2:C19)$

Table 1.13 Example 8: Calculation of the portfolio beta

	A	B	C
1	Month	Monthly portfolio	Monthly benchmark performance
2	07/2012	6.10 %	6.01 %
3	08/2012	5.50 %	5.45 %
4	09/2012	4.70 %	4.63 %
5	10/2012	−5.00 %	−6.99 %
6	11/2012	−5.10 %	−4.16 %
7	12/2012	6.70 %	7.07 %
8	01/2013	6.03 %	5.97 %
9	02/2013	−3.23 %	−2.95 %
10	03/2013	5.12 %	4.66 %
11	04/2013	5.21 %	4.91 %
12	05/2013	−4.10 %	−4.01 %
13	06/2013	−4.50 %	−3.87 %
14	07/2013	1.75 %	−2.95 %
15	08/2013	3.71 %	4.52 %
16	09/2013	−4.20 %	−3.93 %
17	10/2013	4.26 %	4.99 %
18	11/2013	−4.00 %	−3.84 %
19	12/2013	5.10 %	4.99 %
21		$\sigma_{Pf,Bm}^{monthly}$ =	0.00226
23		$\sigma_{Bm}^{monthly}$ =	4.91 %
25		$\beta_{Pf}^{monthly}$ =	0.9345

Source: Own, for illustrative purposes only

- Monthly portfolio beta $\beta_{Pf}^{monthly}$ vs. benchmark in cell $C25$:
  $0.9345 = COVARIANCE.S(B2 : B19, C2 : C19)/STDEV.S(C2 : C19)\char`^2$

End of Example 8

1.3.5.2 Note

β is often understood as a sort of correlation. But although the formulas look the same, β can have values that are less than -1 and greater than 1. For example, a beta of two states a higher sensitivity to the market, i.e., on average the portfolio earns a return twice as high as the market return.

In the long run, the beta of a portfolio or single security versus a comparable index is positive. However, there may be periods with negative beta. A prominent example is Volkswagen in 2008, when it was engaged in take-over and merger discussions with Porsche. Although the German stock index DAX showed primarily negative returns in 2008, the Volkswagen stock increased, resulting in a negative beta over the time period October 23, 2007–October 24, 2008, see Fig. 1.22. However, if the time period only changes slightly like in Fig. 1.23, the stock beta is not negative any more.

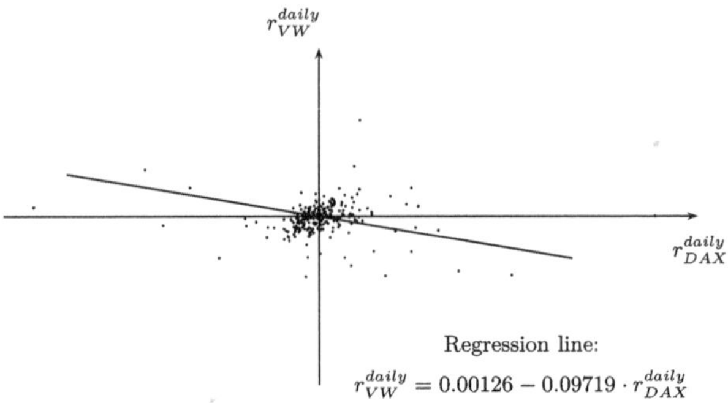

Fig. 1.22 Example of a negative stock beta over the time period October 23, 2007–October 24, 2008. *Source*: Bloomberg (*Tickers*: VOW for Volkswagen and DAX for DAX Index)

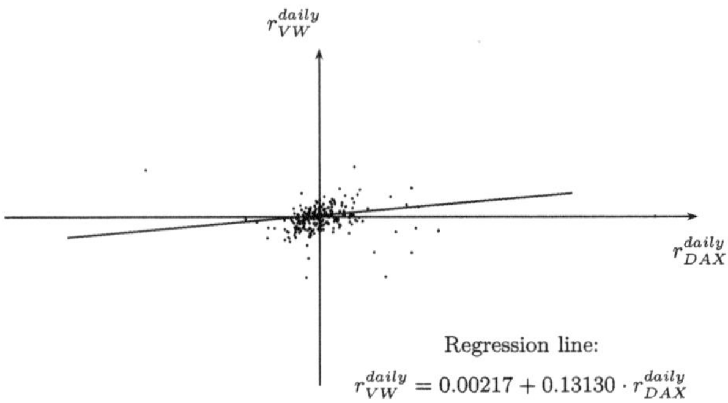

Fig. 1.23 Example of a positive stock beta over the time period November 1, 2007–October 7, 2008. *Source*: Bloomberg (*Tickers*: VOW for Volkswagen and DAX for DAX Index)

1.3.5.3 Conclusion

Beta is an important risk measure for assessing the performance of a portfolio versus the market represented by an index. The value of beta depends on the chosen subperiod length, for example, daily or monthly return data. As seen in the Volkswagen example, the value of beta is very sensitive to the chosen historical time period. However, if a security reacts differently in up and down markets, this cannot be captured by beta. For this, we would need to look at a more specialized concept of beta, the bull and bear market beta.

1.3.6 Bull and Bear Market Beta

If beta is only measured in bull markets the value will be different than when calculated in bear markets for the same security.[44] This observation lead to the concept of the bull and bear market beta whose importance is shown by the fact that investment houses regularly publish separate betas over bull and bear markets for a range of securities to show differing levels of upside potential and downside risk.

[44]For more information see, for example, Woodward and Anderson (2009).

Definition: Bull and Bear Market Beta
Let us split time interval $[0, T]$ into N equidistant subintervals which are usually days or months. Let further

$$r_{Pf}^1, r_{Pf}^2, \dots, r_{Pf}^N$$

be the subperiod percentage returns of the portfolio and

$$r_{Bm}^1, r_{Bm}^2, \dots, r_{Bm}^N$$

the subperiod percentage returns of the benchmark. Using the notation from above, ${}^{bull}\beta_{Pf}$ and ${}^{bear}\beta_{Pf}$ can mathematically be calculated in a similar way as the overall beta β_{Pf}. However, the overall covariance and volatility have to be replaced with conditional measures.

For this let N^{bull} be the number of subperiods in $[0, T]$ with a positive benchmark return and N^{bear} the number of subperiods in $[0, T]$ with a zero or negative benchmark return. Then, we obviously have $N^{bull} + N^{bear} = N$.

Covariance ${}^{bull}\sigma_{Pf,Bm}$ between the portfolio and the benchmark in case of bull markets is then calculated using Eq. (1.45) but applying only N^{bull} subintervals with a positive benchmark return.

Similarly, covariance ${}^{bear}\sigma_{Pf,Bm}$ between the portfolio and the benchmark in case of bear markets is calculated using Eq. (1.45) but applying only N^{bear} subintervals with a zero or negative benchmark return.

As a special case we get the bull and bear variances for bull and bear markets:

$$ {}^{bull}\sigma_{Bm}^2 = {}^{bull}\sigma_{Bm,Bm} \quad \text{and} \quad {}^{bear}\sigma_{Bm}^2 = {}^{bear}\sigma_{Bm,Bm}. \tag{1.72}$$

The bull and bear market beta can be calculated as

$$ {}^{bull}\beta_{Pf} = \frac{{}^{bull}\sigma_{Pf,Bm}}{{}^{bull}\sigma_{Bm}^2} \quad \text{and} \quad {}^{bear}\beta_{Pf} = \frac{{}^{bear}\sigma_{Pf,Bm}}{{}^{bear}\sigma_{Bm}^2} \tag{1.73}$$

These definitions can be done using daily subintervals or monthly subintervals. We then get

$$ {}^{bull}\sigma_{Pf,Bm}^{daily} \quad \text{and} \quad {}^{bull}\sigma_{Pf,Bm}^{monthly}$$

as the bull market covariance,

$$ {}^{bear}\sigma_{Pf,Bm}^{daily} \quad \text{and} \quad {}^{bear}\sigma_{Pf,Bm}^{monthly}$$

as the bear market covariance,

$$ {}^{bull}\sigma_{Bm}^{2,daily} \quad \text{and} \quad {}^{bull}\sigma_{Bm}^{2,monthly}$$

as the bull market variance,

$$ {}^{bear}\sigma_{Bm}^{2,daily} \quad \text{and} \quad {}^{bear}\sigma_{Bm}^{2,monthly} $$

as the bear market variance,

$$ {}^{bull}\beta_{Pf}^{daily} \quad \text{and} \quad {}^{bull}\beta_{Pf}^{monthly} $$

as the bull market beta and

$$ {}^{bear}\beta_{Pf}^{daily} \quad \text{and} \quad {}^{bear}\beta_{Pf}^{monthly}. $$

as the bear market beta.

1.3.6.1 Interpretation

While β_{Pf} measures the sensitivity of a portfolio to its benchmark over all N subintervals in time period $[0, T]$, the bull market beta ${}^{bull}\beta_{Pf}$ is more specific and considers only subintervals with a positive benchmark return. On the other hand, the bear market beta ${}^{bear}\beta_{Pf}$ measures this sensibility relative to subintervals with a zero or negative benchmark performance. The interpretation of the bear-beta and bull-beta is, therefore, straightforward:

- If ${}^{bear}\beta_{Pf} \geq 1$, then the portfolio return drops faster than the benchmark in a downward trending market.
- If $0 \leq {}^{bear}\beta_{Pf} \leq 1$, then the portfolio return drops, but less than the benchmark in a declining market.
- If ${}^{bear}\beta_{Pf} \leq 0$, then the performance of the asset rises, if the benchmark's performance drops.

Therefore, in a bear market, the best is to have a negative bear beta: the portfolio wins even when the benchmark loses. The interpretation of the bull market beta ${}^{bull}\beta_{Pf}$ is as follows:

- If ${}^{bull}\beta_{Pf} \geq 1$, then the portfolio return rises faster than the benchmark in an upward trending market.
- If $0 \leq {}^{bull}\beta_{Pf} \leq 1$, then the portfolio return rises, but less than the benchmark in a rising market.
- If ${}^{bull}\beta_{Pf} \leq 0$, then the performance of the portfolio drops, if the benchmark rises.

Obviously, it is advantageous to have a bull beta above 1 in rising markets. In combination, an ideal portfolio has a below 1 or, even better, negative bear beta and a rather high bull beta (at a minimum above 1).

Example 9

The following example analyzes a portfolio versus its benchmark over a period of 18 months, i.e., $T = 1.5$ years and $N = 18$ months. The monthly performance data of portfolio and benchmark are provided in Table 1.14 on page 62 which displays the same performance figures shown in Table 1.13 for Example 8. To calculate the bull and bear market beta, new columns D to G were added. In these columns, an Excel® IF-function is used to obtain the positive and negative values of the benchmark returns only.

- Positive monthly benchmark return $r_{Bm}^{monthly}$ in cell $D2$:
  $6.01\,\% \;=\; IF(C2 > 0, C2, \text{""})$
- Negative monthly benchmark return $r_{Bm}^{monthly}$ in cell $F2$:
  $= \; IF(C2 < 0, C2, \text{""})$
- Monthly portfolio return $r_{Pf}^{monthly}$, if $r_{Bm}^{monthly}$ is positive, in cell $E2$:
  $6.10\,\% \;=\; IF(D2 <> \text{""}, B2, \text{""})$
- Monthly portfolio return $r_{Pf}^{monthly}$, if $r_{Bm}^{monthly}$ is negative, in cell $G2$:
  $= \; IF(F2 <> \text{""}, B2, \text{""})$
- Bull market covariance ${}^{bull}\sigma_{Pf,Bm}^{monthly}$ for positive benchmark returns in cell $D20$:
  $0.000064 \;=\; COVARIANCE.S(D2 : D19, E2 : E19)$
- Bear market covariance ${}^{bear}\sigma_{Pf,Bm}^{monthly}$ for negative benchmark returns in cell $F20$:
  $0.000146 = \; COVARIANCE.S(F2 : F19, G2 : G19)$
- Bull market variance ${}^{bull}\sigma_{Bm}^{2,monthly}$ for positive benchmark returns in cell $D21$:
  $0.000066 = \; VAR.S(D2 : D19)$
- Bear market variance ${}^{bear}\sigma_{Bm}^{2,monthly}$ for negative benchmark returns in cell $F21$:
  $0.00159 = \; VAR.S(F2 : F19)$

  Then, the bull and bear market beta can be calculated via Excel® as follows:

- Monthly portfolio bull market beta ${}^{bull}\beta_{Pf}^{monthly}$ in cell $D22$:

  $0.9726 = \; D20/D21$

- Monthly portfolio bear market beta ${}^{bear}\beta_{Pf}^{monthly}$ in cell $F22$:

  $0.9188 = \; F20/F21$

Table 1.14 Example 9: Calculation of bull and bear market beta

	A	B	C	D	E	F	G
1	Month	Monthly portfolio performance	Monthly benchmark performance	Months with positive benchmark performance	Corresponding portfolio performance	Months with negative benchmark performance	Corresponding portfolio performance
2	07/2012	6.10 %	6.01 %	6.01 %	6.10 %		
3	08/2012	5.50 %	5.45 %	5.45 %	5.50 %		
4	09/2012	4.70 %	4.63 %	4.63 %	4.70 %		
5	10/2012	−5.00 %	−6.99 %			−6.99 %	−5.00 %
6	11/2012	−5.10 %	−4.16 %			−4.16 %	−5.10 %
7	12/2012	6.70 %	7.07 %	7.07 %	6.70 %		
8	01/2013	6.03 %	5.97 %	5.97 %	6.03 %		
9	02/2013	−3.23 %	−2.95 %			−2.95 %	−3.23 %
10	03/2013	5.12 %	4.66 %	4.66 %	5.12 %		
11	04/2013	5.21 %	4.91 %	4.91 %	5.21 %		
12	05/2013	−4.10 %	−4.01 %			−4.01 %	−4.10 %
13	06/2013	−4.50 %	−3.87 %			−3.87 %	−4.50 %
14	07/2013	1.75 %	−2.95 %			−2.95 %	1.75 %
15	08/2013	3.71 %	4.52 %	4.52 %	3.71 %		
16	09/2013	−4.20 %	−3.93 %			−3.93 %	−4.20 %
17	10/2013	4.26 %	4.99 %	4.99 %	4.26 %		
18	11/2013	−4.00 %	−3.84 %			−3.84 %	−4.00 %
19	12/2013	5.10 %	4.99 %	4.99 %	5.10 %		
20			${}^{bull}\sigma_{Pf,Bm}^{monthly} =$	0.000064	${}^{bear}\sigma_{Pf,Bm}^{monthly} =$	0.000146	
21			${}^{bull}\sigma_{Bm}^{2,monthly} =$	0.000066	${}^{bear}\sigma_{Bm}^{2,monthly} =$	0.000159	
22			${}^{bull}\beta_{Pf}^{monthly} =$	0.9726	${}^{bear}\beta_{Pf}^{monthly} =$	0.9188	

Source: Own, for illustrative purposes only

All calculation results are provided in Table 1.14. As can be seen in cell $D22$, the bull market beta stands at 0.9726, i.e., if the benchmark increases by 1 %, the portfolio, on average, increases only by 0.9726 %. The bear market beta of 0.9191 indicates that when the benchmark decreases by 1 %, the fund decreases only by 0.9191 %. Such a portfolio would behave nicely during up- and downward movements of the market: the bull beta should be much higher than the bear beta which is here the case. Ideally, the bull beta should be higher than 1 and the bear beta should be close to 0 or even negative, but the latter is difficult to achieve with long-only portfolios.

End of Example 9

1.3.6.2 Conclusion

The bull and bear beta are plausible concepts that build on the general beta concept. Investors can more thoroughly analyze a portfolio's behavior compared to the market development. Therefore, it makes sense to not only look at the overall beta but also to consider the bull and bear market beta.

1.3.7 Sharpe Ratio

In order to analyze a portfolio we have presented various return and risk measures. However, how do you compare two portfolios with different risks? This calls for adjusting return by risk: The portfolio's *efficiency* has to be measured, i.e., the achieved return per unit of risk. Mathematically, this means that we divide a return measure by a corresponding risk measure.

Definition: Sharpe Ratio

The Sharpe ratio or, as it is also sometimes called, the *reward-to-variability ratio* is the measure of a portfolio's percentage return per unit of absolute risk (volatility), i.e., a *risk-adjusted return ratio*. It applies to a portfolio and therefore, especially also to a single asset.

The portfolio return can be expressed in two ways. It may be indicated by simply using the percentage return of the portfolio alone, but it can also be formulated using the portfolio's percentage return over a risk-free interest rate, the so-called *risk premium*.

To formally define the Sharpe ratio SR, let $[0, T]$ be the analyzed historical time period with $T \geq 1$ year. A time period of this length is required, since the Sharpe ratio is based on annualized data. There are two definitions for the Sharpe ratio of the portfolio and the benchmark[45]:

(continued)

[45]Lhabitant (2004, p. 65).

$$SR_{Pf} = SR_{Pf}^{p.a.} = \frac{r_{Pf}^{p.a.}}{\sigma_{Pf}^{p.a.}} \tag{1.74}$$

or

$$SR_{Pf} = SR_{Pf}^{p.a.} = \frac{r_{Pf}^{p.a.} - r_{rf}^{p.a.}}{\sigma_{Pf}^{p.a.}}. \tag{1.75}$$

Hereby:

$r_{Pf}^{p.a.}$ = annualized return of the portfolio,

$r_{rf}^{p.a.}$ = annual risk-free interest rate,

$\sigma_{Pf}^{p.a.}$ = annualized volatility of the portfolio.

1.3.7.1 Notes

- The Sharpe ratio is a risk-adjusted return measure in the absolute world, i.e., no benchmark is considered.[46] It was developed by William Sharpe[47] in 1966.[48]
- The definition using the risk premium is a key part of modern portfolio theory when deriving the efficient frontier and the CAPM as will be done in Chap. 2. But this definition of the Sharpe ratio requires the specification of the risk-free rate. In order to avoid this, the first definition is often used in risk and performance measurement practice.
- Also for portfolios which are actively managed against a benchmark it makes sense to calculate the Sharpe ratio. In this situation, Eqs. (1.74) and (1.75) are applied and the two Sharpe ratios are compared. In active portfolio management one would expect the Sharpe ratio of the portfolio to be higher than the benchmark's Sharpe ratio over the same time period using the same subintervals.

1.3.7.2 Interpretation

If we analyze a portfolio that is managed against a benchmark, it does not make sense to only look at the portfolio's Sharpe ratio. In this situation we have to compare the Sharpe ratios of the portfolio and the benchmark. For passively managed

---

[46] If we include a benchmark, the risk-adjusted return measure is a relative measure versus a benchmark called *information ratio*. This concept is explained in Sect. 1.3.8.

[47] William F. Sharpe was born on June 16, 1934, in Boston, MA/USA. He is the STANCO 25 Professor of Finance, Emeritus at Stanford University's Graduate School of Business and the winner of the 1990 Nobel Memorial Prize in Economic Sciences. Sharpe was one of the originators of the capital asset pricing model (CAPM) and created the Sharpe ratio for risk-adjusted investment performance analysis. He contributed to the development of the binomial method for the valuation of options, the gradient method for asset allocation optimization, and returns-based style analysis for evaluating the style and performance of investment funds.

[48] Sharpe (1966, p. 123).

Table 1.15 Example 10: Calculation of the Sharpe ratio for portfolio and benchmark

	A	B	C
1	Month	Monthly portfolio performance	Monthly benchmark performance
2	07/2012	6.10 %	6.01 %
3	08/2012	5.50 %	5.45 %
4	09/2012	4.70 %	4.63 %
5	10/2012	−5.00 %	−6.99 %
6	11/2012	−5.10 %	−4.16 %
7	12/2012	6.70 %	7.07 %
8	01/2013	6.03 %	5.97 %
9	02/2013	−3.23 %	−2.95 %
10	03/2013	5.12 %	4.66 %
11	04/2013	5.21 %	4.91 %
12	05/2013	−4.10 %	−4.01 %
13	06/2013	−4.50 %	−3.87 %
14	07/2013	1.75 %	−2.95 %
15	08/2013	3.71 %	4.52 %
16	09/2013	−4.20 %	−3.93 %
17	10/2013	4.26 %	4.99 %
18	11/2013	−4.00 %	−3.84 %
19	12/2013	5.10 %	4.99 %
20	Annual. return	15.79 %	13.02 %
21	Annual. volatility	16.49 %	17.02 %
22	Sharpe ratio	0.96	0.76

Source: Own, for illustrative purposes only

portfolios we expect these two to be (almost) identical. For actively managed portfolios we expect the portfolio's Sharpe ratio to be higher.

If we look at an absolute return portfolio like a hedge fund, the Sharpe ratio has explanatory power on a stand-alone basis: the higher its value, the better.

Example 10

This example, displayed in Table 1.15, shows how the Sharpe ratio can be calculated for a portfolio which is managed against a benchmark. Column A lists the month, column B the monthly portfolio return, column C the monthly benchmark return. To calculate the Sharpe ratio, we choose the formula without the risk-free rate to avoid additional calculations which are unnecessary for the interpretation. Further, $T = 1.5$ years and $N = 18$ months.

The lower part of the table shows the performance, risk and risk-adjusted performance of the portfolio and benchmark. The Excel® calculations behind these results are as follows:

- Annualized portfolio return $r_{Pf}^{p.a.}$ in cell $B20$:

$$15.79\,\% \quad = \quad \{PRODUCT(1 + (B2 : B19))\textasciicircum(12/18) - 1\}$$

- Annualized benchmark return $r_{Bm}^{p.a.}$ in cell $C20$:

  $13.02\,\% \;=\; \{PRODUCT(1 + (C2:C19))\hat{}(12/18) - 1\}$

- Annualized portfolio volatility $\sigma_{Pf}^{p.a.}$ in cell $B21$:

  $16.49\,\% \;=\; SQRT(12) * STDEV.S(B2:B19)$

- Annualized benchmark volatility $\sigma_{Bm}^{p.a.}$ in cell $C21$:

  $17.02\,\% \;=\; SQRT(12) * STDEV.S(C2:C19)$

- Sharpe ratio of portfolio SR_{Pf} in cell $B22$:

  $0.96 \;=\; B20/B21$

- Sharpe ratio of benchmark SR_{Bm} in cell $C22$:

  $0.76 \;=\; C20/C21$

The portfolio has a higher Sharpe ratio than the benchmark, i.e., for one unit of risk (volatility) the investor in the portfolio receives a higher additional return than when passively investing in the index.

End of Example 10

1.3.7.3 Conclusion

The Sharpe ratio is a key ratio when evaluating a portfolio. It is an absolute risk-adjusted return measure as it does not look at portfolio returns versus a benchmark, but rather at the portfolio returns by themselves. In order to calculate the Sharpe ratio, either the portfolio's percentage return or the return premium (return minus risk-free rate) can be used. The associated risk is always volatility. In practice, the risk-free rate should be omitted to avoid questions on the choice of the risk-free rate or its computation. For benchmark-oriented portfolios, the Sharpe ratio should be calculated both for the portfolio and the benchmark in order to find out if the portfolio or the benchmark delivers a better risk-adjusted return in absolute terms.

However, what risk-adjusted return measure would we need to analyze a portfolio versus the benchmark performance? In principle, a *relative risk-adjusted return measure* is very similar to the *absolute risk-adjusted return measure* Sharpe ratio. We only need to replace the absolute return by the relative return (alpha) and the absolute risk measure (volatility) by the relative risk measure (tracking error). This new risk-adjusted return measure is called *information ratio*.

1.3.8 Information Ratio

Similar to Sect. 1.3.7 where we defined the Sharpe ratio as the risk-adjusted return measure in an absolute (i.e., benchmark-agnostic) world, we now introduce the information ratio as its *relative* counterpart.

Definition: Information Ratio
Let $[0, T]$ with $T \geq 1$ year be the analyzed historical time period. The information ratio IR measures the annualized relative return (alpha) of a portfolio generated per unit of annualized relative risk (tracking error) for a portfolio that is managed against a benchmark[49]:

$$IR = IR^{p.a.} = \frac{\alpha^{p.a.}}{TE^{p.a.}} = \frac{r_{Pf}^{p.a.} - r_{Bm}^{p.a.}}{TE^{p.a.}}. \tag{1.76}$$

Hereby:

$r_{Pf}^{p.a.}$ = annualized portfolio return using data from time period $[0, T]$,

$r_{Bm}^{p.a.}$ = annualized benchmark return using data from time period $[0, T]$,

$\alpha^{p.a.}$ = annualized excess return of portfolio over its benchmark, using data from time period $[0, T]$,

$TE^{p.a.}$ = annualized tracking error of the portfolio vs. its benchmark, using data from time period $[0, T]$.

Ideally, a high information ratio is achieved by having a high positive alpha and a low tracking error.

1.3.8.1 Note

Since at least 1 year of performance data is needed to calculate tracking error, the calculation of the information ratio for time periods below 1 year is meaningless. When using monthly return data, ideally 3 years of data should be available to get a meaningful value. Tracking error and alpha have to be annualized. For the calculation of the information ratio it is important to use the same underlying return data, especially, with the same data frequency (for example, monthly or daily).

1.3.8.2 Interpretation

For example, a typical actively managed equity fund, the information ratio can (based on practical experience) broadly be categorized in the following way, if the underlying time series is sufficiently long (ideally five or more years):

- $IR \approx 1.5$ and higher: top portfolio
- $IR \approx 0.8 - 1$: very good portfolio
- $IR \approx 0.5$: average portfolio

[49] Lhabitant (2004, p. 67).

- $IR \approx 0.2$: poor portfolio
- IR negative: bad portfolio since it has a negative alpha

This interpretation is only valid using *gross-of-fee* return data, (i.e., without considering fees for management, which can be substantial, for example, for portfolios of emerging markets equity). Therefore, it would be more precise to talk about a *gross-of-fee information ratio*. In contrast, if the fees were already subtracted from the return (*net-of-fee*), the interpretation for such a *net-of-fee information ratio* would look slightly different and depend on the charged fees. The information ratio is *the* risk-adjusted return measure for actively managed portfolios irrespective of if the investment approach is fundamental, quantitative or a hybrid of both.

Example 11

Table 1.16 is an extension of Table 1.7 to calculate the information ratio. Use the following Excel® formulas to obtain the information ratio:

Table 1.16 Example 11: Calculation of a portfolio's information ratio

	A	B	C	D
1	Month	Monthly portfolio performance (%)	Monthly benchmark performance (%)	Monthly alpha (%)
2	07/2012	6.10 %	6.01 %	0.09 %
3	08/2012	5.50 %	5.45 %	0.05 %
4	09/2012	4.70 %	4.63 %	0.07 %
5	10/2012	−5.00 %	−6.99 %	1.99 %
6	11/2012	−5.10 %	−4.16 %	−0.94 %
7	12/2012	6.70 %	7.07 %	−0.37 %
8	01/2013	6.03 %	5.97 %	0.06 %
9	02/2013	−3.23 %	−2.95 %	−0.28 %
10	03/2013	5.12 %	4.66 %	0.46 %
11	04/2013	5.21 %	4.91 %	0.30 %
12	05/2013	−4.10 %	−4.01 %	−0.09 %
13	06/2013	−4.50 %	−3.87 %	−0.63 %
14	07/2013	1.75 %	−2.95 %	4.70 %
15	08/2013	3.71 %	4.52 %	−0.81 %
16	09/2013	−4.20 %	−3.93 %	−0.27 %
17	10/2013	4.26 %	4.99 %	−0.73 %
18	11/2013	−4.00 %	−3.84 %	−0.16 %
19	12/2013	5.10 %	4.99 %	0.11 %
20			$TE^{monthly} =$	1.29 %
21			$TE^{p.a.} =$	4.48 %
22	$r_{Pf}^{p.a.} =$	15.79 %	$r_{Bm}^{p.a.} =$	13.02 %
23	$IR =$	0.62		

Source: Own, for illustrative purposes only

- Annualized portfolio return $r_{Pf}^{p.a.}$ in cell $B22$:
  $$15.79\,\% \quad = \quad \{(PRODUCT(1 + B2 : B19))\hat{}(12/18) - 1\}$$
- Annualized benchmark return $r_{Bm}^{p.a.}$ in cell $D22$:
  $$13.02\,\% \quad = \quad \{(PRODUCT(1 + C2 : C19))\hat{}(12/18) - 1\}$$
- Monthly tracking error $TE^{monthly}$ in cell $D20$:
  $$1.29\,\% \quad = \quad STDEV.S(D2 : D19)$$
- Annualized tracking error $TE^{p.a.}$ in cell $D21$
  $$4.48\,\% \quad = \quad SQRT(12) * STDEV.S(D2 : D19)$$
- Information ratio IR of the portfolio in cell $B23$:
  $$0.62 \quad = \quad (B22 - D22)/D21$$

An information ratio of 0.62 is rather average. However, a meaningful interpretation of this value is difficult as only 18 months of performance data are available.

End of Example 11

1.3.8.3 Conclusion

The information ratio is used to evaluate the added value (alpha) of an actively managed portfolio and, thereby, the quality of its portfolio manager on a risk-adjusted basis. The higher the information ratio, the better. In the long run, an information ratio of 1 is very good: It indicates that for each additional percentage of tracking error the active manager can generate an additional percentage point of alpha.

1.3.9 Treynor Ratio

The final traditional risk-adjusted return measure that will be presented here is the *Treynor ratio* introduced by Jack Treynor[50] in 1985 as the *reward-to-volatility ratio*.[51] It is based on the differentiation of risk (volatility) into systematic and unsystematic risk. This idea was already presented in Sect. 1.3.5 and graphically shown in Fig. 1.20. The key question here is: why use the overall risk as a risk measure for calculating a risk-adjusted return, when part of this overall risk is the

[50] Jack Treynor was born 1930 and is a U.S. financial engineer and portfolio manager. He studied mathematics at Haverford College and completed the MBA program at Harvard Business School 1955. In 2007, the International Association of Financial Engineers (IAFE) named Treynor as the 2007 IAFE/SunGard Financial Engineer of the Year (FEOY), recognizing him for his preeminent contributions to financial theory and practice, particularly the essence of the CAPM.

[51] Jorion (2001, p. 395).

systematic risk of the index? If volatility is used as the only risk measure, systematic and unsystematic risk are mixed together, and the portfolio's beta is hidden.

While the Sharpe ratio uses systematic and unsystematic (idiosyncratic) risk together in the form of volatility as the risk measure, the Treynor ratio uses the systematic risk component of a portfolio.

Definition: Treynor Ratio

The definition of the Treynor ratio *TR* is similar to the definition of the Sharpe ratio in Eq. (1.74). However, the Treynor ratio uses beta as risk measure. To formally define the Treynor ratio, let $[0, T]$ be the analyzed historical time period with $T \geq 1$ year. Then[52]:

$$TR = TR^{p.a.} = \frac{r_{Pf}^{p.a.} - r_{rf}^{p.a.}}{\beta_{Pf}^{p.a.}}. \tag{1.77}$$

Hereby:

$$r_{Pf}^{p.a.} = \text{annualized return of the portfolio,}$$

$$r_{rf}^{p.a.} = \text{annual risk-free interest rate,}$$

$$\beta_{Pf}^{p.a.} = \text{beta of the portfolio versus benchmark.}$$

1.3.9.1 Interpretation

The Treynor ratio measures the excess return of a portfolio over the risk-free interest rate versus the engaged systematic risk measured by the portfolio beta versus the index. Therefore, the Treynor ratio is a symmetrical and relative risk-adjusted return measure. It shows how much excess return over the risk-free rate the portfolio manager can generate per unit of beta.

Equation (1.77) of the Treynor ratio can also be used for the benchmark instead of the portfolio. In this case the different Treynor ratios would be indicated by a subscript. However, in this case the benchmark beta is 1 and the return difference is simply the benchmark return minus the risk-free rate. The latter is known as the *equity risk premium* which plays a significant role in the models presented in Chap. 2.

Example 12

Table 1.17 shows various ratios for two portfolios A and B managed versus a benchmark. The data of portfolio A and the benchmark are the data used in the

[52]Lhabitant (2004, p. 75).

Table 1.17 Example 12: Calculation of Treynor ratio and comparison, $r_{rf}^{monthly} = 0.2\,\%$

	A	B	C	D
1	Month	Monthly performance of portfolio A	Monthly performance of portfolio B	Monthly performance of benchmark
2	07/2012	6.10 %	7.71 %	6.01 %
3	08/2012	5.50 %	6.38 %	5.45 %
4	09/2012	4.70 %	4.83 %	4.63 %
5	10/2012	−5.00 %	−7.70 %	−6.99 %
6	11/2012	−5.10 %	−4.90 %	4.16 %
7	12/2012	6.70 %	5.32 %	7.07 %
8	01/2013	6.03 %	6.48 %	5.97 %
9	02/2013	−3.23 %	−3.95 %	−2.95 %
10	03/2013	5.12 %	3.80 %	4.66 %
11	04/2013	5.21 %	5.20 %	4.91 %
12	05/2013	−4.10 %	−3.00 %	−4.01 %
13	06/2013	−4.50 %	−2.01 %	−3.87 %
14	07/2013	1.75 %	−3.54 %	−2.95 %
15	08/2013	3.71 %	4.90 %	4.52 %
16	09/2013	−4.20 %	−1.62 %	−3.93 %
17	10/2013	4.26 %	6.05 %	4.99 %
18	11/2013	−4.00 %	−4.10 %	−3.84 %
19	12/2013	5.10 %	6.20 %	4.99 %
20	Annualized return	15.79 %	17.11 %	13.02 %
21	Annualized volatility	16.49 %	17.63 %	17.02 %
22	Beta	0.93	1.01	1
23	Sharpe ratio	0.96	0.97	0.76
24	Treynor ratio	0.14	0.15	0.11

Source: Own, for illustrative purposes only

previous tables. Looking at the Sharpe ratio and the absolute returns, portfolio B looks better than portfolio A. The Treynor ratio which is slightly higher for portfolio B than portfolio A supports this assessment, but portfolio B contains more systematic risk than portfolio A.

Below we show the calculations needed to obtain the results in Table 1.17:

- Annualized return $r_A^{p.a.}$ of portfolio A in cell $B20$:

  $15.79\,\% \;=\; \{PRODUCT(1 + B2 : B19)\hat{}(12/18) - 1\}$

- Annualized return $r_B^{p.a.}$ of portfolio B in cell $C20$:

  $17.11\,\% \;=\; \{PRODUCT(1 + C2 : C19)\hat{}(12/18) - 1\}$

- Annualized return $r_{Bm}^{p.a.}$ of the benchmark in cell $D20$:

  $13.02\,\% \;=\; \{PRODUCT(1 + D2 : D19)\hat{}(12/18) - 1\}$

- Annualized volatility $\sigma_A^{p.a.}$ of portfolio A in cell $B21$:
  $16.49\,\% \quad = \quad SQRT(12) * STDEV.S(B2 : B19)$
- Annualized volatility $\sigma_B^{p.a.}$ of portfolio B in cell $C21$:
  $17.63\,\% \quad = \quad SQRT(12) * STDEV.S(C2 : C19)$
- Annualized volatility $\sigma_{Bm}^{p.a.}$ of the benchmark in cell $D21$:
  $17.02\,\% \quad = \quad SQRT(12) * STDEV.S(D2 : D19)$
- Beta $\beta_A^{monthly}$ of portfolio A versus benchmark in cell $B22$:
  $0.93 \quad = \quad COVARIANCE.S(B2 : B19, D2 : D19)/VAR.S(D2 : D19))$
- Beta $\beta_B^{monthly}$ of portfolio B versus benchmark in cell $C22$:
  $1.01 \quad = \quad COVARIANCE.S(C2 : C19, D2 : D19)/VAR.S(D2 : D19))$
- Beta $\beta_{Bm}^{monthly}$ of the benchmark versus the benchmark in cell $D22$:
  $1 \quad = \quad COVARIANCE.S(D2 : D19, D2 : D19)/VAR.S(D2 : D19))$
- Sharpe ratio SR_A of portfolio A in cell $B23$:
  $0.96 \quad = \quad B20/B21$
- Sharpe ratio SR_B of portfolio B in cell $C23$:
  $0.97 \quad = \quad C20/C21$
- Sharpe ratio SR_{Bm} of the benchmark in cell $D23$:
  $0.76 \quad = \quad D20/D21$
- Treynor ratio TR_A of portfolio A in cell $B24$:
  $0.14 \quad = \quad (B20 - ((1 + 0.2\,\%)\wedge 12 - 1))/B22$
- Treynor ratio TR_B of portfolio B in cell $C24$:
  $0.15 \quad = \quad (C20 - ((1 + 0.2\,\%)\wedge 12 - 1))/C22$
- Treynor ratio TR_{Bm} of the benchmark in cell $D24$:
  $0.11 \quad = \quad (D20 - ((1 + 0.2\,\%)\wedge 12 - 1))/D22$

End of Example 12

1.3.9.2 Note

According to Spremann, in most cases the values of the Treynor ratio range between -0.3 and $+0.5$.[53] Important to note is also that the Treynor ratio can only be

[53]Spremann (2008, p. 381).

meaningfully applied to portfolios whose betas are calculated against the same index.

1.3.9.3 Conclusion

The Treynor ratio is a symmetrical, relative risk-adjusted return measure for the portfolios managed against a benchmark. It measures the added value per unit of beta, i.e., per unit of systematic risk. Thereby, the Treynor ratio does not consider the idiosyncratic risk of a portfolio.

1.4 Advanced Risk and Risk-Adjusted Return Measures

Until now, we have presented the typical risk and risk-adjusted return measures which are symmetrical measures: they treat a gain and a loss of the same magnitude in the same way. In addition, a neat interpretation for both of these risk measures is based on the assumption that the underlying return data are normally distributed. While this assumption is questionable in general, it became even more dubious when the most recent financial crisis started in August 2007 with the so-called *quant crisis*.

In August 2007, over-leveraged hedge funds started to aggressively unwind their positions and continued to do so through the last few months of 2007. Increased risk aversion and immediate margin calls led to what might be described as clumsy, reactionary trading during August. To meet capital requests from liquidations, traders at these hedge funds had between one and three months, depending on the notice period of the fund, to structure more sophisticated, non-linear trades to both mask and hedge the impact and size of the trades. This was evident, to some extent, in August 2007 and had a highly significant influence on the market and quant asset managers in general in November 2007.

Hedge funds tried to hide their liquidity redemption needs through carefully structured trades. This becomes clear from an analysis of the quant equity strategies that have performed poorly around the world: most of the underperforming strategies were from the developed markets, where hedge funds are prominent. Regions such as Pacific ex-Japan and the emerging markets, where hedge funds are scarcer, did not suffer such extremes in relative performance vs. index.

Although quant models of quantitative asset managers vary greatly in terms of their specific inputs, quant factors probably had some overlap and correlation with the reactionary trades carried out by hedge funds in August. The large variance of intra-month alpha in August 2007 shows that liquidity had a significant impact on excess returns. Whilst not as dramatic, excess returns in November 2007 were still subject to liquidity pressures: In November, growing concerns over inflationary risks undermined expectations that Western central banks would cut interest rates in order to mitigate the impact of the credit crisis and the slowing economic growth.

The crisis, often called *subprime crisis*, continued in 2008 and deepened with the default of Lehman Brothers in September 2008. After a strong recovery in the second, third and fourth quarter of 2009 and good returns in 2010, the crisis came back in 2011 because of the euro problems. Figure 1.24 shows the changes of

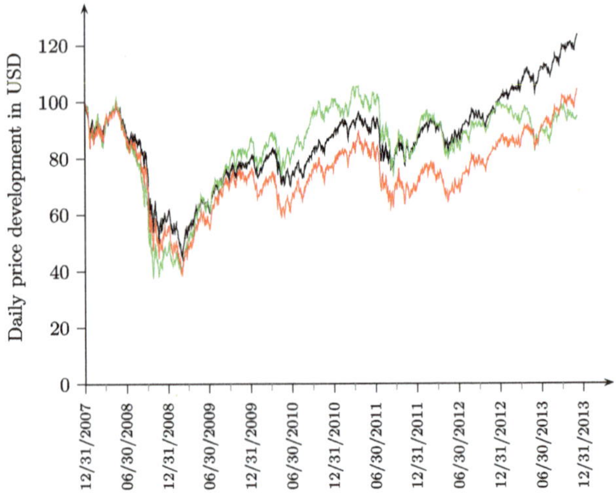

Fig. 1.24 Daily price development of the MSCI World Index (in *black*), the MSCI Emerging Markets Index (in *green*) and the MSCI Europe Index (in *red*). All indices are net dividends reinvested and in USD from December 31, 2007, (base value = 100) to December 31, 2013. *Source*: Factset (*Tickers*: 990100, 891800 and 990500)

various key indices: the MSCI World Equity Index, the MSCI Emerging Markets Equity Index and the MSCI Europe Index, based on an investment of $100 on December 31, 2007. Figure 1.25 shows the corresponding drawdown diagrams[54] for the indices displayed in Fig. 1.24.

The extreme movements in 2008 have been called the *Great Recession* echoing the *Great Depression* that started 1929. The crisis entered a new phase when, in spring 2010, the debt of countries like Greece and the resulting euro crisis took center stage.[55]

In the financial markets, movements with high and highest volatility as well as so-called *Black Swan* events have occurred frequently since 2008. A Black Swan is an unexpected event of large magnitude, an extreme outlier which compared to the normal distribution assumption should not or only rarely occur.[56] A common feature of the equity markets worldwide, be it developed or emerging markets, were the extremely high volatility levels: market downturns, especially end of 2008 and early 2009, alternated with strong market upturns, for example, between April and

[54]A drawdown diagram shows the development of an index whereby positive subinterval returns are only taken into account if the index value is below 1. If a subinterval return is positive while the index value is already 1, the index value will remain at 1. A drawdown diagram shows the losses an index suffers and how long it takes to regain these losses and reach the initial investment of 1.

[55]Portugal, Italy, Ireland, Greece and Spain were often referred to as the PIIGS countries. Common to all of them were the increasing debt ratios that in many cases already had been high even before the crisis.

[56]See page 77 for more details.

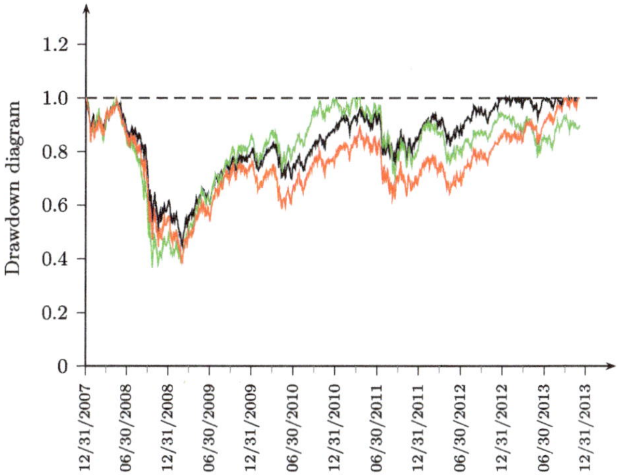

Fig. 1.25 Corresponding drawdown diagrams of all indices displayed in Fig. 1.24: MSCI World Index in *black*, Emerging Markets Index in *green* and MSCI Europe Index in *red*. All indices are net dividends reinvested and in USD from December 31, 2007 to December 31, 2013. *Source*: Factset (*Tickers*: 990100, 891800 and 990500)

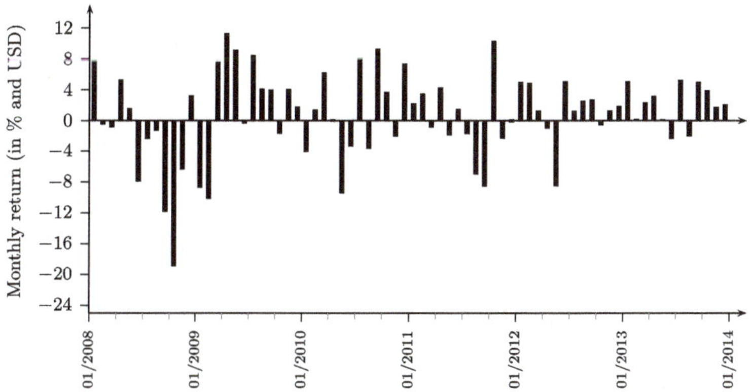

Fig. 1.26 Monthly returns of the MSCI World Index (net dividends reinvested) in USD from January 2008 to December 2013. *Source*: Factset (*Ticker*: 990100)

December 2009. The instability of the markets is visible in Fig. 1.26 which shows the monthly returns of the MSCI World Index (net dividends reinvested) in USD from January 2008 to December 2013.

In addition, Fig. 1.27 displays the VIX Index which measures the implied volatility of the S&P 500 Index, covering the same time period from January 1, 2008 to December 31, 2013.

Obviously, the VIX Index increased significantly in the middle of 2008 and spiked during September 2008 when Lehman Brothers went bankrupt. Thereafter and over the course of 2009 the implied volatility fell significantly before steeply

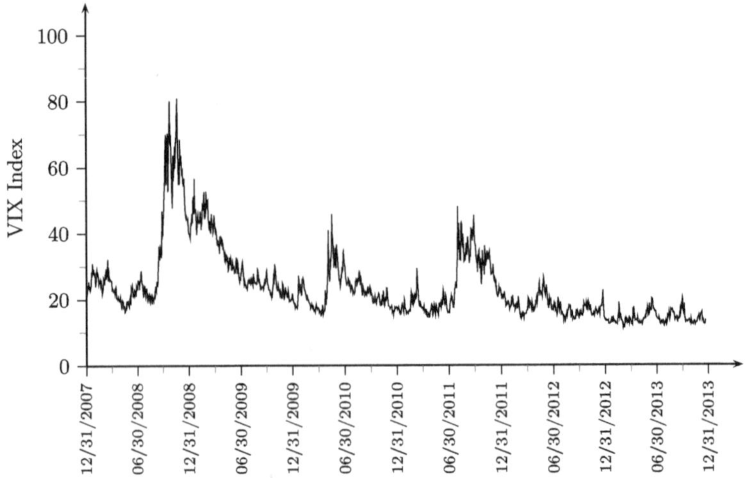

Fig. 1.27 Daily development of the VIX Index from January 1, 2008 to December 31, 2013. *Source*: Factset (*Ticker*: VIX)

Table 1.18 Black Swans: Extreme daily S&P 500 Index returns in recent history

Day	S&P 500 Index return (in USD)
October 19, 1987	−17.13 %
October 25, 1987	−8.26 %
April 14, 2000	−6.63 %
August 31, 1998	−6.59 %
October 27, 1997	−6.53 %
January 8, 1997	−5.54 %
October 13, 1989	−5.34 %
September 17, 2001	−5.07 %
October 16, 1987	−4.72 %
September 11, 1986	−4.35 %

Source: Bloomberg

rising again early 2010 when it seemed as if the crisis of 2008 would return. A similar situation could be observed middle of 2011. Thereafter, the VIX Index returned to pre-Lehman levels as shown in the figure.

Such extreme market movements with corresponding spikes in market-implied volatility could already be observed in previous crises. Table 1.18 provides an overview of various extreme daily returns of the S&P 500 Index in the U.S.

Taleb[57] calls such events *Black Swans*[58] and described them in detail in Taleb (2010). In practice, most return distributions are neither normal nor even symmetrical. An even more troublesome feature of real life return distributions are so-called *fat tails*. This term describes the fact that extreme positive and extreme negative returns are much more likely than a normal distribution assumption implies. Obviously, fat tails on the positive side are not harmful but rather wanted. However, the trouble stems from fat tails on the negative side. Events like in 2007 and 2008 contributed greatly to fat tails on the left hand side of the return distributions.

The idea of confidence intervals can be used to interpret the extreme market events cited in Table 1.18. For example, there is a 0.001 % probability of the outcome of a 6σ (standard deviation) event, i.e., it should occur only once in 10,000 trading days (roughly 40 years). However, such events were much more frequent in the past 40 years, resulting in asymmetrical return distributions.

It is questionable how relevant the dispersion of returns around an average (as measured by symmetrical risk measures) is from an investor's standpoint. Indeed, investors are more averse to negative deviations from the mean value than pleased with positive ones of the same magnitude. Most only perceive risk as a failure to achieve a specific goal (for example, achieving a risk-free rate of return or the benchmark return) or as losing part of the initial investment. This is why we now need to look for tools that allow to measure, loosely speaking, the left hand side of the return distribution.

Table 1.19 shows the month-end dates of 60-month market lows and declines from peak values during previous 59 month-end dates as measured by the CRSP[59] value-weighted market index (December 1925–March 2010). Month-end dates of 60-month lows are based on S&P 500 Index values. Declines from peak values are calculated using the broader value-weighted index provided by the CRSP inclusive of dividend reinvestment.

Table 1.20 shows the 1- and 5-year market returns following extreme month-end market lows. The returns are calculated using the value-weighted market index provided by the CRSP and include dividend reinvestment.

[57]Nassim N. Taleb (born January 1, 1960, in Amioun, Lebanon) is an essayist, scholar and former practitioner of mathematical finance. He is best known as the author of the book *The Black Swan*. Taleb has pursued three distinct careers. Firstly, he is a bestselling author with about 3 millon copies sold in over 30 languages. Secondly, he is a university professor in risk engineering, a scholar, an epistemologist and a philosopher of science. Finally, he is a former senior Wall Street trader, risk expert, and practitioner of mathematical finance. *The Black Swan* has been described by *The Times* as one of the 12 most influential books since World War II.

[58]The Black Swan theory or *theory of Black Swan events*, was developed by Nassim N. Taleb. It explains the disproportionate role of high-impact, hard-to-predict, and rare events that are beyond the realm of normal expectations in history, science, finance and technology. Unlike the earlier philosophical *Black Swan problem*, Taleb's *Black Swan theory* refers only to unexpected events of large magnitude and consequence and their dominant role in history. Such events, considered extreme outliers, collectively play vastly larger roles than regular occurrences.

[59]CRSP is the Center for Research in Security Prices at the University of Chicago.

Table 1.19 Extreme stock market lows

Episode	Month-end date (yyyy-mm)	CRSP value-weighted market index	Index decline from peak value during previous 59 months (in %)	S&P 500 market index
1	1931-09	1.18	−68.30	9.71
1	1931-11	1.16	−68.82	9.50
1	1931-12	1.01	−72.98	8.12
1	1932-01	1.00	−73.27	7.89
1	1932-03	0.94	−74.85	7.31
1	1932-04	0.77	−79.35	5.83
1	1932-05	0.610	−83.64	4.47
1	1932-06	0.607	−83.72	4.43
2	1942-03	1.91	−34.16	8.01
2	1942-04	1.82	−37.00	7.66
3	1970-05	59.80	−29.72	76.55
3	1970-06	56.74	−33.30	72.72
4	1974-08	60.47	−39.53	72.15
4	1974-09	53.84	−46.16	63.54
5	2002-09	1748.88	−44.88	815.29
6	2008-10	2510.65	−37.51	968.75
6	2008-11	2296.74	−42.84	896.24
6	2009-01	2164.36	−46.13	825.88
6	2009-02	1945.84	−51.57	735.09

Source: Reinganum (2010, p. 1)

Table 1.20 One- and five-year market returns following extreme month-end market lows, calculated using the CRSP value-weighted market index (including dividend payments)

Episode	Month-end date (yyyy-mm)	Subsequent 1-year market return (total and in %)	Subsequent 5-year market return (total and in %)	Subsequent 5-year market return (annualized and in %)
1	1932-06	156.35	337.82	34.36
2	1942-04	61.02	151.62	20.27
3	1970-06	44.30	49.12	8.32
4	1974-09	38.72	144.66	19.59
5	2002-09	27.59	123.94	17.50
6	2009-02	58.22	N/A	N/A

Source: Reinganum (2010, p. 2)

Table 1.21 shows corrections after initial market rallies that follow extreme stock market lows. The exact dates of stock market lows are determined using the value-weighted market index provided by the CRSP. The return advances during market rallies and return declines in the subsequent corrections include dividend reinvestment as calculated by the CRSP. The market correction return for episode

Table 1.21 Corrections after initial market rallies that follow extreme stock market lows, calculated using the CRSP value-weighted market index (including dividend payments)

Episode	Date of market low (mm/dd/yyyy)	Initial market rally (in %)	Length of rally in calendar days	Subsequent market correction (in %)	Length of correction in calendar days
1	07/08/1932	+102.10	61	−36.15	173
2	04/28/1942	+81.96	442	−10.44	138
3	07/07/1970	+54.18	295	−12.86	209
4	10/03/1974	+61.17	285	−13.56	78
5	10/09/2002	+21.16	49	−13.2	104
6	03/09/2009	+79.93	410	−12.52	42

Source: Reinganum (2010, p. 3)

6 is calculated using the S&P 500 Index (without dividend reinvestment) over the period from April 23, 2010, through June 4, 2010.

All these tables give a feeling for how volatile the market can get and how quickly a market drawdown can occur. Especially, Table 1.21 already indicates what will be the topic of Chap. 3: the bubble before the crash. Therefore, we need to look at measures for such negative events. Depending on the definition of *negative event* various asymmetrical risk measures can be defined.

1.4.1 Maximum Absolute Drawdown

Definition: Maximum Absolute Drawdown

The *absolute drawdown*[60] measures the cumulative loss of a portfolio that sustains consecutive negative returns. It is calculated using the absolute returns of the portfolio. The largest cumulative loss is called the *maximum absolute drawdown MADD.*

Example 13

Although the maximum absolute drawdown can be defined mathematically, we omit the mathematical definition here because of its complexity. The idea, however, can be easily explained with an example. As usual, we split interval $[0, T]$ into N equidistant subintervals. For the observed time period $[0, T]$, the value of T is 1.5 years with $N = 18$ representing the 18 monthly subintervals. Figure 1.28 explains the situation graphically, spanning the 18 months from July

[60] Lhabitant (2004, pp. 55–56).

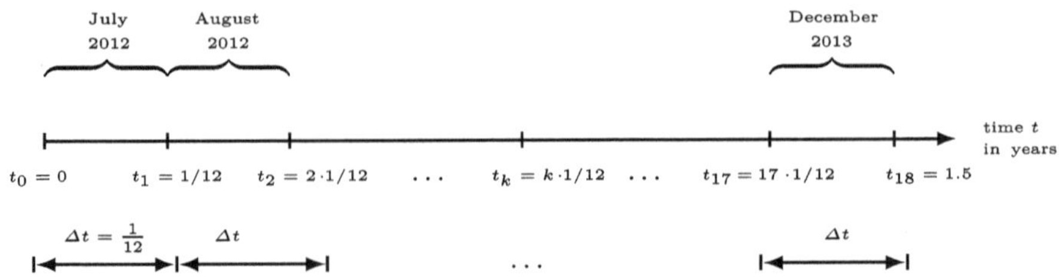

Fig. 1.28 Notation concerning *MADD*. *Source*: Own, for illustrative purposes only

Table 1.22 Example 13: Calculation of the maximum absolute drawdown *MADD*

	A	B	C	D
1	Month	Monthly portfolio performance	Absolute drawdown (yes/no)?	Cumulative return in drawdown period
2	07/2012	6.10 %	No	
3	08/2012	5.50 %	No	
4	09/2012	4.70 %	No	
5	10/2012	−5.00 %	Yes	−5.00 %
6	11/2012	−5.10 %	Yes	−9.85 %
7	12/2012	6.70 %	No	
8	01/2013	6.03 %	No	
9	02/2013	−3.23 %	Yes	−3.23 %
10	03/2013	5.12 %	No	
11	04/2013	5.21 %	No	
12	05/2013	−4.10 %	Yes	−4.10 %
13	06/2013	−4.50 %	Yes	−8.42 %
14	07/2013	1.75 %	No	
15	08/2013	3.71 %	No	
16	09/2013	−4.20 %	Yes	−4.20 %
17	10/2013	4.26 %	No	
18	11/2013	−4.00 %	Yes	−4.00 %
19	12/2013	5.10 %	No	
20			*MADD* =	−9.85 %

Source: Own, for illustrative purposes only

2012 until December 2013. For each of these subintervals we note the absolute return of a portfolio. The data which are identical with Example 12 above are listed in Table 1.22 below, together with the calculations.

Figure 1.29 shows that there are five time periods that qualify as *drawdown periods* with consecutive negative returns. Such a drawdown period ends when it is followed by a subperiod (in our case a month) with a positive portfolio return. A drawdown period can consist of one single subperiod only. The choice of N, i.e., the choice of the subperiods (daily, weekly, monthly, etc.) plays a critical role in the calculation of the maximum absolute drawdown.

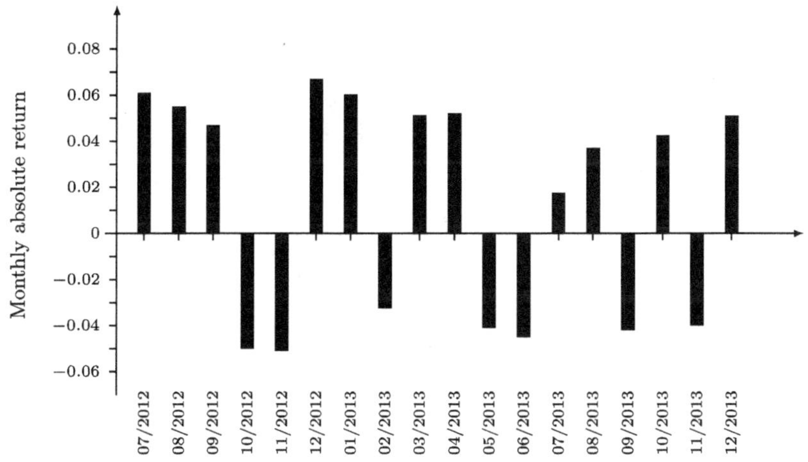

Fig. 1.29 Graphical illustration of the monthly returns in Example 13. *Source*: Own, for illustrative purposes only

Drawdown periods in our example:

- Drawdown period 1: October and November 2012, 2 months
- Drawdown period 2: February 2013, 1 month
- Drawdown period 3: May and June 2013, 2 months
- Drawdown period 4: September 2013, 1 month
- Drawdown period 5: November 2013, 1 month

The stated length in this list is the respective *drawdown duration*. After having determined the drawdown periods with the consecutive negative absolute returns we have to calculate the magnitude of the negative returns for each of the five drawdown periods using the idea of Eq. (1.4) from page 8. This yields the following results for each of the five drawdown periods in the analyzed time frame:

- Drawdown period 1: cumulative absolute return of −9.85 %
- Drawdown period 2: cumulative absolute return of −3.23 %
- Drawdown period 3: cumulative absolute return of −8.42 %
- Drawdown period 4: cumulative absolute return of −4.20 %
- Drawdown period 5: cumulative absolute return of −4.00 %

Therefore, the worst drawdown period was the first one spanning October and November 2012 with a drawdown duration of 2 months and a cumulative absolute return of −9.85 % which is the *MADD*. Figure 1.30 shows the same situation again, displaying the cumulative returns on the vertical axis. This allows to better spot the time point when the loss of a drawdown period is recovered. This so-called *recovery period* is often sought after by investors who

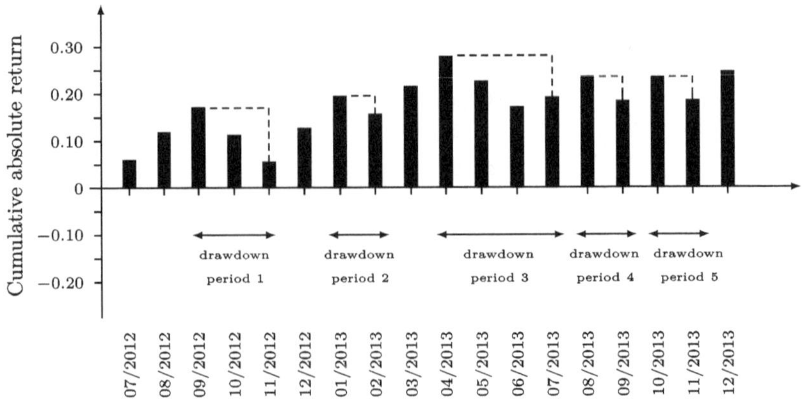

Fig. 1.30 Graph of the cumulative absolute returns in Example 13 which allows to spot the drawdown periods and recovery periods in a *MADD* situation. *Source*: Own, for illustrative purposes only

after consecutive losses want to know how long it takes (based on historical experience) to recover the losses.

As Fig. 1.30 shows, it only took 2 months after the *MDD* period to recover all the suffered losses. Table 1.22 below displays the data used in order to create Figs. 1.29 and 1.30.

The required Excel®-functions for calculating *MADD* are as follows:

- Occurrence of a drawdown (i.e., negative monthly return) in cell $C5$:

  $yes \;=\; IF(B5 < 0,\text{"yes","no"})$

- Cumulative portfolio return r_{Pf}^{cum} in drawdown period in cell $D5$:

  $-5.00\,\% \;=\; IF(B5 < 0, (1 + D4) * (1 + B5) - 1, 0)$

- Maximum absolute drawdown *MADD* in cell $D20$:

  $-9.85\,\% \;=\; MIN(D2 : D19)$

End of Example 13

Conclusion

The maximum drawdown is a typical downside risk measure. It looks at consecutive negative losses and states the worst loss over a historical time period, called maximum absolute drawdown *MADD*. Other things being equal, maximum absolute drawdowns will be greater if the length of the subperiods increases. It is much easier to have 2 months of consecutive negative monthly returns than having 40 (business) days of consecutive negative daily returns. Other things being equal, an *MADD* is also likely to be greater for longer historical time periods. Therefore, a portfolio with a longer track record will tend to have a more severe *MADD* as it will also include more periods of negative portfolio performance. In consequence, an *MADD* should be given with stating the observed historical time period and the considered time subperiods. Especially, when calculating the *MADD* for two different portfolios, it is mandatory to apply the *MADD* calculation to the same historical time period and the same subintervals. Otherwise the two calculated *MADD*s are not comparable.

1.4.2 Maximum Relative Drawdown

When looking at active portfolio management, however, clients—when evaluating a portfolio manager—are not interested in the absolute return or risk, but in the relative return or risk versus the assigned benchmark. They do not care about the absolute losses discussed above, but focus on the relative losses versus the index, i.e., on consecutive negative alphas. They simply want to know the worst drawdown relative to the benchmark and the time the asset manager needed in the past to recover these consecutive negative alphas.

Definition: Maximum Relative Drawdown
The *relative drawdown* measures the sustained cumulative relative loss of a portfolio versus its benchmark, i.e., consecutive negative alphas. The worst of these relative drawdowns is called *maximum relative drawdown MRDD*.

In this section, we will use an example—Example 14—which expands Example 13 from the previous section by including the benchmark that was already used in other examples before. Like *MADD*, the maximum relative drawdown can be defined mathematically, but we omit the mathematical definition due to complexity. The idea, however, can again be easily explained when looking at a picture, i.e., Fig. 1.31 below.

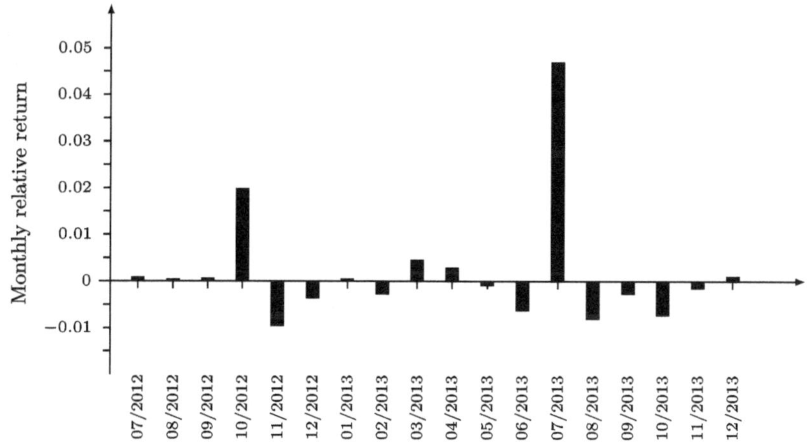

Fig. 1.31 Graphical illustration of the monthly excess returns in Example 14. *Source*: Own, for illustrative purposes only

1.4.2.1 Interpretation

We have dealt with *MADD* by looking at the drawdown of a portfolio alone. However, if we analyze an active portfolio managed against a benchmark, this absolute drawdown measure is irrelevant. What has to be analyzed is its relative counterpart *MRDD*. This measure provides information about the active asset manager's consecutive negative alphas and allows to analyze the asset manager's ability to regain positive alphas.

Example 14

Figure 1.31 shows the monthly excess returns of a portfolio versus its benchmark between July 2012 and December 2013. The data used are identical to Example 13 for *MADD* displayed in Table 1.22. However, we have now added the benchmark using data provided in Example 12, see Table 1.23. For the observed time period $[0, T]$ the value of T is 1.5 years with $N = 18$ representing the 18 subintervals (here: months).

In Fig. 1.31 we can identify four drawdown periods with consecutive negative alphas. Such a relative drawdown period ends when it is followed by a subperiod (in our case a month) with a positive alpha. This especially means that a drawdown period with consecutive negative alphas can also be one single subperiod only. It also shows that the choice of N, i.e., the choice of the subperiods (daily, weekly, monthly, etc.) plays a critical role in the *MRDD* calculation. Drawdown periods in our example are:

- Drawdown period 1: November and December 2012, 2 months.
- Drawdown period 2: February 2013, 1 month.
- Drawdown period 3: May and June 2013, 2 months.
- Drawdown period 4: August–November 2013, 4 months.

The stated length in the list above is the respective *drawdown duration*. After having determined the drawdown periods with the consecutive negative alphas we have to calculate the magnitude of the negative returns for each of the four drawdown periods. This is shown graphically in Fig. 1.32.

Using the idea of Eq. (1.4) from page 8 we get:

- Drawdown period 1: cumulative relative return of $-1.36\,\%$.
- Drawdown period 2: cumulative relative return of $-0.28\,\%$.
- Drawdown period 3: cumulative relative return of $-0.69\,\%$.
- Drawdown period 4: cumulative relative return of $-1.93\,\%$.

Therefore, the worst relative drawdown occurred in period 4 with $MRDD = -1.93\,\%$ spanning the 4 months from August to November 2013. The calculations of the relative drawdowns are shown in Table 1.23 on the next page. The necessary formulas for the cells $D2$ to $H2$ are:

Table 1.23 Example 14: Calculation of the maximum relative drawdown *MRDD*

	A	B	C	D	E	F	G	H
1	Month	Monthly portfolio performance	Monthly benchmark performance	Monthly alpha	Relative drawdown (yes/no)?	Cumulative portfolio return in the drawdown period	Cumulative benchmark return in the drawdown period	Alpha of the cumulated returns in the drawdown period
2	07/2012	6.10 %	6.01 %	0.09 %	No			
3	08/2012	5.50 %	5.45 %	0.05 %	No			
4	09/2012	4.70 %	4.63 %	0.07 %	No			
5	10/2012	−5.00 %	−6.99 %	1.99 %	No			
6	11/2012	−5.10 %	−4.16 %	−0.94 %	Yes	−5.10 %	−4.16 %	−0.94 %
7	12/2012	6.70 %	7.07 %	−0.37 %	Yes	1.26 %	2.62 %	−1.36 %
8	01/2013	6.03 %	5.97 %	0.06 %	No			
9	02/2013	−3.23 %	−2.95 %	−0.28 %	Yes	−3.23 %	−2.95 %	−0.28 %
10	03/2013	5.12 %	4.66 %	0.46 %	No			
11	04/2013	5.21 %	4.91 %	0.30 %	No			
12	05/2013	−4.10 %	−4.01 %	−0.09 %	Yes	−4.10 %	−4.01 %	−0.09 %
13	06/2013	−4.50 %	−3.87 %	−0.63 %	Yes	−8.42 %	−7.72 %	−0.69 %
14	07/2013	1.75 %	−2.95 %	4.70 %	No			
15	08/2013	3.71 %	4.52 %	−0.81 %	Yes	3.71 %	4.52 %	−0.81 %
16	09/2013	−4.20 %	−3.93 %	−0.27 %	Yes	−0.65 %	0.41 %	−1.06 %
17	10/2013	4.26 %	4.99 %	−0.73 %	Yes	3.59 %	5.42 %	−1.84 %
18	11/2013	−4.00 %	−3.84 %	−0.16 %	Yes	−0.56 %	1.37 %	−1.93 %
19	12/2013	5.10 %	4.99 %	0.11 %	No			
20							*MRDD* =	-1.93 %

Source: Own, for illustrative purposes only

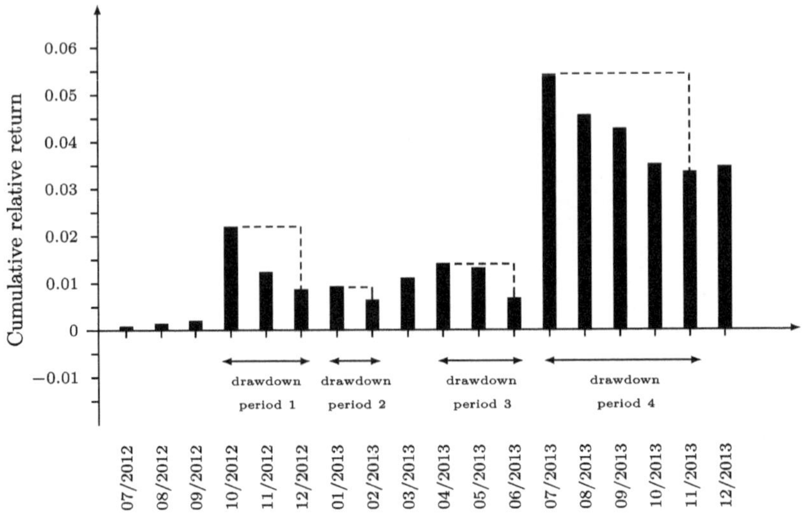

Fig. 1.32 Graph of the cumulative relative returns in Example 14 with drawdown periods and recovery periods. *Source*: Own, for illustrative purposes only

- Monthly portfolio alpha $\alpha^{monthly}$ in cell $D6$:

  $-0.94\,\% \quad = \quad B6 - C6$

- Occurrence of a relative drawdown (i.e., negative monthly alpha) in cell $E6$:

  $yes \quad = \quad IF(D6 < 0, "yes", "no")$

- Cumulative portfolio return r_{Pf}^{cum} in the drawdown period in cell $F6$:

  $-5.10\,\% \quad = \quad IF(D6 < 0, (1 + B6) * (1 + F5) - 1, "")$

- Cumulative benchmark return r_{Bm}^{cum} in the drawdown period in cell $G6$:

  $-4.16\,\% \quad = \quad IF(D6 < 0, (1 + C6) * (1 + G5) - 1, "")$

- Cumulative alpha α^{cum} in the drawdown period in cell $H6$:

  $-0.94\,\% \quad = \quad IF(D6 < 0, F6 - G6, "")$

The maximum relative drawdown *MRDD* in cell $H20$ is then obtained by $MIN(H2 : H19) \quad = \quad -1.93\,\%$.

Therefore, the worst drawdown period was the last one spanning August–November 2013 with a duration of 4 months and a cumulative alpha of $-1.93\,\%$. Figure 1.33 shows the same situation again, now displaying the cumulative alpha on the vertical axis. This allows to better spot the time point when the loss of a drawdown period is recovered. This recovery period is an important measure since it shows the capability of an asset manager to cope with difficult investment periods (based on historical experience).

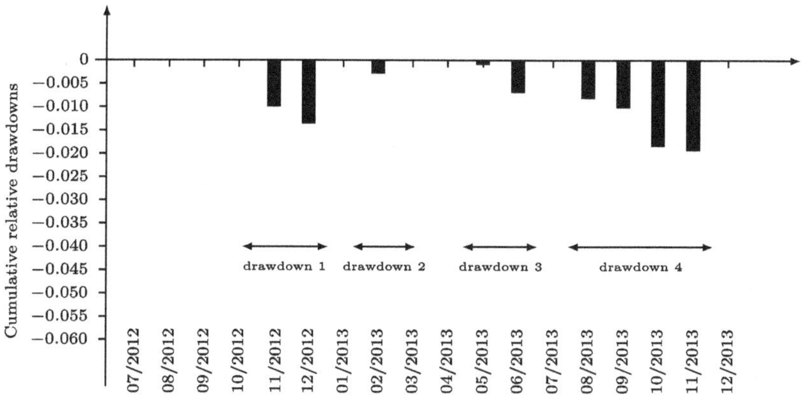

Fig. 1.33 Drawdown diagram for Example 14. *Source*: Own, for illustrative purposes only

To finalize this section, let us look at another graphical representation of the relative drawdown. Figure 1.32 shows the relative drawdowns and the length of the recovery periods, but the magnitude of the drawdown is not obvious to see. Therefore, drawdowns are often represented like in Fig. 1.33 which graphically represents column H of Table 1.23. Here, the magnitude of each single relative drawdown and the *MRDD* can be seen directly (but not the recovery period).

End of Example 14

1.4.2.2 Conclusion

The maximum relative drawdown is a key asymmetrical risk measure in active portfolio management which shows the maximum consecutive negative alpha a portfolio had in the past. In combination with the measured drawdown and recovery period an investor can evaluate the portfolio manager's ability to regain losses versus the benchmark using historical data. As in the case of the absolute drawdown, the length of the historical time period and the choice of the subperiods is key. The shorter the subperiods (for example, days vs. months) and the analyzed historical time period, the less severe the *MRDD* will be.

The historical time period should cover at least one market crash in order to deliver a meaningful *MRDD*. For example, doing an *MRDD* analysis for an active equity portfolio in the beginning of 2008 based on data for 2004–2007 would be irrelevant, since no stock market crash occurred during this time. Here, extending the time period to include the internet bubble 1999–2003 would be necessary, assuming the equity portfolio already existed then.

MRDD is a powerful instrument to compare the drawdown risk and the recovery potential of actively managed funds. The historical time periods and subperiods used for this comparison have to be identical for all analyzed portfolios. Otherwise, an *MRDD* comparison is meaningless.

1.4.3 Semi-deviation and Semi-variance

Definition: Semi-deviation and Semi-variance
Semi-deviation measures, similar to volatility, a dispersion. However, it measures the dispersion *below* a target rate of return and does not consider time periods with above target rate returns. This means that semi-deviation only looks at the left hand side of the return distribution and is, therefore, an asymmetrial risk measure. Using the same notation as before, semi-deviation can mathematically be defined as follows[61]:

$$^{down}\sigma_{Pf} = \sqrt{\frac{1}{N}\sum_{k=1}^{N} d_k^2} \tag{1.78}$$

where

$$\begin{aligned} T &= \text{time point at the end of the whole period } [0, T], \\ N &= \text{number of equidistant time periods in } [0, T], \\ r^* &= \text{target rate of return for any of the } N \\ &\quad \text{equidistant subintervals in } [0, T], \\ r_{Pf}^k &= \text{return of the portfolio in the} \\ &\quad \text{k-th subinterval } [t_{k-1}, t_k],\ 1 \leq k \leq N. \\ d_k &= \left\{ \begin{array}{ll} r^* - r_{Pf}^k & \text{if } r_{Pf}^k \leq r^* \\ 0 & \text{otherwise.} \end{array} \right\} \end{aligned}$$

In this definition, the target rate of return r^* is a constant. However, it can also be dependent of the subinterval k, $1 \leq k \leq N$. For example, $r^* = r^*(k)$ can be defined as the benchmark return in subinterval k. In this case semi-deviation only takes subintervals into account where the achieved return is below the benchmark, i.e., where alpha is negative.

Semi-variance $^{down}\sigma_{Pf}^2$ is the square of semi-deviation. If r^* is not a constant, semi-variance and semi-deviation are also sometimes called *below-target semi-deviation* and *below-target semi-variance*.

If the subintervals are months, we write $^{down}\sigma_{Pf}^{monthly}$ for the monthly semi-deviation and $^{down}\sigma_{Pf}^{2,monthly}$ for the monthly semi-variance.

[61]Lhabitant (2004, p. 51).

1.4.3.1 Interpretation

The interpretation of semi-deviation is not as easy as the interpretation of volatility assuming a normal distribution for the subperiod percentage returns. But the advantage of semi-deviation is that such an assumption is not needed. It can best be interpreted in comparison: The higher the semi-deviation (all else equal), the higher the drawdown risk, i.e., the risk of having below target returns. In the extreme case of a semi-deviation of 0, all subperiods in time period $[0, T]$ have returns equal or above the subperiods' target rates which indicates a very low drawdown risk. However, a semi-deviation of 0 does not imply that there is no drawdown risk in general, since the choice of $[0, T]$ is important for the calculation of the semi-deviation.

Example 15

Let us consider portfolio A from the previous examples over the time period from July 2012 until December 2013. This means $T = 1.5$ years and $N = 18$ subintervals. In each month the return of the portfolio is compared to the target rate of return r^* which is assumed to be the risk-free rate.[62] In this example the monthly risk-free rate is assumed to be 0.2 % like in Example 12. The data and the final results are displayed in Table 1.24. Column C answers the question if the return of a particular month is below the target of 0.2 %. The Excel® formulas for calculating the values in Table 1.24 are as follows:

- The answer ("yes" or "no") in cell $C2$:

  $$no = IF(B2 < 0.002, "yes", "no")$$

- Squared deviation of the portfolio return from the target rate in case of an underperforming portfolio in cell $D2$:

  $$0.00\,\% = IF(B2 < 0.002, (0.002 - B2)\text{^}2, 0)$$

- Monthly portfolio semi-variance ${}^{down}\sigma_{Pf}^{2,monthly}$ in cell $D20$:

  $$0.08\,\% = (1/18) * SUM(D2 : D19)$$

- Monthly portfolio semi-deviation ${}^{down}\sigma_{Pf}^{monthly}$ in cell $D21$:

  $$2.83\,\% = SQRT((1/18) * SUM(D2 : D19))$$

End of Example 15

1.4.3.2 Conclusion

Semi-deviation and semi-variance are the first asymmetrical risk measures presented in this chapter. They only look at subperiods where the achieved return is below a certain target rate (either static or dynamic) and measure the "downside" deviation from this target rate. This allows to see how a portfolio behaves regarding losses, i.e.,

[62] In this example a static target rate of return is used. However, the mechanics remain the same for a non-static return target. Often, the benchmark return is used as the target.

Table 1.24 Example 15: Calculation of semi-deviation and semi-variance, $r^* = 0.2\,\%$

	A	B	C	D
1	Month (k)	Monthly returns of portfolio	Is $r_{Pf}^k < r^*$ (yes/no)?	d_k^2
2	07/2012	6.10 %	No	0.00 %
3	08/2012	5.50 %	No	0.00 %
4	09/2012	4.70 %	No	0.00 %
5	10/2012	−5.00 %	Yes	0.27 %
6	11/2012	−5.10 %	Yes	0.28 %
7	12/2012	6.70 %	No	0.00 %
8	01/2013	6.03 %	No	0.00 %
9	02/2013	−3.23 %	Yes	0.12 %
10	03/2013	5.12 %	No	0.00 %
11	04/2013	5.21 %	No	0.00 %
12	05/2013	−4.10 %	Yes	0.18 %
13	06/2013	−4.50 %	Yes	0.22 %
14	07/2013	1.75 %	No	0.00 %
15	08/2013	3.71 %	No	0.00 %
16	09/2013	−4.20 %	Yes	0.19 %
17	10/2013	4.26 %	No	0.00 %
18	11/2013	−4.00 %	Yes	0.18 %
19	12/2013	5.10 %	No	0.00 %
20			${}^{down}\sigma_{Pf}^{2,monthly} =$	0.08 %
21			${}^{down}\sigma_{Pf}^{monthly} =$	2.83 %

Source: Own, for illustrative purposes only

returns below the target rate. Volatility and variance, the symmetrical counterparts introduced in Sect. 1.3.1, are not able to provide such information.

1.4.4 Shortfall Probability

Definition: Shortfall Probability
Semi-deviation and semi-variance concern time periods where the achieved portfolio return is below the target return, but do not provide information on the probability of their occurrence. This is measured by the *shortfall probability*.

Shortfall probability, also called *shortfall risk*, is a simple tool to evaluate the probability of a portfolio return being below a target return r^*. The target return r^* has to be defined up-front. Using the same notation as before the shortfall risk shortfall_{Pf} of the portfolio can be calculated as[63]

(continued)

[63]Bacon (2008, p. 94).

$$\text{shortfall}_{Pf} = P(r_{Pf}^k \leq r^*), \tag{1.79}$$

where:

$$P(.) = \text{probability function,}$$
$$r_{Pf}^k = \text{portfolio return in time period } [t_{k-1}, t_k],\ 1 \leq k \leq N,$$
$$r^* = \text{target rate of return for any subperiod } k.$$

Like before, the target rate r^* can be constant or dynamic. The shortfall risk only evaluates the probability of a shortfall with respect to the target r^* in a subinterval, but not the potential size of the shortfall. If the subintervals are months we write $\text{shortfall}_{Pf}^{monthly}$.

Clearly, given that the shortfall risk is calculated with historical data, its application for the future is not without shortcomes. The implicit assumption is that the past is representative for the future. However, this can be misleading: If historical time series for 2004–2007 had been used for calculating the shortfall risk, they were of little use in 2008 when the economy slid into the Great Recession. The macroeconomic environment has to be accounted for when applying this risk measure.

1.4.4.1 Interpretation

The interpretation of shortfall risk is straightforward: it is the probability that the portfolio return is below the target return in a certain time period. In practice, it is the relative frequency of a portfolio return being below the target rate within the N subperiods of the historical time period $[0, T]$. This relative frequency is then interpreted as probability assuming the past behavior of the portfolio vs. the target rate continues into the future. For example, a shortfall probability of 50 % means that in the past the portfolio did not reach the target return 50 % of the time. Assuming comparable macroeconomic conditions for this portfolio and given a sufficient length of the time series used to calculate the shortfall probability, we can estimate that in the coming years, the shortfall probability will be close to this calculated number.

Let us now look at shortfall probability through a computational example.

Example 16

We consider a portfolio which is managed against a benchmark and use the same data as in the previous examples. The historical time period is from July 2012 to December 2013, i.e., $T = 1.5$ years using monthly data (i.e., $N = 18$). All data can be found in Table 1.25.

Table 1.25 Example 16: Calculation of shortfall probability

	A	B	C	D
1	Month	Monthly portfolio performance	Monthly benchmark performance	Below target?
2	07/2012	6.10 %	6.01 %	No
3	08/2012	5.50 %	5.45 %	No
4	09/2012	4.70 %	4.63 %	No
5	10/2012	−5.00 %	−6.99 %	No
6	11/2012	−5.10 %	−4.16 %	Yes
7	12/2012	6.70 %	7.07 %	Yes
8	01/2013	6.03 %	5.97 %	No
9	02/2013	−3.23 %	−2.95 %	Yes
10	03/2013	5.12 %	4.66 %	No
11	04/2013	5.21 %	4.91 %	No
12	05/2013	−4.10 %	−4.01 %	Yes
13	06/2013	−4.50 %	−3.87 %	Yes
14	07/2013	1.75 %	−2.95 %	No
15	08/2013	3.71 %	4.52 %	Yes
16	09/2013	−4.20 %	−3.93 %	Yes
17	10/2013	4.26 %	4.99 %	Yes
18	11/2013	−4.00 %	−3.84 %	Yes
19	12/2013	5.10 %	4.99 %	No
22			Number of shortfalls	9
23			$\text{shortfall}_{Pf}^{monthly}$	50 %

Source: Own, for illustrative purposes only

To obtain the shortfall probability, we use the Excel® formulas below. Column C shows if within the subperiods there is a shortfall (i.e., a portfolio return below the benchmark) or not. Column D can be filled out using the IF-function in Excel®.

- Occurrence of a shortfall in cell $D2$:

  $$no = IF(B2 < C2,\text{"}yes\text{"},\text{"}no\text{"})$$

- Number of shortfalls over the observed time period in cell $D22$:

  $$9 = COUNTIF(D2 : D19,\text{"}yes\text{"})$$

- Monthly portfolio shortfall risk $\text{shortfall}_{Pf}^{monthly}$ in cell $D23$:

  $$50\,\% = D22/COUNT(D2 : D19)$$

Thus, cell $D22$ contains the number of shortfalls and cell $D23$ the number of shortfalls (9) divided by the number of subperiods in total (18), yielding the percentage (relative frequency) of months with a shortfall. As elaborated above, we now interpret this figure as the shortfall probability going forward. This implicitly assumes that the analyzed historical time period is representative for the time to come such that we speak in general of a *shortfall probability of* 50 %.

End of Example 16

1.4.4.2 Conclusion

The shortfall probability, or shortfall risk, states (based on historical data) the probability of a portfolio return falling short of a target return. However, this measure does not indicate the magnitude of a downside deviation.

1.4.5 Sortino Ratio

Definition: Sortino Ratio

The Sortino ratio (*SoR*) is a relative risk-adjusted return ratio that uses an asymmetrical risk measure.[64] It measures the excess return of a portfolio over a so-called *minimum acceptable rate of return* per unit of asymmetrical risk measured by semi-volatility with respect to the minimum acceptable rate of return.

In order to mathematically define the Sortino ratio *SoR*,[65] time interval $[0, T], T \geq 1$, is split, as always, into N equidistant subintervals. These are usually days or months. The minimum acceptable rate *MAR*, a constant, is the return that is the minimum aimed for in each subinterval. Let us then calculate the portfolio's semi-volatility ${}^{down}\sigma_{Pf}$ according to Eq. (1.78) from page 88 with $r^* = MAR$, i.e., ${}^{down}\sigma_{Pf}$ is the portfolio's semi-volatility with respect to the minimum acceptable rate of return for a subperiod.

Let further r_{Pf}^k be the return of the portfolio during the k-th subperiod $[t_{k-1}, t_k], 1 \leq k \leq N$. According to Eq. (1.4) from page 8 and Eq. (1.7) from page 9, the geometric mean portfolio return $\hat{r}_{Pf}$ of a subinterval within time period $[0, T]$ can be calculated as

$$\hat{r}_{Pf} = \prod_{k=1}^{N}(1 + r_{Pf}^k)^{\frac{1}{N}} \;-\; 1. \tag{1.80}$$

Then, the Sortino ratio is defined as

$$SoR = \frac{\hat{r}_{Pf} - MAR}{{}^{down}\sigma_{Pf}}. \tag{1.81}$$

---

[64]The Sortino ratio was devised by Brian M. Rom, founder and president of the software development company Investment Technologies, in 1983. The ratio is named after Dr. Frank A. Sortino, an early popularizer of downside risk optimization.

[65]Sortino and Price (1994).

1.4.5.1 Interpretation

As with the other risk-adjusted return ratios presented above, the higher the value of the Sortino ratio, the better the risk-adjusted performance of the portfolio. The key in the Sortino ratio is its use of semi-volatility versus a minimum acceptable return, i.e., it is a downside risk measure. Therefore, the Sortino ratio can also be called a downside-adjusted return ratio where the return is the excess return of the portfolio over the minimum acceptable return. It is a very important measure for comparing two portfolios in terms of their ability to generate returns by taking their downside risk into consideration.

Example 17

Using the data from Example 16 and assuming an *MAR* of 0.2 % per month, we can calculate the Sortino ratio as done in Table 1.26.

The portfolio's semi-variance ${}^{down}\sigma_{Pf}^{2,monthly}$ in cell $D20$ and semi-volatility ${}^{down}\sigma_{Pf}^{monthly}$ in cell $D21$ are calculated in exactly the same way as in Example 16. Other Excel® formulas include:

- Using Eq. (1.80) for the geometric mean $\hat{r}_{Pf}$ in cell $B20$:
  $1.23\,\% = \{PRODUCT(1 + B2 : B19)\hat{}(1/18) - 1\}$

Table 1.26 Example 17: Calculation of the Sortino ratio, $MAR = 0.2\,\%$

	A	B	C	D
1	Month (k)	Monthly returns of portfolio	Is $\mathbf{r}_{Pf}^{k} < \mathbf{r}^{*}$?	$\mathbf{d}_{k}^{2}$
2	07/2012	6.10 %	No	0.00 %
3	08/2012	5.50 %	No	0.00 %
4	09/2012	4.70 %	No	0.00 %
5	10/2012	−5.00 %	Yes	0.27 %
6	11/2012	−5.10 %	Yes	0.28 %
7	12/2012	6.70 %	No	0.00 %
8	01/2013	6.03 %	No	0.00 %
9	02/2013	−3.23 %	Yes	0.12 %
10	03/2013	5.12 %	No	0.00 %
11	04/2013	5.21 %	No	0.00 %
12	05/2013	−4.10 %	Yes	0.18 %
13	06/2013	−4.50 %	Yes	0.22 %
14	07/2013	1.75 %	No	0.00 %
15	08/2013	3.71 %	No	0.00 %
16	09/2013	−4.20 %	Yes	0.19 %
17	10/2013	4.26 %	No	0.00 %
18	11/2013	−4.00 %	Yes	0.18 %
19	12/2013	5.10 %	No	0.00 %
20	$\hat{r}_{Pf} =$	1.23 %	${}^{down}\sigma_{Pf}^{2,monthly} =$	0.08 %
21	$SoR =$	0.36	${}^{down}\sigma_{Pf}^{monthly} =$	2.83 %

Source: Own, for illustrative purposes only

- Sortino ratio *SoR* in cell *B*21:

  $0.36 = \quad (B20 - 0.2\,\%)/D21$

End of Example 17

1.4.5.2 Conclusion

The Sortino ratio can be considered as an improvement of the Treynor ratio. The risk-free rate is replaced by a chosen minimum acceptable return and the risk is evaluated by the risk of losing money, a downside risk measure. Therefore, the Sortino ratio is a key ratio when evaluating the risk-return relationship of a portfolio in times of crisis, i.e., when downside risk is the risk measure to look at and not simply the volatility.

1.5 Portfolio Return and Volatility

Until now, we have introduced various measures of return and risk for an investment which usually would be a portfolio. What we have not done yet (but what will be needed in the next chapter) is to calculate the portfolio return and risk if we only know the return and risk figures of the portfolio's constituents. This is the topic of this section where we specify risk as volatility.

Let us look at an example first: If we invest 50 % in security 1 which has an expected return of 5 % and the other 50 % in security 2 which has an expected return of 10 %, then the expected return of our portfolio lies exactly in the middle at 7.5 %.

What if security 1 has a volatility of 10 % and security 2 has a volatility of 20 %? Does the portfolio necessarily have a volatility of 15 %? No! The reason is that the expected value is linear [see Eq. (1.82) below] but volatility is not [see Eq. (1.83) below]. To calculate the volatility, we also need to know the covariance between the returns of the securities 1 and 2. Let us formulate this mathematically.

Let w_i be the *weight* of security i in the portfolio with M securities during a fixed time interval where the proportion of the portfolio invested in security i, $1 \leq i \leq M$, remains unchanged. Using R_i as the random variable representing the return of security i and R_{Pf} as the random variable representing the portfolio return over this time period, we calculate the expected return of the portfolio as[66]

$$\mathbb{E}(R_{Pf}) = \mathbb{E}\left(\sum_{i=1}^{M} w_i \cdot R_i\right) = \sum_{i=1}^{M} w_i \cdot \mathbb{E}(R_i). \tag{1.82}$$

Calculating the volatility of the portfolio is trickier.[67] We need to consider the covariance of the various portfolio securities, which is the reason for what in Chap. 2

[66] For the formula, see Reilly and Brown (1997, p. 254).

[67] Reilly and Brown (1997, p. 261).

will be called *diversification*. Here, we focus on the formula for calculating the portfolio volatility. For this let σ_i be the volatility of security i, $1 \leq i \leq M$, and $\sigma_{i,j}$ the covariance between security i $(1 \leq i \leq M)$ and security j $(1 \leq j \leq M)$. Then, the portfolio volatility can be calculated as

$$\begin{aligned}
\sigma_{Pf} &= \sqrt{Var(R_{Pf})} \\
&= \sqrt{Var\left(\sum_{i=1}^{M} w_i \cdot R_i\right)} \\
&= \sqrt{\sum_{i=1}^{M} w_i^2\sigma_i^2 + \sum_{i=1}^{M}\sum_{j=i+1}^{M} 2 \cdot w_i \cdot w_j \cdot \sigma_{i,j}}. \qquad (1.83)
\end{aligned}$$

In our example from above, we have $w_1 = w_2 = 50\,\%$. Then, according to Eq. (1.82), the expected return of the portfolio is

$$\begin{aligned}
\mathbb{E}(R_{Pf}) &= w_1 \cdot \mathbb{E}(R_1) + w_2 \cdot \mathbb{E}(R_2) \\
&= 0.5 \cdot 5\,\% + 0.5 \cdot 10\,\% = 7.5\,\% \qquad (1.84)
\end{aligned}$$

and, according to Eq. (1.83), the volatility is [using $\sigma_{1,2} = \rho_{1,2}\sigma_1\sigma_2$ from Eq. (1.43)]

$$\begin{aligned}
\sigma_{Pf} &= \sqrt{w_1^2\sigma_1^2 + w_2^2\sigma_2^2 + 2w_1w_2\sigma_{1,2}} \\
&= \sqrt{w_1^2\sigma_1^2 + w_2^2\sigma_2^2 + 2w_1w_2\rho_{1,2}\sigma_1\sigma_2} \\
&= \sqrt{0.5^2 \cdot 0.1^2 + 0.5^2 \cdot 0.2^2 + 2 \cdot 0.5 \cdot 0.5 \cdot \rho_{1,2} \cdot 0.1 \cdot 0.2} \\
&= \sqrt{0.0125 + 0.01 \cdot \rho_{1,2}}. \qquad (1.85)
\end{aligned}$$

The volatility of the portfolio depends on the correlation $\rho_{1,2}$. The lower the correlation, the lower the volatility of the portfolio, the better the diversification benefits. Let us assume the correlation to be $\rho_{1,2} = -0.44$. Then Eq. (1.85) yields

$$\sigma_{Pf} = \sqrt{0.0125 + 0.01 \cdot (-0.44)} = \sqrt{0.0081} = 9\,\%. \qquad (1.86)$$

The volatility of the portfolio is lower than the volatilities of both assets 1 and 2.

To conclude this chapter, we return to our business case and calculate the volatility of the portfolio.

Business Case (cont.)
Let us continue to look at our business case using the Delta (DAL) stock and oil. Oil has a volatility of 42.03 % [Eq. (1.65)], Delta has a volatility of 75.91 % [Eq. (1.63)], and the correlation is −0.379 [Eq. (1.67)]. If we invest 70 % in oil and 30 % in Delta, how much is the volatility of the portfolio?
We use Eq. (1.85) to calculate the portfolio volatility, where oil is the first asset and Delta stock is the second asset. Then

$$\begin{aligned} \sigma_{Pf} &= \sqrt{\begin{matrix} 0.7^2 \cdot 0.420^2 + 0.3^2 \cdot 0.759^2 + \\ 2 \cdot 0.7 \cdot 0.3 \cdot (-0.379) \cdot 0.420 \cdot 0.759 \end{matrix}} \\ &= 29.59\,\%. \end{aligned} \tag{1.87}$$

This portfolio has a lower volatility than its individual components oil and Delta, because the correlation between oil and Delta is negative. The diversification works very well.

1.6 Summary

In this chapter, after having briefly introduced the basic principles of calculating cumulative and annualized returns for an investment, we have described various measures of risk and distinguished two scenarios. There are portfolios managed on an absolute return basis, i.e., with absolute return targets which are not tied to an index as a benchmark. There are also portfolios managed against a benchmark with relative return targets known as excess returns or alphas. For both scenarios we have presented different risk measures which can be classified as symmetrical and asymmetrical risk measures. Symmetrical risk measures look at both sides of the return distribution. In times of market stress, however, more investors care about downside risk measures and look for protection against tail risk.

Before 2008, asymmetrical risk measures were not widespread in traditional asset management but common in hedge fund management where returns are almost always asymmetrical. This has significantly changed as we will show in Chap. 6. The foundations laid in Chap. 1 will now be used in Chap. 2 to introduce the key principles of modern portfolio theory.

References

Bacon, C. R. (2008). *Practical portfolio performance measurement and attribution* (2nd ed.). Chichester: Wiley Finance.

Chincarini, L. B., & Kim, D. (2006). *Quantitative equity portfolio management*. New York: McGraw-Hill.

Clarke, R. G., de Silva, H., & Thorley, S. (2002, September/October). Portfolio constraints and the fundamental law of active management. *Financial Analysts Journal, 58*, 48–66.
DeFusco, R. A., McLeavey, D. W., Pinto, J. E., & Runkle, D. E. (2004). *Quantitative methods for investment analysis* (2nd ed.). Charlottesville: CFA Institute.
Esch, L., Kieffer, R., & Lopez, T. (2005). *Asset and risk management*. Hoboken: Wiley Finance.
Glavin, B., & Reinganum, M. R. (2013). Get active: Two approaches to actively managing equities. *State Street Global Advisors*.
Grinold, R. C. (1989, Spring). The fundamental law of active management. *Journal of Portfolio Management, 15*(3), 30–37.
Hooker, M., & Schulmerich, M. (2008, March). *Aktives quantitatives Aktienmanagement in 2008*. Germany: DPN.
Hull, J. C. (2009). *Options, futures, and other derivatives* (7th ed.). Upper Saddle River: Pearson Prentice Hall.
Jorion, P. (2001). *Financial risk manager handbook* (1st ed.). New York: John Wiley & Sons.
Lhabitant, F. S. (2004). *Hedge funds quantitative insights*. Chichester: John Wiley & Sons.
Lowenstein, R. (2002). *When genius failed: The rise and fall of Long-Term Capital Management*. London: Fourth Estate.
Markowitz, H. M. (1959). *Portfolio selection*. New Haven: Yale University Press.
Markowitz, H. M. (2008). *Portfolio selection*. München: FinanzBuch Verlag.
Natenberg, S. (1994). *Option volatility and pricing*. New York: McGraw-Hill.
Neftci, S. N. (2000). *An introduction to the mathematics of financial derivatives* (2nd ed.). San Diego: Academic Press.
Øksendal, B. (1995). *Stochastic differential equations* (4th ed.). New York: Springer.
Reilly, F. K., & Brown, K.C. (1997). *Investment analysis and portfolio management*. Chicago: The Dryden Press.
Reinganum, M. R. (2010). Overcrowding or dynamic missteps? Why active quant equity performance was challenged. *State Street Global Advisors*.
Schulmerich, M. (1997, September 9). *Statistische Verfahren für Diffusionsprozesse mit Anwendung auf stochastische Zinsmodelle der Finanzmathematik*. Master's thesis in Mathematics.
Schulmerich, M. (2005). *Real options valuation* (1st ed.). New York: Springer.
Schulmerich, M. (2008a). *Statistische Verfahren für Diffusionsprozesse mit Anwendung auf stochastische Zinsmodelle der Finanzmathematik*. GRIN Verlag. Master's thesis in Mathematics from 1997 published 2008, School of Mathematics, Johannes Gutenberg-University Mainz, Germany.
Schulmerich, M. (2008b, January 3). Quantitative Anlagemodelle in einem unsicheren Umfeld. *Neue Zürcher Zeitung*.
Schulmerich, M. (2009, November). Quantitative investing: Seizing opportunities in the European equities market. *IPE Magazine*.
Schulmerich, M. (2010a). *Real options valuation* (2nd ed.). Berlin: Springer.
Schulmerich, M. (2010b, June/July). *Warum quantitative Anlagen? Wir sind nicht voreingenommen!*. Germany: DPN.
Schulmerich, M. (2013). Can the Black-Litterman framework improve asset management outcomes? *State Street Global Advisors. SSgA Capital Insights*.
Schulmerich, M., & Hooker, M. (2008). Quantitative and fundamental active equities in institutional portfolios. In *SSgA Investment Quarterly, State Street Global Advisors* (pp. 1–3).
Schulmerich, M., Andres, P., Baur, B., Bornemann, H.-O., Kempf, A., Kleeberg, J., et al. (2009, April/May). *Roundtable discussion: Quant management*. Germany: DPN.
Schulmerich, M., Hooker, M., McGoldrick, H., & Mallik, G. (2008). Comparing quantitative and fundamental approaches to active equity investing. *State Street Global Advisors*.
Sharpe, W. F. (1966, January). Mutual fund performance. *Journal of Business, 39*(1), 119–138 [Part 2: Supplement on Security Prices].
Sortino, F. A., & Price, L. N. (1994, Fall). Performance measurement in a downside risk framework. *Journal of Investing, 3*(3), 59–64.

Spremann, K. (2008). *Portfoliomanagement* (4th ed.). München: Oldenbourg Verlag.
Taleb, N. N. (2010). *The black swan* (2nd ed.). New York: Random House Trade Paperback.
Woodward, G., & Anderson, H. (2009). Does beta react to market conditions? Estimates of bull and bear betas using a nonlinear market model with endogenous threshold parameter. *Quantitative Finance, 9*(8), 913–924.
Zagst, R., Goldbrunner, J., & Schlosser, A. (2010). *Zu nah an der Sonne*. München: FinanzBuch Verlag.

Modern Portfolio Theory and Its Problems 2

2.1 Introduction

A quant revolution started on Wall Street in 1952, when Harry M. Markowitz established the *modern portfolio theory* (MPT) which applies mathematical concepts to finance. Based on his work, the *capital asset pricing model* (CAPM) was developed one decade later. Today, the results of the CAPM (and its extended versions) are widely used for describing the risks and returns of portfolios and for performance measurement. This chapter is devoted to the theoretical part of asset management and shows the key tests that compare this theory to practice.

Using the mathematical prerequisites on risk and return measurement from the first chapter, we will look at modern portfolio theory in detail. Before we present MPT, Sect. 2.2 provides a review of regression analysis as needed for this chapter. The concept of regression will be shown by extending our business case from the first chapter. The capital asset pricing model is thereafter introduced in Sect. 2.3, a model which allows us to estimate the required return of a risky asset (or a portfolio) based on its β, i.e., its sensitivity to market movements. This model will be presented together with its assumptions and empirical tests. This section will discuss the validity of the CAPM based on the empirical tests and also present some critical views on the simplistic assumptions of the CAPM.

Although the CAPM has found useful applications, for example, in performance measurement, in explaining the benefits of diversification (by introducing the notion of systematic and unsystematic risk), and as a tool for finding undervalued/overvalued securities, empirical evidence suggests that the model describes capital markets returns only incompletely. Some returns can be influenced by other risk factors than the exposure to the market, some by stock market anomalies. Stock market anomalies will be described in detail in Chap. 3, and historical stock market crashes will be summarized in Chap. 4. Anomalies and crashes are not captured by the CAPM but rather by taking psychological factors into account. The irrational

M. Schulmerich et al., *Applied Asset and Risk Management*,
Management for Professionals, DOI 10.1007/978-3-642-55444-5_2

behavior of market participants heavily influences market prices and the returns of assets. This is researched by *behavioral finance*, which will be covered in Chap. 5.

Empirical evidence suggests that the returns of a risky asset/portfolio are not only driven by market movements, but also by other risk factors which are not included in the CAPM. In 1992, Eugene Fama[1] and Kenneth French[2] developed an extension of the CAPM, the Fama–French three-factor model (FF3M), introduced in Sect. 2.3.6, which incorporates two of these factors: the size of companies and the book-to-market ratio. Like in the discussion about the CAPM, empirical tests of the FF3M will be presented, together with some critical views of this model.

As said before, this chapter looks at the theory while Chaps. 3 and 4 look at the reality of financial markets. Chapter 3 discusses stock market anomalies, i.e., market irregularities which distort the price-return relationship of assets and contradict traditional finance theory. For example, calendar effects have an impact on asset returns while standard finance theory does not distinguish between the holiday season and spring. That is, according to standard finance theory, a calendar effect should not exist. Chapter 4 then looks at behavioral finance, the psychology of investing. Here we will see how psychology plays an important role in finance which again, according to standard finance theory, should not be the case. Yet, behavioral factors were of critical importance for what happened in the worldwide stock markets in 2000–2003, 2008–2009 or 2011.

2.2 A Quick Review of Regression Analysis

Regression analysis is a widely used technique in finance, especially in applications of the CAPM and of the Fama–French three-factor model.[3] The goal of regression analysis is to model a variable Y as a function of different input parameters $X_1, \ldots, X_K$. For our purposes, the asset return is the variable Y, and the different factors which contribute to the asset returns serve as the X-parameters, for example,

[1] Eugene F. Fama, born in 1939, is an American economist. He is known for his work on portfolio theory and asset pricing, both theoretical and empirical. He won the Nobel Prize in Economics in 2013 together with Robert J. Shiller and Lars Peter Hansen, see Reinganum (2013). His Ph.D. thesis (one of his supervisors was Nobel Prize winner Merton Miller) concluded that stock price movements are unpredictable and follow a random walk. Fama (1970) proposes the groundbreaking concept of *efficient-markets* (see Sects. 3.1 and 5.1) such that Fama is most often thought of as the father of the efficient-market hypothesis. Currently, he is a professor of finance at the University of Chicago, Booth School of Business.

[2] Kenneth R. French, born in 1954, is an American economist and professor of finance at Dartmouth College, Tuck School of Business. He has previously been a faculty member at MIT, the Yale School of Management, and the University of Chicago Booth School of Business. He obtained his Ph.D. in Finance in 1983 from the University of Rochester. French is an expert on the behavior of security prices and investment strategies and is most famous for his work on asset pricing with Eugene Fama and the Fama–French three-factor model (1992) as an extension of the CAPM (see Sect. 2.4).

[3] This paragraph is based on Lhabitant (2004, pp. 147–175).

return on the market, return on oil, GDP growth, etc. While the correlation quantifies how consistently two variables vary together, regression analysis describes the specific relationship between two (or more) variables. Therefore, regression analysis can be seen as an extension of the correlation/covariance concept.

2.2.1 Simple Linear Regression

The most basic type of regression is the simple linear regression. In order to understand the concept, let us go back to our business case from Chap. 1 and continue to analyze the stock returns of US Airways (LCC), Delta Airlines (DAL) and the oil price. On page 52 we found that the stock return of US Airways is positively related to the stock return of Delta Airlines, but negatively correlated to oil. The goal of simple linear regression is to find a linear relationship. How much has the price of US Airways changed when oil went up 1 % or when Delta Airlines went up 1 %? And how accurate is this linear relationship?

Let us analyze the period January 2008–June 2010, divided into 30 monthly subperiods. For example, we describe the dependency of US Airways' monthly returns r^k_{LCC} (in the kth month) to oil monthly returns r^k_{Oil} in the form

$$r^k_{LCC} = \underbrace{a + b \cdot r^k_{Oil}}_{\text{linear model}} + \underbrace{\varepsilon_k}_{\text{error term}} . \tag{2.1}$$

The linear relationship, described by a *regression line* of the form

$$r^{monthly}_{LCC} = a + b \cdot r^{monthly}_{Oil} \tag{2.2}$$

is what we are looking for, and the numbers a and b have to be calculated. The monthly return of oil is used to *explain* the monthly returns of Delta, this is why we call $r^{monthly}_{Oil}$ an explanatory variable. The error term

$$\varepsilon_k = r^k_{LCC} - (a + b \cdot r^k_{Oil}) \tag{2.3}$$

accounts for the deviation from our model due to other factors which explain the returns on US Airways, but are not related to oil. Take a look at Fig. 2.1 which is the same as Fig. 1.19 on page 48, but with the regression line (2.2) added (which still has to be calculated). Given the monthly returns $r^1_{LCC}, r^2_{LCC}, \ldots, r^{30}_{LCC}$ of US Airways and $r^1_{Oil}, r^2_{Oil}, \ldots, r^{30}_{Oil}$ of oil, the graph plots the points

$$(r^1_{Oil}, r^1_{LCC}), (r^2_{Oil}, r^2_{LCC}), \ldots, (r^{30}_{Oil}, r^{30}_{LCC}).$$

The line (2.2) is drawn such that the error terms ε_k, i.e., the distances of the points (r^k_{Oil}, r^k_{LCC}) from the line, are kept small on average. We could search for the line which minimizes the average distance from the line, but in practice, the squared sum

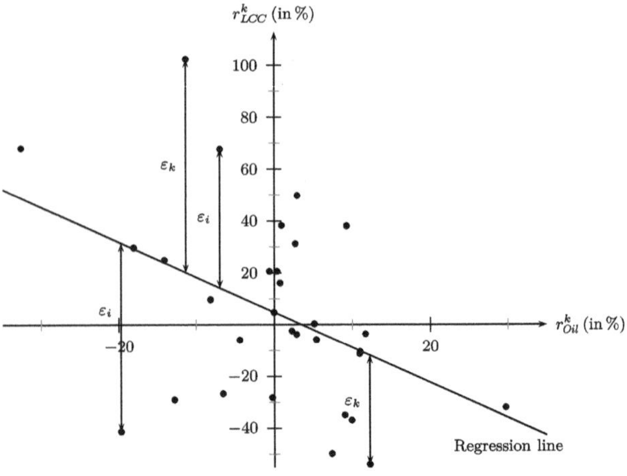

Fig. 2.1 Scatter plot of monthly returns of US Airways (LCC) against oil (Crushing, OK Crude Oil Futures Contract) for the period January 2008–June 2010, together with regression line. *Sources*: Yahoo! Finance and U.S. Energy Information Administration

of the error terms is used as the term to be minimized which puts greater weight to big outliers.

We get the numbers a and b for the regression line (2.2)

$$r_{LCC}^{monthly} = a + b \cdot r_{Oil}^{monthly}$$

by solving the following problem: Find the numbers a and b such that the sum of squared errors

$$\sum_{i=k}^{30} \varepsilon_k^2 = \sum_{k=1}^{30} (r_{LCC}^k - (a + b \cdot r_{Oil}^k))^2 \tag{2.4}$$

is minimal. We will present the solution to this problem after we have introduced the general problem and the general solution.

Given data points (X_1, Y_1), $(X_2, Y_2), \ldots, (X_N, Y_N)$, in the general problem we want to find the *regression line* (or *best-fit line*)

$$Y = a + b \cdot X \tag{2.5}$$

which describes best the relationship between X_k and Y_k.[4] In this regression equation, we call X the *independent variable*, Y the *dependent variable*, a the *intercept* (where the line crosses the Y-axis) and b the *slope coefficient*.[5] We refer to a and b as the *regression coefficients*.

[4]For an introduction to linear regression, see also DeFusco, McLeavey, Pinto, and Runkle (2004, pp. 395–420).

[5]The terms are from DeFusco et al. (2004, p. 395).

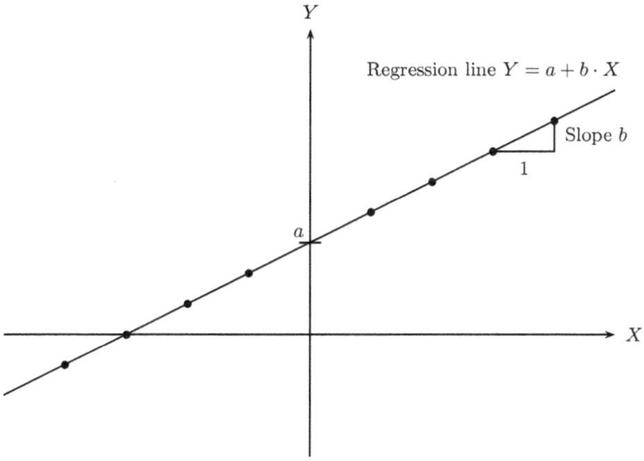

Fig. 2.2 Regression line for perfect linear relationship between Y and X. *Source*: Own, for illustrative purposes only

X is also called *explanatory variable*[6] because X is used to explain the variable Y. The larger the absolute value of b, the greater the *explanatory power* of X. If b is close to zero, then changes in X cannot explain changes in Y, and X has no explanatory power.

The graph in Fig. 2.2 shows a scatter plot where all data points (X,Y) lie on the line $Y = a + b \cdot X$. In practice, you usually have a scatter plot like in Figs. 2.3 or 2.4 where the data points are scattered around a line and where you can still see a linear tendency, but in the latter plot, the linear relationship is weak.

For any data point (X_k, Y_k), the error term ε_k is the difference between the actual Y_k value and the Y value predicted by the line (2.5), hence

$$\varepsilon_k = Y_k - (a + b \cdot X_k). \tag{2.6}$$

The numbers a and b for the regression line $Y = a + b \cdot X$ are chosen such that the sum of the squared error terms

$$\sum_{k=1}^{N} \varepsilon_k^2 = \sum_{k=1}^{N} (Y_k - (a + b \cdot X_k))^2 \tag{2.7}$$

is minimized. For that reason, this is also called the ordinary least squares (OLS) method.

[6]For example, the term *explanatory variable* is used in Fama and MacBeth (1973, p. 618).

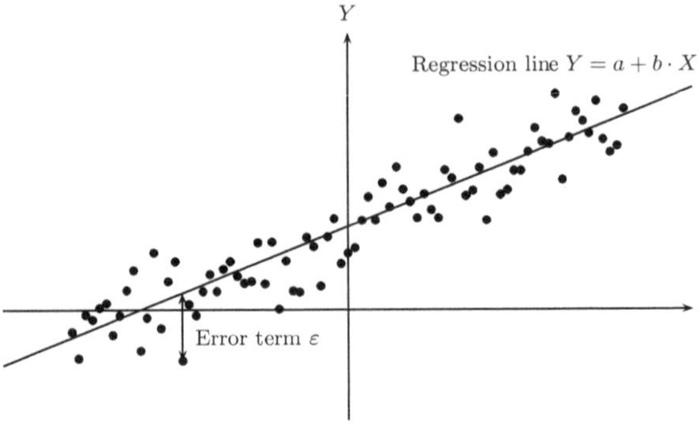

Fig. 2.3 Regression line for an approximate linear relationship between Y and X. Example for a high coefficient of determination R^2. *Source*: Own, for illustrative purposes only

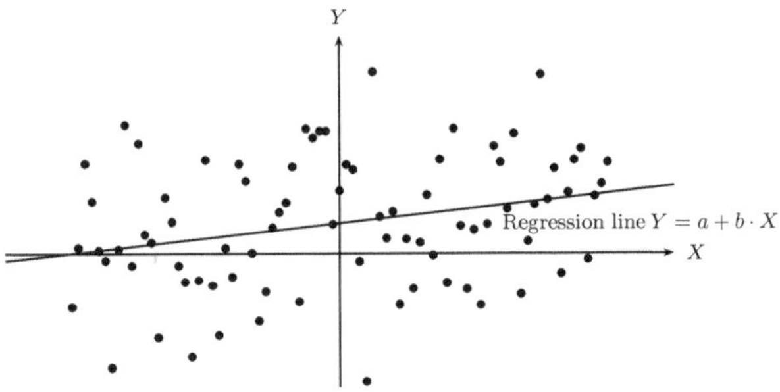

Fig. 2.4 The regression line provides only a weak description of the relationship between Y and X. Example for a low coefficient of determination R^2. *Source*: Own, for illustrative purposes only

Let

$$\bar{X} = \frac{1}{N}\sum_{k=1}^{N} X_k \quad \text{and} \quad \bar{Y} = \frac{1}{N}\sum_{k=1}^{N} Y_k.$$

Then, the solutions for the parameters a and b are

$$b = \frac{\sum_{k=1}^{N} (X_k - \bar{X}) \cdot (Y_k - \bar{Y})}{\sum_{k=1}^{N} (X_k - \bar{X})^2} \tag{2.8}$$

and

$$a = \bar{Y} - b \cdot \bar{X}. \tag{2.9}$$

For any plot one can find a regression line, but it does not always describe the relationship between Y and X well. Figure 2.3 shows one example where the data points are close to the regression line, and Fig. 2.4 shows one example where they are far off the line.

In order to describe the quality of the regression $Y = a + b \cdot X$, we use the R^2 measure, also known as the *coefficient of determination* R^2. Let $\hat{Y}_k$ be the predicted $\hat{Y}$ value based on the X_k value and the regression equation. In the case of the simple linear regression,

$$\hat{Y}_k = a + b \cdot X_k. \tag{2.10}$$

Then R^2 is the ratio between the variation explained by our regression model

$$\sum_{k=1}^{N} (\hat{Y}_k - \bar{Y})^2$$

and the total variation

$$\sum_{k=1}^{N} (Y_k - \bar{Y})^2.$$

In other words:

$$R^2 = \frac{\sum_{k=1}^{N} (\hat{Y}_k - \bar{Y})^2}{\sum_{k=1}^{N} (Y_k - \bar{Y})^2}. \tag{2.11}$$

In our case of simple linear regression where $Y = a + b \cdot X$, R^2 can also be calculated as[7]

$$R^2 = \frac{\left(\sum_{k=1}^{N} (X_k - \overline{X}) \cdot (Y_k - \overline{Y})\right)^2}{\sum_{k=1}^{N} (X_k - \overline{X})^2 \sum_{k=1}^{N} (Y_k - \overline{Y})^2}. \tag{2.12}$$

Please note:

- R^2 lies between zero and one. The larger the value of R^2, the more accurate the regression. R^2 is 1 if and only if all data points lie on the regression line $Y = a + b \cdot X$.
- The value of R^2 provides the percentage of the variation of the variable Y which is explained by the variation of the variable X. For example, $R^2 = 0.75$ means that the variable X explains 75 % of the variation of the variable Y.

[7] DeFusco et al. (2004, p. 403): For simple linear regression, R^2 is the square of the correlation between X and Y.

- In Fig. 2.3, R^2 is large (but not 1) because the data points lie close to the regression line, which describes the relation between Y and X well.
- In Fig. 2.4, R^2 is small because the regression line only vaguely represents the relationship between Y and X.

Business case (cont.)

Let us get back to our business case from Chap. 1 (an Excel® file with the calculations can be downloaded at http://www.pecundus.com/publications/springer-solutions [username: *solutions*; password: *springer-book-sle*]). We want

$$r_{LCC}^{monthly} = a + b \cdot r_{Oil}^{monthly}.$$

The equation for the slope coefficient b from (2.8) can be rewritten as

$$b = \frac{\sigma_{LCC,Oil}}{\sigma_{Oil}^2}, \tag{2.13}$$

using formulas for monthly volatility (1.22) and monthly covariance (1.45). We use the known results (1.56) and (1.62) for calculation.

$$b = \frac{\sigma_{LCC,Oil}^{monthly}}{\left(\sigma_{Oil}^{monthly}\right)^2} = \frac{-0.019952}{0.12133^2} = -1.355. \tag{2.14}$$

The intercept a is calculated using Eq. (2.9), together with the results from Eqs. (1.53), (1.52) and (2.14):

$$\begin{aligned} a &= \bar{r}_{LCC} - b \cdot \bar{r}_{Oil} \\ &= 4.73\,\% - (-1.356 \cdot (-0.04\,\%)) = 4.68\,\%. \end{aligned} \tag{2.15}$$

The resulting regression line is

$$r_{LCC}^{monthly} = 4.68\,\% - 1.356 \cdot r_{Oil}^{monthly}. \tag{2.16}$$

The R^2 formula (2.12) can be rewritten as a square of the correlation, using formulas for volatility (1.20) and monthly covariance (1.45) and correlation (1.43):

$$R^2 = \frac{\sigma_{Oil,LCC}^2}{\sigma_{Oil}^2 \cdot \sigma_{LCC}^2} = \rho_{Oil,LCC}^2. \tag{2.17}$$

(continued)

Use the result (1.68) to calculate

$$R^2 = \rho^2_{LCC,Oil} = (-0.435)^2 = 0.189. \qquad (2.18)$$

Our interpretation of the result is: For every 1 % increase (decrease) of the oil price in a given month, the US Airways stocks tend to decrease (increase) by 1.356 %. 18.9 % of the variation of US Airways can be explained by the movements of oil prices.

Let us calculate the regression line for the monthly returns $r_{LCC}^{monthly}$ of US Airways (LCC) against the monthly returns $r_{DAL}^{monthly}$ of Delta Airlines (DAL), based on the period January 2008–June 2010 (see Table 1.11). The regression equation is

$$r_{LCC}^{monthly} = a + b \cdot r_{DAL}^{monthly}. \qquad (2.19)$$

The slope b is calculated using Eq. (2.8), together with the results from Eqs. (1.54) and (1.60):

$$b = \frac{\sigma_{DAL,LCC}^{monthly}}{\left(\sigma_{DAL}^{monthly}\right)^2} = \frac{0.065368}{0.21914^2} = 1.361. \qquad (2.20)$$

The intercept a is calculated using Eq. (2.9), together with the results from Eqs. (1.52), (1.51) and (2.20):

$$\begin{aligned} a &= \overline{r}_{LCC} - b \cdot \overline{r}_{DAL} \\ &= 4.73\,\% - 1.361 \cdot 1.59\,\% = 2.57\,\%. \end{aligned} \qquad (2.21)$$

The resulting regression line is

$$r_{LCC}^{monthly} = 2.57\,\% + 1.361 \cdot r_{DAL}^{monthly}. \qquad (2.22)$$

R^2 is calculated from Eq. (2.12) using (1.66):

$$R^2 = \rho^2_{DAL,LCC} = 0.789^2 = 0.623. \qquad (2.23)$$

Our interpretation of the result is: For every 1 % increase (decrease) of Delta Airlines stocks in a given month, the US Airways stocks tend to increase (decrease) by 1.361 %. 62.3 % of the variation of the US Airways stocks can be explained by movements of Delta Airlines stocks.

(continued)

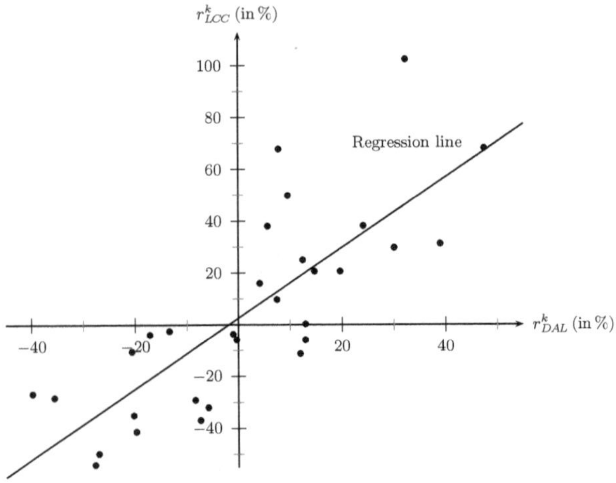

Fig. 2.5 Scatter plot of monthly returns of US Airways (LCC) against Delta (DAL) for the period January 2008–June 2010, together with regression line. *Source*: Yahoo! Finance

Figure 2.5 on the following page shows the scatter plot of monthly returns of US Airways against Delta Airlines for the period January 2008–June 2010, together with the regression line from Eq. (2.22).

When we compare the scatter plots in Figs. 2.5 and 2.1, we can see that the data points of the former are much closer to the regression line than those of the latter. This can be explained with R^2 being much greater in the former case [0.623, see Eq. (2.23)] than in the latter case [0.189, see Eq. (2.18)].

2.2.2 Multi-Linear and Non-Linear Regression

We can extend the concept of simple linear regression to multi-linear (or non-linear) regression to study the dependency of a variable on multiple variables (or to study non-linear dependencies).[8]

An example for multi-linear regression is

$$Y = a + b \cdot X + c \cdot W + d \cdot Z. \tag{2.24}$$

An example for non-linear regression is

$$Y = a + b \cdot X + c \cdot X^2 + d \cdot Z. \tag{2.25}$$

[8]For an introduction to multi-linear regression, see also DeFusco et al. (2004, pp. 441–494).

For any observed data point, the *error term* ε_k is the difference between the actual observed Y_k value and the value $\hat{Y}_k$ predicted by the regression equation, i.e.,

$$\varepsilon_k = Y_k - (a + b \cdot X + c \cdot W + d \cdot Z) \tag{2.26}$$

in our multi-linear example and

$$\varepsilon_k = Y_k - (a + b \cdot X + c \cdot X^2 + d \cdot Z) \tag{2.27}$$

in our non-linear example. We call X, W and Z the *independent variables*, Y the *dependent variable*, a the *intercept* (where the line crosses the Y-axis) and b the *slope coefficient*.[9] We refer to a and b as the *regression coefficients*.

In analogy to the simple linear regression, we get the regression equation by finding the *regression coefficients* a, b, c, d (and we may have more parameters for other examples) such that the sum of squares $\sum_{k=1}^{N} \varepsilon_k^2$ is minimal.

The variables X, W and Z are also called *explanatory variables* because they serve as variables to explain changes in Y. The larger the slope coefficients b, c and d, the stronger the *explanatory power* of the corresponding explanatory variables X, W and Z in describing changes in Y. If a slope coefficient (for example, d) turns out to be close to zero, then the corresponding explanatory variable (Z) is rather useless for describing the variable Y.

The R^2 measure from the simple linear regression [see Eq. (2.11)] is also used here to measure the quality of the regression. Multi-linear regressions also use R^2_{adj}, the *adjusted* R^2,[10] to describe the quality of the regression:

$$R^2_{adj} = 1 - \frac{N-1}{N-I-1}(1 - R^2) \tag{2.28}$$

where N is the number of observations and I is the number of independent variables.

We now have all prerequisites in order to understand the basic concepts of MPT as presented in the following section. We will start with the key idea of diversification and then continue and introduce the mean-variance efficient portfolio which will lead us to the famous capital asset pricing model (CAPM). When looking at empirical tests of the CAPM we will see that the CAPM often does not hold in reality. Therefore, we will present an extension of the CAPM: the Fama–French three-factor model.

[9]The terms are from DeFusco et al. (2004, p. 443).

[10]DeFusco et al. (2004, p. 457).

2.3 The Capital Asset Pricing Model (CAPM)

2.3.1 Introduction

Harry Markowitz[11] laid down the foundation of *modern portfolio theory* (MPT)[12] in his article Markowitz (1952) and his book *Portfolio Selection: Efficient Diversification of Investments*. A decade later, Treynor in 1961 (see French 2002), Sharpe (Sharpe 1964), Lintner (Lintner 1965a) and Mossin (Mossin 1966) built on his work to develop the capital asset pricing model (CAPM).

In Sect. 2.3 of this book, Sect. 2.3.2 presents the assumptions which are necessary for the CAPM to hold. Although these assumptions will turn out to be too idealized and unrealistic which is why the CAPM is not exactly true, the main implications and theories are still valid. Section 2.3.3 introduces the capital asset pricing model. Using a hypothetical *market portfolio*, i.e., a portfolio containing all risky assets, the set of optimal portfolios turns out to be a combination of the risk-free asset and the market portfolio. As a result, the expected return of a portfolio can be calculated by the CAPM equation as the sum of the risk-free rate plus the excess return (market return minus risk-free rate) multiplied by β, the sensitivity of the portfolio to market movements. This equation implies that the market rewards investors who take risk in terms of β. A brief summary of important aspects in Sects. 2.3.2 and 2.3.3 can be found in Schulmerich (2012a) or Schulmerich (2013b), *The Efficient Frontier in Modern Portfolio Theory: Weaknesses and How to Overcome Them (white paper).*[13]

Various empirical tests of the CAPM have been conducted with mixed results. The tests are presented in Sect. 2.3.4, followed by a discussion about the empirical validity of the CAPM in Sect. 2.3.5. The CAPM is often criticized for its unrealistic assumptions which may lead to questionable results. Section 2.3.6 will discuss the assumptions and show that by relaxing them the results will be slightly changed without changing the main implications.

---

[11]Harry M. Markowitz, born August 24, 1927, in Chicago, Illinois, is an American economist who become famous for his pioneering work in modern portfolio theory. He earned his Bachelor of Philosophy from the University of Chicago in 1947 and received his Master and Doctor of Economics at the same university in 1950 and 1954, respectively. He held various positions with RAND corporation (1952–1963), Consolidated Analysis Centers, Inc. (1963–1968), the University of California, Los Angeles (1968–1969), Arbitrage Management Company, (1969–1972), and IBM's T.J. Watson Research Center (1974–1983) before becoming a professor of finance at Baruch College of the City University of New York. He joined the University of California, San Diego in 1994 as a research professor of economics where he became a finance professor at the Rady School of Management in 2006. He received the John von Neumann Theory Prize in 1989 and the Nobel Memorial Prize in Economic Sciences in 1990.

[12]Although he is known as the father of MPT, Markowitz credits Andrew D. Roy with half of the honor in Markowitz (1999).

[13]An extension of these articles on MPT including different risk measures can be found in Schulmerich (2012b), *Extending Modern Portfolio Theory: Efficient Frontiers for Different Risk Measures (white paper)*. A third white paper in this MPT trilogy is *Can the Black–Litterman Framework Improve Asset Management Outcomes?*, see Schulmerich (2013a).

2.3.2 Assumptions

Before we introduce the CAPM, we need to talk about its assumptions first. These are[14]:

- **Assumption 1 (A1)**:
  All investors have homogeneous expectations, i.e., they expect the same probability distribution of returns.
- **Assumption 2 (A2)**:
  All investors want to invest in an optimal portfolio based on Markowitz's mean-variance framework, i.e., for a given expected return, they target the portfolio with the lowest volatility.
- **Assumption 3 (A3)**:
  All investors can lend and borrow any amount of money at the risk-free rate.
- **Assumption 4 (A4)**:
  All investors have the same one-period horizon.
- **Assumption 5 (A5)**:
  All assets are infinitely divisible.
- **Assumption 6 (A6)**:
  There are no taxes and transaction costs.
- **Assumption 7 (A7)**:
  There is no inflation or any change in interest rates, or inflation is fully anticipated.
- **Assumption 8 (A8)**:
  Capital markets are *efficient*, i.e., they are in equilibrium.

These assumptions are idealized and unrealistic. Does this mean that the CAPM is useless because it is based on "wrong" assumptions? There are two important points to note[15]: Many of these assumptions can be relaxed to get closer to the real world, which is discussed in Sect. 2.3.5. This will lead to slight modifications of the CAPM without changing the main conclusions. The other point is that "*a theory should never be judged on the basis of its assumptions, but rather on how well it explains and helps us predict behavior in the real world*". [16]

[14]Reilly and Brown (1997, p. 279). This work offers a good introduction and is used as a standard reference work in the CFA curriculum. The sections in this book about the CAPM and its assumptions are based on this reference.

[15]Reilly and Brown (1997, p. 279).

[16]Reilly and Brown (1997, p. 279).

2.3.3 The Model

After we have defined the set of assumptions of the capital asset pricing model, we now turn to the model itself. Of course, the simplistic assumptions in Sect. 2.3.2 will lead to a rather simplistic model. This model should be thought of as a fundament for financial analysis which helps us to understand the relation between risk and return. It helps us to understand the real world, but does not represent the real world.

You may ask why you should care about an oversimplistic model. Maybe you remember the Bohr[17] atom model from high school physics. In this model, the atom consists of a heavy positively charged nucleus surrounded by electrons which travel around the nucleus on circular orbits. You can compare this model to our solar system, where the planets orbit around the sun. Although this model was a breakthrough to understand atomic physics, it is very simplistic and was later substituted by the theory of quantum mechanics. The point is, that like the Bohr model, the CAPM should just be treated as a starting point to understand mathematical finance. You have to understand the basic model before you can deal with more complex questions.

Like models in physics, models in finance have to be positively tested until they are accepted. The most important tests of the CAPM will be presented in Sect. 2.3.3, followed by a discussion about the empirical validity of the CAPM in Sect. 2.3.4. Section 2.3.5 will then deal with relaxed assumptions for the CAPM which lead to a slightly modified model.

Let us assume a future 1-year horizon $[0, 1]$, for which we have available data about the expected (annual) returns of individual securities, their risks in terms of (annual) volatility, and the correlations between the securities. For every single portfolio, we can calculate the expected annual return $\mathbb{E}[R_{Pf}]$ [see Eq. (1.82)] and its annual volatility σ_{Pf} [see Eq. (1.83)], as outlined in Sect. 1.5.

Please note:

- The future horizon for the CAPM does not have to be 1 year, but we assume this for simplicity. You could also have one quarter year, one month, one week, etc. as a future horizon, and if not dealing with annual returns, the CAPM makes the same statements about quarterly, monthly and weekly returns.
- We use capital letters for random variables, for example the future annual return of a portfolio R_{Pf} which is uncertain, and lower letters are used for numbers like historical returns which we know with certainty.
- For annual returns on the time horizon $[0, 1]$ we use the same notations as for annualized returns because they are the same.

[17]Niels H.D. Bohr (October 7, 1885–November 18, 1962) was a Danish physicist who made foundational contributions to understanding atomic structure and quantum theory, for which he received the Nobel Prize in Physics in 1922.

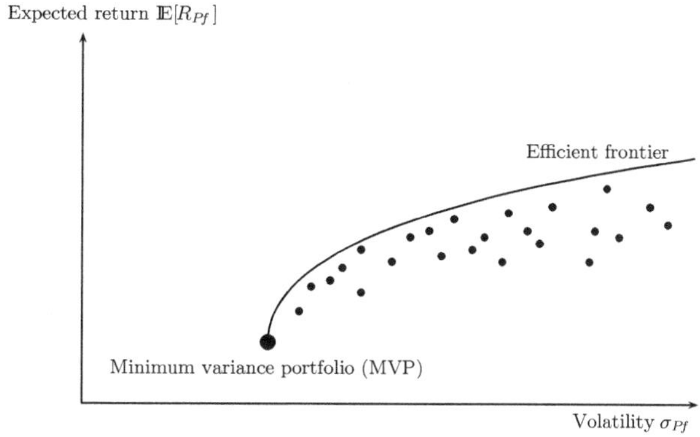

Fig. 2.6 Mean-variance diagram with efficient frontier. *Source*: Own, for illustrative purposes only

The rational investor who is the presumed CAPM protagonist, wants to earn a certain return and tries to identify a portfolio of minimal risk which satisfies this goal. Following this purpose, we plot all possible portfolios of risky assets in a mean-variance diagram,[18] as shown in Fig. 2.6 where the points represent the expected returns $\mathbb{E}[R_{Pf}]$ (vertical axis) and the volatilities σ_{Pf} (horizontal axis) of the portfolios.

A portfolio is called *mean-variance efficient* (or just *efficient*), if for a given volatility there is no portfolio with a higher return. As Merton[19] notes, the set of efficient portfolios in the mean-variance diagram is called the *efficient frontier* and has the shape of a hyperbola.[20] It is the upper boundary of all portfolios in the mean-variance diagram from Fig. 2.6. This is exactly the set of portfolios, that the rational investor in our framework is looking for: They maximize the expected return for a given risk, and they minimize the risk for a given return.

---

[18]Technically a *volatility-mean* diagram, it is called *mean-variance* diagram for historical reasons because when Harry Markowitz introduced MPT in Markowitz (1952, 1999), he plotted the variance against the mean return, although today it is more common to plot mean return against volatility.

[19]Robert C. Merton (born July 31, 1944), an American economist and professor at the MIT Sloan School of Management, is known for his significant contributions to continuous-time finance, especially the first continuous-time option pricing model, the Black–Scholes–Merton formula. Together with Myron Scholes, Merton received the Alfred Nobel Memorial Prize in Economic Sciences in 1997 for expanding the Black–Scholes–Merton formula. He earned his Doctor of Economics from the MIT in 1970 under the guidance of Paul Samuelson. He then joined the faculty of the MIT Sloan School of Management where he taught until 1988 before moving to Harvard Business School, where he stayed until 2010. In 2010 he rejoined the MIT Sloan School of Management as the School of Management Distinguished Professor of Finance. He is the past President of the American Finance Association.

[20]Merton (1972, p. 1856).

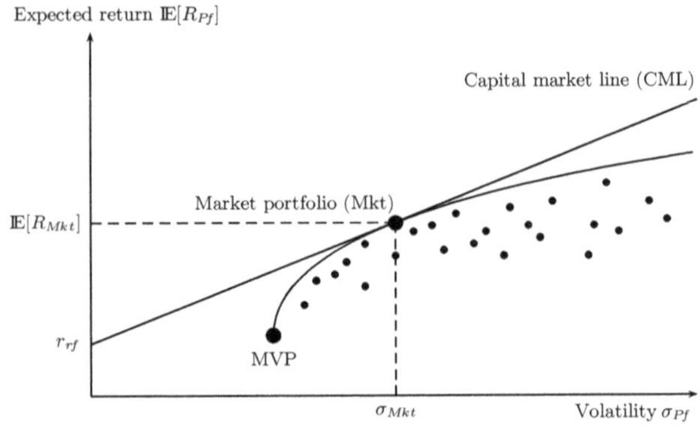

Fig. 2.7 The mean-variance diagram shows the capital market line (CML) which connects the risk-free asset with the tangency point on the efficient frontier, the market portfolio. The efficient frontier is based on portfolios without the risk-free asset. *Source*: Own, for illustrative purposes only

The portfolio on the efficient frontier with the lowest volatility is called *minimum-variance portfolio (MVP)*.[21] If a *risk-free asset*[22] exists, that is, an asset with *zero volatility*, then the set of mean-variance efficient portfolios, formed with risk-free and risky assets, is a line from the risk-free asset to the tangency point on the efficient frontier, as shown in Fig. 2.7. We call the tangency point and the corresponding portfolio "Mkt", and the risk-free asset is labeled "rf". The reason why we get a line is that the volatility σ_{rf} and also the covariance $\sigma_{rf,Mkt}$ are zero (see definition of covariance in Eq. (1.42). We use $r_{rf} = \mathbb{E}[R_{rf}]$), so the volatility of the portfolio, with weight w_{Mkt} invested in "Mkt" and $1 - w_{Mkt}$ invested in the risk-free asset, is [see Eq. (1.83)]

$$\begin{aligned}\sigma_{Pf} &= \sqrt{w_{Mkt}^2\sigma_{Mkt}^2 + (1-w_{Mkt}^2)\underbrace{\sigma_{rf}^2}_{=0} + 2w_{Mkt}(1-w_{Mkt})\underbrace{\sigma_{rf,Mkt}}_{=0}} \\ &= w_{Mkt}\cdot\sigma_{Mkt}.\end{aligned} \tag{2.29}$$

Then

$$\begin{aligned}\mathbb{E}[R_{Pf}] &= w_{Mkt}\mathbb{E}[R_{Mkt}] + (1-w_{Mkt})r_{rf} \\ &= r_{rf} + w_{Mkt}(\mathbb{E}[R_{Mkt}] - r_{rf}) \\ &= r_{rf} + \sigma_{Pf}\cdot\frac{\mathbb{E}[R_{Mkt}] - r_{rf}}{\sigma_{Mkt}}.\end{aligned} \tag{2.30}$$

[21] Bodie, Kane, and Marcus (2009, p. 211).

[22] Risk-free assets are discussed below on page 118.

This gives us the equation for our line which is called the *capital market line* (CML). All portfolios on the CML are a combination of the tangency portfolio and the risk-free asset. Since it is assumed that capital markets are in equilibrium, every investor holds a portfolio on the CML and therefore a part of the same tangency portfolio, the latter has to be the *market portfolio*.

Black,[23] Jensen[24] and Scholes[25] were among the first to use the term *market portfolio*.[26] The market portfolio is a theoretical portfolio which is central to the CAPM. It contains every risky asset in the market,[27] including stocks, bonds, options, real estate, coins, stamps, art, antiques[28] and also human capital.[29] The assets are weighted according to their market value.

The reason for this is simple[30]: In our framework, every investor invests in the same risk-free asset and the same risky portfolio. For example, if all investors have 1 % of their risky portfolio invested in Apple Inc., then Apple comprises 1 % of the

---

[23]Fischer S. Black (January 11, 1938–August 30, 1995), an American economist, is one of the authors of the famous Black–Scholes equation. He graduated from Harvard College in 1959 and received a Ph.D. in Applied Mathematics from Harvard University in 1964. He was initially expelled from the Ph.D. program due to his inability to settle on a thesis topic, having switched from physics to mathematics, then to computers and artificial intelligence. In 1971, he began to work at the University of Chicago but later left to work at the MIT Sloan School of Management. In 1984, he joined Goldman Sachs where he worked until his death in August 1995 from throat cancer.

[24]Michael C. Jensen (born November 30, 1939 in Rochester, Minnesota, U.S.) is an American economist working in the area of financial economics. He is currently the managing director in charge of organizational strategy at Monitor Group, a strategy consulting firm, and the Jesse Isidor Straus Professor of Business Administration, Emeritus, at Harvard University. He received his B.A. in Economics from Macalester College in 1962 and both his M.B.A. (1964) and Ph.D. (1968) degrees from the University of Chicago Booth School of Business, notably working with Professor Merton Miller, the 1990 co-winner of the Nobel Prize in Economics. Jensen is also the founder and editor of the Journal of Financial Economics. The Jensen Prize in corporate finance and organizations research is named in his honor.

[25]Myron S. Scholes (born July 1, 1941), a Canadian-born American financial economist, is one of the authors of the Black–Scholes equation. In 1968, after finishing his dissertation under the supervision of Eugene Fama and Merton Miller, Scholes took an academic position at the MIT Sloan School of Management where he met Fischer Black and Robert C. Merton, who joined MIT in 1970. For the following years Scholes, Black and Merton undertook groundbreaking research in asset pricing, including the work on their famous option pricing model. In 1997 he shared the Nobel Prize in Economics with Robert C. Merton "for a new method to determine the value of derivatives". Fischer Black, who co-authored with them the work that was awarded, had died in 1995 and thus was not eligible for the prize. In 1981 Scholes moved to Stanford University, where he remained until he retired from teaching in 1996. Since then he holds the position of Frank E. Buck Professor of Finance Emeritus at Stanford.

[26]Black, Jensen, and Scholes (1972, p. 444).

[27]Reilly and Brown (1997, p. 284).

[28]This list of examples is mentioned in Reilly and Brown (1997, p. 284).

[29]Richard Roll mentioned human capital as part of the market portfolio in Roll (1977, p. 131 and p. 155).

[30]This reasoning is from Bodie, Kane, and Marcus (1999, p. 253).

whole market, and vice versa. Every investor holds the same risky portfolio which excludes the possibility that Apple stock represents a 1.5 % portion in the portfolio of one investor and a 0.5 % portion in a another one.

The market portfolio cannot only contain stocks or bonds, it has to contain everything. If it did not include any Picasso paintings, then by our theory nobody would own Picasso paintings, because everyone's risky portfolio is supposed to be the same. There would be no demand for those, and they would not be worth millions. According to our theory, every investor who makes a risky investment holds some tiny fractions of Picasso paintings. Of course, this is not realistic, and this clearly shows the limitations of CAPM.

It is common practice to use Treasury bills as *the* risk-free asset and its yield as the risk-free rate r_{rf}[31]: As a short-term fixed income investment it is insensitive to interest rate fluctuations. It is backed by the U.S. government and has virtually no default risk. In practice, all *money market instruments* can be treated as risk-free assets because their short maturities make them virtually free of interest rate risk. They are fairly safe in terms of default or credit risk, and as one of the most liquid asset types their liquidity risks are low. Money market instruments are short-term debt instruments that have original maturities of 1 year or less.[32] They include U.S. Treasury bills, commercial paper, some medium-term notes, bankers acceptances, federal agency discount paper, most certificates of deposit, repurchase agreements, federal funds, and short-lived mortgage- and asset-backed securities.

After having discussed the optimal portfolios which are situated on the CML, we want to take a closer look at the individual securities within the portfolio. Since the rational investor does not hold one single, but many securities in his portfolio, we have to assess the risk of these securities in the portfolio context and not just the volatility which describes the risk of holding only one security alone. Let us take a look at the mean-variance diagram in Fig. 2.8, and analyze security A. If you are invested only in this asset and nothing else, then the volatility σ_A is the risk which matters. But if you hold security A as part of a portfolio on the CML, then some of the risk, the unsystematic risk $\sigma_{\varepsilon,A}$, is diversified away, and what is left is the systematic risk $\beta_A \cdot \sigma_{Mkt}$ (β_A is defined in Eq. (2.31)). In this case, security A becomes equivalent to a portfolio B on the CML with the same expected return as portfolio A, but with a lower risk $\beta_A \cdot \sigma_{Mkt}$.

This risk cannot be diversified away. Dividing it by σ_{Mkt} produces the *standardized measure of systematic risk, beta* β_A. The market portfolio has beta 1. Since every rational investor is assumed to hold a portfolio on the CML, we only have to take account of the systematic risk, i.e., β.

In equilibrium, *all* securities and *all* portfolios are on the security market line (SML), as shown in Fig. 2.9 which plots the expected return of a portfolio against its *beta*. The beta of the risk-free asset is 0, the beta of the market portfolio is 1, and the SML is the line which connects these portfolios.

---

[31]Bodie et al. (1999, p. 181).

[32]Fabozzi, Mann, and Choudhry (2002, p. 1).

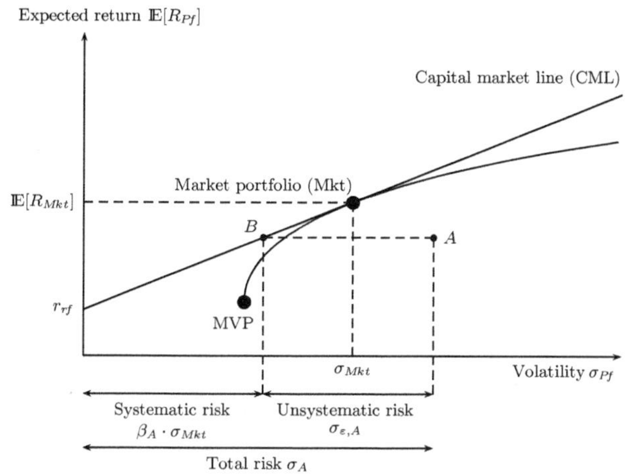

Fig. 2.8 The mean-variance diagram shows the capital market line (CML) which connects the risk-free asset with the tangency point on the efficient frontier, the market portfolio. *Source*: Own, for illustrative purposes only

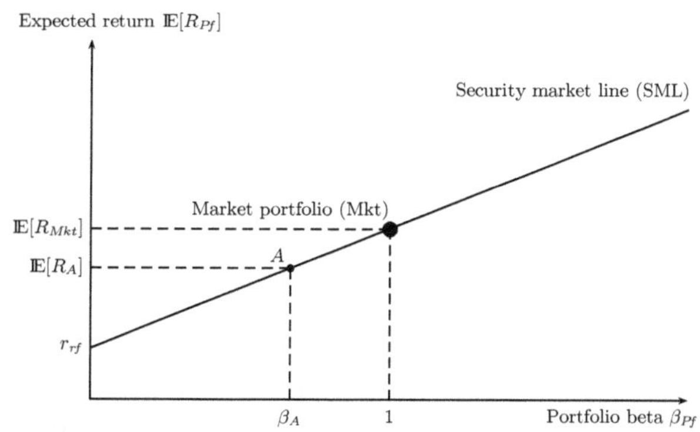

Fig. 2.9 The diagram shows the security market line (SML) which connects the risk-free asset with the market portfolio. In equilibrium, all assets and portfolios plot on the SML. *Source*: Own, for illustrative purposes only

The *beta* (β_{Pf}) of a portfolio is the covariance between the portfolio return and the market return divided by the variance of the market return:

$$\beta_{Pf} = \frac{\sigma_{Pf,Mkt}}{\sigma_{Mkt}^2}. \tag{2.31}$$

In an efficient market, the expected value $\mathbb{E}$ of the portfolio return R_{Pf} satisfies the CAPM equation for any possible portfolio

$$\mathbb{E}[R_{Pf}] = r_{rf} + \beta_{Pf} \cdot (\mathbb{E}[R_{Mkt}] - r_{rf}), \tag{2.32}$$

where the portfolio beta is β_{Pf}, the risk-free return is r_{rf}, and the market return is r_{Mkt}. The difference $\mathbb{E}[R_{Mkt}] - r_{rf}$ is called the *market risk premium*. This equation also shows that the investor is directly rewarded for taking risk in terms of β_{Pf}, but not for volatility σ_{Pf}.

There are big differences between the two risk measures beta and volatility:

- While the volatility of a portfolio is an absolute measure which depends only on its returns, beta is a relative measure, since it measures how the portfolio moves relative to the market. A β_{Pf} of 0.8 means that a 1 % increase in the market return increases the portfolio return by 0.8 %. Contrary to the volatility, β_{Pf} could be negative which would mean that market returns and portfolio returns tend to move in opposite directions.
- While computing the volatility σ_{Pf} of a portfolio with M securities (with portfolio weights $w_1, \ldots, w_M$), one has to take into account the pairwise covariances

$$\sigma_{i,j} \quad (1 \leq i, j \leq M, i \neq j)$$

among these M securities besides their volatilities

$$\sigma_i \quad (1 \leq i \leq M),$$

see Eq. (1.83). The relation for beta is linear:

$$\beta_{Pf} \;=\; w_1 \cdot \beta_1 + w_2 \cdot \beta_2 + \ldots + w_M \cdot \beta_M. \tag{2.33}$$

According to the CAPM, all optimal portfolios are a combination of the risk-free portfolio and the market portfolio, so every investor—independent of his risk-preference—will always hold a combination of those two portfolios. If w_{Mkt} is the portfolio weight invested in the market portfolio and $1-w_{Mkt}$ the proportion invested in the risk-free asset, then the portfolio beta β_{Pf} is w_{Mkt}. In order to construct individually tailored optimal portfolios, two tasks have to be completed, which can be done separately (this is also called *Tobin separation*[33]):

---

[33] Spremann (2008, pp. 227–228), initially discussed in Tobin (1958, p. 66). In Markowitz (1999, p. 10), Harry Markowitz refers to the Tobin (1958) article as the *first CAPM*, and summarizes Tobin's work. In Sect. 3.6. of Tobin's article, the author presented his seminal result, today known as the Tobin separation theorem. Tobin assumed a portfolio selection model with M risky assets and one riskless asset: cash. Because these assets were monetary, i.e., "*marketable, fixed in money value, free of default risk*" (see Tobin 1958, p. 66), the risk was market risk, not default risk. Holdings had to be nonnegative. Borrowing was not permitted. Implicitly, Tobin assumed that the covariance matrix for risky assets is nonsingular. The result was that "*the proportionate composition of the non-cash assets is independent of their aggregate share of the investment balance. This fact makes it possible to describe the investor's decisions as if there were a single non-cash asset, a composite formed by combining the multitude of actual non-cash assets in fixed proportions.*" (see Tobin 1958, p. 84).

Table 2.1 Example for calculating the expected return of a portfolio using CAPM

i	Weight w_i	Beta β_i
1	15 %	1.2
2	20 %	0.6
3	25 %	0.5
4	15 %	2.4
5	25 %	−0.4

Source: Own, for illustrative purposes only

- Construct the risk-free portfolio (with money market instruments) and the market portfolio. This is the task of the portfolio manager.
- Determine the correct risk profile of the client and find the optimal allocation between risk-free and risky investments. This is the role of the investment advisor.

In order to find the optimal portfolio, one would have to know the composition of the market portfolio. In theory, this would also include non-tradable assets like human capital, rarely traded assets like art collections, and assets like real estate which are difficult to value. Their respective market capitalization has to be known in order to set up the market portfolio. But given that human capital makes up a large part of the market portfolio with estimates of its proportion within the portfolio ranging from 50 to 90 %,[34] the market portfolio cannot be measured as long as the valuation of human capital remains questionable.[35]

Example

Let us consider the following example: We want to calculate the expected return of a portfolio containing five different assets. Table 2.1 shows the weights of the assets in our portfolio. Assume that the risk-free rate r_{rf} is 5 % and the expected market return $\mathbb{E}[R_{Mkt}]$ is 13 %.
The first step is to calculate the portfolio beta β_{Pf} using Eq. (2.33) with $M = 5$:

$$\begin{aligned}\beta_{Pf} &= w_1 \cdot \beta_1 + \ldots + w_5 \cdot \beta_5 \\ &= 0.15 \cdot 1.2 + 0.20 \cdot 0.6 + 0.25 \cdot 0.5 + 0.15 \cdot 2.4 + 0.25 \cdot (-0.4) \\ &= 0.685. \end{aligned} \tag{2.34}$$

We use this result to calculate the expected return $\mathbb{E}[R_{Pf}]$ of the portfolio from the CAPM equation—Eq. (2.32):

[34] In the studies Kendrick (1974, 1976, 1994) and Eisner (1985, 1989) which used the cost-based approach, the size of human capital was about the size of non-human capital. In the study Jorgenson and Fraumeni (1989, 1992) which used the income-based approach, the share of human capital in total wealth was over 90 %.

[35] A good summary about the market portfolio is provided in Le, Gibson, and Oxley (2003).

$$\begin{aligned}\mathbb{E}[R_{Pf}] &= r_{rf} + \beta_{Pf} \cdot (\mathbb{E}[R_{Mkt}] - r_{rf}) \\ &= 5\,\% + 0.685 \cdot (13\,\% - 5\,\%) = 10.48\,\%. \end{aligned} \tag{2.35}$$

The expected return of our portfolio is 10.48 %.

End of Example

Remark. We have derived the CAPM equation—Eq. (2.32)—for a 1-year period and for annualized returns. But the CAPM also holds for arbitrary periods, for example, quarters, months, weeks, days, and you also get CAPM equations with quarterly, monthly, weekly and daily returns. For example, for monthly returns, the CAPM formula is

$$\mathbb{E}[R_{Pf}^{monthly}] = r_{rf}^{monthly} + \beta_{Pf} \cdot (\mathbb{E}[R_{Mkt}^{monthly}] - r_{rf}^{monthly}). \tag{2.36}$$

2.3.4 Empirical Tests

We have described the CAPM after starting with very simplistic and unrealistic assumptions in Sect. 2.3.1 and made two points about the usefulness of the theory[36]: First, many assumptions can be relaxed to approximate real-world conditions. This would slightly modify the CAPM without changing the main implications. Second, a theory should "*not be judged on the basis of it assumptions, but on how well it explains relationships that exist in the real world*".[37]

Therefore, in this section, we will discuss the empirical tests of the CAPM. This section is based on the paper Fama and French (2004) and Reilly and Brown (1997, pp. 310–311).

2.3.4.1 The Testable Hypotheses

Let us recall the CAPM equation (2.32) which should be tested:

Given the risk-free return r_{rf} and the market return R_{Mkt}, the expected return $\mathbb{E}[R_{Pf}]$ of the portfolio depends only on the systematic risk, the portfolio beta β_{Pf} (i.e., the covariance of the portfolio *Pf* with the market portfolio *Mkt* divided by the variance of the market portfolio), satisfying the equation

$$\mathbb{E}[R_{Pf}] = r_{rf} + \beta_{Pf} \cdot (\mathbb{E}[R_{Mkt}] - r_{rf}). \tag{2.37}$$

The first question which arises is if beta is a stable measure of systematic risk.[38] Can historical betas be used as an estimate for future betas? If betas change too

[36] Reilly and Brown (1997, p. 279).

[37] Reilly and Brown (1997, p. 310).

[38] Reilly and Brown (1997, p. 310).

much over time, then they are no useful measures for future risks and for estimating expected returns with the CAPM formula.

Equation (2.37) has four testable implications which are to be tested[39]:

- **Hypothesis 1 (C1)—*Linearity*:**
  Expected returns of all assets are linearly related to their betas.
- **Hypothesis 2 (C2)—*No systematic non-beta risks*:**
  Beta is a complete measure of risk for an asset in the market portfolio. No other variable has marginal explanatory power to explain returns.
- **Hypothesis 3 (C3)—*Positive beta premium*:**
  The beta premium,[40] i.e., the difference between the expected return of the market portfolio and the expected return of assets uncorrelated to the market, is greater than zero. In other words, the expected return of the market portfolio exceeds the expected return of assets uncorrelated to the market.
- **Hypothesis 4 (C4)—*Risk-free return on zero-beta assets*:**
  Assets uncorrelated to the market portfolio have expected returns equal to the risk-free rate.

2.3.4.2 Regressions

There are two different simple linear regressions which are used to test Eq. (2.37):

- The *cross-sectional regressions* test the CAPM across assets with different betas. The returns of different assets are regressed over the betas.
- The *time-series regressions* test the CAPM equations for each individual asset over time. The excess portfolio returns in the subperiods are regressed over the respective excess return of the market portfolio.

Cross-Sectional Regression

The cross-sectional regression is the main tool used to test Eq. (2.37) over a period $[0, T]$ of T years. CAPM tests do not exactly test Eq. (2.37) because it includes *expected returns* which are not measureable. Instead, the question is if the relation holds for *realized returns*, i.e., if the relation has been correct in the past. The cross-sectional regression tests the CAPM equation by regressing the (arithmetic) average annual return $\overline{r}_A$ of an asset A over beta.

Figure 2.10 shows a typical cross-sectional regression. For every asset A, the data point (β_A, r_A) is plotted with β_A its beta and $\overline{r}_A$ as its (arithmetic) average annual return. We then get the regression line which has the form

---

[39]These are from Fama and French (2004, p. 30), and Fama and MacBeth (1973, p. 610 and p. 613).

[40]Literally, the beta premium is the premium per unit of beta. The CAPM implies that the beta premium is the excess market return, i.e., the difference between the expected return on the market portfolio and the risk-free rate. But this equality is equivalent to the fourth hypothesis, that zero-beta assets expect to earn the risk-free rate. The *positive beta premium* does not test the equality with the excess market return, only the positivity.

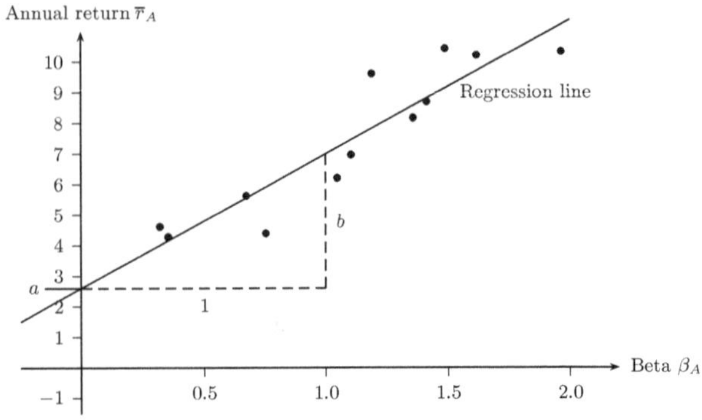

Fig. 2.10 Cross-sectional regression. Annualized returns of asset returns are regressed over their respective betas. *Source*: Own, for illustrative purposes only

$$\overline{r}_A = a + b \cdot \beta_A. \tag{2.38}$$

This diagram looks very similar to Fig. 2.9 where the expected returns of assets are plotted against their betas. They lie on the security market line (SML) which is the line *predicted* by the CAPM, whereas the regression line is the *empirical* one we observe, based on historical data. The empirical test of the CAPM examines if both lines are equal, or in other words, if the data fits the predicted line, the SML. The intercept a in Eq. (2.38) and the slope coefficient b (the beta premium) correspond to the risk-free rate r_{rf} in Eq. (2.37) and the market risk premium $R_{Mkt} - r_{rf}$, respectively. With this regression, we can check hypotheses (C3) and (C4) from the list on page 123:

- *Positive beta premium*: The beta premium b is positive.
- *Risk-free return on zero-beta assets*: The intercept a equals the risk-free rate r_{rf}.

Given the market data, we still need the betas β_A to plot a diagram like Fig. 2.10 and to perform our regression analysis. We get the betas from a time-series regression.

Time-Series Regression

The time-series regression tests the CAPM equation for each individual asset separately on N equidistant subperiods of a time horizon $[0, T]$. The excess asset return in each subperiod k is regressed over the excess market return from the same subperiod. Figure 2.11 shows a typical time-series regression. Given an asset A, for every subperiod k, the data point $\tilde{r}^k_{Mkt}, \tilde{r}^k_A$ is plotted where $\tilde{r}^k_{Mkt} = r^k_{Mkt} - r^k_{rf}$ is the excess return of the market portfolio in subperiod k, i.e., the difference between the market return and the return on a risk-free asset, and $\tilde{r}^k_A = r^k_A - r^k_{rf}$ is the excess return of the asset A in the subperiod k.

We get a regression line of the form

$$\tilde{r}_A^{subperiod} = a + b \cdot \tilde{r}_{Mkt}^{subperiod} \tag{2.39}$$

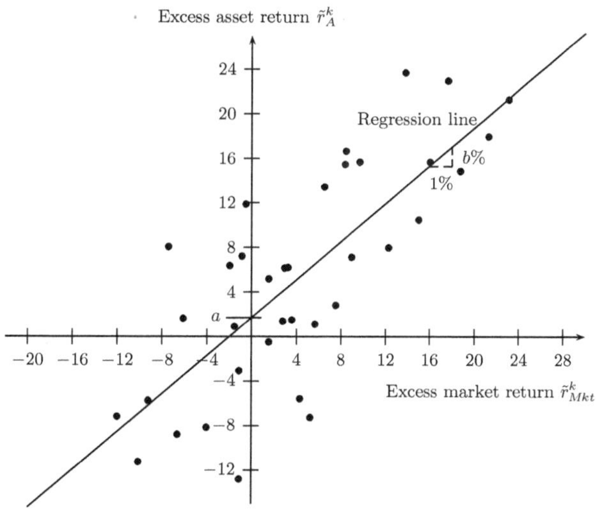

Fig. 2.11 Time-series regression. Excess returns of an asset A in subperiod i are regressed over the excess market return. *Source*: Own, for illustrative purposes only

where *subperiod* can be *quarterly, monthly, weekly, daily,* etc. and refers to the length of the subperiods. For example, if we choose monthly subperiods, then we have a regression on monthly returns:

$$\underbrace{\tilde{r}_A^{monthly}}_{\text{monthly excess return}} = a + b \cdot \tilde{r}_{Mkt}^{monthly}. \tag{2.40}$$

Let us recall the CAPM equation using monthly returns (Eq. (2.36)):

$$\mathbb{E}[R_A^{monthly}] = r_{rf}^{monthly} + \beta_A \cdot (\mathbb{E}[R_{Mkt}^{monthly}] - r_{rf}^{monthly}).$$

Rewrite this as

$$\underbrace{\mathbb{E}[R_{Pf}^{monthly}] - r_{rf}^{monthly}}_{\text{expected monthly excess return}} = \beta_A \cdot (\mathbb{E}[R_{Mkt}^{monthly}] - r_{rf}^{monthly}), \tag{2.41}$$

and we can see that in the regression equation—Eq. (2.40)—a should be zero and b is the beta β_A. For any asset A, we call the intercept a from the above regression the *alpha* of A. In case of monthly subperiods, we write $\alpha_A^{monthly}$, similarly to other subperiods. The alpha represents the incremental rate of return which exceeds the theoretical rate of return implied by the CAPM. It is introduced in Jensen (1967) as the intercept of the linear time-series regression and as a performance measure to evaluate a portfolio manager who invests in portfolio A.[41]

[41] Jensen (1967, p. 8).

Regarding the CAPM hypotheses on page 123, the time-series regression can be used to test hypothesis (C4):

Risk-free return on zero-beta assets: The alphas of all assets are zero.

Note that for our regressions, we do not use the real market portfolio *Mkt* because it is hardly observable. Instead, we use a proxy like a stock market index to mimic the market return.

2.3.4.3 Beta Stability

In the CAPM, the only relevant risk measure is the systematic risk, beta. For any asset with historical data, we can calculate the historical beta with a time-series regression. But is the historical beta reliable enough to assess future risk? Figure 2.12 depicts the following scenario for an investor: Assume that you are looking for a rather conservative investment with a low beta of around 0.5, and you have found a portfolio with a historical beta of 0.5. Your natural question to ask is if the portfolio's beta will only change little over time, i.e., whether it will remain stable.

There have been lots of studies on this topic, with similar results. They are summarized in Reilly and Brown (1997, pp. 310–311), and this section about beta stability is based on this book. The two researchers tested the correlation between two consecutive time periods, as illustrated in Fig. 2.13: The earlier period is called *estimation period*, the latter is termed *subsequent period*.[42] If the correlation between the betas of both periods is high, then beta is stable, and the investor can rely on the historical beta to provide a good estimate of the future beta. The correlation of betas of consecutive time periods is therefore used as a measure for beta stability.

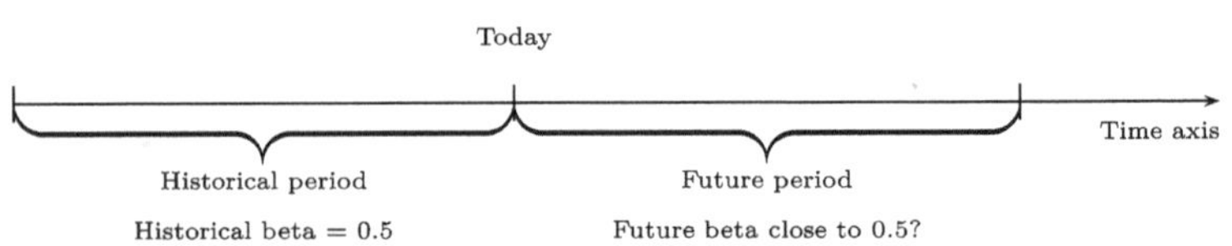

Fig. 2.12 Illustration of historical beta vs. future beta. *Source*: Own

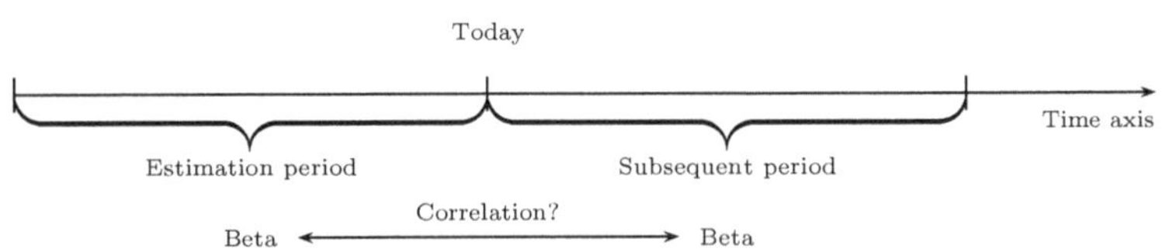

Fig. 2.13 Illustration of two consecutive periods: estimation period and subsequent period. *Source*: Own

[42] This terminology is from Roenfeldt, Griepentrog, and Pflaum (1978).

Marshall E. Blume studied the variation of the beta measure for all NYSE common stocks (the number of stocks ranged from 415 to 890)[43] in the period July 1926–June 1968 in his paper Blume (1971). He split this 42-year period into seven consecutive 6-year periods for which he calculated the betas (based on monthly returns) of portfolios with 1, 2, 4, 7, 10, 20, 35, 50, 75 and 100 stocks. He then calculated the correlations of betas between consecutive periods, as illustrated on the timeline in Fig. 2.14.

His result is that the correlations of betas are low for one single stock (around 0.6), but these numbers increase with the number of stocks in the portfolio. With 50 stocks, the correlation becomes very high (0.98). The beta is unstable for individual stocks, but it is very stable for a larger portfolio with 50 stocks.[44]

Robert A. Levy did a similar study in Levy (1971). Here, the time periods were much shorter: 13 weeks, 26 weeks and 52 weeks. The betas were calculated based on weekly returns. Short-term betas are more important for portfolio managers with shorter time horizons.[45] Again, the beta of the single stock was unstable. The longer the time period (at least 26 weeks) and the more stocks (at least 25 stocks) in the portfolios, the more stable the beta of the portfolio.

R.B. Porter and John R. Ezzell argue in Porter and Ezzell (1975) that the high correlation of betas in Blume's study is only due to his portfolio selection methodology. Blume selected the portfolios in his study based on the betas[46]: The stocks with the lowest betas are assigned to portfolio 1, the next smallest ones to portfolio 2, etc. Porter and Ezzell repeated his study with randomly selected portfolios and found that in contrast to Blume's results, the stability of beta is "*relatively light and* [...] *totally unrelated to the number of securities in the portfolio*".[47]

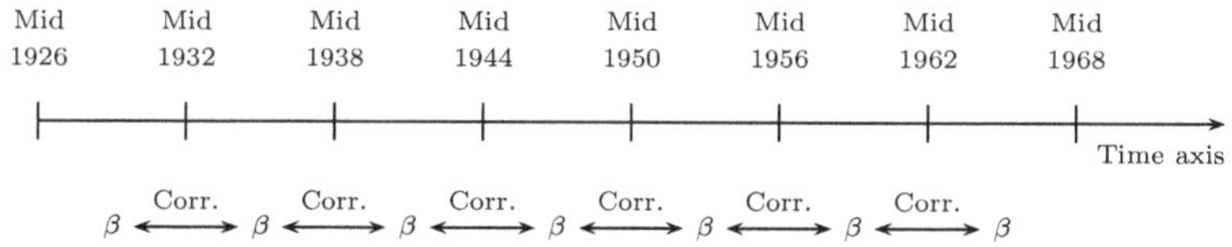

Fig. 2.14 Illustration of the timeline used in Blume (1971) to study the stability of beta. The time period July 1926–June 1968 is split in seven equidistant subperiods for which the betas are measured. The correlation between the betas of consecutive subperiods are calculated and used as a measure for beta stability. *Source*: Own

[43] Blume (1971, p. 4).

[44] Blume (1971, p. 7).

[45] Levy (1971, p. 57).

[46] Blume (1971, p. 6).

[47] Quote from Porter and Ezzell (1975, p. 369). See p. 370, Table 2, which compares Blume's to Porter's & Ezzell's correlation numbers.

T.M. Tole examined the standard deviation of betas as another measure of beta stability in Tole (1981). He found a significant decrease in the standard deviation (and therefore an increase in beta stability) with increasing portfolio size,[48] even beyond 100 stocks.[49]

To which extent does beta stability depend on the lengths of the estimation and subsequent periods? In Baesel (1974), the beta stability of single U.S. stocks is examined by using different lengths for the two periods. Baesel found that an increase in these lengths also increases the beta stability.[50] Altman, Jacquillat, and Levasseur (1974) presents the same results for French stocks. While both periods were equally long in these two studies, Roenfeldt et al. (1978) compares the beta of a 4-year estimation period to the betas of subsequent 1-year, 2-year, 3-year and 4-year periods. They found that estimated betas based on the 4-year period were more reliable as a forecast for subsequent 2-year, 3-year and 4-year periods than for a subsequent 1-year period.[51]

According to Reilly and Brown (1997), the overall conclusion of the empirical tests is that individual betas are generally volatile, whereas portfolio betas are stable. The estimation period should be at least 36 months to forecast beta.[52]

2.3.4.4 Tests of the Main Hypotheses

Now we discuss tests of the main hypotheses which were formulated on page 123. Because the beta stability was low for individual securities and high for large portfolios (as discussed in the Section *"Beta Stability"* above), the CAPM was only tested for portfolios. This is plausible, since the CAPM is only interesting for portfolios where the beta is supposed to be the significant risk measure, in contrast to an individual security. An application to single securities does not make sense.

A few years after the CAPM was developed in Sharpe (1964), Lintner (1965a) and Mossin (1966), the first empirical tests were performed. We want to focus on three groundbreaking empirical studies which shaped the discussion about the empirical CAPM and led to modifications of this model: Black et al. (1972), Fama and MacBeth (1973) and Fama and French (1992).

Black, Jensen and Scholes (1972): *The Capital Asset Pricing Model: Some Empirical Tests*

The early empirical cross-sectional tests in Douglas (1969) and Miller and Scholes (1972) showed that zero-beta assets earned more than the risk-free rate and that the beta premium was lower than the market excess return, in violation with hypothesis (C3).[53] Friend and Blume (1970) also showed that low-beta assets earned positive alpha and high-beta assets earned negative alpha. Black et al. (1972) provided

---

[48]Tole (1981, p. 47, Exhibit 1).

[49]This is pointed out in Reilly and Brown (1997, p. 310).

[50]Conclusion in Baesel (1974, p. 1493). Compare Table 1 on p. 1492 with Table 2 on p. 1493.

[51]Conclusion in Roenfeldt et al. (1978, p. 120 and Table 1 on p. 119).

[52]Reilly and Brown (1997, p. 311)

[53]Fama and French (2004, p. 32).

additional time-series and cross-sectional tests in support of these results and, as a consequence, postulated a modified version of the CAPM, the *zero-beta CAPM*, which accommodated zero-beta returns above the risk-free rate.

The data used in the tests to be described were taken from the University of Chicago Center for Research in Security Prices monthly price relative file, which contains monthly price, dividend, and adjusted price and dividend information for all securities listed on the New York Stock Exchange in the period January 1926–March 1966.[54] The monthly returns on the market portfolio $R_{Mkt}^{monthly}$ were defined as the returns that would have been earned on a portfolio consisting of an equal investment in every security listed on the NYSE at the beginning of each month.[55] The risk-free rate was defined as the 30-day rate on U.S. Treasury bills for the period 1948–1966. For the period 1926–1947, where Treasury bill rates were not available, the dealer commercial paper rate was used.

Portfolio selection process[56]: The tests used portfolios with a wide range of betas. This was done by ranking individual securities by their betas and then assigning the ones with the highest beta to portfolio 1, the next highest to portfolio 2, etc. But if these betas were used as an input for the regression analysis, this would cause a bias.[57] This is why the analysis is split in two phases:

1. Portfolio formation period: In this period, the betas of individual securities are calculated and ranked. Based on the rankings, the portfolios are formed. The betas from this period are not used in the test period.
2. Test period: The returns and betas of portfolios which are used for regression analysis are computed. The beta is calculated only on the basis of the test period.

For the beginning of January 1931, only stocks with at least 24 months of available data were considered, and their individual betas were calculated based on the monthly returns in the 5-year period January 1926–December 1930 (or from a shorter period if less than 5 years of data were available). These securities were ranked according to the calculated betas and then assigned to ten portfolios, with an equal investment in each security[58]: the 10 % with the largest beta to the first portfolio, the next 10 % to the second portfolio, etc. The return in each of the next 12 months for each of the ten portfolios was calculated.

After 12 months, for the beginning of January 1932, the ten portfolios were rebalanced using the same procedure: Only stocks with at least 24 months of previous monthly returns were considered, their betas were calculated based on the

[54]Black et al. (1972, p. 10).

[55]Black et al. (1972, p. 8).

[56]Black et al. (1972, p. 11).

[57]Black et al. (1972, p. 9).

[58]Black et al. (1972, p. 8): An equal investment is indicated by using the average return and average risk (beta) of all securities in a portfolio.

Table 2.2 Summary of portfolio formation period and testing period

Period	1	2	...	35
Portfolio formation period	1926–1930	1927–1931	...	1960–1964
Testing period	1931	1932	...	1965

Source: Black et al. (1972, p. 11)

Table 2.3 Summary of regression statistics for the time-series regression for ten different portfolios based on monthly returns in the period January 1931–December 1965

Portfolio #	1	2	3	4	5	6	7	8	9	10
β	1.561	1.384	1.248	1.163	1.057	0.923	0.853	0.753	0.629	0.499
α (in %)	−0.083	−0.194	−0.065	−0.017	−0.054	0.059	0.046	0.081	0.197	0.201

Source: Black et al. (1972, Table 2, p. 14)

5-year period January 1927–December 1931 (or on a shorter period if less than 5 years of data were available), the stocks were ranked and assigned to the ten portfolios based on the rankings.

This process was repeated for the beginnings of January 1933, January 1934, etc. through January 1965. Table 2.2 summarizes the periods:

The total number of stocks in the portfolios ranged from 582 to 1,094. For each portfolio, we get 35 years of monthly data.

Time-series regression was run on the ten portfolios based on 35 years of monthly returns in the period January 1931–December 1965 (420 observations). The results are shown in Table 2.3.

The table shows the alphas and betas of the ten portfolios, calculated with a linear time-series regression. Portfolio number 1 contains the highest-beta securities, portfolio number 10 consists of the lowest-beta securities. The portfolio betas range from 0.499 to 1.561. The critical result is that the high-beta portfolios ($\beta > 1$) consistently show negative alphas while low-beta portfolios ($\beta < 1$) show positive alphas. In other words, high-beta portfolios yield lower returns and low-beta portfolios yield higher returns than predicted by the CAPM. This rejects the CAPM hypothesis (C4) (see the list of hypotheses on page 123) that alpha should be zero. Furthermore, the study shows that the incremental return per unit of risk β (the beta premium) is smaller than implied by the CAPM.

Black et al. (1972) also performed a cross-sectional regression analysis with the same ten portfolios for the period January 1931–December 1965, which is illustrated in Fig. 2.15.

For each of the ten portfolios, the average excess monthly return[59] (vertical axis) and the beta (horizontal axis) are plotted (represented by dots), together with the (bold) regression line. The square represents the market portfolio and has beta 1.

[59]The excess return of a portfolio is the difference between the return of a portfolio and the return on a risk-free asset.

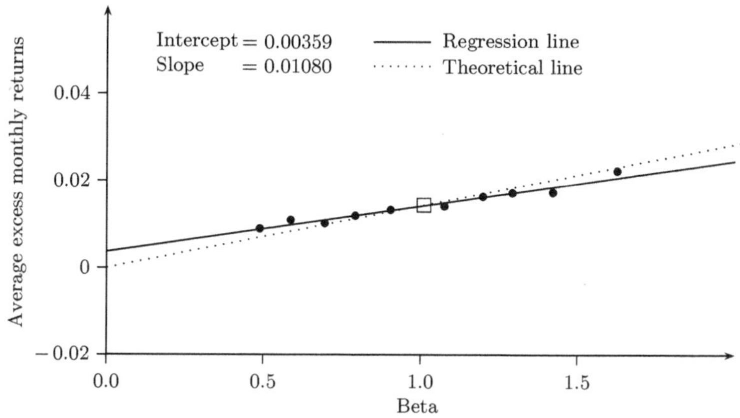

Fig. 2.15 Average excess monthly returns of ten portfolios (denoted by *dots*) and the market portfolio (denoted by the *square*) are plotted against their betas for the 35-year period 1931–1965. The *bold line* is the regression line, the *dotted line* is the theoretical line implied by the CAPM. *Source*: Black et al. (1972, p. 21)

The dotted line is the theoretical CAPM line (i.e., the SML) which goes through $(0, 0)$ and the market portfolio is also drawn to compare it with the regression line. The regression line has a flatter slope than the theoretical line and a higher intercept which is significantly greater than zero. The latter contradicts[60] the CAPM hypothesis (C4) (see the list of hypotheses on page 123). On the other hand, the ten data points representing the portfolios plot close to the line, indicating a linear relationship between return and beta (in support of (C1)).

Black et al. (1972) proceeded by examining the time-dependency of the regression line and its variation over time. A 35-year period was divided in four equal subperiods of 105 months length, and the same cross-sectional regression analysis was done for these subperiods. Figure 2.16 shows the regression for the period January 1931–September 1939. The intercept was significantly less than zero and the regression line was steeper than the theoretical line. Figure 2.17 shows the regression for October 1939–June 1948. The intercept is significantly greater than zero, and the regression line is flatter than the theoretical line. Figure 2.18 shows how the regression line became even flatter in the period July 1948–March 1957. From April 1957 to December 1965, the regression line in Fig. 2.19 even shows a negative slope, implying a negative beta premium which contradicts hypothesis (C3).

[60]In the general cross-sectional regression (2.38) that we described earlier, the average return, not the average *excess* return, is regressed on beta. The only difference is that we have to shift the graph from Fig. 2.16 down by the average risk-free rate r_{rf}. So the second hypothesis stated on page 124, i.e., *risk-free return on zero-beta assets*, translates into: "The intercept a is zero."

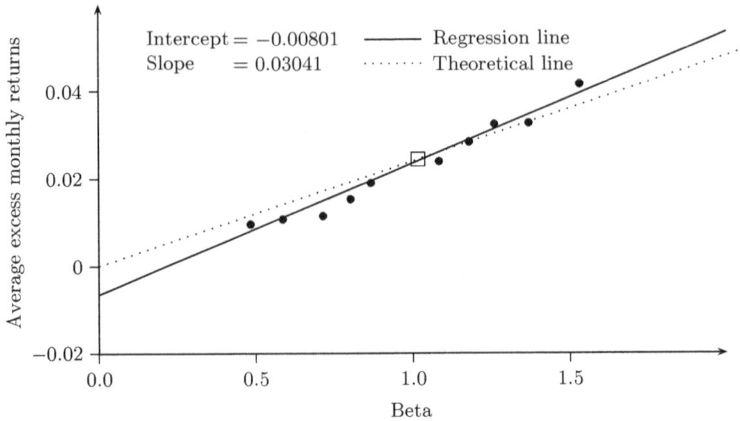

Fig. 2.16 Average excess monthly returns of ten portfolios (denoted by *dots*) and the market portfolio (denoted by the *square*) are plotted against their betas for the 105-month period January 1931–September 1939. The *bold line* is the regression line, the *dotted line* is the theoretical line implied by the CAPM. *Source*: Black et al. (1972, p. 24)

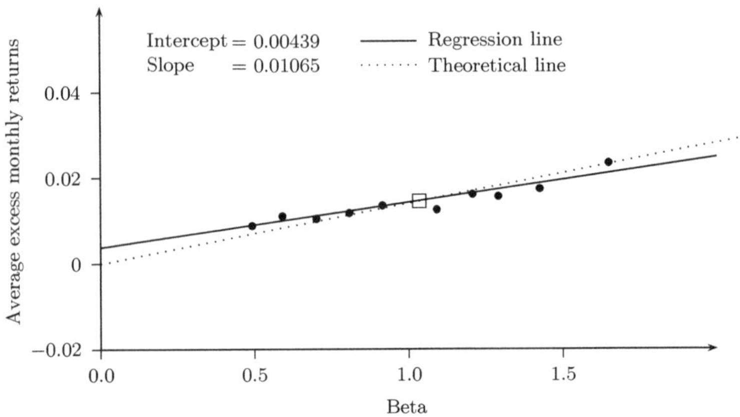

Fig. 2.17 Average excess monthly returns of ten portfolios (denoted by *dots*) and the market portfolio (denoted by the *square*) are plotted against their betas for the 105-month period October 1939–June 1948. The *bold line* is the regression line, the *dotted line* is the theoretical line implied by the CAPM. *Source*: Black et al. (1972, p. 24)

On the other hand, in all four subperiods which show different behaviors of the regression line (the slope was greater than the theoretical line in one subperiod and negative in an other subperiod), the data points plot close to the regression line, supporting hypothesis (C1) about the linearity between return and beta.

The overall conclusion of the empirical tests is that linearity between return and beta is supported by the data, but the hypothesis that zero-beta assets earn the

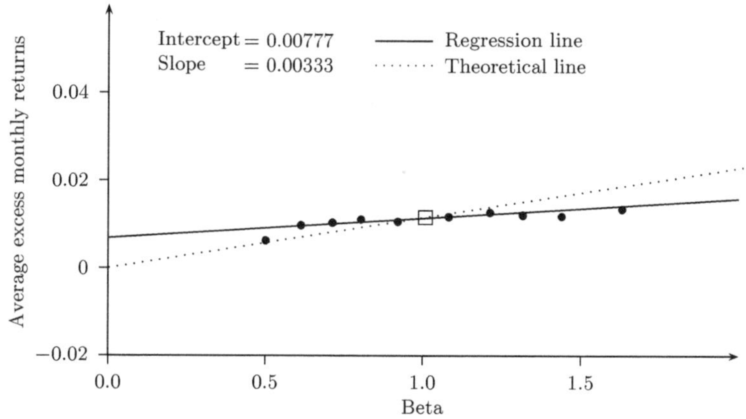

Fig. 2.18 Average excess monthly returns of ten portfolios (denoted by *dots*) and the market portfolio (denoted by the *square*) are plotted against their betas for the 105-month period July 1948–March 1957. The *bold line* is the regression line, the *dotted line* is the theoretical line implied by the CAPM. *Source*: Black et al. (1972, p. 24)

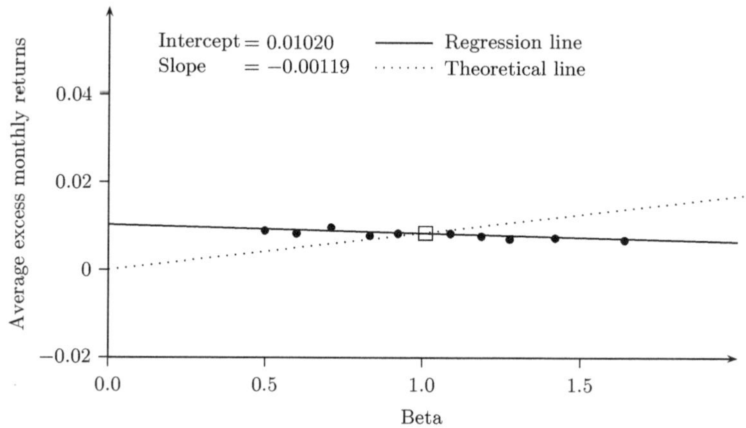

Fig. 2.19 Average excess monthly returns of ten portfolios (denoted by *dots*) and the market portfolio (denoted by the *square*) are plotted against their betas for the 105-month period April 1957–December 1965. The *bold line* is the regression line, the *dotted line* is the theoretical line implied by the CAPM. *Source*: Black et al. (1972, p. 24)

risk-free rate is rejected. The paper therefore suggests a modified CAPM without the assumption of risk-free borrowing and lending, the *zero-beta CAPM* (see also Sect. 2.3.5):

Let R_Z be the return on an asset Z which has zero beta. Then

$$R_{Pf} = \mathbb{E}[R_Z] + \beta_{Pf} \cdot (\mathbb{E}[R_{Mkt}] - \mathbb{E}[R_Z]). \tag{2.42}$$

The zero-beta CAPM was introduced in Black (1972), where it was shown that Eq. (2.42) holds when all CAPM assumptions from Sect. 2.3.1 are true except the assumption of risk-free borrowing and lending (A3). The empirical tests from Black et al. (1972) are consistent with this model.[61]

Fama and MacBeth (1973): ***Risk, Return, and Equilibrium: Empirical Tests***
Before Eugene F. Fama and James D. MacBeth published their study in 1973, empirical tests of the CAPM were focused on the hypothesis that zero-beta assets earn the risk-free rate (C4, see page 123).[62] This hypothesis was rejected in several papers, including Douglas (1969), Friend and Blume (1970), Miller and Scholes (1972) and Black et al. (1972). This led to the introduction of the zero-beta CAPM (2.42).

Fama and MacBeth (1973) made the next step: They formulated[63] the list of testable hypotheses (C1)–(C4) implied by the CAPM as shown on page 123 and tested these hypotheses by extending the cross-series regression with beta as the only explanatory variable to a multi-linear regression with three explanatory variables:

- **Beta**:
  As applied in the cross-sectional regressions above.
- **Beta squared**:
  Beta squared is used to test linearity (C1). If it turns out to have explanatory power in explaining returns, then the relation between return and beta cannot be linear. Beta squared serves as an explanatory variable for possible low-beta or high-beta tilts.[64] If the slope coefficient of beta squared is positive, then high-beta securities have higher expected returns and low-beta securities have lower expected returns than predicted by the CAPM. If the slope coefficient is negative, then it is the other way round.
- **Unsystematic risk**:
  Total risk (in terms of volatility) is the sum of systematic and unsystematic risk which are not correlated with each other.[65] Unsystematic risk is used as an explanatory variable in the regression to test (C2), i.e., if non-beta risks also explain returns. If the CAPM is valid, then unsystematic risk should have no explanatory power.

The data used in the study are the monthly returns (including dividends and capital gains with the appropriate adjustments for capital changes such as splits and stock dividends) for all common stocks traded on the New York Stock Exchange

[61] Black et al. (1972, p. 25).
[62] Fama and MacBeth (1973, p. 614).
[63] Fama and MacBeth (1973, p. 610 and p. 613).
[64] Fama and MacBeth (1973, p. 614).
[65] Fama and MacBeth (1973, p. 616).

during the period January 1926–June 1968.[66] The data are taken from the Center for Research in Security Prices of the University of Chicago. *Fisher's arithmetic index* is chosen as the market portfolio, an equally-weighted index on all stocks listed on the New York Stock Exchange.[67] The monthly risk-free rate is taken to be the yield on the 1-month Treasury bills.[68]

The portfolio selection process for the study is similar to Black et al. (1972), as described on page 129. 20 portfolios are selected based on the beta rankings of the securities, with equal investment in all securities after a *portfolio formation period* of 7 years (except for the first period which is 4 years long, see Table 2.4). Data which is used for beta rankings is not reused for regression analysis because this introduces statistical biases.[69] The next 5 years of data in the *initial estimation period* is used to estimate the betas and the unsystematic risk (the average of the unsystematic risks of all securities in the portfolio) of the portfolios. After that, in the *testing period*, monthly returns of the 20 portfolios are measured. For every month in the testing period, the monthly returns of the 20 portfolios are regressed over beta, beta squared and unsystematic risk which are measured based on the preceding 5 years of data. Then the regression coefficients are averaged to get the result. Table 2.4 shows the different periods.

Let us illustrate the methodology for period 1: First, the betas of the securities are measured based on the period 1926–1929 and ranked. The 5 % of the securities with the lowest betas go to portfolio 1, the next lowest 5 % to portfolio 2, etc., until 20 portfolios are established. The first month in the test period is illustrated in Fig. 2.20: In January 1935, the monthly returns of the 20 portfolios are measured, and they are cross-sectionally regressed over beta, beta squared and unsystematic risk which were measured based on the preceding 60 months of data (January 1930–December 1934).

Table 2.4 Portfolio formation, estimation and testing period

Period	1	2	3	4	5
Portfolio formation period	1926–1929	1927–1933	1931–1937	1935–1941	1939–1945
Initial estimation period	1930–1934	1934–1938	1938–1942	1942–1946	1946–1950
Testing period	1935–1938	1939–1942	1943–1946	1947–1950	1951–1954

Period	6	7	8	9
Portfolio formation period	1943–1949	1947–1953	1951–1957	1955–1961
Initial estimation period	1950–1954	1954–1958	1958–1962	1962–1966
Testing period	1955–1958	1959–1962	1963–1966	1967–1968

Source: Fama and MacBeth (1973, Table 1, pp. 618–619)

[66]Fama and MacBeth (1973, p. 614).

[67]Fama and MacBeth (1973, p. 614).

[68]Fama and MacBeth (1973, p. 626).

[69]Fama and MacBeth (1973, p. 615).

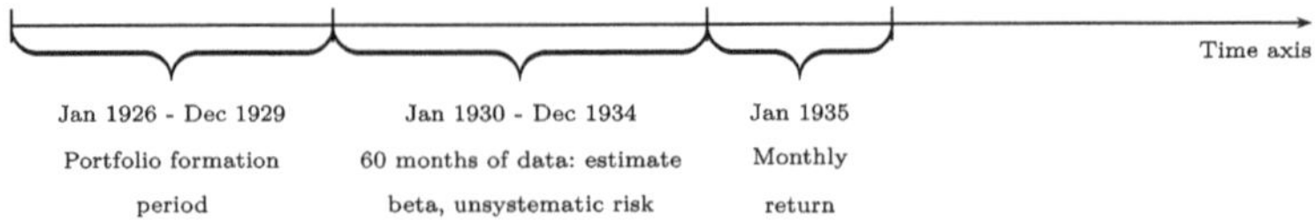

Fig. 2.20 Fama–MacBeth regression illustrated: 1st month of period 1. *Source*: Own

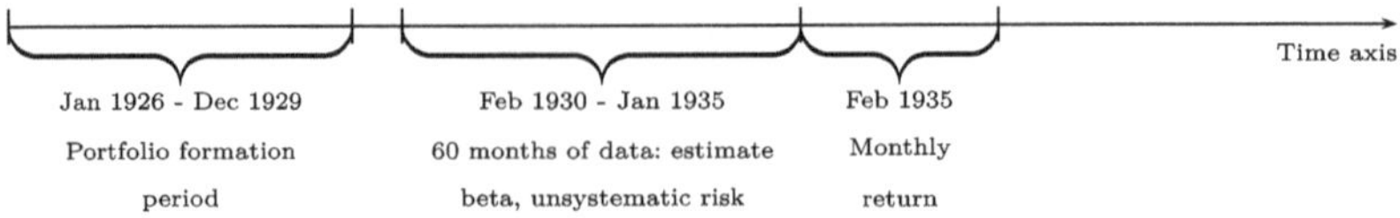

Fig. 2.21 Fama–MacBeth regression illustrated: 2nd month of period 1. *Source*: Own

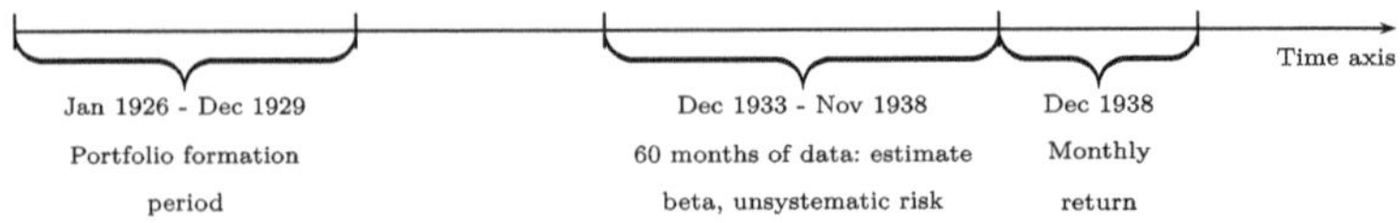

Fig. 2.22 Fama–MacBeth regression illustrated: last month of period 1. *Source*: Own

The second month in the test period is illustrated in Fig. 2.21: In February 1935, the monthly returns of the 20 portfolios are measured, and they are cross-sectionally regressed over beta, beta squared and unsystematic risk which were measured based on the preceding 60 months of data (February 1930–January 1935).

This procedure is continued until the last month of period 1, illustrated in Fig. 2.22: In December 1938, the monthly returns of the 20 portfolios are measured, and they are cross-sectionally regressed over beta, beta squared and unsystematic risk which were measured based on the preceding 60 months of data (December 1933–November 1938).

For each of the 48 months in the testing period 1935–1938, a cross-sectional regression is performed and regression coefficients are calculated. These are averaged for the whole period 1935–1938. This process is repeated for the other eight testing periods (see Table 2.4), the regression coefficients are averaged over all months of the period 1935–1968, and the result for the entire period is obtained.

This regression method, which is also known as the *Fama–MacBeth regression*, has become standard in the literature.[70] Fama and MacBeth performed their tests on the 20 portfolios for the period January 1935–June 1968 and its subperiods (see Table 2.4). The results of Fama's and MacBeth's tests were the following[71]:

- The beta premium is positive for the overall period, which validates hypothesis (C3) that beta is positively related to return.[72]

[70]Fama and French (2004, p. 31).

[71]Fama and MacBeth (1973, p. 624).

[72]The beta premium turned negative during a short subperiod (1956–1960), but this does not invalidate the long-term result.

- Beta squared showed no explanatory power for returns. Therefore, the test supports hypothesis (C1) that the relation between return and beta is linear.
- Unsystematic risk showed no explanatory power for returns, which supports hypothesis (C2) that no variable other than beta explains returns.

The results support the zero-beta CAPM (2.42). In addition, the cross-sectional regression analysis with beta as the only explanatory variable shows that the returns on zero-beta assets are higher than the risk-free rate, thus invalidating hypothesis (C4).

There is one big difference between the cross-sectional regressions used in the tests in Black et al. (1972) and Fama and MacBeth (1973): In Black et al. (1972), monthly returns and betas are measured in the *same* testing period. While this is a test of the CAPM, this is not a test if estimated betas can be used for predicting returns because beta is unknown before the returns are measured. The Fama–MacBeth regression is an improvement because betas are measured *before* the monthly returns are realized. So this empirical test also supports the assertion that estimated betas can be used for making predictions and decisions.[73] Another difference is that while the early cross-sectional tests treated the zero-beta returns as constant (because the CAPM assumes a constant and deterministic risk-free rate), the Fama–MacBeth regression uses time-varying zero-beta returns which come closer to the stochastic nature of zero-beta returns, as postulated in the zero-beta CAPM.

Other tests like Blume and Friend (1973) and Stambaugh (1982) also positively tested the linearity assumption (C1) while rejecting (C4).

Fama and French (1992): *The Cross-Section of Expected Stock Returns*

After the CAPM was developed, some researchers started to doubt the relation between beta and return. Reinganum (1981) found no significant relationship between the beta and the returns of NYSE/AMEX stocks in the period 1964–1979. Lakonishok and Shapiro (1986) arrived at the same result for NYSE stocks in the period 1962–1981.

While more and more doubts were raised about beta as a basis to predict returns, other factors like size and price ratios were discovered to play a role as well.[74] Basu (1977) showed that high P/E stocks earned lower returns and low P/E stocks earner higher returns than predicted by the CAPM. Banz (1981) showed a size effect: small firm stocks earned a higher return and large firm stocks earned a lower return than indicated by the CAPM. Bhandari (1988) found a positive relation between leverage (in terms of debt-equity ratio, the book value of debt over market value of equity) and returns, controlling for beta and size.

Rosenberg, Reid, and Lanstein (1985) showed that U.S. stocks with high book-to-market equity ratios (the book value of common stock over its market value)

[73]Fama and MacBeth (1973, p. 618).

[74]This summary is from Fama and French (2004, p. 35).

earned higher returns than stocks with lower book-to-market ratios, controlling for betas. Chan, Hamao, and Lakonishok (1991) came to the same conclusion for Japanese stocks. Note that all of these new factors (size, P/E, leverage) contain the stock price which reflects the expected future returns,[75] so the fact that these factors explain returns should not be seen as a big surprise.

Fama and French (1992) provided an extensive test on the role of beta, size, book-to-market (B/M) ratio, leverage, earnings-to-price (E/P) and their combinations in explaining average returns. The study covered the period July 1963–December 1990 and confirmed the results of the other studies mentioned above.

The authors used market data of non-financial stocks traded on the NYSE, AMEX and NASDAQ which they obtained from the Center for Research in Security Prices (CRSP).[76] The accounting data was taken from COMPUSTAT, also maintained by CRSP. The 1962 start date reflects the fact that book values were not generally available before. The market portfolio is the value-weighted portfolio of all NYSE, AMEX and NASDAQ stocks.[77]

For examining the effect of beta and size on returns, 12 portfolios were constructed based on pre-ranking betas and sizes. They were rebalanced for July in every year. All portfolios were equally weighted.[78]

Construction of Beta-Based Portfolios

To set up the portfolios for the 1-year period July 1963–June 1964, the securities were ranked based on the beta estimates of the preceding 60 months (July 1958–June 1962). The distribution scheme of the portfolios is shown in Table 2.5, starting with the lowest betas on the left (1A) and ending with the highest betas on the right (10B).

The 5 % of the securities with the lowest beta are in portfolio 1A, the next 5 % in 1B, the following 10 % in portfolio 2, etc.

Table 2.5 Distribution scheme for β-based portfolios from low beta (1A) to high beta (10B) which is also used for size-based portfolios

Portfolio	1A (low)	1B	2	3	4	5	6	7	8	9	10A	10B (high)
Distribution	5 %	5 %	10 %	10 %	10 %	10 %	10 %	10 %	10 %	10 %	5 %	5 %

Source: Fama and French (1992, Table II, pp. 436–437)

[75]This argument was initially used for the E/P ratio in Ball (1978). But it was generalized to other factors in Fama and French (1992, p. 428).

[76]Fama and French (1992, p. 429). Financial stocks were excluded because the high leverage that is normal for these firms probably does not have the same meaning for non-financial firms, where high leverage more likely indicates distress.

[77]Fama and French (1992, p. 431).

[78]Fama and French (1992, p. 431 and p. 433).

Table 2.6 Portfolios formed on pre-ranking β from lowest β (1A) to highest β (10B) with average monthly returns and post-ranking β in the period July 1963–December 1990

Portfolio	1A (low)	1B	2	3	4	5	6	7	8	9	10A	10B (high)
Return	1.20	1.20	1.32	1.26	1.31	1.30	1.30	1.23	1.23	1.33	1.34	1.18
β	0.81	0.79	0.92	1.04	1.13	1.19	1.26	1.32	1.41	1.52	1.63	1.73

Source: Fama and French (1992, Table II, pp. 436–437)

After 1 year, the portfolios for the period July 1964–June 1965 were reconstructed, again based on the beta estimates of the preceding 60 months (July 1959–June 1963). We continued this procedure for every year until 1990 to get the 12 portfolios for the period July 1963–December 1990. Table 2.6 lists the average monthly returns of the portfolios, together with their post-ranking betas which are based on the whole period July 1963–December 1990.[79]

Construction of Size-Based Portfolios

A similar procedure was applied to construct 12 size-based portfolios. To set up the portfolios for the 1-year period July 1963–June 1964, the securities were ranked based on their size (in terms of market equity, or ME, i.e., stock price times number of shares outstanding) at the end of June 1963. The distribution scheme of the size-based portfolios was the same as for the beta-based portfolios shown in Table 2.5, starting with the lowest size on the left (1A) and ending with the highest size on the right (10B).

After 1 year, the portfolios for the period July 1964–June 1965 were reconstructed, based on the sizes at the end of June 1964. This procedure was continued for every year until 1990 to establish the 12 portfolios for the period July 1963–December 1990. Table 2.7 lists the average monthly returns of the portfolios, together with their post-ranking betas which are based on the whole period July 1963–December 1990.[80]

In Table 2.7, we can observe that when forming portfolios based on size, the average return decreases with increasing size (portfolio 1A contains the smallest companies, 10B the largest), from 1.64 % per month for the smallest-size portfolio to 0.90 for the largest. This shows the *size effect*, i.e., size is negatively related to return. We can also see that with increasing size, β decreases like the return, from 1.44 for the smallest-size portfolio to 0.90 for the largest, suggesting a strong positive relationship between return and β. But for portfolios based on pre-ranked betas, Table 2.6 shows a rather flat relationship between return and β: While the

---

[79]The betas are calculated based on the monthly returns which were realized by the portfolios after they were set up. They have nothing to do with the beta used for the pre-ranking.

[80]The beta is calculated based on the monthly returns which were realized by the portfolios after they were set up. They have nothing to do with the beta used for the pre-ranking.

Table 2.7 Portfolios formed on size, from lowest size (1A) to highest size (10B) with average monthly returns and post-ranking β in the period July 1963–December 1990

Portfolio	1A (low)	1B	2	3	4	5	6	7	8	9	10A	10B (high)
Return	1.64	1.16	1.29	1.24	1.25	1.29	1.17	1.07	1.10	0.95	0.88	0.90
β	1.44	1.44	1.39	1.34	1.33	1.24	1.22	1.16	1.08	1.02	0.95	0.90

Source: Fama and French (1992, Table II, p. 436)

betas range from 0.81 to 1.73 (and this range of betas is larger than the range of betas for the size-based portfolios), the range of average returns is quite small, from 1.18 to 1.34 % per month. The lowest-beta portfolio (1A) even shows a higher return (1.20 %) than the highest-beta portfolio (10B) (1.18 %). So a variation in β which is tied to size is positively related to return, while a variation in β alone does not explain a variation of returns.

To examine the pure effect of β on returns without any size effects on β, we form 100 size-β portfolios.

Construction of Size-β Portfolios

For the end of June 1963, we rank the stocks based on size (market equity). We divide the stocks in ten size groups based on data for the end of June 1963, i.e., ME-1 contains the smallest 10 % stocks, ME-2 the next smallest 10 %, etc. Next, for each of the size groups we form ten portfolios based on the pre-ranking β (calculated on the basis of the preceding 60 months, i.e., July 1958–June 1963). For example, within the size group ME-2, we put the 10 % lowest-beta stocks into portfolio ME-2/$\beta - 1$, the next 10 % into ME-2/$\beta - 2$, etc. We hold these portfolios for the period July 1963–June 1964.

For the end of June 1964 the portfolios are rebalanced: We rank the stocks based on size at the end of June 1964, divide them in ten size groups, and for each size group form ten portfolios based on the pre-ranking β (calculated on the basis of the preceding 60 months, i.e., July 1959–June 1964). We hold these portfolios for the period July 1964–1965.

We repeat this process year after year and set up 100 portfolios (ten in each size group) with different size-beta characteristics for the period June 1963–December 1990. Since the ten portfolios from the same size group have a wide range of betas, but similar stock sizes, we can always use the ten portfolios from the same size group to test the effects of β which are unrelated to size. Table 2.8 shows the post-ranking betas and Table 2.9 shows the average monthly returns (in %) of the 100 size-β portfolios. Each row in these tables represents a certain size group, and when we read the numbers within a row from left to right, we can observe the results for increasing betas within a size group.

We can see that the betas in Table 2.8 increase from left to right: This is how we constructed the portfolios. In Table 2.9 when reading each row from left to right, we can see how increasing beta—while keeping size constant—effects returns:

Table 2.8 Post-ranking betas of the size-β portfolios formed on size (down) and then β (across) in the period July 1963–December 1990

	β-1 (low)	β-2	β-3	β-4	β-5	β-6	β-7	β-8	β-9	β-10 (high)
	Post-ranking βs									
ME-1 (small)	1.05	1.18	1.28	1.32	1.40	1.40	1.49	1.61	1.64	1.79
ME-2	0.91	1.15	1.17	1.24	1.36	1.41	1.43	1.50	1.66	1.76
ME-3	0.97	1.13	1.13	1.21	1.26	1.28	1.39	1.50	1.51	1.75
ME-4	0.78	1.03	1.17	1.16	1.29	1.37	1.46	1.51	1.64	1.71
ME-5	0.66	0.85	1.12	1.15	1.16	1.26	1.30	1.43	1.59	1.68
ME-6	0.61	0.78	1.05	1.16	1.22	1.28	1.36	1.46	1.49	1.70
ME-7	0.57	0.92	1.01	1.11	1.14	1.26	1.24	1.39	1.34	1.60
ME-8	0.53	0.74	0.94	1.02	1.13	1.12	1.18	1.26	1.35	1.52
ME-9	0.58	0.74	0.80	0.95	1.06	1.15	1.14	1.21	1.22	1.42
ME-10 (large)	0.57	0.71	0.78	0.89	0.95	0.92	1.02	1.01	1.11	1.32

Source: Fama and French (1992, Table I, p. 435)

Table 2.9 Average monthly returns of the size-β portfolios formed on size (down) and then β (across) in the period July 1963–December 1990

	β-1 (low)	β-2	β-3	β-4	β-5	β-6	β-7	β-8	β-9	β-10 (high)
	Average monthly return (in %)									
ME-1 (small)	1.71	1.57	1.79	1.61	1.50	1.50	1.37	1.63	1.50	1.42
ME-2	1.25	1.42	1.36	1.39	1.65	1.61	1.37	1.31	1.34	1.11
ME-3	1.12	1.31	1.17	1.70	1.29	1.10	1.31	1.36	1.26	0.76
ME-4	1.27	1.13	1.54	1.06	1.34	1.06	1.41	1.17	1.35	0.98
ME-5	1.34	1.42	1.39	1.48	1.42	1.18	1.13	1.27	1.18	1.08
ME-6	1.08	1.53	1.27	1.15	1.20	1.21	1.18	1.04	1.07	1.02
ME-7	0.95	1.21	1.26	1.09	1.18	1.11	1.24	0.62	1.32	0.76
ME-8	1.09	1.05	1.37	1.20	1.27	0.98	1.18	1.02	1.01	0.94
ME-9	0.98	0.88	1.02	1.14	1.07	1.23	0.94	0.82	0.88	0.59
ME-10 (large)	1.01	0.93	1.10	0.94	0.93	0.89	1.03	0.71	0.74	0.56

Source: Fama and French (1992, Table I, p. 434)

The relationship between average monthly returns and betas is flat. More surprisingly, in each size group, the lowest-beta portfolio (β-1) earned a higher return than the highest-beta portfolio (β-10)! For example, within the size group ME-5, the lowest-beta portfolio (ME-5/β-1) earned 1.34 % per month and outperformed the highest-beta portfolio (ME-5/β-10) with 1.08 % per month.

The overall conclusion about beta is "*that variation in beta that is tied to size is positively related to return, but variation in beta unrelated to size is not compensated in the average returns in 1963–1990*".[81] Fama and French did the same test on NYSE stocks in the 50-year period 1941–1990 and found the same result: There is a "*reliable size effect* [...] *but little relation between beta and average return*". On the other hand, there was a positive relation between beta and average return in the period 1941–1965, which is exactly the period which was used in the early CAPM tests. But "*even for the 1941–1965 period, however, the relation between beta and average return disappears when we control for size.*"[82] Due to this striking discovery that beta appears to have no relation to average returns, the paper raised the question "*Can β Be Saved?*"[83]

Construction of B/M-based Portfolios

By constructing size-based portfolios, we have observed a negative relation between size and average return. We apply the same approach to examine the relation between the book-to-market ratio and average returns. To set up portfolios for the 1-year period July 1963–June 1964, we rank the stocks based on the book-to-market value at the end of December 1962 using accounting data from the latest fiscal year.[84] The stocks are assigned to the portfolios based on the scheme explained in Table 2.5, from smallest B/M (1A) to highest (10B). After 1 year, the portfolios are rebalanced for the period July 1964–June 1965 based on the book-to-market value at the end of December 1963. This procedure is continued for every year until 1990.

According to Table 2.10, the average monthly return of portfolios based on B/M increases with increasing B/M ratio, from 0.30 % per month for the smallest B/M ratio to 1.83 % per month for the largest B/M ratio. The difference between the largest and the smallest B/M portfolio is 1.53 % and even twice as large as the difference between the largest- and smallest-size portfolio of 0.74 % (see Table 2.7). Note that beta stays flat across different B/M ratios, so the difference in return cannot be explained by beta!

Table 2.10 Portfolios formed on book-to-market (B/M) ratio, from lowest B/M (1A) to highest B/M (10B) with average monthly returns and post-ranking β of the portfolios

Portfolio	1A (low)	1B	2	3	4	5	6	7	8	9	10A	10B (high)
Return	0.30	0.67	0.87	0.97	1.04	1.17	1.30	1.44	1.50	1.59	1.92	1.83
β	1.36	1.34	1.32	1.30	1.28	1.27	1.27	1.27	1.27	1.29	1.33	1.35

Source: Fama and French (1992, Table IV, p. 442)

[81] Fama and French (1992, p. 433).

[82] Fama and French (1992, p. 440).

[83] Fama and French (1992, p. 439).

[84] Fama and French (1992, Table IV, p. 442).

Fama–MacBeth Regression

The Fama–MacBeth regression serves to determine the effects of beta, size, leverage and E/P ratio (as single variables and also in combination) on average return. Fama and French (1992) describes the regression of monthly returns of stocks over β, $\ln(ME)$ (ME = market value of common equity in millions of dollars, as a proxy for size), $ln(BE/ME)$ ratio (BE = book value of common equity), $ln(A/ME)$ and $ln(A/BE)$ (as proxies for leverage, A = value of total assets) and E/P ratio. The regression coefficients were averaged over the period July 1963–December 1990. In particular, for each of the 12 months in the period July 1963–June 1964, the monthly returns of the stocks were regressed over

- the post-ranking beta of the size-β portfolio to which the stock was assigned at the end of June 1963. This was used as an approximation of the stock's beta because it is much more stable than the individual stock's beta.
- the size (measured in $\ln(ME)$) where the market value was taken as of end of June 1963.
- leverage, BE/ME and E/P ratio: the accounting variables were taken from the fiscal year ending in December 1962.

After 12 months, the data was updated. For the period July 1964–June 1965, the monthly returns of the stocks were regressed over post-ranking beta at the end of June 1964, size at the end of June 1964, and leverage, $ln(BE/ME)$ and P/E ratio from the fiscal year ending in December 1963. This process was repeated for every year and the result was averaged. The conclusions of the Fama–MacBeth regression are that

- beta[85] does not help explaining average stock returns in 1963–1990, neither alone nor in combination with size, leverage and E/P ratio.
- as single explanatory variables, size[86] and total assets-to-book ratio[87] are negatively related to average returns, whereas book-to-market ratio,[88] total assets-to-market ratio[89] and earnings-to-price ratio[90] are positively related to average returns.
- when combining all explanatory variables, the size and book-to-market effect turn out to be most significant, making the other variables redundant. The book-to-market effect is even more powerful than the size effect.[91]

---

[85]Fama and French (1992, p. 438).

[86]Fama and French (1992, p. 438).

[87]Fama and French (1992, pp. 441–442).

[88]Fama and French (1992, p. 441).

[89]Fama and French (1992, pp. 441–442).

[90]Fama and French (1992, pp. 442–443).

[91]Fama and French (1992, p. 440).

Other Empirical Tests and the Revival of Beta

As a response to the studies which showed that the relationship between return and beta is flat, Pettengill, Sundaram, and Mathur (1995) argued that previous tests of the CAPM were flawed.[92] They offered a modification: Although the CAPM predicts that the *expected* return should be positively related to beta, this should not be true for all *realized* returns. When markets go up, then high-beta assets overperform low-beta assets. But when the market goes down, then assets with high beta (and therefore with higher risk) should underperform conservative low-beta assets. This kind of behavior of high-beta stocks illustrated why they carry higher risk than low-beta stocks. Previous tests like Fama and French (1992) which only looked for a general positive relationship between beta and returns did not take this implication of the CAPM into account. The empirical test by Pettengill et al. (1995) which covered U.S. stocks (which were available in the Center for Research in Security Prices (CRSP) monthly returns file) in 1936–1990 used a slightly different approach: They tested the relationship between beta and return for up markets (i.e., months where the monthly return of the market portfolio exceeded the monthly risk-free rate) and down markets (i.e., months where the monthly returns of the market portfolio was below the monthly risk-free rate) separately. The sensitivity of returns to up markets/down markets is also known as *bull beta* and *bear beta*, see also Sect. 1.3.6.

The empirical result is that during the period 1936–1990 and also during the subperiods 1936–1950, 1951–1970, 1971–1990, the relationship between beta and return was positive in up markets and negative in down markets.[93] When running the usual CAPM test without distinguishing down and up markets, Pettengill et al. (1995) found a positive relationship between beta and return only in the subperiod 1936–1950, but a flat relationship in the subperiods 1951–1970 and 1971–1990. The authors suggested that these "*results are biased due to the aggregation of positive and negative market excess return periods*".[94] The separate CAPM tests for up markets and down markets became increasingly popular and this method was used in several other publications (see also Table 2.11).

2.3.4.5 Roll's Critique: The Market Portfolio Problem

Researchers have done several empirical tests which have supported and rejected implications of the CAPM. But in Roll (1977), Richard Roll[95] questions their validity because the tests only used stocks (or stock market indices like the S&P

---

[92]Pettengill et al. (1995, pp. 102–104) for the argument.

[93]Pettengill et al. (1995, p. 110).

[94]Pettengill et al. (1995, p. 109).

[95]Richard Roll (born October 31, 1939) is an American economist, best known for his work on portfolio theory and asset pricing, both theoretical and empirical. In 1968, he received his Ph.D. from the Graduate School of Business at the University of Chicago in economics, finance, and statistics. In 1976, Roll joined the faculty at UCLA, where he remains as Japan Alumni Chair Professor of Finance. In 1987, Roll was elected President of the American Finance Association.

500) to represent the whole market. However, the market portfolio of the CAPM should contain every risky asset in the market, including stocks, bonds, options, real estate, coins, stamps, art, antiques,[96] and also human capital.[97] Since a lot of assets have been ignored, Roll (1977) argues that the CAPM has never been tested.

The trouble is that the zero-beta CAPM

$$R_{Pf} = \mathbb{E}[R_Z] + \beta_{Pf} \cdot (\mathbb{E}[R_m] - \mathbb{E}[R_Z]) \tag{2.43}$$

holds for *any* mean-variance efficient portfolio m and vice versa.

The zero-beta CAPM is tautological to the market portfolio Mkt being mean-variance efficient, and so are the implications (C1)–(C3) from page 123 (linearity in beta, no non-beta risk, positive beta premium) which were empirically tested. So the hypotheses (C1)–(C3) are not independently testable, and it just comes down to the question if the market portfolio is mean-variance efficient or not. This cannot be answered by using stock-only market proxies, and the market is unobservable. Roll argues in Roll (1977, p. 130): "*The theory is not testable unless the exact composition of the true market portfolio is known and used in the tests. This implies that the theory is not testable unless all individual assets are included in the sample.*"[98]

The equation

$$R_{Pf} = \mathbb{E}[R_Z] + \beta_{Pf} \cdot (\mathbb{E}[R_m] - \mathbb{E}[R_Z]) \tag{2.44}$$

holds for *any* mean-variance efficient portfolio m.[99] This is a mathematical fact without any model assumptions.[100] Even if the CAPM hypotheses (C1)–(C3) are positively tested for a market proxy m, this only implies that the market proxy m is mean-variance efficient, but the true market portfolio might be still inefficient (therefore rendering CAPM invalid), leading to the wrong validation of CAPM. Or the CAPM hypotheses (C1)–(C3) are negatively tested with a wrong proxy m (which is inefficient), while the true market portfolio is efficient, leading to a wrong rejection of CAPM. Therefore, CAPM tests with wrong market proxies are inconclusive.

2.3.5 Evaluation of the CAPM

Let us summarize the results of the empirical tests: Early tests (Black et al. 1972; Douglas 1969; Miller and Scholes 1972) rejected the hypothesis (C4) that zero-beta

[96]This list of examples is mentioned in Reilly and Brown (1997, p. 284).

[97]Richard Roll mentions human capital as part of the market portfolio in Roll (1977, p. 131 and p. 155).

[98]Roll (1977, p. 136).

[99]β_{Pf} is the portfolio beta measured relative to the portfolio m. The zero-beta asset Z has zero beta relative to m.

[100]Roll (1977, p. 130 and p. 136).

Table 2.11 Empirical tests of the capital asset pricing model

Study	Data used	Results
Douglas (1969): *Risk in the Equity Markets: An Empirical Appraisal of Market Efficiency*	1954–1963: All 301 stocks from the S&P 425, for which all the data were in the broader study where available for all years.[a]	The test showed that • zero-beta return was greater than risk-free rate. • beta premium was positive, but lower than market excess return. • the residual from the regression which was used to estimate beta had a significant positive impact on return.
Friend and Blume (1970): *Measurement of Portfolio Performance under Uncertainty*	January 1960–June 1968: All stocks from the New York Stock Exchange (NYSE)	For the whole period January 1960–June 1968, the rate of return was positively related to beta, but the Jensen alpha and other performance measures like Sharpe ratio and Treynor ratio were negatively related to beta. However, in the subperiod April 1964–June 1968 the performance measures and beta were positively related.
Miller and Scholes (1972): *Rates of Return in Relation to Risk: A Reexamination of Some Recent Findings*	1954–1963: 631 stocks from the New York Stock Exchange (NYSE)	The test showed that • zero-beta return was greater than risk-free rate. • beta premium was positive, but lower than market excess return. • the residual from the regression which was used to estimate beta had a significant positive impact on return. The authors mention two possible explanations for that: positive association between systematic risk and residual risk, skewness of the underlying distribution of returns.
Black et al. (1972): *The Capital Asset Pricing Model: Some Empirical Tests*	January 1931–December 1965: All stocks from the New York Stock Exchange (NYSE)	In the period January 1931–December 1965, high-beta portfolios ($\beta > 1$) earned negative alphas while low-beta portfolios ($\beta < 1$) earned positive alphas. In the same period, there was also a positive linear relation between beta and the rate of return. This period was divided in four equidistant subperiods for further study: In January 1931–September 1939, the rate of returns for zero-beta

		assets was below the risk-free rate and the empirical CAPM line was steeper than the theoretical line. Over time, the empirical CAPM line became flatter and even had a negative slope in the subperiod April 1957–December 1965.
Fama and MacBeth (1973): *Risk, Return, and Equilibrium: Empirical Tests*	January 1926–June 1968: All stocks from the New York Stock Exchange (NYSE)	The test showed that • beta premium was positive. • beta squared showed no explanatory power, therefore supporting the linearity between beta and return. • unsystematic risk showed no explanatory power, therefore supporting the hypothesis that no variable other than beta explains returns.
Blume and Friend (1973): *A New Look at the Capital Asset Pricing Model*	January 1955–June 1968: All stocks from the New York Stock Exchange (NYSE)	The test showed that the returns of zero-beta portfolios assets were different from the risk-free rate. It confirmed the linearity between beta and the rate of return by showing that beta squared had no explanatory power.
Basu (1977): *The Investment Performance of Common Stocks in Relation to their Price-Earnings Ratios: A Test of the Efficient Market Hypothesis*	April 1957–March 1971: All stocks from the New Stock Exchange (NYSE) from firms with December as the fiscal-year end	High P/E stocks had lower returns, low P/E stocks had higher returns than predicted by the CAPM.
Reinganum (1981): *A New Empirical Perspective on the CAPM*	January 1964–December 1979: All stocks from the New York Stock Exchange (NYSE) and American Stock Exchange (AMEX)	No significant relationship between beta and return.

(continued)

Table 2.11 (continued)

Study	Data used	Results
Banz (1981): *The Relationship Between Returns and Market Value of Common Stock*	January 1936–December 1975: All stocks from the New York Stock Exchange (NYSE)	Small size firms had higher returns, large size firms had lower returns than predicted by the CAPM.
Stambaugh (1982): *On the Exclusion of Assets from Tests of the Two-Parameter Model*	February 1953–December 1976: All stocks from the New York Stock Exchange, four preferred stocks and five bond portfolios	Confirmed linearity between beta and return. Beta premium is positive, zero-beta return is different from risk-free rate. The empirical test was done with different market proxies, yielding similar results and the same conclusions about CAPM.
Rosenberg et al. (1985): *Persuasive Evidence of Market Inefficiency*	January 1973–September 1984: 1,400 of the largest U.S. companies in the COMPUSTAT database, containing largely stocks from the New York Stock Exchange (NYSE), and a few from other exchanges like the AMEX and NASDAQ	High P/E stocks had higher returns than low P/E stocks, controlling for betas.
Lakonishok and Shapiro (1986): *Systematic Risk, Total Risk and Size as Determinants of Stock Market Returns*	January 1962–December 1981: All stocks from the New York Stock Exchange (NYSE)	Among the three variables—systematic risk, total risk and size—only the size matters when describing returns. If January data are excluded, even size loses its significance. When considering up markets/down markets only, the average return was positively/negatively related to beta.
Bhandari (1988): *Debt/Equity Ratio and Expected Common Stock Returns*	January 1948–December 1979: All stocks from the New York Stock Exchange (NYSE)	Debt/equity (more precisely: debt/market equity) ratio was positively related to returns, even when controlling for size and beta. The study was also performed on manufacturing firms to remove biases from high-leverage industries like finance. Here, the positive effect of the debt/equity ratio on return

		was even larger. Size had a negative impact on return, but only in January. The relationship between return and beta was positive only in January.
Chan et al. (1991): *Fundamentals and Stock Returns in Japan*	January 1971–December 1988: Stocks from the Tokyo Stock Exchange (TSE)	Four explanatory variables for returns were tested: size, book/market ratio, earnings/price ratio, cash flow/price ratio. Book/market and cash flow/price ratios had a positive impact on returns. Earnings/price ratio showed little impact after controlling for other variables. The impact of size depended on the regression model used.
Fama and French (1992): *The Cross-Section of Expected Stock Returns*	July 1963–December 1990: Non-financial stocks which were traded on the New York Stock Exchange (NYSE), American Stock Exchange (AMEX) and NASDAQ	The following explanatory variables for returns were tested: beta, size, book/market ratio, leverage (total assets/book equity and total assets/market equity ratio), earnings/price ratio. Conclusions: • Beta does not explain returns. • As single explanatory variable, size and total assets/book equity are negatively related to average returns, whereas book/market, total assets/market equity and earnings/price ratios are positively related to average returns. • Of all explanatory variables, size and book/market ratio turn out to be most significant, making the other variables redundant. Book/market ratio has a stronger effect on returns than size.
Herrera and Lockwood (1994): *The Size Effect in the Mexican Stock Market*	January 1987–December 1992: All stocks from the Mexican stock exchange and similar stocks from the NASDAQ. To create the sample of NASDAQ stocks, one NASDAQ stock with similar industry characteristics was selected for each Mexican stock.	For Mexican stocks and the NASDAQ sample, a positive beta premium and a negative size premium were found, as single explanatory variables and when both beta and size are examined together.

(continued)

Table 2.11 (continued)

Study	Data used	Results
Pettengill et al. (1995): *The Conditional Relation between Beta and Returns*	January 1936–December 1990: U.S. stocks from the Center for Research in Security Prices (CRSP) monthly returns file	In up markets/down markets, the relation between beta and average returns was positive/negative for the periods 1936–1950, 1951–1970 and 1971–1990. Without distinguishing between up markets and down markets, a positive relationship between beta and average returns was found in 1936–1950, while it was flat in the periods 1951–1970 and 1971–1990.
Strong and Xu (1997): *Explaining the Cross-Section of U.K. Expected Stock Returns*	July 1960–June 1992: U.K. stocks from the London Share Price Database (LSDP). For the period in July 1973–June 1992 where accounting data were also used, only stocks from companies on both Exstat (which contains accounting data) and LSDP databases were included.	For July 1960–June 1992, the study found a positive beta premium and negative size premium (as single explanatory variables). For the time period of July 1973–June 1992 the same effects for beta and size were observed. Furthermore, as single explanatory variables, the premium on total assets/book equity ratio was negative, whereas the premiums for book/market and total assets/market equity were positive. The earnings/price ratio had a positive, albeit not significant, effect on price. The book/market ratio and leverage (total assets/book equity and total assets/market equity ratio) had significant explanatory power in combination with other variables.
Fletcher (1997): *An Examination of the Cross-Sectional Relationship of Beta and Return: U.K. Evidence*	January 1975–December 1994: U.K. stocks from the London Stock Exchange and Unlisted Securities Market, from the London Business School Share Price Database (LBS)	In 1975–1994 and in both subperiods 1975–1984 and 1985–1994, the relationship between beta and return was flat. When looking at up markets/down markets only, there was a positive/negative relationship between beta and return in 1975–1994 and in both subperiods 1975–1984 and 1985–1994. The negative beta premium in down markets was higher than the positive beta premium in up markets.

Chui and Wei (1998): *Book-to-market, Firm Size, and the Turn-of-the-Year Effect: Evidence from Pacific-Basin Emerging Markets*	Test on the stock markets of five countries: Malaysia, Taiwan (both July 1981–June 1993), Korea (July 1982–June 1993), Hong Kong (July 1984–June 1993) and Thailand (July 1988–June 1993)	The regressions on portfolios showed that the relationship between beta and return was flat for all markets. The negative size effect was significant only for Korea. Book/market ratio and return had a significantly positive relationship only for Hong Kong, Korea and Malaysia. The regressions on individual stocks showed the same conclusion regarding beta and book/market ratio. The negative size effect was significant for Hong Kong, Korea, Malaysia and Thailand. The tests showed opposite seasonal patterns of the size effect in Hong Kong and Korea: While in Hong Kong the size premium in January was positive (i.e., larger firms earned higher returns) and negative in the other months, the negative size effect in Korea was concentrated in January.
Heston, Rouwenhorst, and Wessels (1999): *The Role of Beta and Size in the Cross-Section of European Stock Returns*	January 1978–December 1995: 2100 stocks from 12 European countries: Austria, Belgium, Denmark, France, Germany, Italy, Netherlands, Norway, Spain, Sweden, Switzerland and U.K., covering 60–90 % of each country's market capitalization	Average return is positively related to beta, but this is partly due to the fact that high-beta countries overperform low-beta countries. Within countries, this positive relation vanishes and can only be observed in January. The size effect can be observed in the Europe-wide test and in 5 (out of 12) countries. The value-weighted MSCI index of the 12 sample countries was used as a proxy for the market portfolio.
Fletcher (2000): *On the Conditional Relationship between Beta and Return in International Stock Returns*	January 1970–July 1998: Morgan Stanley Capital International (MSCI) World Equity Index and 18 MSCI developed markets country indices	The CAPM was tested on the 18 country portfolios of the developed markets,[b] represented by the respective MSCI country portfolios. The MSCI World Index was used as a proxy for the market portfolio. The test shows no significant relationship between beta and average return in January 1970–July 1998. But when testing up market and down market months only, beta shows a significant positive relationship with average returns in up markets and negative relationship in down markets.

(continued)

Table 2.11 (continued)

Study	Data used	Results
Hodoshima, Garza-Gómez, and Kunimura (2000): *Cross-Sectional Regression Analysis of Return and Beta in Japan*	January 1956–December 1995: Japanese stocks from the first section (large-cap) of the Tokyo Stock Exchange	Empirical tests were run for the entire period 1956–1995 and for the subperiods 1956–1965, 1966–1975, 1976–1985 and 1986–1995. In the entire period and all subperiods, the relation between beta and return is flat. But in up markets/down markets only, there is a positive/negative relation between beta and return, for 1956–1995 and for all subperiods. When considering beta together with size and book/market ratio in the period July 1962–December 1995, then size is the only significant variable with negative premium. In up markets, beta has a positive, size has a negative effect on return, book/market ratio is insignificant. In down markets, the size effect becomes insignificant while beta has a negative effect on return, book/market ratio is insignificant. In January, all three explanatory variables size (negative), beta and book/market (positive) are significant in explaining returns, while in non-January months, none of these are significant.
Faff (2001): *A Multivariate Test of a Dual-Beta CAPM: Australian Evidence*	January 1974–January 1995: Australian stocks from the Price Relatives File of the Centre for Research in Finance (CRIF) at the Australian Graduate School of Management	The relation between beta and return was flat. In up markets/down markets only, the relation between beta and returns was positive/negative. There is minimal evidence of a difference between up market beta and down market beta.

Elsas, El-Shaer, and Theissen (2003): *Beta and Returns Revisited: Evidence from the German Stock Market*	January 1968–December 1995: Stocks from Germany. Sample size ranges from 211 to 316 stocks.	Over the period 1968–1995, the relation between beta and return was flat. When considering up markets/down markets only, the relation between beta and return was positive/negative.
Wang and Xu (2004): *What Determines Chinese Stock Returns?*	July 1996–June 2002: A-shares[c] (which are available only to domestic investors, for non-financial companies, as opposed to B-shares which are for international investors) from the Chinese stock market, traded on the Shanghai Stock Exchange or the Shenzhen Stock Exchange, from the China Stock Market and Accounting Research Database	The relationship between beta and return was flat. Size had a significant negative effect on return. Book/market ratio was found to be insignificant. The paper also tested free-float (percentage of shares which are tradable in the stock markets) as a proxy for good governance which is an important aspect in a country where companies are mostly government-controlled. The study found a positive relationship between free-float and return, when taking into account the other variables beta, size and book/market ratio.

[a] Lintner (1965b, p. 611) for the data used in the study originally done by J. Lintner
[b] Listed on Fletcher (2000, p. 238): The 18 countries are Australia, Austria, Belgium, Canada, Denmark, France, Germany, Hong Kong, Italy, Japan, Netherlands, Norway, Singapore, Spain, Sweden, Switzerland, U.K. and USA
[c] See also Wang and Xu (2004, pp. 65–66) for the definition of A-shares and B-shares

returns equal the risk-free rate. This led to the introduction of a modified version of the CAPM, the zero-beta CAPM (see Black 1972).

Fama and MacBeth (1973) tested the linearity between return and beta and non-beta risk as explanatory variable for returns. Unsystematic risk was provided as a proxy for non-beta risk. The Fama–MacBeth regression, which provided an important framework for future tests, supported the linearity (C1) and beta as the only variable with explanatory power for returns (C2). The linearity assumption was also supported in other tests, like Blume and Friend (1973) and Stambaugh (1982).

Early studies declared a positive beta effect on expected returns (C3) in the U.S. stock market, as shown in Black et al. (1972) for the period 1931–1965 and in Fama and MacBeth (1973) for the period 1935–1968. But Black et al. (1972) also revealed that the relation between average return and beta flattened over time. Doubts about the role of beta in explaining returns were articulated in Reinganum (1981), when no significant relationship between beta and the returns of NYSE/AMEX stocks was found in the period 1964–1979. The same result was obtained for NYSE stocks in the period 1962–1981 in Lakonishok and Shapiro (1986). Instead of beta, other variables like P/E ratio (Basu 1977), size (Banz 1981), leverage (Bhandari 1988), and book-to-market ratio (Rosenberg et al. 1985) were found to play significant roles in explaining returns.

Fama and French (1992) raised the question *"Can β be Saved?"* when the authors discovered that beta played no role in explaining average returns for the 50-year period 1941–1990. They showed that even in the 1941–1965 period, where authors of the early CAPM tests had agreed that the beta premium was significant, the relationship between beta and return vanished when controlling for size.

Roll (1977) argued that the empirical tests were not a test of the CAPM which requires the true market portfolio. The hypotheses (C1)–(C3) are true for any mean-efficient portfolio used as a market proxy, because of a mathematical tautology and without any model assumptions.

The conclusion of the tests is that the CAPM "*never has been an empirical success*".[101] The zero-beta CAPM had some success until other variables like size and price ratios were discovered which explained average returns. Fama and French (2004, p. 43), states: "*The problems are serious enough to invalidate most applications of the CAPM*".[102] CAPM estimates for high-beta stocks are too high, and estimates for low-beta stocks turned out to be too low. The risk-adjusted performance measure α, introduced in Jensen (1967), which is the return in excess of the CAPM-implied return, turns out to be larger for small-beta portfolios and smaller for large-beta portfolios. Funds could simply concentrate on stocks with low beta, small size and high B/M and earn positive alpha without special stock picking skills.[103]

[101]Fama and French (2004, p. 43).

[102]Fama and French (2004, pp. 43–44).

[103]Fama and French (2004, p. 44).

Pettengill et al. (1995) revived the importance of beta, suggesting that previous tests of the relation between beta and return have been biased due to the aggregation of up markets and down markets. By testing up markets and down markets separately, they found significant relationships (i.e., positive for up markets and negative for down markets) between beta and return.

Table 2.11 shows a summary of the empirical tests.

2.3.6 A Critical View of the CAPM Assumptions

The CAPM has been criticized for its unrealistic assumptions (listed in Sect. 2.3.1). We will now discuss some relaxations of these assumptions which lead to slight modifications of the model. This section is based on Reilly and Brown (1997, pp. 305–309).

2.3.6.1 Zero-Beta CAPM

The zero-beta CAPM was derived in Black (1972). It was postulated in Black et al. (1972) after empirical tests rejected the hypothesis that the return of a zero-beta asset is the risk-free rate.

The zero-beta CAPM makes all the assumptions from the original CAPM except (A3), i.e., that all investors can lend and borrow any amount of money at the risk-free rate.

By combining different assets (since shorting is allowed), several portfolios can be created with zero beta. Let Z be the zero-beta portfolio with minimal variance. Then all portfolios plot on a security market line (SML) which connects the portfolios Z and Mkt (market portfolio), as shown in Fig. 2.23.

The equation for the zero-beta CAPM is

$$\mathbb{E}[R_{Pf}] = \mathbb{E}[R_Z] + \beta_{Pf} \cdot (\mathbb{E}[R_{Mkt}] - \mathbb{E}[R_Z]). \qquad (2.45)$$

2.3.6.2 Different Borrowing and Lending Rates

The CAPM assumption (A3) that all investors can borrow and lend any amount for the risk-free rate is unrealistic. While investors can lend any amount at the risk-free rate r_{rf} by buying Treasury bills, most investors usually have to pay a higher rate r_b for borrowing. The effect of the different borrowing and lending rates is illustrated in Fig. 2.24. It shows the mean-variance diagram with the efficient frontier, together with the risk-free rate r_{rf} and the borrowing rate r_b. The line segment between $r_{rf} - Mkt$ represents all portfolios which are combinations of the market portfolio and the risk-free asset (i.e., lending at r_{rf}). If it were possible to borrow at r_{rf}, then this line segment would extend beyond the point Mkt. The point K is the tangency point from r_b to the efficient frontier, and this tangency line ends at G. The line segment $K - G$ represents all investment opportunities where the investors borrow at the rate r_b and invest in the portfolio K. The segment $Mkt - K$ on the efficient frontier does not involve any borrowing or lending. The CML, the set of all optimal

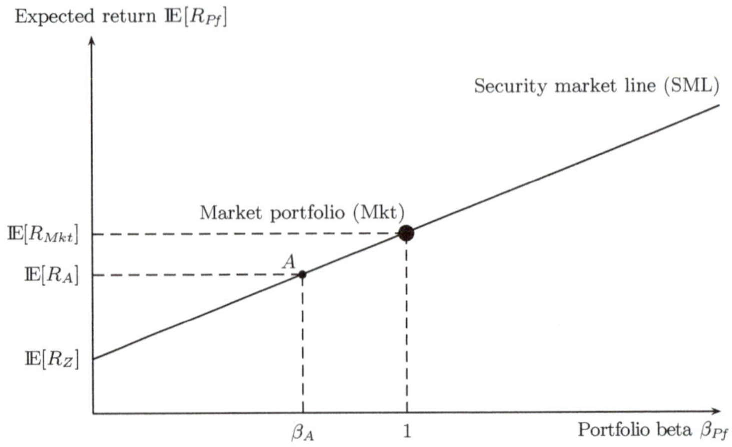

Fig. 2.23 The diagram shows the security market line (SML) without the risk-free asset. The SML connects the market portfolio with the zero-beta portfolio Z. In equilibrium, all assets plot on the SML. *Source*: Own, for illustrative purposes only

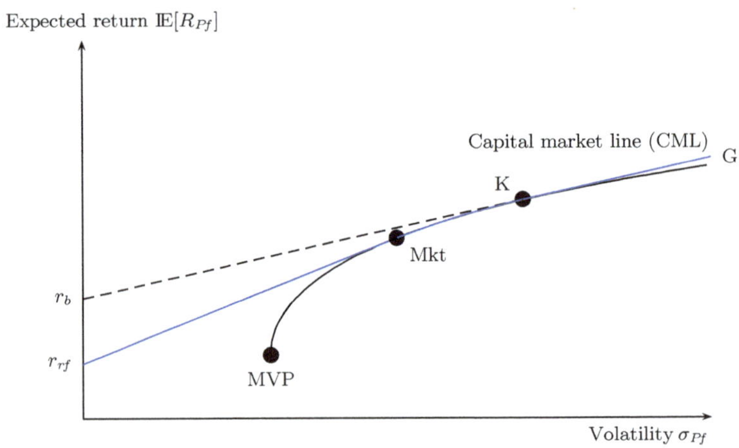

Fig. 2.24 The mean-variance diagram shows the CML when borrowing cost (r_b) is greater than the risk-free rate (r_{rf}). The CML (marked in *blue*) is made up of $r_{rf} - Mkt - K - G$, i.e., of a line segment $r_{rf} - Mkt$, a curve segment $Mkt - K$ and a line segment $K - G$. *Source*: Own, for illustrative purposes only

investment opportunities, is made of $r_{rf} - Mkt - K - G$, i.e., of a line segment $r_{rf} - Mkt$, a curve segment $Mkt - K$ and a line segment $K - G$.

2.3.6.3 Transaction Costs

The CAPM assumes no transaction costs (A6), which means that investors will even buy or sell securities when they are only slightly mispriced. If a security plots above the SML, then its expected return is higher than its theoretical return implied by the

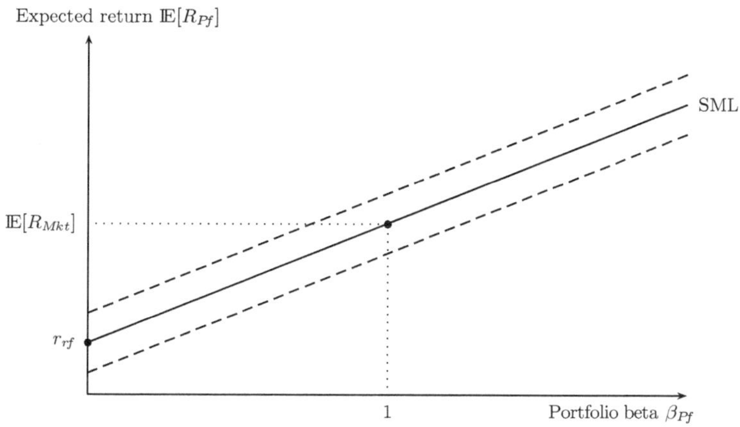

Fig. 2.25 Security market line (SML) with transaction costs. *Source*: Own, for illustrative purposes only

CAPM, i.e., the security price is underpriced. Investors will buy this security and bid up the price until it is fairly valued, i.e., it plots on the SML. If a security plots below the SML, then investors short it until it plots on the SML. With transaction costs, investors will not correct small mispricings when the costs of buying and selling eat up the potential gains. So the SML will rather be a band, as illustrated in Fig. 2.25, and the greater the transaction costs, the wider the band becomes.

2.3.6.4 Heterogeneous Expectations, Investment Horizons and Taxes

The CAPM assumes homogeneous expectations (A1), while in reality, investors have different expectations about risk and returns. If we assume that every investor has his own beliefs, then each one would have a unique CML and/or SML. The composite graph would be a band, and its breadth would reflect the divergence of opinions.

The result is similar when we allow for different investment horizons. The CAPM presupposes the same investment horizon (A4) for all investors, but in reality you have short-time investors who need their money in 1 month and long-term investors saving for their retirement in 30 years. The CAPM is a one-period model, but the investor with a 1-month investment horizon has a different CML/SML from the investor with a 30-year planning period.

The CAPM does not account for taxes (A6) , but investors pay different taxes on capital gains and dividends, and rational investors will consider their after-tax returns. Since taxes have an impact on the after-tax return, different taxes will cause different CML/SML among investors.[104]

[104]For a detailed discussion on taxes, see Black and Scholes (1974) and Litzenberger and Ramaswamy (1979).

2.4 The Fama–French Three-Factor Model

2.4.1 Introduction

After the development of the capital asset pricing model (CAPM) in the 1960s (Treynor in 1961 (see French 2002), Sharpe 1964, Lintner 1965a and Mossin 1966), many empirical tests were developed. Early tests (Black et al. 1972; Douglas 1969; Miller and Scholes 1972) rejected the CAPM and led to a modification of the model, the zero-beta CAPM (Black 1972). Empirical tests in the 1970s (Black et al. 1972; Fama and MacBeth 1973) validated that model, but later on significant doubts were raised about the beta premium (Lakonishok and Shapiro 1986; Reinganum 1981). On the other hand, other factors like P/E ratio (Basu 1977), size (Banz 1981), leverage (Bhandari 1988), and book-to-market ratio (Rosenberg et al. 1985) were found to play significant roles in explaining returns.

In Fama and French (1992), various factors were tested (as single explanatory variables and in combinations). The size and book-to-market ratio were found to be the most significant ones for describing returns. These variables were incorporated into the *Fama–French three-factor model (FF3M)* which is a modification of the CAPM. The big difference between the two is that the CAPM was derived from market portfolio theory with a huge list of idealized assumptions, whereas FF3M is a model developed as a modification of the CAPM to better fit the empirical data.

2.4.2 The Model

The Fama–French three-factor model was introduced in Fama and French (1993) as a modification of the CAPM and included the size and book-to-market ratio as additional factors describing returns. These were found in Fama and French (1992) to be the most significant ones: Small-caps outperformed large-caps and high-B/M stocks outperformed low-B/M stocks.

For the model, we first have to specify a certain stock market. For a stock A, according to the Fama–French three-factor model,[105] the monthly return $R_A^{monthly}$ is

$$\begin{aligned} R_A^{monthly} &= \alpha + r_{rf}^{monthly} + \beta_{1,A} \cdot (R_{Mkt}^{monthly} - r_{rf}^{monthly}) \\ &\quad + \beta_{2,A} \cdot SMB + \beta_{3,A} \cdot HML, \end{aligned} \tag{2.46}$$

where

- $r_{rf}^{monthly}$ is the monthly risk-free rate.
- $R_{Mkt}^{monthly}$ is the monthly return on the portfolio of all stocks (in the specified stock market).

[105]For the equation, see Fama and French (1993, p. 24, Table 6). It is also on p. 37, Table 9a, as one of the regression equations studied in Fama and French (1993).

- *SMB* ("small minus big") is the difference between small-cap and large-cap returns (defined below).
- *HML* ("high minus low") is the difference between high-B/M and low-B/M returns (defined below).
- α is the component of the return not described by the factors and should be insignificant.

The model would also apply to annual, quarterly, weekly, daily returns, etc., but Fama and French (1993) originally used monthly returns when FF3M was formulated and studied.

In the Fama–French model from Fama and French (1993), the method to calculate *SMB* and *HML* is as follows[106]:

All stocks are ranked according to size, and the largest 50 % are put into the *big* group, the smaller 50 % into the *small* group. Independent from that, the stocks are also ranked according to B/M ratio. The highest 30 % end up in the *high*, the middle 40 % in the *medium*, and the lowest 30 % in the *low* group.

Six portfolios are formed from the combinations of these groups (*small/high, small/medium, small/low, big/high, big/medium, big/low*).

SMB is the difference between the arithmetic average of the monthly returns on the three small-stock portfolios (*small/high, small/medium, small/low*) and the arithmetic average of the monthly returns on the three big-stock portfolios (*big/high, big/medium, big/low*). *SMB* describes the difference between a large-cap portfolio and a small-cap portfolio with similar book-to-market equity.

HML is defined similarly: the difference between the arithmetic average of the monthly returns on the high-B/M portfolios (*small/high, big/high*) and the arithmetic average of the monthly returns on the low-B/M portfolios (*small/low, big/low*). *HML* describes the difference between a high-B/M and a low-B/M portfolio with similar size.

SMB and *HML* are usually positive, since small-caps tend to outperform large-caps and high-B/M portfolios tend to outperform low-B/M portfolios.

The variables β_1, β_2 and β_3 are determined by a time-series regression[107]:

- β_1 is the market sensitivity of the stock, controlling for size and B/M. Note that this is usually different from the CAPM-β which accounts only for the market sensitivity of the stock.
- β_2 is the size coefficient. Smaller companies tend to have larger β_2s than larger companies.
- β_3 is the B/M coefficient. High-B/M companies tend to have larger β_3s than low-B/M companies.

---

[106]Fama and French (1993, pp. 8–9). Although the original method in Fama and French (1993) was slightly more complicated, the method we present here will more easily convey the idea behind it.

[107]The time-series regression was introduced on page 124 for one independent variable, but can easily be extended to multiple variables.

2.4.3 Theoretical Explanations of the Fama–French Three-Factor Model

The main difference between the capital asset pricing model and the Fama–French three-factor model is that the former has a theoretical foundation while the latter is an ad hoc model which was just introduced because it better fits the empirical data. The reason why Fama and French (1993) introduced FF3M is because a higher return for small-size and high-B/M firms was observed. The theoretical justification from Fama and French (1993) of size and B/M as factors to drive returns was that firms with high B/M or low size tend to have higher earnings, therefore describing risk factors.[108]

Other explanations of the value premium (i.e., the effect that high-B/M firms outperform low-B/M firms) stem from behavioral finance[109]:

Lakonishok, Shleifer, and Vishny (1994) offers the view that the value premium is due to the overreaction of the markets.[110] They overreact to good news and therefore overbuy *glamour* stocks which have performed well in the past. These stocks are overpriced and therefore have low B/M. On the other hand, investors also overreact to bad news and oversell badly performing stocks. These *value* stocks are underpriced and have high B/M. When the market corrects the overreaction, value stocks overperform glamour stocks. De Bondt and Thaler (1985) validates the overreaction hypothesis by pointing out that "losers" overperform "winners".

One other behavioral view is that investors simply like growth stocks (low B/M, strong firms) and dislike value stocks (high B/M, weak firms).[111] In this case, the value premium would not be due to risk but to characteristics of value stocks which have the effect of turning investors away. Daniel and Titman (1997) formulates and presents the theory and evidence that the similar characteristics of high-B/M firms rather than the factor loadings from the FF3M drive their returns and explain the high correlation of returns among those.[112]

We summarize the theoretical arguments for the size effect from van Dijk (2011, pp. 11–23):

The argument from Daniel and Titman (1997) that characteristics and not factor loadings describe returns also applies to size, contrary to the view in Fama and French (1993) that size describes a risk factor related to financial distress.

Another reason for the size effect is that it includes transaction costs and liquidity risks: Stoll and Whaley (1983) argues that dealers require bigger bid-ask spreads for trading in small firms because of the infrequent trading activity and higher risk.[113]

---

[108]Fama and French (1993, pp. 7–8).

[109]These explanations can be found in Davis, Fama, and French (2000, pp. 389–390).

[110]Lakonishok et al. (1994, p. 1542).

[111]Davis et al. (2000, p. 390).

[112]Daniel and Titman (1997, pp. 3–4).

[113]Stoll and Whaley (1983, p. 58).

For NYSE stocks from 1960–1978, the abnormal returns are shown to be eliminated for investment horizons of less than 1 year.[114] Amihud (2002) finds an *illiquidity premium*, i.e., that expected market illiquidity positively affects stock returns.[115] Since small firms are more affected by illiquidity, this explains part of the size effect.

Pástor and Stambaugh (2003) explains that sensitivity to market liquidity is priced[116] as a risk factor and that the smallest firms tend to be most affected by market illiquidity because their shares are illiquid.[117] But on the other hand, the paper contends that the relation between the liquidity of stocks and market liquidity is not straightforward. A drop in market liquidity causes many investors to move from stocks to bonds, therefore selling the most liquid (large-cap) stocks to save transaction costs.[118]

Last, but not least, there are also explanations of the size effect from behavioral finance.

The overreaction hypothesis which explains the value premium might also be applicable to size, and Chan and Chen (1991) indicates that many small-size firms have performed poorly and lost market value in the past,[119] but it seems to be unexplored whether overreaction is a driving factor of the size effect.[120]

One behavioral view of the value premium is that investors like growth stocks and dislike value stocks. The same argument also applies to large-cap stocks vs. small-cap stocks. Gompers and Metrick (2001) found that institutional investors increased the demand for liquid, large-cap stocks in the U.S equity markets in 1980–1996. The shift from individual to institutional ownership led to a disappearance of the size effect in that period.[121] This is an example for the theory of Daniel and Titman (1997) that size as a characteristic and not as a risk factor drives returns.

Market frictions are also found to have impact on returns: Hou and Moskowitz (2005) holds that the delay, i.e., the time needed until the price reflects information, requires a premium. Small firms are most affected by the delay which captures part of the size premium.[122]

After we have explored the theoretical explanations of the FF3M, we turn to the empirical tests.

[114]Stoll and Whaley (1983, p. 58).

[115]Amihud (2002, p. 31).

[116]Pástor and Stambaugh (2003)

[117]Pástor and Stambaugh (2003, p. 677).

[118]Pástor and Stambaugh (2003, p. 677).

[119]Chan and Chen (1991, pp. 1467–1468).

[120]van Dijk (2011, p. 16).

[121]Gompers and Metrick (2001, p. 17).

[122]Hou and Moskowitz (2005, p. 981).

2.4.4 Empirical Tests

2.4.4.1 Fama and French (1993): *Common Risk Factors in the Returns on Stocks and Bonds*

After Fama and French (1992) found that size and book-to-market equity were the most significant drivers of returns, the Fama–French three-factor model was developed in Fama and French (1993) which included these factors besides beta. An empirical study was also done in the same paper.

The study uses the same data as Fama and French (1992)[123]: The market data were from non-financial stocks which were traded on the NYSE, AMEX and NASDAQ and taken from the Center for Research in Security Prices (CRSP).[124] The accounting data was taken from COMPUSTAT, also maintained by the CRSP. The 1962 start date reflects the fact that book values were not generally available before. The market portfolio is the value-weighted portfolio of all NYSE, AMEX and NASDAQ stocks.[125] The risk-free rate is the 1-month treasury bill rate.[126]

Calculation of SMB and HML

SMB and *HML* are calculated by dividing the U.S. stocks in *small/big* and *high/medium/low*.[127] We do these calculations year after year, from July 1963 to June 1964:

At the end of June 1963, all NYSE stocks on CRSP are ranked on size. The median NYSE size splits the NYSE/AMEX/NASDAQ stocks into the groups *small* and *big*. To form the B/M-groups, we determine the breakpoints for top 30 % (*high*), middle 40 % (*medium*) and bottom 30 % (*low*) based on the B/M ratios of the NYSE stocks. We use the accounting data from January 1962 to December 1962 from COMPUSTAT for book equity. We divide it by the market equity at the end of December 1963 to get B/M. Firms with negative book value are excluded from the study.

SMB (small minus big) is the difference between the arithmetic average of the monthly returns on the three small-stock portfolios (*small/high*, *small/medium*, *small/low*) and the arithmetic average of the monthly returns on the three big-stock portfolios (*big/high*, *big/medium*, *big/low*).

HML (high minus low) is the difference between the arithmetic average of the monthly returns on the high-B/M portfolios (*small/high*, *big/high*) and the arithmetic average of the monthly returns on the low-B/M portfolios (*small/low*, *big/low*).

---

[123]Fama and French (1993, p. 4) which mentions that it extends the study of Fama and French (1992, using p. 429 for the data).

[124]Financial stocks were excluded because the high leverage that is normal for these firms probably does not have the same meaning for non-financial firms, where high leverage more likely indicates distress.

[125]Fama and French (1992, p. 431).

[126]Fama and French (1993, p. 10).

[127]Methodology from Fama and French (1993, pp. 8–9).

For July 1964–June 1965, we repeat the same method: At the end of June 1964, we form the groups *big/small* based on size and the groups *high/medium/low* based on B/M. To calculate B/M, the book equity data at the end of December 1964 is used. Then *SMB* and *HML* are calculated as above. The whole process is repeated for every year until December 1991.

Construction of Size-B/M Portfolios

The construction of the 25 size-B/M portfolios is similar to the construction of the *small/big* and *high/medium/low* portfolios above, but we use five different size groups and B/M groups (based on the NYSE size and B/M quintiles). By intersecting them we form the 25 size-B/M combinations.[128] In Table 2.12, you can see the average monthly excess returns (i.e., return minus risk-free rate) of the 25 size-B/M portfolios. They range from 0.32 to 1.05 %. The table shows the negative effect of size and the positive effect of B/M on returns. For all sizes, the returns rise with B/M. For all but the lowest B/M, the returns tend to decrease with size.

Time-Series Regression

Different time-series regressions were done on each of the 25 size-B/M portfolios to compare the explanatory power of the regression of the CAPM against the FF3M. Let

$$\tilde{r}_{Pf}^{monthly} = r_{Pf}^{monthly} - r_f^{monthly} \tag{2.47}$$

be the monthly excess return of the portfolio Pf over the risk-free rate $r_f^{monthly}$. The regression which is shown in Table 2.13 explains a big part of the variation of returns, although there is still some room for improvement[129]: The R^2 value is around 0.9 for the big-stock and low-B/M portfolios, but most of the portfolios have an R^2 of about 0.7 to 0.8.

Table 2.12 Average monthly excess returns of the 25 size-B/M portfolios in the period July 1963–December 1991

	B/M-1 (low)	B/M-2	B/M-3	B/M-4	B/M-5 (high)
	Average monthly return (in %)				
Size-1 (small)	0.39	0.70	0.79	0.88	1.01
Size-2	0.44	0.71	0.85	0.84	1.02
Size-3	0.43	0.66	0.68	0.81	0.97
Size-4	0.48	0.35	0.57	0.77	1.05
Size-5 (big)	0.40	0.36	0.32	0.56	0.59

Source: Fama and French (1993, Table 2, pp. 14–15)

[128]Fama and French (1993, p. 8).

[129]Fama and French (1993, p. 19).

Table 2.13 Regression of excess stock returns of the 25 size-B/M portfolios on excess market return in the period July 1963–December 1991

	Regression $\tilde{r}_{Pf}^{monthly} = \alpha + \beta \cdot \tilde{r}_{Mkt}^{monthly}$				
	B/M-1 (low)	B/M-2	B/M-3	B/M-4	B/M-5 (high)
	β				
Size-1 (small)	1.40	1.26	1.14	1.06	1.08
Size-2	1.42	1.25	1.12	1.02	1.13
Size-3	1.36	1.15	1.04	0.96	1.08
Size-4	1.24	1.14	1.03	0.98	1.10
Size-5 (big)	1.03	0.99	0.89	0.84	0.89
	R^2				
Size-1 (small)	0.67	0.70	0.68	0.65	0.61
Size-2	0.79	0.79	0.76	0.76	0.71
Size-3	0.84	0.84	0.80	0.79	0.74
Size-4	0.89	0.90	0.87	0.80	0.76
Size-5 (big)	0.89	0.92	0.84	0.79	0.69

Source: Fama and French (1993, Table 4, p. 20)

Table 2.14 shows the results of a linear regression of monthly excess stock returns over monthly excess market return, *SMB* and *HML* over the time period July 1963–December 1991. Compared to the CAPM-β in Table 2.13, the market sensitivities (β_1) change once other factors like *SMB* and *HML* are included in the regression. The β_1s are much closer to 1 than the βs from Table 2.13. In Table 2.14, we can see that β_2 increases for decreasing size and β_3 tends to increase for increasing B/M. This is exactly how the coefficients are supposed to behave according to the FF3M: Small-caps earn higher returns than large-caps, high-B/M firms earn higher returns than low-B/M firms. Most of the R^2 values from Table 2.14 are greater than 90 %, which means that the data fits the model very well. The R^2 values are much higher than those from the CAPM regression in Table 2.13, indicating a significant improvement.

There were studies which have questioned the significance of the CAPM-β,[130] and when looking at Table 2.14, one may think that β_1 plays an insignificant role. The variation is small, having barely any effect in describing returns. However, the market sensitivity β_1 in the Fama–French three-factor model does play a significant role. Fama and French (1993) ran a regression of monthly excess returns on size and B/M alone, and the results can be seen in Table 2.15. The R^2 values range from 0.04 to 0.65. For small-caps, the R^2 value is around 0.6, but it decreases for larger sizes.

Compared to the CAPM regression in Table 2.13, the R^2 values of the regression on only size and B/M are low. So although the values of β_1 differ very little across

[130] See Sect. 2.3.3 about empirical tests.

Table 2.14 Regression of excess stock returns of the 25 size-B/M portfolios on excess market return, *SMB* and *HML*, in the period July 1963–December 1991

	Regression $\tilde{r}_{Pf}^{monthly} = \alpha + \beta_1 \cdot \tilde{r}_{Mkt}^{monthly} + \beta_2 \cdot SMB + \beta_3 \cdot HML$				
	B/M-1 (low)	B/M-2	B/M-3	B/M-4	B/M-5 (high)
	β_1 (coefficient of market excess return)				
Size-1 (small)	1.04	1.02	0.95	0.91	0.96
Size-2	1.11	1.06	1.00	0.97	1.09
Size-3	1.12	1.02	0.98	0.97	1.09
Size-4	1.07	1.08	1.04	1.05	1.18
Size-5 (big)	0.96	1.02	0.98	0.99	1.06
	β_2 (coefficient of *SMB*)				
Size-1 (small)	1.46	1.26	1.19	1.17	1.23
Size-2	1.00	0.98	0.88	0.73	0.89
Size-3	0.76	0.65	0.60	0.48	0.66
Size-4	0.37	0.33	0.29	0.24	0.41
Size-5 (big)	−0.17	−0.12	−0.23	−0.17	−0.05
	β_3 (coefficient of *HML*)				
Size-1 (small)	−0.29	0.08	0.26	0.40	0.62
Size-2	−0.52	0.01	0.26	0.46	0.70
Size-3	−0.38	0.00	0.32	0.51	0.68
Size-4	−0.42	0.04	0.30	0.56	0.74
Size-5 (big)	−0.46	0.00	0.21	0.57	0.76
	R^2				
Size-1 (small)	0.94	0.96	0.97	0.97	0.96
Size-2	0.95	0.96	0.95	0.95	0.96
Size-3	0.95	0.94	0.93	0.93	0.93
Size-4	0.94	0.93	0.91	0.89	0.89
Size-5 (big)	0.94	0.92	0.88	0.90	0.83

Source: Fama and French (1993, Table 6, pp. 24–25)

portfolios in the three-factor regression, the market sensitivity is needed as a factor together with size and B/M to make the data fit very well.[131]

After we have analyzed the sensitivities β_1, β_2, β_3 and the coefficient of determination R^2, we will look at the intercept α and test if it is close to zero. Table 2.16 shows the results for the CAPM, the FF3M and the two-factor model with size and B/M only. The usual CAPM regression (1) shows mainly positive intercepts. This was usually observed in CAPM tests and led to an early rejection of the original CAPM. The intercept decreases with increasing size (except for the lowest B/M quintile) and increases with increasing B/M.

Regression (2) with size and B/M as factors shows positive large intercepts. Nineteen of the portfolios have an intercept greater than 0.5 %, and four have

[131] Fama and French (1993, p. 21).

Table 2.15 Regression of excess stock returns of the 25 size-B/M portfolios on *SMB* and *HML* in period July 1963–December 1991

	Regression $\tilde{r}_{Pf}^{monthly} = \alpha + \beta_2 \cdot SMB + \beta_3 \cdot HML$				
	B/M-1 (low)	B/M-2	B/M-3	B/M-4	B/M-5 (high)
	β_2 (coefficient of *SMB*)				
Size-1 (small)	1.93	1.73	1.63	1.59	1.67
Size-2	1.52	1.46	1.35	1.18	1.40
Size-3	1.28	1.12	1.05	0.93	1.16
Size-4	0.86	0.82	0.77	0.72	0.95
Size-5 (big)	0.28	0.35	0.22	0.29	0.44
	β_3 (coefficient of *HML*)				
Size-1 (small)	−0.95	−0.57	−0.35	−0.18	0.01
Size-2	−1.23	−0.66	−0.38	−0.16	0.00
Size-3	−1.09	−0.65	−0.31	−0.11	−0.01
Size-4	−1.11	−0.65	−0.36	−0.11	−0.01
Size-5 (big)	−1.07	−0.65	−0.42	−0.06	0.08
	R^2				
Size-1 (small)	0.65	0.60	0.60	0.60	0.59
Size-2	0.59	0.53	0.49	0.42	0.44
Size-3	0.51	0.43	0.37	0.31	0.35
Size-4	0.43	0.30	0.24	0.18	0.23
Size-5 (big)	0.34	0.18	0.08	0.04	0.06

Source: Fama and French (1993, Table 5, p. 22)

an intercept between 0.4 and 0.5 %. This reveals that size and B/M alone do not "*explain the average premium of stock returns over 1-month bill returns.*"[132]

When adding the excess market return to the previous regression, the intercepts get closer to zero. In absolute terms, the intercepts of 16 portfolios are smaller than 0.1 %, 22 are smaller than 0.2 %. "*Intercepts close to 0 say that the regressions that use [monthly excess returns], SMB and HML to absorb common time-series variation in returns do a good job explaining the cross-section of average stock returns.*"[133]

The conclusion is that *SMB* and *HML* explain the variation of returns across stocks, whereas the market factor explains why stock returns are on average higher than the risk-free rate.[134]

By using the Gibbons, Ross, and Shanken (1989) test,[135] Fama and French (1993) rejected the assertion that all intercepts in the regression (1), (2) and (3) from

[132]Fama and French (1993, p. 35).
[133]Fama and French (1993, p. 38).
[134]Fama and French (1993, p. 38).
[135]Fama and French (1993, p. 40).

Table 2.16 Intercepts α from the regression of excess stock returns on excess market return, *SMB* and *HML*, in the period July 1963–December 1991

	Intercept α				
	B/M-1 (low)	B/M-2	B/M-3	B/M-4	B/M-5 (high)
	(1) Regression $\tilde{r}_{Pf}^{monthly} = \alpha + \beta_1 \cdot \tilde{r}_{Mkt}^{monthly}$				
Size-1 (small)	−0.22	0.15	0.30	0.42	0.54
Size-2	−0.18	0.17	0.36	0.39	0.53
Size-3	−0.16	0.15	0.23	0.39	0.50
Size-4	−0.05	−0.14	0.12	0.35	0.57
Size-5 (big)	−0.04	−0.07	−0.07	0.20	0.21
	(2) Regression $\tilde{r}_{Pf}^{monthly} = \alpha + \beta_2 \cdot SMB + \beta_3 \cdot HML$				
Size-1 (small)	0.24	0.46	0.49	0.53	0.55
Size-2	0.52	0.58	0.64	0.58	0.64
Size-3	0.52	0.61	0.52	0.60	0.66
Size-4	0.69	0.39	0.50	0.62	0.79
Size-5 (big)	0.76	0.52	0.43	0.51	0.44
	(3) Regression $\tilde{r}_{Pf}^{monthly} = \alpha + \beta_1 \cdot \tilde{r}_{Mkt}^{monthly} + \beta_2 \cdot SMB + \beta_3 \cdot HML$				
Size-1 (small)	−0.34	−0.12	−0.05	0.01	0.00
Size-2	−0.11	−0.01	0.08	0.03	0.02
Size-3	−0.11	0.04	−0.04	0.05	0.05
Size-4	0.09	−0.22	−0.08	0.03	0.13
Size-5 (big)	0.21	−0.05	−0.13	−0.05	−0.16

Source: Fama and French (1993, Table 9a, pp. 36–37)

Table 2.16 are zero.[136] The Fama–French three-factor model performed best in the test and just failed it slightly. The reason for the rejection of FF3M is the data for the low-B/M portfolios: The return on the small-size portfolio was too low (intercept −0.34) and the return on the big-size portfolio was too high (intercept 0.21).[137] In other words, the size effect was missing in the lowest-B/M quintile. But despite the marginal rejection of the FF3M based on the Gibbons et al. (1989) test, Fama and French (1993) finds that the FF3M "*does a good job on the cross-section of average stock returns.*"[138] The R^2 values are very high (see Table 2.14) and intercepts are all close to zero except for the smallest-size lowest-B/M portfolio.[139]

2.4.4.2 Other Empirical Tests

In Sect. 2.4.4.1 we looked as an example at an empirical test conducted by Fama and French in 1993. However, many other tests of the Fama-French three-factor model exist. Table 2.17 shows a summary of these empirical tests.

136 Fama and French (1993, p. 39).

137 Fama and French (1993, pp. 40–41).

138 Fama and French (1993, p. 41).

139 Fama and French (1993, p. 41).

Table 2.17 Empirical tests of the Fama–French three-factor model

Study	Data used	Results
Fama and French (1993): *Common Risk Factors in the Returns on Stocks and Bonds*	July 1963–December 1991: Non-financial stocks which were traded on the New York Stock Exchange (NYSE), American Stock Exchange (AMEX) and NASDAQ	Marginal rejection of the FF3M by the GRS test[a] because the size effect did not occur for the low-book/market portfolios.
Griffin (2002): *Are the Fama and French Factors Global or Country Specific?*	January 1981–December 1995: Stocks from USA, Canada, Japan and U.K.	For the countries USA, Canada, Japan and U.K., the domestic Fama–French three-factor model was compared to a world version of the FF3M. The domestic versions fit the data much better than the world Fama–French model. But the GRS test rejected the (domestic/world) FF3M for all countries except Japan.
Faff (2004): *A Simple Test of the Fama and French Model Using Daily Data: Australian evidence*	May 1996–April 1999: Stocks from the Australian Stock Exchange (ASX)	Supports the FF3M equation using the MacKinlay–Richardson test[b] but finds a reversed size effect, i.e., negative *SMB*, which means that big firms outperform small firms.
Bahl (2006): *Testing the Fama and French Three-Factor Model and its Variants for the Indian Stock Returns*	July 2001–June 2006: 79 stocks from the BSE-100 Index (Bombay Stock Exchange)	GRS test supported FF3M
Rogers and Securato (2007): *Comparative Study of CAPM, Fama and French Model and Reward Beta Approach in the Brazilian Market*	July 1995–June 2006: Non-financial stocks from the Sao Paulo Stock Exchange (BOVESPA)	Wald test supported FF3M, but the book/market factor was not found to be significant.

(continued)

Table 2.17 (continued)

Chen and Fang (2009): *Uniform Testing and Portfolio Strategies for Single and Multifactor Asset Pricing Models in the Pacific Basin Markets*	Test on seven different stock markets: Japan (1976–2000), Hong Kong (1981–1999), South Korea (1982–1999), Malaysia (1979–1999), Thailand (1983–1999), Indonesia (1990–1998) and Singapore (1976–1997)	GRS test supported FF3M in all seven stock markets. Compared to CAPM, the average adjusted R^2 improves by 29 %. Size matters more than book/market.
Artmann, Finter, Kempf, Koch, and Theissen (2012): *The Cross-Section of German Stock Returns: New Data and New Evidence*	July 1962–December 2006: Over 900 non-financial stocks from the Frankfurt Stock Exchange	The FF3M was rejected by the GRS test. The highest R^2 value for a test portfolio was 81.6 %, but most of the R^2 values were below 80 %.
Bauer, Cosemans, and Schotman (2010): *Conditional Asset Pricing and Stock Market Anomalies in Europe*	February 1986–June 2002: Over 2000 stocks from exchanges of Austria, Belgium, Denmark, Finland, France, Germany, Greece, Ireland, Italy, Netherlands, Norway, Portugal, Spain, Sweden, Switzerland and the U.K., covering about 80 % of the European market capitalization	GRS test rejected FF3M: The small-cap low-book/market portfolio had big abnormal returns because it contained many start-ups which soared during the tech bubble. For the test portfolios, the adjusted R^2 ranged from 73 to 89 %. Book/market is insignificant when explaining returns. For European stocks, a significant size effect was observed, while for a sample of U.S. stocks, the size effect was insignificant.

[a] GRS test stands for the test statistic developed in Gibbons et al. (1989)

[b] The test used here is the generalized method of moments (GMM) developed by MacKinlay and Richardson (1991)

2.5 Summary

In this chapter, we have presented the basics of modern portfolio theory as introduced by Markowitz in the 1950s. The ideas of diversification and the efficient frontier are key when investing today. The capital asset pricing model allows investors to detect over- and undervalued securities, but it shows weaknesses when it is subjected to empirical tests. Its extension, the Fama–French three-factor model, uses more input parameters than the CAPM when determining investment returns, but also exhibits shortcomings in practice. Both models are theoretically plausible, but there is a discrepancy with reality which, in times of crisis, is significant. Moreover, traditional finance theory cannot explain market situations like crashes and stock market anomalies. The latter will be the topic of the next chapter.

References

Altman, E. I., Jacquillat, B., & Levasseur, M. (1974, December). Comparative analysis of risk measures: France and the United States. *Journal of Finance, 29*(5), 1495–1511.

Amihud, Y. (2002). Illiquidity and stock returns: Cross-section and time-series effects. *Journal of Financial Markets, 5*(1), 31–56.

Artmann, S., Finter, P., Kempf, A., Koch, S., & Theissen, E. (2012, January). The cross-section of German stock returns: New data and new evidence. *Schmalenbach Business Review (SBR), 64*, 20–43.

Baesel, J. B. (1974, December). On the assessment of risk: Some further considerations. *Journal of Finance, 29*(5), 1491–1494.

Bahl, B. (2006, September). *Testing the Fama and French three-factor model and its variants for the Indian stock returns.* U.K.: Department of Economics, University of Warwick.

Ball, R. (1978). Anomalies in relationship between securities' yields and yield-surrogates. *Journal of Financial Economics, 6*, 103–126.

Banz, R. (1981). The relationship between returns and market value of common stock. *Journal of Financial Economics, 9*, 3–18.

Basu, S. (1977). Investment performance of common stocks in relation to their price-earnings ratios: A test of the efficient market hypothesis. *Journal of Finance, 32*(3), 663–682.

Bauer, R., Cosemans, M., & Schotman. P. C. (2010). Conditional asset pricing and stock market anomalies in Europe. *European Financial Management, 16*(2), 165–190.

Bhandari, L. C. (1988). Debt/equity ratio and expected common stock returns: Empirical evidence. *Journal of Finance, 43*(2), 507–528.

Black, F. S. (1972). Capital market equilibrium with restricted borrowing. *Journal of Business, 45*(3), 444–455.

Black, F. S., Jensen, M. C., & Scholes, M. S. (1972). The capital asset pricing model: Some empirical tests. In *Studies in the theory of capital markets* (pp. 79–121). New York: Praeger Publishers.

Black, F. S., & Scholes, M. S. (1974, May). The effect of dividend yield and dividend policy on common stock prices and returns. *Journal of Financial Economics, 1*(1), 1–22.

Blume, M. E. (1971, March). On the assessment of risk. *Journal of Finance, 26*(1), 1–10.

Blume, M. E., & Friend, I. (1973). A new look at the capital asset pricing model. *Journal of Finance, 28*(1), 19–33.

Bodie, Z., Kane, A., & Marcus, A. J. (1999). Investments (4th ed.). New York: Irwin/McGraw-Hill.

Bodie, Z., Kane, A., & Marcus, A. J. (2009). Investments (8th ed.). Boston: Irwin/McGraw-Hill.

Chan, K. C., & Chen, N.-F. (1991, September). Structural and return characteristics of small and large firms. *Journal of Finance, 46*(4), 1467–1484.
Chan, L. K. C., Hamao, Y., & Lakonishok, J. (1991, December). Fundamentals and stock returns in Japan. *Journal of Finance, 46*(5), 1739–1764.
Chen, A., & Fang, S. (2009). Uniform testing and portfolio strategies for single and multifactor asset pricing models in the Pacific Basin markets. *Applied Economics, 41*(15), 1951–1963.
Chui, A. C. W., & Wei, K. C. J. (1998, August). Book-to-market, firm size, and the turn-of-the-year effect: Evidence from Pacific-Basin emerging markets. *Pacific-Basin Finance Journal, 6*(3–4), 275–293.
Daniel, K., & Titman, S. (1997). Evidence on the characteristics of cross-sectional variation in stock returns. *Journal of Finance, 52*(1), 1–33.
Davis, J. L., Fama, E. F., & French, K. R. (2000). Characteristics, covariances, and average returns: 1929 to 1997. *Journal of Finance, 55*(1), 389–406.
De Bondt, W. F. M., & Thaler, R. H. (1985). Does the stock market overreact? *Journal of Finance, 40*(3), 793–805.
DeFusco, R. A., McLeavey, D. W., Pinto, J. E., & Runkle, D. E. (2004). *Quantitative methods for investment analysis* (2nd ed.). Charlottesville: CFA Institute.
Douglas, G. W. (1969). Risk in the equity markets: An empirical appraisal of market efficiency. *Yale Economic Essays, 9*, 3–45.
Eisner, R. (1985). The total incomes system of accounts. *Survey of Current Business, 65*(1), 24–48.
Eisner, R. (1989). *The total incomes system of accounts*. Chicago: University of Chicago Press.
Elsas, R., El-Shaer, M., & Theissen, E. (2003, February). Beta and returns revisited: Evidence from the German stock market. *Journal of International Financial Markets, Institutions and Money, 13*(1), 1–18.
Fabozzi, F. J., Mann, S. V., & Choudhry, M. (2002). *The global money markets*. Hoboken: Wiley.
Faff, R. (2001). A multivariate test of a dual-beta CAPM: Australian evidence. *Financial Review, 36*(4), 157–174.
Faff, R. (2004). A simple test of the Fama and French model using daily data: Australian evidence. *Applied Financial Economics, 14*(2), 83–92.
Fama, E. F. (1970). Efficient capital markets: A review of theory and empirical work. *Journal of Finance, 25*(2), 383–417.
Fama, E. F., & French, K. R. (1992). The cross-section of expected stock returns. *Journal of Finance, 47*(2), 427–465.
Fama, E. F., & French, K. R. (1993). Common risk factors in the returns on stocks and bonds. *Journal of Financial Economics, 33*, 3–56.
Fama, E. F., & French, K. R. (2004, Summer). The capital asset pricing model: Theory and evidence. *Journal of Economic Perspectives, 18*(3), 25–46.
Fama, E. F., & MacBeth, J. D. (1973). Risk, return, and equilibrium: Empirical tests. *Journal of Political Economy, 81*(3), 607–636.
Fletcher, J. (1997, May–June). An examination of the cross-sectional relationship of beta and return: U.K. evidence. *Journal of Economics and Business, 49*(3), 211–221.
Fletcher, J. (2000, Autumn). On the conditional relationship between beta and return in international stock returns. *International Review of Financial Analysis, 9*(3), 235–245.
French, C. W. (2002). Jack Treynor's 'Toward a theory of market value of risky assets'. Available at SSRN: papers.ssrn.com/sol3/papers.cfm?abstract_id=628187.
Friend, I., & Blume, M. E. (1970). Measurement of portfolio performance under uncertainty. *American Economic Review, 60*(4), 561–575.
Gibbons, M. R., Ross, S. A., & Shanken, J. (1989, September). A test of the efficiency of a given portfolio. *Econometrica, 57*(5), 1121–1152.
Gompers, P. A., & Metrick, A. (2001, February). Institutional investors and equity prices. *Quarterly Journal of Economics, 116*, 229–259.
Griffin, J. M. (2002, Summer). Are the Fama and French factors global or country specific? *Review of Financial Studies, 15*(3), 783–803.

Herrera, M. J., & Lockwood, L. J. (1994, September) The size effect in the Mexican stock market. *Journal of Banking and Finance, 18*(4), 621–632.
Heston, S. L., Rouwenhorst, K.G., & Wessels, R. E. (1999). The role of beta and size in the cross-section of European stock returns. *European Financial Management, 5*(1), 9–27.
Hodoshima, J., Garza-Gómez, X., & Kunimura, M. (2000, November–December) Cross-sectional regression analysis of return and beta in Japan. *Journal of Economics and Business, 52*(6), 515–533.
Hou, K., & Moskowitz, T. J. (2005, Autumn) Market frictions, price delay, and the cross-section of expected returns. *Review of Financial Studies, 18*(3), 981–1020.
Jensen, M. C. (1967). The performance of mutual funds in the period 1945–1964. *Journal of Finance, 23*(2), 389–416.
Jorgenson, D., & Fraumeni, B. M. (1989). The accumulation of human and nonhuman capital, 1948–84. In R. E. Lipsey & H. S. Tice (Eds.), *The measurement of saving, investment, and wealth*. Chicago: University of Chicago Press. http://www.nber.org/chapters/c8121.pdf?new_window=1. Accessed 8 Feb 2014.
Jorgenson, D., & Fraumeni, B. M. (1992). The output of the education sector. In Z. Griliches (Ed.), *Output measurement in the service sectors*. Chicago: University of Chicago Press. http://www.nber.org/chapters/c7238.pdf?new_window=1. Accessed 8 Feb 2014.
Kendrick, J. W. (1974) The accounting treatment of human investment and capital. *Review of Income and Wealth, 20*(4), 439–468.
Kendrick, J. W. (1976). Conceptual foundations of the study. In *The formation and stocks of total capital* (pp. 1–21). New York: National Bureau of Economic Research (NBER).
Kendrick, J. W. (1994) Total capital and economic growth. *Atlantic Economic Journal, 22*(1), 1–18.
Lakonishok, J., & Shapiro, A. C. (1986, December). Systematic risk, total risk and size as determinants of stock market returns. *Journal of Banking and Finance, 10*(1), 115–132.
Lakonishok, J., Shleifer, A., & Vishny, R. W. (1994, December) Contrarian investment, extrapolation, and risk. *Journal of Finance, 49*(5), 1541–1578.
Le, T., Gibson, J., & Oxley, L. (2003). Cost- and income-based measure of human capital. *Journal of Economic Surveys, 17*(3), 271–307.
Levy, R. A. (1971, November–December) On the short-term stationarity of beta coefficients. *Finance Analysts Journal, 27*(6), 55–62.
Lhabitant, F. S. (2004). *Hedge funds quantitative insights*. Hoboken: Wiley.
Lintner, J. (1965a). The valuation of risk assets and selection of risky investments in stock portfolios and capital budgets. *Review of Economics and Statistics, 47*(1), 13–37.
Lintner, J. (1965b, December). Security prices, risk, and maximal gains from diversification. *Journal of Finance, 20*(4), 587–615.
Litzenberger, R., & Ramaswamy, K. (1979, June). The effect of personal taxes and dividends on capital asset prices: Theory and empirical evidence. *Journal of Financial Economics, 7*(2), 163–195.
MacKinlay, A. C., & Richardson, M. P. (1991, June). Using generalised method of moments to test mean-variance efficiency. *Journal of Finance, 46*(2), 511–527.
Markowitz, H. M. (1952). Portfolio selection. *Journal of Finance, 7*(1), 77–91.
Markowitz, H. M. (1999, July/August) The early history of portfolio theory: 1600–1960. *Financial Analysts Journal, 55*(4), 5–16.
Merton, R. C. (1972, September). An analytic derivation of the efficient portfolio frontier. *Journal of Financial and Quantitative Analysis, 7*(4), 1851–1872.
Miller, M. H., & Scholes, M. S. (1972). Rates of return in relation to risk: A re-examination of some recent findings. In M. C. Jensen (Ed.), *Studies in the theory of capital markets* (pp. 47–78). New York: Praeger Publishers.
Mossin, J. (1966, October). Equilibrium in a capital asset market. *Econometrica, 34*(4), 768–783.
Pástor, L., & Stambaugh, R. F. (2003). Liquidity risk and expected stock returns. *Journal of Political Economy, 111*(3), 642–685.
Pettengill, G. N., Sundaram, S., & Mathur, I. (1995, March). The conditional relation between beta and returns. *Journal of Financial and Quantitative Analysis, 30*(1), 101–116.

Porter, R. B., & Ezzell, J. R. (1975, October) A note on the predictive ability of beta coefficients. *Journal of Business Research, 3*(4), 365–372.

Reilly, F. K., & Brown, K. C. (1997) *Investment analysis and portfolio management*. Chicago: Dryden Press.

Reinganum, M. R. (1981, December) A new empirical perspective on the CAPM. *Journal of Financial and Quantitative Analysis, 16*(4), 439–462.

Reinganum, M. R. (2013). *2013 Nobel Prize in economic sciences recognizes the important contributions of quant equity*. State Street Global Advisors.

Roenfeldt, R. L., Griepentrog, G. L., & Pflaum, C. C. (1978, March). Further evidence on the stationarity of beta coefficients. *Journal of Financial and Quantitative Analysis, 13*(1), 117–121.

Rogers, P., & Securato, J.R. (2007, November). *Comparative study of CAPM, Fama and French model and reward beta approach in the Brazilian market*. University of Vicosa and University of Sao Paulo.

Roll, R. (1977). A critique of the asset pricing theory's tests. *Journal of Financial Economics, 4*(2), 129–176.

Rosenberg, B., Reid, K., & Lanstein, R. (1985). Persuasive evidence of market inefficiency. *Journal of Portfolio Management, 11*(3), 9–16.

Schulmerich, M. (2012a). *The efficient frontier in modern portfolio theory: Weaknesses and how to overcome them*. State Street Global Advisors. SSgA Capital Insights.

Schulmerich, M. (2012b). *Extending modern portfolio theory: Efficient frontiers for different risk measures*. State Street Global Advisors. SSgA Capital Insights.

Schulmerich, M. (2013a). *Can the Black-Litterman framework improve asset management outcomes?* State Street Global Advisors. SSgA Capital Insights.

Schulmerich, M. (2013b, July/August). The efficient frontier in modern portfolio theory: Weaknesses and how to overcome them. In *Investments & Wealth Monitor* (pp. 27–32). Greenwood Village: IMCA - Investment Management and Consultants Association.

Sharpe, W. F. (1964). Capital asset prices: A theory of market equilibrium under conditions of risk. *Journal of Finance, 19*(3), 425–442.

Spremann, K. (2008). *Portfoliomanagement* (4th ed.). München: Oldenbourg Verlag.

Stambaugh, R. F. (1982). On the exclusion of assets from tests of the two-parameter model. *Journal of Financial Economics, 10*, 237–268.

Stoll, H. R., & Whaley, R. E. (1983). Transaction costs and the small firm effect. *Journal of Financial Economics, 12*(1), 57–79.

Strong, N., & Xu, X. G. (1997, March). Explaining the cross-section of U.K. expected stock returns. *British Accounting Review, 29*(1), 1–23.

Tobin, J. (1958, February). Liquidity preference as behavior towards risk. *Review of Economic Studies, 25*(2), 65–86.

Tole, T. M. (1981, Winter). How to maximize stationarity of beta. *Journal of Portfolio Management, 7*(2), 45–49.

van Dijk, M. A. (2011, September). Is size dead? A review of the size effect in equity returns. *Journal of Banking and Finance*, 1–40. http://papers.ssrn.com/sol3/papers.cfm?abstract_id=879282#. Accessed 3 Jan 2014.

Wang, F., & Xu, Y. (2004, November–December) What determines Chinese stock returns? *Financial Analysts Journal, 60*(6), 65–77.

Stock Market Anomalies 3

3.1 Introduction

Most of the people who studied finance were taught that stock markets are efficient. According to Fama, there are three general forms of efficiency[1]:

- **Weak form:**
  Only historical price data is reflected in today's stock price.
- **Semi-strong form:**
  All publicly available information is reflected in today's stock price.
- **Strong form:**
  All publicly available information plus all insider knowledge is reflected in today's stock price.

The efficient market hypothesis states that financial markets are semi-strong efficient or *informationally efficient*. Consequently, an investor cannot consistently achieve returns in excess of average market returns on a risk-adjusted basis. This is a key part of traditional finance theory (presented in the previous chapter) which assumes that the investor is rational (*Homo economicus*).

However, theoretical implications from MPT and the efficient markets hypothesis do not allow for stock market anomalies that have been increasingly observed since the 1980s. A stock market anomaly is a market situation that cannot be explained by traditional finance theory. To some degree it is a persistent situation and *not* an arbitrage opportunity which, as soon as spotted, disappears because everyone aims to exploit it. A stock market anomaly persists although people trade on it and can make consistent gains.

[1] Fama (1970).

M. Schulmerich et al., *Applied Asset and Risk Management*,
Management for Professionals, DOI 10.1007/978-3-642-55444-5_3

This chapter provides a summary of the most important stock market anomalies that have occurred to date. One can classify them in at least four categories.

The first one is based on fundamentals. In other words, anomalies can be noticed through the study of accounting data and possibly exploited. One example of such anomalies would be the P/E (price-to-earnings) ratio effect.[2] The second kind of stock market anomalies refers to the calendar. Stocks seem to perform differently depending on the time of the year, holidays, end of the month, etc. The third type are structure-related anomalies. For instance, unfair competition, regulations or market transparency can be the origin of exploitable anomalies. The fourth kind are behavior-based anomalies. For example, economic agents like arbitrageurs or other investment experts may generate trading patterns that affect the market and can thus be exploited.

The aim of this chapter is to provide the reader an overview of the most important market anomalies:

- Weekend effect (Sect. 3.2)
- January effect (Sect. 3.3)
- Turn-of-the-month effect and holiday effect (Sect. 3.4)
- S&P 500 effect (Sect. 3.5)
- Trading by insiders (Sect. 3.6)
- Momentum of industry portfolio (Sect. 3.7)
- Home bias (Sect. 3.8)
- Value Line enigma (Sect. 3.9)
- Expiry of IPO lockups (Sect. 3.10)

Each section describes the anomaly and supplies evidence and explanations on the issue when available. For a summary of stock market anomalies see Schulmerich (2014).

3.2 Weekend Effect

The weekend effect is one of the oldest existing stock market anomalies, given that it has existed for many decades. In this section, we define the anomaly, present the evidence and discuss its persistence.

[2]Basu has shown that low P/E stocks tend to outperform both the market and high P/E stocks. In *What Works on Wall Street*, O'Shaughnessy found that the P/E ratio is particularly relevant for large stocks. However, he argued that the price-to-sales ratio is an even better indicator of excessive returns. Fama and French find that market and size factors in earnings help explain the P/E ratio effect. See O'Shaughnessy (1998, p. 16), Basu (1977) and Fama and French (1995).

3.2.1 Description

The weekend effect can be defined as a Friday's return minus the following Monday's return for a single security or a portfolio of securities.[3] Under normal conditions, there should be no substantial difference between each day of the week through a long time span. However, in 1980, French analyzed daily returns of stocks for the period 1953–1977 and found that there is a tendency for returns to be negative on Mondays whereas they are positive on the other days of the week.[4] He writes that these negative returns are *"caused only by the weekend effect and not by a general closed-market effect."* A general closed-market effect would mean that most of the market agents would sell on Friday and buy on Monday. However, it seems that abnormal positive stock returns occur on Friday and the contrary happens on Mondays. So buying regularly on Monday and selling on Friday may bring abnormal returns.

3.2.2 Evidence

In Table 3.1, all the common shares traded on the NYSE, the AMEX and NASDAQ have been regrouped and divided into two categories. The first category represents the stocks traded on the equally-weighted (EW) index, i.e., each stock has the same weight in the index. The second category includes the stocks traded on the value-weighted (VW) index where the weight for each share is proportional to the capitalization of the company relative to the capitalization of the whole traded index. The mean daily returns in percent of these shares have been computed for the period starting from July 1962 to July 2001.

The weekend effect reaches 0.34 % if and only if the index is equally-weighted. The difference of returns among the weekdays from July 1962 to July 2001 can be seen at a glance in the chart in Fig. 3.1.

It is evident that there is a performance shift between Fridays and Mondays which leads to abnormal returns. It is important to note that the difference in returns through the week is fairly low, namely 0.33 %, yet considerable in comparison with other week days. Researchers also found that the weekend effect changed over time to such a point that during the period 1990–2001 there were negative returns in an equally-weighted index. They also observed that the effect was strengthened during long weekends while it was weakened when investors could trade on Saturdays.[5]

The discovery of the weekend effect lead Kamara to study the S&P 500 from 1962 to 1993.[6] He found no significant Monday effect after April 1982 except for a portfolio of smaller U.S. stocks where the Monday effect remained undiminished

[3]Chen and Singal (2003, p. 80).

[4]French (1980, p. 56).

[5]Singal (2006, p. 45).

[6]Kamara (1997).

Table 3.1 Daily average return in percent of common stocks on the NYSE, AMEX and NASDAQ from July 1962 to July 2001

	Monday	Tuesday	Wednesday	Thursday	Friday	Weekend (Fri.–Mon.)
Overall period						
1962–2001						
EW	−0.093	0.000	0.133	0.125	0.246	0.339
VW	−0.055	0.044	0.099	0.047	0.098	0.153
By decades						
1962–1970						
EW	−0.105	−0.008	0.176	0.074	0.218	0.326
VW	−0.124	0.022	0.145	0.028	0.131	0.255
1971–1980						
EW	−0.082	−0.019	0.112	0.115	0.245	0.327
VW	−0.100	0.035	0.098	0.049	0.111	0.211
1981–1990						
EW	−0.173	−0.038	0.108	0.123	0.231	0.403
VW	−0.078	0.062	0.112	0.049	0.109	0.187
1991–2001						
EW	−0.021	0.056	0.144	0.174	0.283	0.304
VW	0.063	0.053	0.057	0.057	0.050	−0.013

Source: Singal (2006, p. 42)

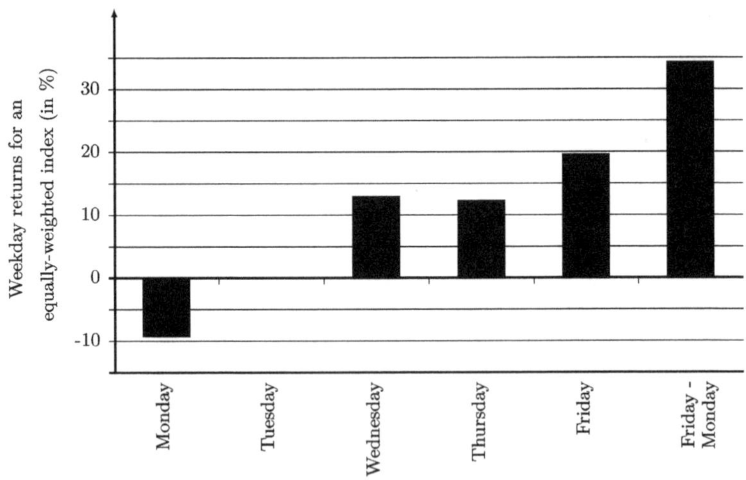

Fig. 3.1 Weekday returns in percent for an equally-weighted index over the 1962–2001 period. *Source*: Singal (2006, Fig. 3.1, p. 44)

until 1993. Apparently, the weekend effect was tightly linked to small capitalization and has a higher impact on equally-weighted indices than on value-weighted indices.

The following research focused on the international scale. Did the weekend effect occur outside U.S. markets? In 1994, Agrawal and Tandon found significant negative returns on Mondays in nine countries and on Tuesdays in eight countries, while large and positive returns could be observed on Fridays in 17 of the 18 countries studied.[7] Unfortunately, their data did not extend beyond 1987.

A quite interesting study was conducted in 2001: Steeley noticed that the weekend effect in the U.K. had disappeared in the 1990s,[8] while in 2010 Benjamin Liu and Bin Li could still note its existence on the Australian stock exchange, yet with varying strength among shares and industries.[9]

3.2.3 Explanations

It was only in 2003 that Chen and Singal provided a satisfying explanation for this long-lasting anomaly which seemed to curiously disappear in the 1990s in a few developed countries. The weekend effect can be mostly attributed to short sellers.[10] There are two kinds of short selling: hedging and speculation. In the first case, short selling is used to offset an existing position, resulting in a neutral position. On the other hand, speculative short selling consists in selling an asset without having it, and then re-buying it later. The difference generates the profit or loss. The key aspect is to sell something before owning it.

To do so, investors borrow the asset and keep it as long as the initial owner does not want it back. This speculative strategy is quite risky because if the initial owner reclaims his asset at a time when the price is not beneficial to the short seller, he has to buy back the sold asset so as to then give it back to the initial owner. He may thus incur a loss. This happened before organized derivatives markets existed, and happens also now. The second drawback of short selling is that the expected loss is virtually unlimited. This is even more true for futures and options, which are financial products whose prices and returns are derived from assets like stocks. For example, if a stock costs $20, and drops to $19, the loss is not overwhelming. But the loss for the derivative may amount to, say, $1,000. Moreover, while the maximum loss of the share is $20 (the share has then no value), the loss on the derivatives market could reach $20,000, while only $2,000 were initially invested.

After these remarks on short selling, we can focus on Chen and Singal's explanations. Short sellers go for a quick profit and prefer not to hold their position too long, because of the high volatility and the cost of borrowing. Taking a long

[7] Agrawal and Tandon (1994, p. 101)

[8] Steeley (2001).

[9] Liu and Li (2010).

[10] Chen and Singal (2003).

and a short position, namely in derivatives, is costly. If these operations happen too frequently the profit suffers. A single day strategy would be too short for short sellers, as a daily renewal of their positions would be too costly. However, weekends represent close to 50 h without trading. They are a landmark for buying back what was short sold at the beginning of the week. A buy order on a stock is then executed on a Friday to settle the short sell order of a Monday. An increase in asset prices is then observed on stock exchanges on Fridays and a decrease is observed on Mondays. The same effect occurs before and after holidays.[11]

Since options are easily available and less expensive for large stocks, the weekend effect for the value-weighted index began to disappear in recent years. Researchers noticed these changes after the introduction of organized option markets. On the other hand, for an average stock, options are either non-existent or too expensive to trade. That is why the weekend effect for the equally-weighted index remained fairly unchanged through time.[12] Furthermore, the weekend effect tends to be stronger if institutional investors engage in the market.[13]

Additional explanations emerged from other researchers:

- Measurement errors
- Specialist-related biases in prices
- Timing of corporate releases after the stock exchanges close on Fridays
- Reduced institutional trading and greater individual trading on Mondays
- Daylight saving time changes for two weekends of the year
- Delay in the settlement of trades and bid-ask bounce[14]

3.2.4 Persistence

The weekend effect was observed for a wide spectrum of assets, indices and time spans. It has been assumed that the timing of news releases was the main reason for the observed abnormal returns on worldwide indices. However, news releases are not restricted to a given type of firm, and explain no more than 3.4 % of the weekend effect.[15] So the timing of the delivery of bad news may not be a sufficient reason to explain this anomaly. Daylight saving time changes did not seem to impact the weekend effect in 1989,[16] neither did they later between 2001 and 2010.[17] The

[11]Singal (2006, p. 48).

[12]Singal (2006, p. 48).

[13]Sias and Starks (1995, p. 66).

[14]The bid-ask bounce is the process that on Fridays, the asset is traded at Friday's ask price at the close of trading, whereas on Mondays, it trades at Fridays' bid price at the start of the trading session.

[15]Damodaran (1989, p. 607).

[16]Damodaran (1989, p. 616).

[17]Patel (2012, p. 109).

delay in settlement is assumed to represent around 17 % of the weekend effect, while the bid-ask bounce would explain 32 and 10 % of the observed market anomaly in an equally-weighted and value-weighted index, respectively.[18]

The effect can still be noticed nowadays. It has lasted so long for three major reasons:

- It was not understood before 2003.
- It is tightly linked to specific index and share types.
- Its magnitude is small.

3.2.5 Summary

Before 2003, the reasons of the weekend effect were hardly understood. It could thus disappear without notice. It was also impossible to optimize and refine a strategy to capture the highest available profit. This anomaly is close to non-existent in a value-weighted index, and it is hard to benefit from an equally-weighted index. Illiquid markets for small-caps and high trading costs can make it prohibitive to surf on the weekend effect. It is even more complicated to generate a profit since the average returns do not exceed 0.33 %.[19] That is why the best an investor can do would be to just buy stocks on Mondays and sell them on Fridays.

3.3 January Effect

The January effect is a particularly interesting anomaly because it has not disappeared, despite being well known since the mid-1970s. According to arbitrage theory, any anomaly should disappear as traders attempt to exploit it. How was it discovered and what are its properties?

3.3.1 Description and Evidence

The January effect is an abnormal return on a given set of stocks achieved in January compared to other months of the year. In 1976, Rozeff and Kinney discovered that abnormal higher returns were achieved during the first days of January as compared to other months of the year.[20] They found that the average return on the New York Stock Exchange small-caps between 1904 and 1974 reached 3.48 % during this

[18]Singal (2006, p. 47).

[19]Singal (2006, p. 42).

[20]Rozeff and Kinney, Jr. (1976, p. 349).

month versus 0.42 % during other months.[21] In 1983, Donald Keim[22] who was then a graduate student at the University of Chicago, observed that stock prices increase during the month of January in almost every year.[23] Furthermore, he found that such abnormal returns could not be traced back to price information. This contradicts the efficient market hypothesis (EMH).

This anomaly has persisted through time, notably in more recent years. Bhardwaj and Brooks as well as Eleswarapu and Reinganum proved its occurrence in the periods 1977–1986 and 1961–1990, respectively.[24] These findings allowed Robert Haugen and Philippe Jorion to note that *"the January effect is, perhaps, the best-known example of anomalous behavior in security markets throughout the world."*[25]

Additional properties of the January effect were unveiled by Chang and Pinegar in 1986. They actually discovered that it also occurred in bonds market.[26] In 1998, Maxwell additionally showed that this anomaly is strong for non-investment grade bonds, but not for investment grade bonds.[27]

3.3.2 Explanations

According to this and further research, the January effect is attributed to the rebound of stocks after the year-end tax selling period. Actually, stocks depressed near year-end are more likely to be sold for tax purposes. On an accounting basis, any stock which lost value during the former year yields a tax credit and the investor has less to pay to the state for the past or future period, according to accounting standards. Yet, Chen and Singal,[28] among others, have also identified a *December Effect*, which seems to stem from the requirement that many funds report holdings at this time of the year as well as from investors who buy in advance of potential January increases.[29] Fund managers have to follow investment objectives. Given

[21] Rozeff and Kinney, Jr. (1976, p. 349).

[22] Donald Keim received his Ph.D. from the University of Chicago in 1983. He is very well known and widely cited for the discovery of the January effect. Currently, he is teaching at Wharton University as John B. Neff Professor of Finance.

[23] Keim (1983).

[24] Bhardwaj and Brooks (1992) and Eleswarapu and Reinganum (1993).

[25] Haugen and Jorion (1996, p. 27).

[26] Chang and Pinegar (1986).

[27] Investment grade bonds are bonds audited by rating agencies like Moody's or Fitch Rating. An investment grade is any rate between AAA and BBB- or Aaa and Baa. Non-investment grade (also known as *junk*) spans from BB (or Ba) to D (default). They help to evaluate the default risk associated with a bond and are used by investors to assess the credit worthiness of a corporate or sovereign bond. See Maxwell (1998).

[28] Honghui Chen is an assistant professor at the University of Central Florida, Orlando. Vijay Singal, CFA, is J. Gray Professor of Finance at Pamplin College of Business, Virginia Tech, Blacksburg.

[29] Chen and Singal (2003).

their aggregated weight and knowledge of the global direction of selected stocks thanks to fundamental or quantitative analysis, they can have an observable impact on the market as a whole. Finally, Bhabra, Dhillon, and Ramirez (1999) document a *November Effect*,[30] which has occurred only after the Tax Reform Act of 1986. They also found that the January effect has become stronger since then. Taken together, their results point toward a tax-loss selling explanation of the January effect.

3.3.3 Persistence

The January effect persists because it is not possible to arbitrage the anomaly. It is also important to note that the stocks affected by the January effect are rather small. It reaches only the bottom 20 % of all stocks that trade on organized stock exchanges and NASDAQ: From the overall 6,500 stocks, the anomaly concerns only 1,300 stocks with a median market capitalization of about $25 mn.[31] Because of high trading costs, investors will not be able to benefit from this anomaly. Even index futures, options, and mutual funds are useless for diminishing trading costs, because their capitalization size is too small to take benefit of it. On the other end, the December and November effects are not as clearly understood. That is why experts estimate that these effects should disappear through time.[32] Meanwhile, buying stocks at the beginning of December or 6 days before year end and selling them on the last trading day of the year could generate consistent abnormal returns.

In addition to the January effect, a comparable anomaly occurs at each end of the month: *the turn-of-the-month effect*.

3.4 Turn-of-the-Month and Holiday Effect

The turn-of-the-month effect and the holiday effect are two typical seasonal anomalies. They are explained here together because they share similar properties.

3.4.1 Description

The turn-of-the-month effect is a typical seasonal stock market anomaly. Studies have shown that stocks offer higher returns on the last and first days of every month relative to the other days. This effect is called *turn-of-the-month effect* and has been well documented over time and across countries. Depending on researchers, the turn of the month is defined as the three to five trading days at the end of the month and

[30]Bhabra et al. (1999).

[31]Singal (2006, p. 33).

[32]Singal (2006, p. 37).

at the beginning of the next month.[33] The holiday effect is similar in that returns are on average higher on the day before a holiday, compared to other trading days.[34]

3.4.2 Evidence

Lakonishok and Smidt (1988) shows that U.S. stock returns are significantly higher during the turn-of-the-month period.[35] Ariel (1987) points out that returns tend to be higher on the last day of the month.[36]

In 1991, Ziemba discovered a turn-of-the-month effect for Japan when the turn of the month is defined as the last five and the first two trading days of the month.[37] Hensel and Ziemba found that returns at the turn of the month consistently and significantly exceeded averages during the period 1928–1993 and declared that *"the total return from the S&P 500 over this sixty-five-year period was received mostly during the turn of the month."*[38] Two years later, in 1998, Kunkel and Compton unveiled a turn-of-the-month effect on the S&P 500 Index but for a different time span.[39] Both studies underline how abnormal returns can be earned by exploiting this anomaly. However, in 1992, Cadsby and Ratner detected similar turn-of-the-month effects in some countries like Canada, the U.K., Australia and Switzerland, but they also discovered that it does not occur in other countries.[40] More recently, Frank Russell Company examined the returns of the S&P 500 over a 65-year period. They also found that U.S. large-cap stocks consistently generate higher returns at the turn of the month.[41]

A second very famous seasonal anomaly is the holiday effect. A range of researchers provided evidence that returns are on average higher on the day before a holiday than on other trading days.[42] This effect is present across major U.S. stock exchanges, namely the NYSE, the AMEX, and NASDAQ.[43] It is also noticeable in other parts of the world like Japan and the United Kingdom.[44] Brockman and

[33]Kunkel and Compton (1998, p. 207), Ziemba (1991, p. 119) and Hensel and Ziemba (1996, p. 17).

[34]Scott (2003).

[35]Lakonishok and Smidt (1988).

[36]Ariel (1987).

[37]Ziemba (1991).

[38]Hensel and Ziemba (1996, p. 21).

[39]Kunkel and Compton (1998).

[40]Cadsby and Ratner (1992).

[41]Russell Investment Group website: http://www.russell.com/us/education_center. See also Gonzalez (1996).

[42]Lakonishok and Smidt (1988), Ariel (1990), Cadsby and Ratner (1992).

[43]The acronyms stand for New York Stock Exchange, American Stock Exchange and National Association of Securities Dealers Automated Quotations.

[44]Kim and Park (1994).

Michayluk (1998) add that the pre-holiday effect is one of the oldest and most consistent of all seasonal irregularities.[45]

3.4.3 Explanations

As for now, a set of explanations are proposed for both anomalies. The most plausible one for the turn-of-the-month effect appears to be related to the information delivery process. It seems that these anomalies arise from clustered information on the macroeconomic level. Given that this information is systematically released at specific known dates, market agents and investors seem to ask for a higher risk premium on these dates. On a risk-adjusted basis, the anomaly seems to disappear.

A similar explanation is sometimes given for the holiday effect. However, some researchers also believe that the holiday effect is due to end-of-month cash flows (salaries, mortgages, credit cards, etc.).[46] Singal underlines that part of this effect is also due to short sellers,[47] as short sellers close their position by buying back the stock they have previously short sold.

3.4.4 Persistence

Unfortunately, there is no consensus regarding the explanation of the turn-of-the-month effect or the holiday effect. However, assuming that the effect arises from information delivery, it seems impossible to hedge against or to take advantage of these anomalies unless one can benefit from information which has not been publicly released. But if the effect is assumed to be due to cash flow delivery or even to short selling, then one can buy the stock slightly before the end of the month or the holidays' starting day.

Stock market anomalies are not only seasonal or periodic. They can be structural as in the case of the S&P 500 Index effect.

3.5 S&P 500 Index Effect

The S&P 500 effect is also known as the S&P game. It is quite easy to recognize but less easy to explain. In the following part, we will briefly present the index and the effect linked to it. Then, we provide a series of evidence followed by a list of possible explanations of this conspicuous anomaly which has persisted with its present characteristics.

[45] Brockman and Michayluk (1998, p. 205).

[46] Russel and Torbey (2002).

[47] Singal (2006, p. 47).

3.5.1 Description

Through the 1970s, new indices started to emerge and were increasingly used as benchmarks for trading. Yet, much earlier, in 1957, the S&P 500 Index was founded. This index is a portfolio that tracks five hundred top U.S. shares. The included firms are supposed to represent the leading industries in the U.S. economy, and the index is believed to be a good indicator of the country's market performance. The adjusted market capitalization as of September 2013 was about $15 tn. The included stocks can reach a market capitalization as high as $401.73 bn., and as low as $1.99 bn., for an average of $30.35 bn.[48]

Stocks can be added or withdrawn from the index based on the four following criteria:

- The ownership of the firm must not lie in the hands of only a few shareholders.
- The firm is a market leader with a large market share.
- The firm is profitable and is expected to continue to be so in the long run.
- The stock is traded in sufficient volumes and has a high liquidity on the stock exchange.

But there are many more companies that fulfill these criteria than can be included in the index. As a result, a necessary subjectivity in the process of selecting the index constituents makes the outcome unpredictable. However, market observers keep trying to identify the happy few who are qualified to enter the index. For example, Lehman Brothers, a former leading investment bank in the United States before the subprime crisis, identified nineteen candidates to be included versus ten stocks to be excluded in 2002. It turned out that four of the identified companies were included and two of them were excluded.[49] In any case, on the day of addition/deletion, the S&P 500 Index is always composed of exactly 500 stocks. So, if two stocks are deleted from the index, two new companies' shares are included.

As previously said, the index is used as a benchmark, and fund managers and other investors who are willing to track the index as closely as possible will adapt their portfolio in the case of the addition or deletion of a new stock. To prevent any excess volatility, starting from 1989, such an event will be announced a week before it takes effect.

Now, we can deal with the S&P 500 effect. In 1986, Harris, Gurel and Schleifer found an abnormal increase in stock prices that could reach 3 % because of the announcement of a stock's inclusion into the S&P 500 Index.[50] Their discovery triggered a wide range of research trying to define and explain this observation.

Today, the S&P 500 Index reflects the fact that a newly added stock will yield an abnormal return following its listing. This effect will vanish within a few weeks or

---

[48] *Source*: http://www.spindices.com/indices/equity/sp-500.

[49] Singal (2006, p. 165).

[50] Harris and Gurel (1986, p. 815) and Shleifer (1986, p. 583).

at most a month. A deleted stock will first drop in value, but including an already listed stock into an index does not seem to reveal information. However, it does have an impact on its return and volatility. Therefore it is an anomaly which runs against the idea of the efficient market hypothesis.

3.5.2 Evidence

To be deleted or added to the S&P 500 Index has a significant impact on a company as can be seen in Tables 3.2 and 3.3. Table 3.2 represents the price impact of additions to the S&P 500 Index from 1962 to 2000. Table 3.3 is its equivalent in the case of a deletion. Note that the number of companies included into this sample is significantly smaller than for additions. This is due to the fact that only stocks that were removed from the index by Standard and Poor's are taken into account. Bankruptcy, mergers or acquisitions are not considered so as to minimize the sample selection bias. Also note that AD is an acronym for *announcement date* and ED stands for *effective date*.

Table 3.2 shows different time periods. The first period runs from 1962 to 1976. Until 1976, as noticed before, indices were not a popular tool. Hence, Standard and Poor's made no public announcement. The second period runs from 1976 to 1989. Starting 1976, Standard and Poor's announced changes in their index after market close on Wednesdays, a day before the changes became effective. However, these

Table 3.2 Price impact of additions to the S&P 500 Index from 1962 to 2000

Period	Total addition	Additions in the sample	Abnormal return on AD + 1 (%)	Abnormal return to AD + 1 to ED (%)	Abnormal return to 20 days after ED (%)	Abnormal return to 60 days after ED (%)
7/1962–8/1976	304	285	0.0	N/A	−0.5	0.9
9/1976–9/1989	297	274	3.0	N/A	2.9	3.5
1989–2000	278	224	5.3	8.4	5.6	4.5
1990	13	11	3.1	6.7	3.1	−3.8
1991	13	9	5.8	8.2	4.5	0.3
1992	7	6	4.6	6.4	6.6	2.3
1993	13	9	4.7	7.3	5.6	6.3
1994	18	16	2.2	4.7	1.2	0.3
1995	32	21	4.3	7.2	3.0	2.5
1996	27	20	3.8	7.5	2.9	3.8
1997	28	24	8.1	10.5	7.5	5.8
1998	48	37	5.5	9.3	3.5	0.5
1999	41	38	5.8	8.5	6	7.8
2000	33	29	7.1	10.7	13.5	12.5

Source: Singal (2006, Table 8.1, p. 166)

Table 3.3 Price impact of deletions from the S&P 500 Index from 1962 to 2000

Period	Total addition	Deletions in the sample	Abnormal return on AD + 1 (%)	Abnormal return on AD + 1 to ED (%)	Abnormal return to 20 days after ED (%)	Abnormal return to 60 days after ED (%)
7/1962–8/1976	304	170	−0.3		1.6	3.5
9/1976–9/1989	297	61	−1.6		−3.8	−2.9
1989–2000	278	88	−5.4	−10.3	−3.3	2.6
1990	13	5	−1.6	−4.6	−9.1	43.7
1991	13	4	−19.3	−19.3	20.1	21.4
1992	7	5	−10.2	−32.4	6.0	8.2
1993	13	6	−2.3	−7.2	−4.9	3.3
1994	18	10	−2.8	−5.4	1.4	2.3
1995	32	11	−5.8	−15.7	−10.7	−16.7
1996	27	13	−4.0	−7.4	−0.1	3.0
1997	28	4	−5.2	−7.4	4.9	22.5
1998	48	8	−7.3	−10.5	−12.1	−2.8
1999	42	9	−3.2	−3.3	2.8	4.4
2000	32	12	−5.9	−10.9	−11.9	−7.7

Source: Singal (2006, Table 8.2, p. 167)

notifications were only given to interested investors, for example, fund managers, and were not entirely public information.

Index management became more and more popular. The research on the benefits of indices created a growing interest from institutional and private investors. As a result, on the D-days of introduction or withdrawal, abnormal volatilities, volumes and returns were observed for the newly included or deleted stocks of the S&P 500 Index. That is why, in 1989, the company in charge of this index started to make announcements a week before the effective date of a new addition or deletion. This procedure has not been changed as of March 2014.

Stocks that are added to the S&P 500 are already present on a stock exchange, like the AMEX, NASDAQ or even the NYSE. All public information is already available. Given that apparently no information is added through the listing or delisting process, no abnormal return should be witnessed. However, a clear pattern has emerged which is very different in the case of an addition or a deletion. Let us first analyze Table 3.2, which shows the addition effect on the price.

In the years 1962–1976, in the event of an addition, abnormal returns oscillate between −0.5 and 0.9 % after 20 and 60 days following the introduction, respectively. The day following the announcement, and thus the effective date of the inclusion, is null, in statistical terms it is a non-event. From 1976 to 1989, a shift seems to appear. An abnormal positive return can be observed, varying from 2.9 to 3.5 % for 20 and 60 days after the effective date. A single day after the announcement date, an investor could generate a substantial 3.0 % return. The pattern intensifies from 1989 to 2000 with a 5.6 versus 4.5 % abnormal return after

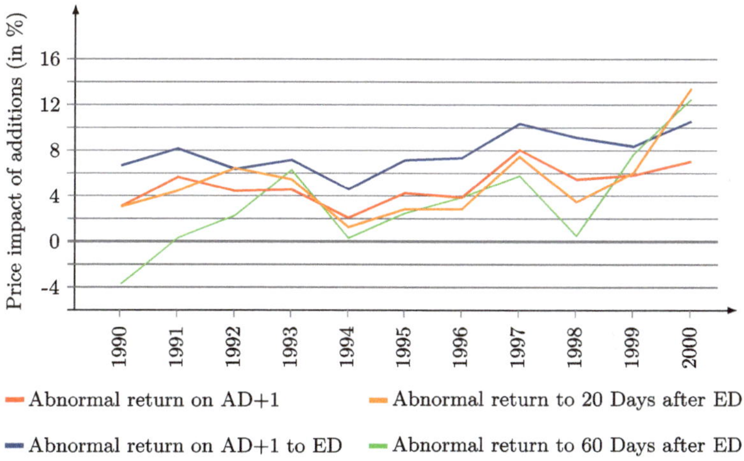

Fig. 3.2 Impact of additions from the S&P 500 Index from 1989 to 2000. *Source*: Own, based on Singal (2006, p. 166)

20 and 60 days following the introduction. Such a figure is even more outstanding if one focuses on the day following the announcement date: It offered a return of 5.3 %. This could sky rocket to 8.4 % if the position was held until the effective inclusion date.

Figure 3.2 also shows that the effect strengthened over time. It can also be observed that in 2000, most abnormal returns were generated 1 day after the announcement date.

In Table 3.3, we analyze the effect of a deletion. In the years 1962–1976, in the event of a deletion, abnormal returns of 1.6 and 3.5 % occur after 20 and 60 days following the effective date. During the period 1976–1989, returns are negative and amount to −3.8 and −2.9 %, respectively. Even more intriguing, from 1989 to 2000, one could expect a −3.3 % abnormal return 20 days after the effective date and yet get a positive 2.6 % return 40 days later.

How is this possible? Generally speaking, on the day following the announcement date, a significant loss occurs and widens from −0.3 in the 1962–1976 period to −5.5 % in the 1989–2000 period. However, Fig. 3.3 below which represents the impact of deletions from the S&P 500 Index from 1989 to 2000, shows that approaching the 2000s, the effect seems to contract around −5.0 to −10.0 %.

It took years for financial researchers to find explanations for these observations made at various times,[51] yet none of them is fully satisfying. We will look at them in the next section.

[51] Arbel (1985, p. 4) and Chen, Noronha, and Singal (2003, pp. 1901–1902).

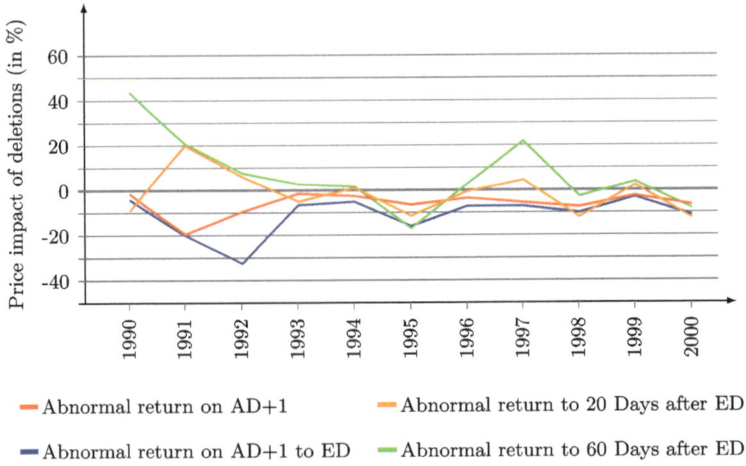

Fig. 3.3 Impact of deletions from the S&P 500 Index from 1989 to 2000. *Source*: Own, based on Singal (2006, p. 167)

3.5.3 Explanations

As for now, five possible explanations co-exist to explain the S&P 500 effect:

- Certification
- Imperfect substitutes
- Liquidity improvement
- Price pressure
- Recognition from investors

3.5.3.1 Certification

The certification explanation proposes that the addition of a stock to the index produces additional information about the firm (i.e., expected long-term profitability),[52] even though Standard and Poor's analysis is based on all publicly available information. According to this hypothesis, the listing or delisting of a stock should lead to positive or negative price movements, respectively. This was definitely the case in recent years. However, in the period 1964–1976, stocks were introduced and withdrawn, yet with no significant effect on the market prices. Furthermore, the certification hypothesis cannot explain the temporary effect of the price drop.[53] If it held true, the negative effect should be permanent since the information about, say, negative expectations about the long-term profitability would not vanish in a matter of weeks.

[52]Denis, McConnell, Ovtchinnikov, and Yu (2003, p. 52).

[53]Shleifer (1986, p. 579).

3.5.3.2 Imperfect Substitutes

The imperfect substitutes hypothesis is a second explanation. First, it assumes that a listing on the S&P 500 does not create any additional information. If it were created, then prices should have varied already before 1976, which was not the case. Moreover, Standard and Poor's bases its analysis on publicly available information. At best, a re-release of information would bring information to the market. Such an effect occurs when the Wall Street Journal publishes its column *Insider Trading Spotlight* but has yet to be proven for the re-composition of the S&P 500.[54] Liquidity improvement is not significant as will be discussed. So the single reason left is the shock in demand created by indexers, i.e., fund managers who replicate an index. They create an upward price shift in the case of an addition and a price drop in the case of a deletion. The key point is the following:

In theory, a firm's share represents very little in comparison to the market as a whole. Hence, it is assumed under the efficient market hypothesis that all shares are perfect substitutes between each other. So, in theory, buying Boeing, Airbus or Lockheed Martin would make no difference, since these companies are all in the aeronautics and defense industry. If a stock rises more than its competitors, then a short sell followed by a buy order on one of the competitors would nullify the abnormal return. However, assuming stocks have imperfect substitutes, the arbitrageurs can profit from the situation and generate abnormal returns. This can be an explanation for the observations made in Tables 3.2 and 3.3.

But this hypothesis, while being a best-fit, is contradicted by two facts. The first one is that no relation so far has been found between indexing and price impact. If the imperfect substitutes hypothesis is true, the greater the demand for the introduced stock, the greater the price impact. But this is not the case.[55] A second issue is that the price loss through deletion should be as permanent as it seems to be for an addition. This is against observations, therefore, another explanation is required.

3.5.3.3 Liquidity Improvement

A third explanation is liquidity improvement. The S&P 500 shares benefit from a very liquid market. Hedge and McDermott observe an improved liquidity for added stocks primarily due to lower transaction costs and an improvement in information flow around the announcement date.[56] For deleted stocks, the liquidity declines over 3 months on average. However, in the meantime, the number of shares available for trading may negatively impact the liquidity of the shares and very few studies show a permanent increase in liquidity.[57] Most of the studies report that liquidity increases or decreases vanish within days in the case of an introduction, and within months in the case of a deletion. This explanation is thus not sufficient to explain the anomaly.

---

[54] Chang and Suk (1998).

[55] Wurgler and Zhuravskaya (2002, p. 583).

[56] Hegde and McDermott (2003).

[57] Singal (2006, p. 171).

3.5.3.4 Price Pressure

A fourth explanation is the price pressure hypothesis. It also relies on imperfect substitutes but focuses more on short-term price changes. Under this hypothesis, adding a new stock to the index generates a burst in demand for the stock. This generates an upward price pressure in the case of an addition, and a downward pressure in the case of a deletion. In the case of an introduction, as soon as the demand is satisfied, the temporary abnormal return reaches a stable new level. This fits with observations stated in Table 3.2 starting from 1989. Kaul, Mehrotra and Morck found equivalent results for the Toronto Stock Exchange in 1996 with no long-term reversal.[58] Harris and Gurel's observations also fit with the hypothesis.[59] However, in their sample which preceded 1986, it seems that the stock that appreciated after an introduction loses its abnormal return in a matter of 2 weeks. It thus seems that a significant shift did occur around 1989, so that the price pressure hypothesis does not hold. The price drift should disappear with time passing by, but this has not been the case anymore starting from 1989 as stated in Table 3.2 and subsequent research.

3.5.3.5 Recognition from Investors

A fifth and last explanation is proposed which is known as investor recognition.[60] The argument is that newly introduced stocks get a bigger visibility, and deleted ones progressively lose it. A firm with higher visibility may get better access to capital markets, even if the stock has already been introduced years ago. Financial analysts and observers will monitor the company much closer. Mechanically, this visibility is also granting them greater access to capital markets since the index is used as a benchmark for many funds. Getting access to the index also grants the newly indexed company access to indirect investors.

As a result, larger projects can be started. Financial institutions may be more willing to lend at lower costs, which leads to a higher debt capacity. The market value of the company, on a fundamental basis, should then see its value increased. On the other hand, the deleted stocks cannot become suddenly unknown to investors. The investor recognition hypothesis therefore assumes an asymmetric effect between the addition and the deletion of a stock. Before 1976, Standard and Poor's did not make any announcement about an introduction or a deletion, and stock prices were not impacted. Starting from 1976, selective announcements were made and price changes occurred with abnormal returns of 3 %. Since 1989, the announcements take place at least a week before taking effect, and the information is widely and publicly delivered.[61] A correlation with a higher abnormal return can be observed in the following decade in Tables 3.2 and 3.3.

---

[58] Kaul, Mehrotra, and Morck (2000).

[59] Harris and Gurel (1986).

[60] Singal (2006, p. 172).

[61] Singal (2006, pp. 171–173).

However, this explanation is tightly linked with another anomaly: the neglected or small firm effect. It states that small-sized firms or firms poorly monitored by outside observers tend to perform better. Newly introduced stocks get known and the stock price moves upward. Newly deleted firms become less well known and are temporarily undervalued. Yet, because they tend to be monitored, an upward movement of the stock price is realized in the following months.

Out of these five possible explanations for the price impact of index changes, the investor recognition hypothesis seems to be the most consistent one. But in the case of a deletion of a stock, this hypothesis does not provide an easy explanation. In this case, the imperfect substitute hypothesis needs to be considered. Let us now focus on the persistence of the S&P 500 effect.

3.5.4 Persistence

The index effect has to be divided into two parts: a permanent and a temporary part. The permanent part seems to be explained thanks to the hypotheses presented above, namely the investor recognition and imperfect substitute hypotheses. Therefore, this part is not exploitable, yet the temporary part is.

When the addition of a stock occurs, both parts co-exist. Starting from the day of the announcement that the stock is going to be included in the index, the permanent S&P 500 Index effect becomes evident. However, the abnormal return generated between the announcement date and the effective date is a temporary effect that is meant to disappear. Standard and Poor's confirms this observation: *"a stock being added to the index would rise about 8.5% between announcement and implementation dates."*[62]

Only the temporary effect is visible when a stock is deleted. All of the abnormal return generated during the period following the effective date of deletion seems to evaporate. This observation is extracted from Fig. 3.3 and corroborates Standard and Poor's year 2000 report: *"The average market decline across all the deletions since 1998, a total of 53, was 11.7%. On average, this decline was nearly fully reversed by the sixth trading day."*[63]

The issue then is to understand the persistence of the temporary effect. Why has it not disappeared after more than 20 years of existence? There are two interconnected reasons for this:

- The index fund evaluation process of index fund managers.
- The announcement of an upcoming re-composition of the index which is exploited by arbitrageurs.

[62]Bos (2000).

[63]Dash (2002).

The evaluation of index fund managers includes the assessment of their ability to reduce tracking error.[64] As seen in Sect. 1.3.2, tracking error can be used to measure how close the passive portfolio manager tracks the actual index. Typically, tracking error shall not exceed 0.1 % for an index tracking portfolio. Therefore, fund managers are likely to buy or sell stocks precisely on the effective date of introduction. This is one part of the game.

But in this scenario, arbitrageurs also play their role. They know how fund managers are evaluated, and in the case of an introduction in the index, they buy the stock around the announcement day and sell it in the days following the effective date. They sell massively and generate an abnormal profit because the demand of the fund managers is large enough to absorb the selling effect. In the case of a deletion, the profit is even more interesting. Arbitrageurs believe in the hypothesis of the imperfect substitute. They short sell the stock to be deleted on the announcement date. The stock price is driven down until the effective date when they repurchase the stock at a lower price to give it back to fund managers. An abnormal profit is generated. Later on, because the price movements are not based on additional information, the stock price is likely to reach its former level.

3.5.5 Summary

To sum up, a listing on the S&P 500 Index is a major event for a firm. It becomes more visible and gains a potentially better access to capital markets. It will also attract investments from fund managers tracking the index. This may increase the liquidity of the stock. Starting from 1989, the investor behavior regarding S&P 500 Index introductions and withdrawals changed. This correlates with Standard and Poor's decision to introduce longer announcement periods in order to limit the demand shock on the effective date, by introducing a longer announcement period. Surprisingly, while the objective definitively was to limit the volatility of a stock by increasing information transparency, the opposite effect was produced. This anomaly is rather structural than seasonal. Let us now focus on another anomaly: trading by insiders.

3.6 Trading by Insiders

An anomaly can be seen as a means which, if exploited, can lead to a significant overperformance over the market. This is also true for the legal trading by insiders. In the following sections, we will take a look at the definition of *insiders* and how they trade. We will present the evidence and the persistence of this stock market anomaly.

[64] Blume and Edelen (2002, p. 1).

3.6.1 Description

The SEC[65] defines an insider or inside investor as the chairman, CEO or president (top executive), or a senior executive (or an officer) of a company, a member of the board of directors (director), an owner of 10 % or more of the firm's shares (a large shareholder), or a close relative of such persons.[66] These people have the right to invest into the company they are managing or to sell their shares as long as they follow the rules set by the SEC. Such rules are strict and trading by insiders can prove to be tricky and complicated. This is why only few people are really interested in this investment possibility.

According to the strong-form efficient market hypothesis, the stock exchanges use all available information, i.e., past, present and insider information. Practitioners believe, however, that stock exchanges follow the semi-strong hypothesis, i.e., that actual prices reflect all publicly available data including past data, but *not* insider knowledge. This is important because in theory, if an investor can understand the investment choices of an insider as defined above, he might be able to continuously generate above-market portfolio returns. It is reasonable to assume that an insider has a better understanding of the company he manages and his industry than an outside observer or an individual trader. With this knowledge, their predictions ought to be more accurate than the predictions or expected future trends based on publicly available information. Since information appears to be essential for price assessment, an investor may be able to generate abnormal returns if he can decipher the information provided by the investments of the insiders.

3.6.2 Evidence and Insider Behavior

On a global level, the evidence for this anomaly is mixed. In general, very little market movement is observed when insiders trade and when they report their trades to the SEC.[67] For example, in 1998 Espen and Smith found no abnormal positive performance by insiders on the Oslo Stock Exchange.[68] If markets are efficient, an anomaly should vanish in a short time if there is any. If there is no important market move, one can assume no anomaly exists.

However, a growing academic literature tends to show that insider trading, motivated by private information, mostly occurs close to corporate announcements. Damodaran and Liu found strong support for the hypothesis that inside traders buy after receiving favorable appraisal news, while they seem to sell after receiving

[65]SEC stands for Unites States Securities and Exchange Commission. Their mission is to monitor and control investment activities.

[66]Singal (2006, p. 135).

[67]Lakonishok and Lee (2001).

[68]Eckbo and Smith (1998).

a negative appraisal news.[69] This leads to significant abnormal returns during the appraisal period. Kahle confirms this observation.[70] She shows that insider sales increase and purchases decrease prior to the issue of information-sensitive securities, i.e., convertible debt and equity. The reason may be the dilution factor that would diminish insiders' wealth. Such an analysis holds true for industrial firms, it does not seem to apply for utility companies. This difference was not accounted for by Kahle's analysis.

According to the analysis of Seyhun who documented the period 1975–1989,[71] the aggregate net number of open market purchases and sales by corporate insiders predicts up to 60 % of the variation in the 1-year-ahead aggregate returns. Lakonishok and Lee conducted an analysis of insider trades on a much larger scale which covered over 20 years of trading on the NYSE, the AMEX and NASDAQ from 1975 to 1995. According to the researchers, if insiders buy in 1 month, there is a 38 % chance that they will buy in the following month, versus a 11 % chance that they will sell.[72] An equivalent pattern seems to exist for insider sales, which is far less obvious though. Furthermore, a stock appears to be more likely to perform better following an insider buying month as compared to following an insider selling month. The difference in returns over the subsequent year varies from 4.8 to 18.7 % according to Singal.[73]

Overall, when insiders actively purchase their own firm shares, these stocks outperform those which insiders have been actively selling. If one uses the monthly trade of insiders as a signal, these stocks gain 24 % on a purchase versus 15.1 % on a sale in the following year.[74] The effect is even more striking as the investment horizon shrinks.

Two theories can explain the positive impact of a buy order from insiders, i.e., the information effect and the stealth trading hypothesis.

3.6.3 Information Effect

The first explanation is based on what is called the *information effect*. Following the new SEC regulations of 1993–1995, share prices react positively to large managerial purchases.[75] These new regulations required inside traders to publish their trading operations in a timelier manner. The price reacts positively with even more strength if the stock belongs to a firm that is either small or undervalued or experiences

[69] Damodaran and Liu (1993).
[70] Kahle (2000).
[71] Seyhun (1992, p. 1303).
[72] Lakonishok and Lee (2001, pp. 89–96).
[73] Singal (2006, p. 139).
[74] Singal (2006, p. 155).
[75] Roth and Saporoschenko (1999).

conflicts between managers and shareholders. Even more interesting, this effect is not reversed within a year and provides a good long-term investment.

3.6.4 Stealth Trading Hypothesis

The second explanation is based on what is called the *stealth trading hypothesis*.[76] Barclay and Warner proposed the hypothesis after they examined the proportion of a stock's cumulative price change in pre-set trade-size categories. Although the majority of trades were small, most of the cumulative stock-price change seemed to be due to medium-size trades. This observation is consistent with the hypothesis that informed trades are concentrated in the medium-size category, and that price movements are mostly related to the private information of inside traders. As a matter of fact, stocks are bought in medium-size sets so as not to alarm the market or in order to comply with legal requirements. The sell orders are realized on a multiple month period. This is why prices will steadily increase through time. This hypothesis is consistent with observations of short- versus long-term investment. Inside traders, due to a strong confidence in their firm's performance, tend to overvalue their company's stock in the long-term,[77] which does not offer a good long-term investment.

Inside traders seem to possess superior knowledge about the market that they use to decide whether to purchase or not their own firm's stock. Hence, stock prices have a high probability of rising following a buy order. A third characteristic of inside traders is the timing. If an insider is willing to buy but expects prices to fall, he is more likely to wait. If he estimates that the stock price will not rise any further, he should be prone to sell. There is a 69 % probability that insiders sell shares once the firm's stocks have increased by more than 10 %, while the probability of a sale drops to 52 % after the share price has fallen by 10 %.[78] According to Singal, these results combined with the continuity of returns and trading patterns indicate that inside traders are able to time their trades.

It seems possible to benefit from insider trading by mimicking the behavior of the traders. However, a perfect mimicking might not be a good choice as we have seen above. Moreover, inside traders tend to have a rather risky profile. According to Lakonishok and Lee insiders are in aggregate contrarian investors, i.e., they bet on a bouncing market when it is falling and vice versa.[79] The effect, however, is driven by the ability of the insiders to predict returns in smaller firms. In addition, the informativeness of insiders' activities is linked to purchases, while insider selling appears to have no predictive ability.[80] An option exercise may also include some

[76]Barclay and Warner (1993).

[77]Kahle (2000).

[78]Singal (2006, p. 142).

[79]Lakonishok and Lee (2001).

[80]Lakonishok and Lee (2001, p. 93).

information, but evidence shows that the predictability of stock sales following an option exercise is very small.[81] Therefore, a cautious investor who tries to follow the trading strategy of an inside trader should rather focus on the buying orders than on the selling orders so as to generate typically abnormal returns between 10 and 15 % per year.[82]

3.6.5 Newspaper and Mimicking

The purpose of this part is not to depict precisely *how* but *if* it is possible to generate abnormal returns out of insider information gained from their purchases and sales, based on academic research.

Bettis, Vickrey and Vickrey underline that, while previous research indicated that corporate insiders can systematically earn abnormal returns, observers could not generate a substantial profit in their attempt to copycat insiders.[83] However, in 1993 the results of their research lead to the general assumption that the market is not semi-strong efficient, and that it is possible to consistently generate abnormal returns.

The question then is what are the tools we dispose of? A good deal of magazines and newspapers publish daily or weekly insider trading information as required by the SEC, at the latest 2 days after operating the transaction. Considered to be complicated, the process of analyzing trading by insiders requires a lot of number crunching. The provided data are already partially processed. Therefore many investors prefer to choose more accessible trading strategies. Still, the Consensus of Information (COI) or the Wall Street Journal *Insider Trading Spotlight* can provide information on inside trading.

In 2002, Friederich, Gregory, Matatko and Tonks tried a mimicking strategy on the London Stock Exchange so as to examine patterns in abnormal returns in the days around inside trader investment decisions.[84] They found what has already been underlined, namely, that directors engaging in short-term market timing[85] have superior predictive content for future returns. This is especially the case for medium-sized trades which are more informative. This is in line with the stealth hypothesis previously presented. The interesting aspect of their research is, however, that even after netting positions, abnormal returns did not disappear. Therefore, one could definitely outperform the market, based on insider trading.

---

[81] Benesh and Pari (1987).

[82] Singal (2006, p. 158).

[83] Bettis, Vickrey, and Vickrey (1997).

[84] Friederich, Gregory, Matatko, and Tonks (2002).

[85] Market timing is the strategy of making buy or sell decisions of financial assets by attempting to predict future market price movements, in this specific case short-term movements.

Ferreira and Brooks uncovered that on re-release[86] dates in the Wall Street Journal, a significant price change could be witnessed for common stocks between 1994 and 1995.[87] On the day of publication, insiders generate a positive abnormal profit in the case of purchases and incur an abnormal loss in the case of sales. But how were the observers' profits? The trading volumes increased significantly compared to the thirty previous days. Yet, more importantly, the researchers discovered a relationship between the abnormal returns and the relative trading volumes on the publication date. In other words, the *Insider Trading Spotlight* column available in the Wall Street Journal seems to provide additional information to the market.

This corroborates evidence found by Bettis, Vickrey and Vickrey, i.e., that the market does not have a semi-strong form efficiency. Therefore, following recommendations from the Wall Street Journal might lead to substantial abnormal returns. Chang and Suk drew equivalent conclusions for the time around the Wall Street Journal publication day.[88] They added that a secondary dissemination of information can affect stock prices if the initial public disclosure attracts only limited attention.

Benesh and Pari conducted a similar study based on a different database: the COI.[89] They noticed that stocks listed in the COI newsletters were characterized by excess positive returns over the 4 months immediately preceding listing. Yet, on the listing date, much of the abnormal return was already gone. Furthermore, users of the COI's recommendations could have earned moderate excess returns by consistently placing a buy order on each recommendation and holding the stock for a year.

Yet, the evaluation of insider trades is not easy. Newsletter performances are not spectacular either. For example, as Singal notes, two newsletters followed by Hulbert Financial Digest underperformed the broader market: The first portfolio earned 118.9 % versus the market's 197.6 % return from January 1985 to June 1992; Market Logic, the second newsletter, had a portfolio that cumulated a 339.2 % versus the market's 432.7 % return.[90]

3.6.6 Persistence

There is no clear explanation of the persistence of this anomaly. The reasons which are evoked either betray a lack of understanding of inside trading strategies or are based on the belief that inside trading is forbidden by the law. The complexity and

---

[86] A re-release is a piece of information that has already been given public. The Wall Street Journal in this case publishes once again an information that is already public on the market.

[87] Ferreira and Brooks (2000).

[88] Chang and Suk (1998).

[89] Benesh and Pari (1987).

[90] Singal (2006, pp. 137–138).

the uncertainty of the method lead practitioners to leave this strategy out of their investment spectrum. Newsletters devoted to this particular kind of strategy have not been very popular or effective either. Yet, the anomaly exist, just as the momentum of industry portfolios.

3.7 Momentum of Industry Portfolios

The momentum of industry portfolios is a topic which is fiercely discussed by academics. Its very existence is contested by many researchers and institutional investors. Observations of the market price behavior seem to prove its existence. In this section, a clear definition of the various industry momenta is provided. Then, evidence of its existence is shown in the following section. Unfortunately, no clear explanation can be provided to justify this anomaly. This is why several hypotheses are proposed, based on current research results. The section is closed by discussing the persistence of this intriguing market anomaly.

3.7.1 Description

A market momentum is a measure of an overall market sentiment, calculated as the change in the value of a market index multiplied by the aggregate trading volume occurring within the index components. In the case of an industry, and not of the market as a whole, the momentum of a stock is a measure of the stock's performance relative to its industry. The momentum of an industry is a measure of the industry's performance over its benchmark, most of the time a well-known index. The concept is also known under the term *relative strength*.

There are various kinds of momentum: stock momentum, intra-industry momentum, cross-industry momentum, etc.[91] The basic concept is the same and can be understood as a lag in price increase following an information event. More generally speaking, it can be depicted as the fact that a firm that had a poor performance in the previous period is likely to pursue its low return path. Likewise, great performers of previous periods tend to bring high returns in the following period. Past winners and losers are expected to be the future winners and losers, respectively.

Economic conditions, market or industry expectations or even the mood of investors create investment shifts from one industry to another. At some point in time, investors may prefer to invest in steel, then they switch to paper or automotive or electronics or aeronautics or the Internet industry. Each time the market has a new

[91]The stock momentum is calculated based on the change in the value of stocks between two dates. The intra-industry momentum is calculated based on the change in the value of stocks in a specific industry index multiplied by the aggregate trading volume occurring within the index components. The cross-industry momentum is calculated based on the change in the value of an industry index multiplied by the aggregate trading volume occurring within the selected industries used as benchmark.

fad, the favored industries perform better than the market as a whole. As a result, selecting these industries may lead to risk-adjusted abnormal returns compared to the market index which is used as a benchmark. The first question is to select which kind of momentum investors should focus on to get the highest and most consistent abnormal return.

Evidence proves that cross-industry momentum, industry momentum versus a benchmark and intra-industry momentum provide excess returns above their benchmark. Let us now focus on this point.

3.7.2 Evidence

September 11, 2001: The World Trade Center collapses under a terrorist attack. Investors understood that the defense and security industry would be favored in the new politico-economic era, but nobody knew what public institutions and private investors were willing to invest to fight the new threat. As the investors' interest shifted toward this industry, stock prices should have mirrored this change, and a price surge should have been witnessed. However, because of the high uncertainty of future prospects stock prices changed only slowly in the security and defense industry. Evidence tends to support that gradual movements in industries and in the returns of industry portfolios are frequent and believed to be normal by investors.[92] Similarly, yet less terrorizing, catastrophic weather conditions in Asia can impact delivery of semi-conductors to IT firms as witnessed in 2004 and 2009, which in turn slows down sales of the firms, and the IT industry does not perform as well as expected. Less investors are then interested in IT stocks whose prices fall across the world since many IT companies have outsourced a significant part of their supply chain to Asia.

The gradual effect is of importance because it may be taken advantage of, even though no absolute information is required or necessary. Abnormal profits can be generated based on historical data. This is in clear contradiction with the efficient market hypothesis, given that the drift takes on average several months to cool down or to revert. The following evidence of momentum is presented in three parts, namely for cross-industry momentum level, industry level and stock level relative to its industry.

Cross-industry momentum defines a relative price strength of certain industries compared to the market, which is typically represented by a benchmark like the Dow Jones Industrial Average Index, the S&P 500 Index, etc. According to Menzly and Ozbas, industries related to each other through the supply chain (upstream or downstream) exhibit strong cross-momentum.[93] Trading strategies that consist of buying industries with large positive returns and selling industries with large

[92] Singal (2006, p. 78).

[93] A company in the upstream part of a supply chain is one of the final customers of the product. If for instance, Goodyear, which produces tires for the car industry, is the upstream company, then

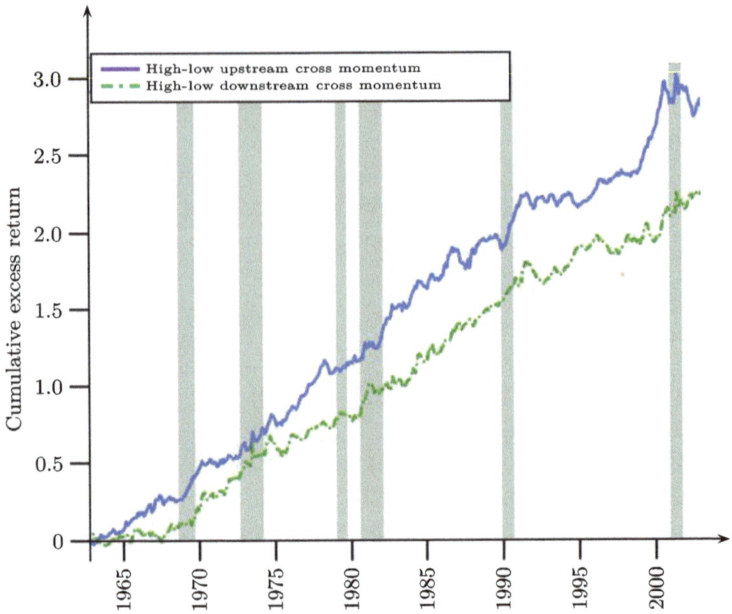

Fig. 3.4 Performance of upstream and downstream strategies from 1964 to 2002. *Source*: Menzly and Ozbas (2004, p. 28)

negative returns over the previous month yield significant profits.[94] As a matter of fact, a zero-investment strategy consisting in buying cross-industry winners and short selling losers should in theory provide abnormal returns, but the study does not take into account trading costs and other related costs.

Figure 3.4 represents the performance of upstream and downstream strategies from 1964 to 2002 in the context of a cross-industry momentum measurement attempt. An upstream strategy means buying a top performer industry and industries to which it sells its raw material or components. A downstream strategy buys a top performer industry and its suppliers. To build this figure, the researchers first ranked industries into five categories, from top performers to poor performers so that there is exactly the same number of stocks in each category. Performance was estimated from the past 1-month period. An equally-weighted and a value-weighted (using industry market capitalization) portfolio were built that bought industries in the highest category and sold industries in the lowest category.

As one can see, the cumulated abnormal returns differ depending on the strategy. They follow a steady path, but the returns are fairly low. Figure 3.5 has a focus on

the corporation owning the trees which supply the raw material would be a downstream company in the supply chain. See Menzly and Ozbas (2004).

[94]Menzly and Ozbas (2004, p. 9).

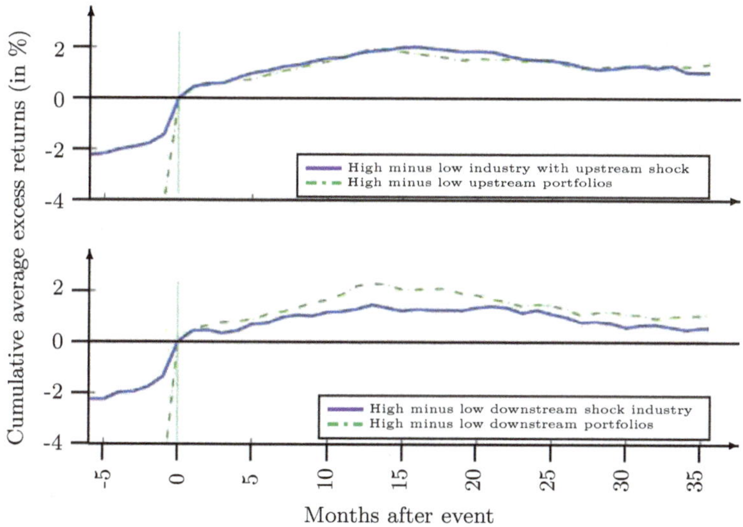

Fig. 3.5 Cross-industry momentum beyond the first month. *Source*: Menzly and Ozbas (2004, p. 27)

the behavior of the cross-industry momentum after implementing the strategy. The implementation date and abnormal return are visible at the junction of the vertical and horizontal line, respectively. While the cumulative return still appears low, the mountain shape of the return is of great interest. Preceding the implementation date of the strategy, a price increase is fairly noticeable. In the long term, an inverted U shape seems to appear, regardless of the employed strategy. This is typical of momentum, be it for a stock, an industry or a cross-industry portfolio. The abnormal return of the equally-weighted strategy reaches 6.8 % per year, against 6.5 % of the value-weighted strategy. Hence, the Sharpe ratio of the equally-weighted strategy is significantly better than that of the value-weighted strategy: 0.852 versus 0.573.[95]

On the *industry momentum level*, Moskowitz and Grinblatt conducted a very important research in 1999.[96] They implemented a strategy of buying the winners of previous periods and selling the losers and found a very strong prevalent momentum effect in industry component stock returns which accounts for a significant part of the stock momentum anomaly. After canceling the industry's momentum, the strategy earns a significantly less profitable return. This is, however, in contradiction with Grundy and Martin who report that neither industry effects nor cross-sectional differences in expected returns are the primary cause of the momentum phenomenon,[97] but rather time varying factor models.

[95]Menzly and Ozbas (2004, p. 12).

[96]Moskowitz and Grinblatt (1999).

[97]Grundy and Martin (2001, pp. 1, 22 and 31).

Nevertheless, Moskowitz and Grinblatt add that their strategy, i.e., exploiting industry momentum, is highly profitable even after controlling for size, book-to-market equity, individual stock momentum, the cross-sectional dispersion in mean returns, and potential micro-structure influences.[98] In other words, even after suppressing a possible size effect, a surprise earnings effect, an abnormal correlation in returns or industry-specific developments, this strategy seems to be interesting. But what abnormal return can an investor expect compared to a benchmark? How long should the portfolio be invested? O'Neal declares that according to latest academic research, the momentum present in U.S. stock returns is, to a large extent, a result of industry momentum if and only if the investment horizon is between 3 and 12 months.[99] He confirms that during this time span, a strong (weak) industry performance is followed by a continuously strong (weak) industry performance. In practice, he bought and held for 6 months top-performing sector funds based on their previous 6-month performance from May 1989 to April 1999.[100] Each fund was invested in a particular industry. This strategy outperformed the S&P 500 and generated a 12 % abnormal return per year according to the paper. Chan, Jegadeesh and Lakonishok found equivalent results for the periods 1973–1993 and 1994–1998.[101]

Jegadeesh and Titman found similar results at the *stock level* compared to the industry.[102] Purchasing stocks that have performed well in the past and selling stocks that have performed poorly generates significant positive returns over a holding period of 3–12 months. More interestingly, according to the researchers, these abnormal returns are not due to systematic risk or to delayed reactions to common factors (for example, earnings announcements, liquidity risk, etc.). Another very important point is that part of the abnormal returns generated in the first year after the portfolio formation disappears in the following 2 years.

A similar observation is made about the earnings announcements of past winners and losers. In other words, a holding period of more than 2 years destroys the momentum anomaly which is possibly initiated by a public information release. Given that this characteristic is shared by all industries, it seems unwise to keep the investment for more than 2 years if the investors wish to take advantage of the stock's momentum. A buy and hold over 6 months generates a compounded interest of 12.01 % per year on average.[103] Such an outperformance does not seem to be the result of systematic risk taking.[104] Jegadeesh and Titman underline, however, that this lag is consistent with a delayed price reaction to firm-specific information.

[98] Moskowitz and Grinblatt (1999).

[99] O'Neal (2000, p. 37).

[100] O'Neal (2000, p. 37).

[101] Chan, Jegadeesh, and Lakonishok (1999).

[102] Jegadeesh and Titman (1993).

[103] Jegadeesh and Titman (1993, p. 89).

[104] Jegadeesh and Titman (1993, p. 89).

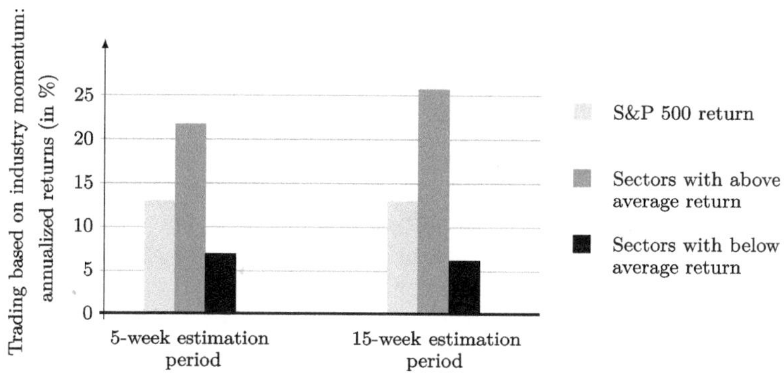

Fig. 3.6 Annualized returns of S&P 500 versus above and below average sectors based on 5-week and 15-week estimation period from 1997 to 2001. *Source*: Singal (2006, p. 99)

This effect seems to disappear on the industry level according to Moskowitz and Grinblatt's report.[105]

Figure 3.6 represents research results for the period from 1997 to 2001. Top performers are compared to poor performers and a benchmark, the S&P 500 Index, based on their respective performance in a 5- and 15-week estimation period, respectively. Returns are annualized.

Hence, top performers effectively outperform by approximately 9 % for a 5-week estimation period followed by a 5-week holding period. This abnormal performance jumps to 13 % above the index in case of a 15-week estimation period followed by a 5-week holding period. Based on equivalent periods, poor performers underperform the index by respectively 6 and 8 %. Apparently, a trading strategy can effectively generate abnormal profits. However, evidence which shows a risk-adjusted overperformance on the cross-industry and industry level seems to be related to additional systematic risk on the stock level. The momentum effect reveals different properties depending on the subject of study (cross-industry, intra-industry, etc.). Yet, no explanation has been provided to date to understand the industry momentum, whatever the level.

3.7.3 Explanations

There are many reasons that could lead to momentum in a stock, an industry or even an index. Possible explanations include:

- Irrationality of the agents
- Lead lags in information dissemination
- Herding behavior

[105]Moskowitz and Grinblatt (1999).

- Market friction and information uncertainty
- Dividend growth rate variations through time

3.7.3.1 Irrationality of the Agents

The first hypothesis explaining industry or stock momentum is the *irrationality of the agents*. According to this explanation, investors are reluctant to change their beliefs quickly in the short term, even in the face of convincing information. Every investor, at some point in time, believes that the information he holds is of superior quality. This may lead to overconfidence as will be discussed in more detail in Sect. 5.3.5, which deals with behavioral finance biases. This overconfidence is progressively undermined by price movements, which do not necessarily follow the expected direction. As more information becomes impounded in prices partially and gradually over time, a price drift is observed through days, weeks or months.[106] In the long term, however, a feedback effect occurs and investors who did not close their losing positions, sell massively, leading to overreactions. So according to this hypothesis, the combination of underreaction (due to the overconfidence of a set of agents) and overreaction leads to momentum. A slow upward price drift should thus be witnessed. For example, investors were well aware about the overvaluation of the Internet stocks in the 2000s, but they still tried to surf on the upward price movement wave.[107]

3.7.3.2 Lead Lags in Information Dissemination

The second hypothesis is about *lead lags in information dissemination*. Large firms or hot industries, like the Internet firms in the 2000s, tend to be in the focus of a large number of analysts, while smaller firms have less visibility. The extent of coverage affects the frequency with which stocks are reviewed. The stocks of large firms or hot industries are reviewed early, which increases the availability of information. Smaller stocks or neglected industries are reviewed later which delays the impact of information on the price.[108] This point, while being controversial, underlines that information dissemination is not immediate. In order to capture the effect of information dissemination, Hong and Stein designed a model with *news watchers* on a side and *momentum traders* on the other.[109] Each news watcher observes some private information, but fails to extract other news watchers' information from prices. If information diffuses gradually across the population, prices underreact in the short run. The underreaction means that the momentum traders can profit by trend chasing. However, if they can only implement simple (i.e., single variable) strategies, their attempts at arbitrage must inevitably lead to overreaction in the long run.[110]

---

[106] Menzly and Ozbas (2004) and Holden and Subrahmanyam (2002).

[107] Singal (2006, p. 83).

[108] Singal (2006, p. 83).

[109] Hong and Stein (1999).

[110] Hong and Stein (1999, p. 2143).

Real estate investment trusts (REITs)[111] provide a good setting to examine this hypothesis because the industry experienced structural changes beginning in the 1990s in the U.S. Chui, Titman and Wei tested predictions that are related to investor overconfidence and the speed of information diffusion.[112]

If overconfidence is the reason for industry momentum, a stronger momentum effect in REITs should have occurred in the post-1990 period compared to the pre-1990 period due to a higher valuation uncertainty in the post-1990 period. On the other hand, if the speed of information diffusion is the key factor, then the momentum effect in REITs should have been stronger in the pre-1990 period than in the post-1990 period due to the higher speed of information diffusion in the post-1990 period. The report established that the evidence tends to support the first prediction. Specifically, while no momentum effect occurred in REITs during the pre-1990 period, a strong and prevalent momentum effect in REITs became visible during the post-1990 period.[113] Therefore, in the case of REITs, the hypothesis does not seem to hold true and is less plausible than the irrationality of the agents.

3.7.3.3 Herding Behavior

A third hypothesis concerns *herding behavior*. Herding is a behavioral finance concept that will be explained later in detail. In short, it means that if the market develops a trend, an individual investor is more likely to irrationally follow it. This may happen even if the investor possesses high quality factual information and is fully aware of his mistake. According to Singal, sell-side equity analysts[114] at times have a tendency to herd towards the consensus estimate when making their quarterly earnings forecasts.[115] Gao argues that such a herding tendency leads to the inefficient aggregation of private information and consequently to the price momentum in stocks.[116] According to his analysis, the price momentum phenomenon presented by Jegadeesh and Titman occurs only during periods when analysts who follow the concerned stocks herd together.[117]

The herding tendency is stronger for smaller stocks, growth stocks, and stocks with higher share turnover ratio and more news media coverage.[118] Finally, Gao states that these findings are distinct from earnings momentum effects, information

[111] A real estate investment trust is a security that sells like a stock on the major exchanges and invests in real estate directly, either through properties or mortgages. REITs receive special tax considerations and typically offer investors high yields, as well as a highly liquid method of investing in real estate.

[112] Daniel, Hirshleifer, and Subrahmanyam (1998, p. 363) and Hong and Stein (1999).

[113] Chui, Titman, and Wei (2003).

[114] Sell-side analysts analyze a small amount of stocks in a specific industry and try to sell their report stating a given expected return in the upcoming period.

[115] Singal (2006, p. 83).

[116] Gao (2006).

[117] Jegadeesh and Titman (1993, p. 90).

[118] Jegadeesh and Titman (1993, p. 65).

uncertainty effects and liquidity risk. Grinblatt, Titman, and Wermers (1995) seem to support his report: their research proves that 77 % of mutual funds were held by momentum investors who exhibit herding behavior.[119]

The difference from the previous topic is subtle. Lead lags in information dissemination refer to the timing of information releases, in particular to the release of timely, independent high quality information from financial analysts which may have the effect to promote hot stocks. Herding behavior occurs, when investors count on each other's beliefs. Depending on the *belief mix*, they adjust their recommendations, but the quality of the information is not at stake. In the case of an asymmetry between the value of the information and the perception of the information, a momentum effect appears. If many fund managers bought a stock which performs poorly, their performance will be less harshly judged since they followed the crowd, and most investors believed that the stock would rise. Therefore, when a few institutional managers begin to buy a particular stock, other managers may feel safe in buying the same stock, which results in momentum. The herding explanation matches with empirical observations and the analysis of the REITs case.[120]

3.7.3.4 Market Friction

The fourth hypothesis is *market friction*. Financial assets are known to reflect all historical market data, but also future expectations discounted at an appropriate rate. The price of real goods, like corn or house prices, reacts more sluggishly under a price movement perspective. This lag between financial markets and the real goods' market creates market friction and is the origin of momentum in a given industry.[121]

The oil market has a futures market. Futures on oil are financial products that derive their price and performance from crude oil. In general, commodity futures reflect the expectations on the underlying good.

For example, when the crude oil spot price (actual price) is stated on the physical market and if the futures oil price on this specific crude oil category is higher, an investor can fairly accurately estimate the demand and offer for oil for a specific time horizon. It is important to note that not all futures markets behave in the same way. For some asset classes, futures prices trade higher than spot prices and are said to move in contango, while for other asset classes, they trade lower than spot prices and are said to move in backwardation. If there are no benefits in holding an asset, the market will move in contango. If there are benefits in holding the asset, the market will move in backwardation.

As a result, the existence of a futures market should diminish the uncertainty on physical markets and the available evidence seems to confirm this.[122] Forecasts are easier, more accurate, and can be retraced on a curve. However, while there

[119] Grinblatt, Titman, and Wermers (1995, pp. 1088 and 1093).

[120] This paragraph was based on Chui et al. (2003).

[121] Singal (2006, p. 87).

[122] Feder, Just, and Schmitz (1980).

Table 3.4 Returns of Fidelity's sector fund for the defense industry (ticker: FSDAX) versus the S&P 500 return from October 2001 to June 2002

Month	FSDAX return (in %)	S&P 500 return (in %)
October 2001	3.6	1.9
November 2001	3.8	7.7
December 2001	4.1	0.9
January 2002	5.8	−1.6
February 2002	2.3	−1.9
March 2002	4.6	3.8
April 2002	2.0	−6.1
May 2002	0.7	−0.7
June 2002	−3.0	−7.1
Total:	23.9	−3.0

Source: Singal (2006, p. 84)

are futures markets for oil or corn, they do not exist for some products classes or industries, for example, semi-conductors. Without a futures market a proper forecast for this industry can only be based on a thorough analysis, but an analyst would at best be able to generate scattered estimates on a given time horizon and not a curve. Under these circumstances, mathematical tools do not offer much help for the trend estimation process.

A simple case study should be sufficient to test the market friction hypothesis. The terrorist attack of September 11, 2001, was a positive signal for defense and security firms and for the likely rise of their stock prices. Table 3.4 retraces the returns of Fidelity's sector fund for the defense industry versus the S&P 500 return from October 2001 to June 2002.

If the market friction hypothesis holds true, two developments should take place. First, the stock prices of defense and security industry as a whole should outperform the market. Second, given that there is no specific futures market, a drift due to future sales of defense products should progressively impact the industry, followed by a necessary downward correction. Table 3.4 shows that from October 2001 to June 2002, the defense and security sector outperforms the benchmark by 26 %. In June 2002, a correction seems to appear in the sample. In this case, the market friction hypothesis matches the observations. However, this hypothesis is event-driven. It does not account for investor fads and cyclical asset allocations witnessed on the market. Neither does it explain properly the abnormal returns generated from buying top-ranked companies and selling poor performers.

3.7.3.5 Dividend Growth Rate Variations Through Time

A last rational explanation arises from the *dividend growth rate variations through time*. According to this hypothesis, the momentum effect does not need to imply investor irrationality, heterogeneous information, or market frictions. Johnson proposes a simple, single firm model with a standard pricing kernel which can produce a short-term underreaction when the expected dividend growth rates vary over time. An enhanced model, where persistent growth rate shocks occur as illustrated in Table 3.4, matches many of the features documented by the empirical research. According to Johnson, the same basic mechanism potentially accounts for underreaction anomalies in general.[123] This is in line with Grundy and Martin's report, which presents a ranking system similar to previous research results.[124] They argue that buying recent winners and shorting recent losers guarantees the returns connected to common risk factors, while protecting investments from those time-varying risk factors during the ranking period. Adjusted for this dynamic risk exposure, momentum profits were remarkably stable across the subperiods of the entire post-1926 era. While factor models can explain 95 % of the positive and negative return variability, this hypothesis cannot explain the mean return component of the industry momentum.[125] In this case, it is not the economic uncertainty that drives the underreaction process, but corporate managers who chose to re-evaluate the dividends.

To sum up, there is no consensus about the reasons underlying the momentum effect, whatever the chosen level. On the industry level at least, it seems highly probable that its emergence is related to herding behavior. Given that this explanation is less mathematical than a dividend yield or a short-term price drift, and given that empirical skepticism about the industry momentum persists despite numerous evidences, further research on the topic might be required to unveil the proper reasons for this effect.

3.7.4 Persistence

Being cross-sectional, industry-only or related to single stocks, the momentum effect has persisted remarkably well for decades,[126] i.e., since 1926. But how to exploit it? A fund that specifically invests in one industry might be the appropriate tool to generate abnormal profits. Dellva, DeMaskey and Smith tried to test the selectivity and timing performance of the Fidelity sector mutual funds during the 1989–1998 time period.[127] They used the S&P 500, the Dow Jones Industry Group Total Return Indexes, and the Dow Jones Subgroup Total Return Indexes as benchmarks. Compared to the Dow Jones Industry benchmarks, the results indicate a positive selectivity and a negative timing ability of many sector fund managers.[128] In other words, part of the performance is lost through the timing process. Secondly, regulations oblige the fund manager to keep part of the invested amount in the fund in cash. The momentum of the industry might not be fully exploited because of

[123] Johnson (2002).

[124] Grundy and Martin (2001).

[125] Grundy and Martin (2001, p. 29).

[126] Grundy and Martin (2001, pp. 1 and 3).

[127] Dellva, DeMaskey, and Smith (2001).

[128] Dellva et al. (2001).

the inactive amount of money, and that even before trading costs are taken under consideration.

Furthermore, in 2004, Lesmond, Schill, and Zhou published a test on the profitability of relative strength or momentum trading strategies (buying past strong performers and selling past weak performers).[129] They found that standard relative strength strategies require frequent trading in disproportionately high cost securities. Trading costs are so high that they prevent the execution of profitable strategies. In the cross-section, the authors found that stocks that generate large momentum returns are precisely stocks with high trading costs. That is why they conclude that the magnitude of the abnormal returns associated with these trading strategies creates an illusion of profit opportunity when, in fact, none exists.

To conclude, the industry momentum supports the concept that abnormal returns can be generated by buying top performing companies and selling poor performers, thanks to an apparent lag in price adjustment. The most plausible explanation for this anomaly is the herding behavior of agents. They prefer to state, compare and evaluate rather subjectively their private information together with other agents. The industry momentum persists although no real profit can be generated out of it. Apparently, stocks that would be the source of abnormal performance appear to be linked to high trading costs, thus killing the possible profit.

However, some limitations have to be added. In 2007, the subprime crisis launched a wave of risk management and investment process re-engineering, which could have altered the structure of the market sufficiently to modify the existence or the nature of momentum. But as of summer 2013, no academic research starting in 2007 or later regarding the momentum effect could be found to clarify possible changes in the characteristics of market, cross-industry or intra-industry momentum.

3.8 Home Bias and International Investing

The home bias is a puzzle for academics. According to MPT, an optimal portfolio should consist of a large number of assets which should include both domestic and foreign stocks. However, empirical research tends to show that investors hold a substantially larger proportion of their wealth portfolios in domestic assets, a phenomenon called *equity home bias*.[130] Evidence supporting the anomaly lead Tesar and Werner to find that existing explanations to the home equity bias are unsatisfactory and to conclude that the issue poses a challenge for portfolio theory.[131]

[129] Lesmond, Schill, and Zhou (2004).

[130] Lewis (1999).

[131] Tesar and Werner (1995).

3.8.1 Description

From all anomalies reviewed until now, the home bias does not specifically defy the efficient market hypothesis but instead the capital asset pricing model itself. The home bias is an anomaly with regard to the model and describes the abnormal overweight in domestic assets in the portfolio asset allocation process compared to what should be observed in the quest of diversification. It shows that if an investor had better diversified his portfolio with stocks of companies traded on foreign exchanges, he would have realized a higher risk-adjusted performance.

3.8.2 Evidence of the Advantages of International Investing

The first evident characteristic when an observer looks at home bias figures is how strong this bias is. In 1991, French and Poterba analyzed and constructed estimates of the international equity portfolio holdings of investors in Britain, Japan and the United States.[132] More than 82 % of the equity portfolio of British investors was held domestically; for Japan and the United States, the figure increased to 94 and 98 %, respectively.[133] In 2001, Jeske provided similar results as depicted in Fig. 3.7. The figure plots the percentage of domestic stocks in equity portfolios and the world market share of the domestic market of eleven industrialized countries in 2000.[134] The home bias of the United States and most of European developed countries in 2013 is shown in Table 3.5.

All countries appear heavily biased toward holding more of their domestic equity than foreign assets. Great Britain is estimated to have portfolios where domestic assets represent 78 % of the total value. Japan keeps a top score with 92 %. The U.S. still has one of the highest domestic shares of about 89 %. The other countries of the study almost reach or narrowly exceed the 80 % mark.

Lintner asserts that in theory, a well-balanced portfolio which attempts to reach the highest risk-adjusted return on an optimal portfolio should include assets from different countries with identical proportions.[135] This assertion, together with market clearing,[136] implies that each country should hold a portfolio where the proportion of domestic assets equals the share of these assets in world market capitalization.

[132] French and Poterba (1991).

[133] French and Poterba (1991, p. 223).

[134] The home share was computed using market capitalization data from the International Federation of Stock Exchanges (FIBV), and the international investment positions were provided by the International Monetary Fund (IMF). See Jeske (2001, p. 33).

[135] Lintner (1965, p. 13).

[136] A market clearing price is the price of a good or service at which the quantity supplied is equal to the quantity demanded. It is sometimes referred to as equilibrium price.

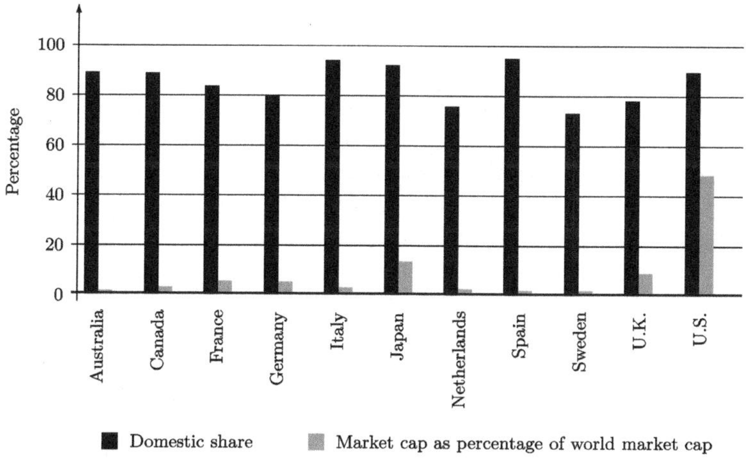

Fig. 3.7 Domestic stocks and their relative market capital allocations in 2000. *Source*: Jeske (2001, chart 1, p. 33)

Table 3.5 Estimates of home bias (in %) in 2012

Country	Market capitalization as % of world	Investment in domestic stocks	Home bias
Australia	2.66	88	> 85
Canada	4.23	88	> 85
China	7.52	88	> 85
France	3.48	83	> 75
Germany	2.63	80	> 70
Italy	0.96	93	> 90
Japan	7.85	91	> 75
Netherlands	1.32	75	> 70
Spain	2.29	95	> 90
Sweden	1.04	72	> 65
United Kingdom	2.67	78	> 65
United States	34.69	90	> 40

Source: Own, based on The World Bank Data as of March 2013. See World Bank (2013)

Accordingly, an investor who follows Lintner's requirements should balance his portfolio with 7.85 % of Japanese stocks and 34.69 % of U.S. stocks, because Japan and the U.S. have this weight in world market capitalization, respectively, according to the World Bank. But the data provided by the World Bank is an estimate of the relative country capitalization compared to the world's total capitalized wealth. Using the MSCI equity index scheme would lead to very different results, especially in the domestic versus emerging markets differentiation.

A second property of the home bias is also related to the domestic market of the investors. French and Poterba developed a simple model of investor preferences and behavior to show that current portfolio patterns imply that investors in each nation expect returns in their domestic equity market to be several hundred basis points higher than returns in other markets.[137] Empirical observations seem to confirm their conclusions.

Jeske conducted a large scale research on the 1991–2000 period.[138] In the U.S., the home bias is almost 150 basis points per year, by far the lowest among all industrialized nations. Australia, Canada, and most European countries display a home bias of between 200 and 500 basis points. The situation gets worse for Italy, Japan, Sweden and Spain: Annual costs are in the range of 700–1,500 basis points. This lack of diversification appears to be the result of investor choices, rather than of institutional constraints.[139] Still, the creation of the European Union and its common currency seems to diminish this spread.

On the investor side, the available research seems to confirm that the behavior of market agents is at least partially responsible for the home bias. The bias might be connected to the reluctance of decision makers to open their domestic stock markets to foreign investors because of possible risks for their domestic market. The question is then: *Is there any risk for countries who receive additional attention from foreign investors from a financial point of view?*

Before focusing on the mathematics of portfolio optimization, which seems to aim at maximizing short-term interests only, let us look at the long-term consequences following an opening-up of a domestic financial market to foreign financing. Kim and Singal conducted a research in regard to emerging markets which opened up their stock exchanges to foreign investment.[140] On a fundamental level, they find that when emerging economies open their stock exchange, the level of stock prices tends to rise without an associated increase in volatility. Hence, more capital becomes available for domestic investment at a lower cost.

The stock markets also appear to become more efficient, which leads to a better allocation of resources. On a macroeconomic level, the inflow of foreign capital does not lead to higher inflation or stronger currencies. The volatilities of inflation or exchange rates do not increase either. Kim and Singal conclude that the experience of emerging countries who opened their markets to overseas investors has been largely beneficial for the host countries. They add that if some countries experience large capital outflows with damaging consequences, the culprits are not the foreign investors, but rather the futile attempt of policymakers to defy market forces and the failure of the domestic economies to put the inflowing capital to productive uses.

To sum up, the opening of the market leads to a higher performance on the stock market without a statistically significant increase in volatility. Inflation and

[137] French and Poterba (1991).

[138] Monthly returns are annualized. Nominal returns for countries' equity markets were taken from MSCI (http://www.msci.com). Returns were then deflated by countries' CPI (consumer price index) data (from International Financial Statistics) and converted into the corresponding country's home currency. See Jeske (2001).

[139] French and Poterba (1991).

[140] Kim and Singal (1997).

currency are not significantly changed either. These developments are even more positive if investment decisions are taken under a longer historical perspective.[141] In the opinion of Kim and Singal, subsequent crashes are caused by the mistakes of policymakers after their loss of power in favor of financial markets. It is important to note, however, that if the money inflows would occur suddenly and on a massive scale the results of the study might differ significantly.

Sarkar and Li identified a complementary characteristic to foreign investing. They examined the international diversification benefits when short selling is not allowed.[142] The benefits remain substantial for U.S. equity investors when they are prohibited from short selling in emerging markets. In contrast, the benefits of investing in developed countries, that are small on average, disappear if short selling is not allowed. Sarkar and Li also pointed out another very important feature. The integration of world equity markets reduces, but does not eliminate, the diversification benefits of investing in emerging markets subject to short-sale constraints. In technical terms, the correlation between international markets increases.[143]

Then what are the benefits for investors to invest in developing or developed countries? In the late 1980s and early 1990s, the emerging stock markets witnessed high rates of economic growth combined with stock market liberalization. In 1993, for example, the composite return of emerging markets (measured in U.S. dollars) was no less than 68 %,[144] which compares very favorably to the S&P 500 performance of still respectable 10 %. However, such high returns are highly volatile. Emerging markets show much more volatile features than developed markets. For instance, the Brazilian stock market fell 63 % in 1987, rose 126 % in 1988, fell 66 % in 1990, and rose 170 % in 1991. The Turkish stock market fell 61 % in 1988, rose 502 % in 1989, fell 42 % in 1991, fell 53 % in 1992, and rose 234 % in 1993. All returns are expressed in U.S. dollars.[145] Let us now look at the annual stock returns broken down by region which are shown in Table 3.6.

From year to year one region performed better than the other. Table 3.6 shows 5 years of impressive market returns in the U.S. Nobody could predict that in the following years investments in any other country were a better choice. While investing in emerging markets was a good choice in 1991–1993 with successive performances of 62.6, 12.1, 76.5 %, EAFA (Europe, Australia and the Far East) was more rewarding during the next 3 years. The objective regarding this table would be to identify a significant part of top performers while diminishing risk. As seen in Sect. 1.5, a basket of two shares has a volatility which comprises the volatility of the two shares, their weights and their correlation. Let us assume the first share is a domestic index, and the second share is a foreign index. In this case,

---

[141] Kim and Singal (1997).

[142] Sarkar and Li (2002).

[143] Sarkar and Li (2002, p. 3).

[144] International Finance Corporation (1997, p. 55).

[145] All the figures presented are extracted from International Finance Corporation (1997, p. 55).

Table 3.6 Annual stock returns by region from 1988 to 2002 (in %)

Year	S&P 500	Europe, Australia and the Far East (EAFE)	Emerging markets (EMF)	All countries except the U.S. (AC-ex-U.S.)	All countries (AC)
1988	17.1	30.0	41.8	30.1	24.7
1989	32.5	12.3	71.2	13.0	18.7
1990	−1.6	−19.9	−6.8	−19.8	−14.5
1991	31.9	14.2	62.6	15.7	21.2
1992	7.9	−11.0	12.1	−10.1	−3.9
1993	10.3	34.8	76.5	36.9	25.6
1994	1.8	8.9	−5.5	7.9	5.7
1995	37.7	12.4	−4.1	10.4	19.9
1996	23.6	6.6	6.7	6.1	13.5
1997	34.9	3.3	−8.9	3.1	16.1
1998	31.4	22.4	−18.8	17.6	18.8
1999	22.0	28.3	70.5	31.4	33.8
2000	−7.9	−13.2	−29.7	−15.2	−13.1
2001	−10.2	−20.1	2.0	−18.2	−14.5
2002	−27.0	−15.4	−5.1	−14.5	−21.8
Average annual return	13.6	6.2	17.6	6.3	8.7

Source: Singal (2006, Table 10.2, p. 237)

the diversification process in each country should have already taken effect. Hence, the domestic investor can only improve on two issues to reach a higher return: the volatility of the foreign index and the correlation between his domestic market and the foreign market. The first can be shrunk thanks to hedging strategies, the latter can be diminished by choosing a foreign index which has a very low correlation with the domestic one. Ideally, a correlation close to zero would lead to a minimum performance loss.

In 1999, Clarke and Tullis looked at the impact of foreign investment exposure on the return and risk of a U.S.-based investment portfolio, based on the relative returns, the currency forward premium and the currency return (and exposure).[146] They developed a framework for analyzing the optimal foreign asset exposure, and modeled a simple portfolio, with a fixed exposure to foreign assets. To test each component, they used performance data from the Morgan Stanley Europe, Australia and Far East Index from 1991 to 1997 to create the 36-month volatility of hedged and unhedged to currency returns against a diversification threshold, returns and cumulative returns.

[146]Clarke and Tullis (1999).

Table 3.7 Optimal portfolios for different periods

Period	U.S. portfolio return (in %)	U.S. portfolio risk (in %)	Portfolio allocation U.S. (in %)	Portfolio allocation EAFE (in %)	Optimal Portfolio return (in %)	Optimal portfolio risk (in %)
1970–2000	12.70	15.30	100	0	19.60	18.60
1970–2000	12.70	15.30	70	30	13.00	14.00
1971–1998	13.40	15.30			13.50	14.30
1980–1992	17.20	15.80	41	59	20.00	14.50
1971–1988	9.90	16.20			13.90	14.50
1971–1988	9.90	16.20		100	18.20	17.50

Source: Singal (2006, Table 10.4, p. 240)

The results show that hedging reduces the volatility of international assets. However, there is more stability in the relative risk (volatility) between markets than between the relative returns. Given that volatility is relatively stable, Clarke and Tullis applied a minimum-variance allocation and analyzed long-run trade-offs between risk and return. Since the equation is fairly easy to compute, the empirical results allow for the inclusion of parameters, i.e., weights and variables can be included in the equation. Clarke and Tullis concluded that a 20–30 % allocation to foreign equity would increase the performance, subject, for example, to whether the forward currency premium is positive or negative. Simply speaking, with 20–30 % of the portfolio dedicated to foreign markets, one can optimally minimize volatilities and achieve maximum returns.[147]

Table 3.7 retraces an optimal portfolio for different periods. This portfolio is a world portfolio containing EAFE[148] and U.S. assets. The optimal weights for the optimal portfolio construction are based on previous research results.[149] The performance and risk of the optimal portfolio is shown in the last two columns.

In the first two lines we oppose a diversified versus a non-diversified worldwide portfolio for a long period. The U.S. portfolio has a 12.70 % return and a 15.30 % volatility during the period 1970–2000. The optimal portfolio shows a higher return and a higher volatility, namely 19.60 and 18.60 %, respectively. If the investor implements the recommendations of Clarke and Tullis, he will optimize his portfolio with a 30 % weight of EAFE stocks, and ends up on the second line of Table 3.7.

Let us now consider the subperiod 1971–1998 which is a special case. During this subperiod, investing in the U.S. yielded nearly the same return as investing in a world portfolio (13.40 versus 13.50 %), but at a higher risk (15.30 versus 14.30 %). Global investing necessarily leads to lower risk characterized by a 1 % volatility diminution.

[147]Clarke and Tullis (1999, p. 33).

[148]The acronym EAFE stands for Europe, Australasia (Australia and New Zealand), and the Far East.

[149]Singal (2006, p. 239).

The subperiod of 1980–1992 in the fourth line is another special case. The annual return is outstanding compared to the 30 years average, namely 17.20 versus 12.70 %. However, even with such impressive returns, the diversification process with a large proportion of EAFE countries (59 % of the portfolio total weight) brings an additional 2.80 % return while reducing the risk from 15.80 to 14.50 %. Please note that starting from 1992 market returns were much lower.

The last two lines show a U.S. stock market average return of 9.90 % for the 1971–1988 subperiod versus a 12.70 % return for the longer period spanning from 1970 to 2000. Until 1988, investors might have been interested in investing all their funds outside of the U.S., but doing so was not appropriate: a nearly doubled return of 18.20 % came at the cost of higher risk, i.e., 17.50 versus 16.20 %.

To sum up, one benefits from diversification when returns are lower than the world average thanks to a lower risk exposure. For the same reason, the investor also takes advantage of foreign investing in the event of equivalent returns for both the domestic and the world market. If returns are high on the domestic market, an even higher return at lower risk can be generated by diversification. Not diversifying under poor domestic conditions can be more risky and less profitable as illustrated in the fifth line of Table 3.7. Finally, investing all funds in a foreign country in the event of high domestic risk and low domestic returns is not necessarily positive. Higher returns could be generated outside of the domestic market, but at the expense of higher risk. Generally, it seems always appropriate for a U.S. based investor to diversify his portfolio at least including stocks from EAFE countries.

As a matter of fact, it seems evident to rational U.S. investors to diversify their portfolio in foreign countries, assuming their decision process is solely based on the return/variance framework stated by the capital asset pricing model. Similar patterns emerge for investors outside of the U.S. One major reason lies in the fact that the market capitalization of U.S. stocks represents approximately a third of world capitalization as of year 2013 and is by itself very diversified domestically compared to France, Germany or the U.K. For an investor with less diversified funds, the benefits of international investing are then necessarily at least similar or higher. However, the home bias, i.e., the abnormal domestic investment relative to an optimal investor portfolio, persists. What can explain such an intriguing anomaly?

3.8.3 Explanations

Many explanations have been proposed to explain home bias. A very large literature supports many hypotheses and explains how they are linked to agents or market structures. The major explanations are:

- Adverse selection problem: a geographic or familiarity preference
- Shadow costs
- Information costs
- Foreign inflation issues

- Optimistic or certain expectations about the domestic market
- Irrational agent behavior

3.8.3.1 Adverse Selection Problem

The *adverse selection problem*[150] arises when an investor faces an issue of information asymmetry, typically found with small firms with little public coverage. According to this hypothesis, investors prefer to choose companies geographically near to them, believing they have a better access to information. Hence, proximity would become a means to overcome an information deficit. In 1999, Coval and Moskowitz documented a strong bias in favor of domestic securities in international investment portfolios.[151] They show that the preference for investing close to home also applies to portfolios of domestic stocks. In particular, U.S. investment managers exhibit a strong preference for locally headquartered firms, particularly small, highly levered firms that produce non-traded goods. The empirical evidence suggests that the information asymmetry between local and non-local investors may drive the preference for geographically proximate investments.

A second explanation of the adverse selection problem refers to the familiarity of investors with domestic shares. As previously said, Coval and Moskowitz show that even in a domestic environment, professionals tend to buy geographically local stocks.[152] Huberman adds that investors have the general tendency to concentrate their portfolio on shares they are familiar with.[153] This attitude leads to very undiversified portfolios: shares of firms which are geographically close to their investors are overrepresented. To support this hypothesis, Huberman shows that shareholders of a Regional Bell Operating Company (RBOC) tend to live in the area which it serves, and that its customers tend to hold its shares rather than an equity of other RBOCs.[154] While the information asymmetry appears to provide a plausible explanation, no existing model can account for the whole home bias effect, nor for its variation across countries.

3.8.3.2 Shadow Costs

The second hypothesis holds that *shadow costs* prevent investors from diversifying their portfolio with foreign shares. Let us assume an investor who is U.S. based

[150] Adverse selection, anti-selection or negative selection is a term used in economics. It refers to a market process in which undesired results occur when buyers and sellers have asymmetric information (access to different information); the *bad* products or services are more likely to be selected.

[151] Coval and Moskowitz (1999).

[152] Coval and Moskowitz (1999).

[153] Huberman (2001).

[154] Regional Bell Operating Companies (RBOC) are the result of what is called *United States v. AT & T*, the U.S. Department of Justice antitrust suit against the former American Telephone & Telegraph Company (later known as AT&T Corp.). On January 8, 1982, AT&T Corp. settled the suit and agreed to divest its local exchange service operating companies. Many local firms emerged from the AT&T split into regional companies.

and wishes to invest in Toyota on the Tokyo Stock Exchange. He might be stunned by the trading costs, which are much higher than for domestic stocks and he may look for other solutions, namely mutual funds,[155] ADRs (American Depository Receipts),[156] iShares[157] or U.S.-traded foreign stocks.[158] But the performance of these investments can also be hampered by high trading costs, like mutual fund managers' fees etc. These costs are called shadow costs. Shadow costs are the perceived annual cost of foreign equity necessary to create a bias away from perfect international risk sharing and toward domestic equity.[159]

To measure how severe home bias is, Jeske introduced a method of quantifying it. He uses a simple asset allocation model to determine the shadow costs of foreign investment. It shows that in most industrialized nations the shadow costs would have to be unrealistically high to account for home bias. For instance, in the United States the home bias is almost 150 basis points per year (1.5 %), by far the lowest among all industrialized nations,[160] while it can climb from 700 to more than 1,500 basis points in countries like Italy, Japan, Sweden or Spain. Therefore, shadow costs cannot explain home bias.

3.8.3.3 Information Costs

The third hypothesis assumes that *information costs* are cheaper for domestic assets than for foreign assets. It would then be irrational to invest in foreign markets: most of the price appraisal would be offset by the acquisition costs or the delays of delivery of the new information. Logically, a home bias would be generated. While this explanation is intuitively plausible, Jeske demonstrates, by using both a

[155] Mutual funds that invest internationally probably will have higher costs than funds that invest only in U.S. stocks. They are also liable to investment style risk: although the fund prospectuses mandate the percentages and limits of where and what to invest in, the latitude can still allow for some wide variances in style and strategy.

[156] An ADR is a registered security issued by a U.S. bank representing shares of a foreign stock. ADRs trade on U.S. stock exchanges and on the over-the-counter market. The price of an ADR corresponds to the price of the foreign stock in its home market, with some adjustments.

[157] iShares are index funds that trade like stocks. They are similar in fashion to ETFs (equity traded funds). Shares are available for both U.S. and international equity indexes. The key difference between iShares and mutual fund index funds is that mutual fund trades are executed at the end of the day (market close). iShares trade throughout the day whenever the market is open.

[158] Although in the U.S. markets most foreign stocks trade as ADRs, some foreign stocks trade in the same form as in their local market. International investing can be more expensive than investing in U.S. companies. In smaller markets, there may be a premium for purchasing shares of popular companies. In some countries, there may be unexpected taxes or transaction costs such as fees or broker commissions. Taxes are often higher than in U.S. markets. Mutual funds that invest abroad often have higher fees and expenses than funds that invest in U.S. stocks, in part because of the extra expense of trading in foreign markets.

[159] Jeske (2001, p. 31).

[160] Jeske (2001, pp. 35–36).

naive model and a rational expectations model, that the theory is unable to account for observed patterns of home bias. He concludes that home bias is still a puzzle.[161]

3.8.3.4 Foreign Inflation

Foreign inflation issues are a much discussed topic to explain home bias. What would be the benefits of diversification if the returns on a foreign market sky-rocket to 65 % with inflation reaching 64 %, while the domestic market shows an average 10 % return for equivalent volatility? Moreover, assuming inflation can be hedged, what are the benefits of hedging if the cost is higher than the profit? To answer both questions, Coen developed an international capital asset pricing model and tested it regarding home bias observed in portfolio choices for nine countries including the United States, Canada, United Kingdom, Japan, Germany, France, Italy, Spain and Sweden from March 1981 to December 1994.[162] First he tested whether the hedge against inflation alone could explain the home bias and showed that this explanation requires very high levels of risk tolerance and negative correlation between asset returns and the domestic inflation rate. By definition, investors are risk-averse, and empirical evidence shows that negative correlation is very seldom. In a second step Coen improves his model with deadweight costs including inflation.[163] Using different levels of risk aversion, he computed the necessary deadweight costs to explain home bias in late December 1994. Again, the calculated costs are too high compared to the real costs carried by international investors. Therefore, inflation issues do not seem to provide an accurate home bias explanation.

3.8.3.5 Optimistic or Certain Expectations About Domestic Markets Versus Foreign Markets

Another intuitive hypothesis holds that domestic investors hold more *optimistic or certain expectations*. Foreign markets are believed to be more risky and less lucrative. In 1991, French and Poterba provided evidence which supported this hypothesis.[164] Nine years later, Hasan and Simaan developed a model that incorporates both the foregone gains from diversification and the informational constraints of international investing.[165] They show that home equity bias is a CAPM parameter issue, i.e., of return and volatility evaluation. In particular, they prove that the risk estimations for international markets can be responsible for this phenomenon. If the expectations of investors about the mean return (currency effect deducted) and the volatility of domestic and foreign markets exceed the observed values, domestic

[161] Jeske (2001, pp. 35–36).

[162] Coën (2001).

[163] Deadweight cost is the extent to which the direct impact of an increase or reduction in tax (or subsidies) is lessened by its indirect effect. For instance, a corporate tax hike will boost government revenue but may also cause companies to go broke, which would have a negative impact on government finances.

[164] French and Poterba (1991).

[165] Hasan and Simaan (2000).

allocation overrules international diversification. To test their hypothesis, French and Poterba examined the returns on eleven international markets during 25 years from the perspective of German, Japanese and U.S. investors. Very impressively, the empirical evidence is consistent with their hypothesis. The key concept underlying home bias would then appear to be related to volatility and return parameters prior to investing in foreign markets.

3.8.3.6 Irrational Behavior and the Overconfidence of Market Agents

The sixth hypothesis is related to the irrational behavior of market agents. The explanation arises from behavioral finance concepts, as overconfidence[166] or herding behavior.[167] Such an attitude, if assumed by a wide set of agents would give rise to a market anomaly. In the present case of home bias, herding would lead investors to overweight their portfolios with domestic poor performing stocks because a majority of investors do it. In 2001, Goetzmann and Kumar tested this hypothesis.[168] They analyzed the diversification choices of more than 60,000 individual investors at a large U.S. discount brokerage house during a 6-year period from 1991 to 1996. This study shows that U.S. individual investors, especially retired people, hold underdiversified portfolios. The level of underdiversification is greater among younger, low-income, less educated, and less sophisticated investors. The level of underdiversification is also correlated with investment choices that are consistent with overconfidence, trend-following behavior (herding), and local bias (geographical bias). Investors who overweight stocks with higher volatility and higher skewness (larger extreme event probability) are also less diversified. In contrast, there is little evidence that portfolio size or transaction costs constrain diversification. While being appealing, the report of Goetzmann and Kumar does not explain the magnitude of the home bias, nor its global prevalence. It shows that poorly educated people who account for a large proportion of the U.S. population might create the shift toward domestic preference.[169] But given that there is a high correlation between income and education, their cumulative weight should not have a strong impact on the market.

3.8.4 Summary

There are currently six popular explanations for home bias, but each one of them is unable to fully and accurately account for this anomaly. The optimistic

[166]Overconfidence arises from the belief that one's knowledge is of great quality in spite of conflicting evidence. An in-depth explanation is proposed in Sect. 5.3.5 which introduces behavioral finance in order to explain stock market crashes.

[167]Herding is an attitude of individuals who follow a trend rather than higher quality information which they possess in the context of finance for instance. See Sect. 5.3.3 for an in-depth explanation.

[168]Goetzmann and Kumar (2001).

[169]Ryan and Siebens (2012, p. 2).

expectations about domestic markets compared to foreign markets seem to be the most rational explanation for the overweight of domestic assets. There is still controversy about human capital limitations, i.e., the amount invested in stocks is not large enough to allow for international allocation. For instance, Goetzmann and Kumar argue that there is little evidence that portfolio size or transaction costs constrain diversification,[170] while Coen affirms that portfolio size does have an impact, yet not big enough to justify the bias.[171]

The persistence of home bias is not at stake, given that it cannot be arbitraged. However, considering the creation of the euro currency and its use in the eurozone, Schoenmaker and Bosch argue that its effect is shrinking in an uneven way among European countries.[172] Generally speaking, the home bias does not bring abnormal profits, since the systematic attempt to diversify portfolios gives rise to a higher correlation between foreign markets, thus hampering the international diversification effect.

3.9 Value Line Enigma

The Value Line enigma is a very interesting anomaly related to stock exchanges. It is also sometimes referred to as the typical implementation pitfall.

It started when a company called Value Line Investment Survey (below: Value Line) launched a fund, after it had ranked a majority of U.S. outstanding shares based on their historical performance. On paper, the strategy generated a return of more than 33,000 % for the period from 1965 to 2012. However, the fund did not perform as well as predicted and performed even worse than the Dow Jones Industrial Average Index.[173] How was this possible? In the following, Value Line and its infamous Ranking System Timeliness are presented and evidence of the estimated performance is provided and explained. The section is closed by discussing the incredible persistence of this anomaly.

3.9.1 Description of Value Line Ranking System Timeliness

Value Line is a company that provides investment services on a large scale of products and strategies. One of its most famous services is a trade mark called Ranking System Timeliness (below: Timeliness), which ranks approximately 1,700 stocks relative to each other according to their price performance. This amount accounts for approximately 90 % of the market capitalization of all stocks traded in the United States. The objective of this service is to predict stock price movements

[170] Goetzmann and Kumar (2001).

[171] Coën (2001).

[172] Schoenmaker and Bosch (2008).

[173] http://www.valueline.com/About/Ranking_System.aspx.

over a 3–6-month period, or 6–12-month period, depending on its version. To compute the ranking, various criteria are applied. The major ones are the 10-year trend of relative earnings and prices, recent earnings and price changes, and earnings surprises. The ranking is computed and published every week.

Stocks are classified into five categories. Category 1 is the highest-ranked category and includes stocks which are expected to reach the highest expected returns. Category 5 is at the bottom end of the ranking.[174] Each category is best described as follows:

- **Rank 1 (Highest):**
  These stocks, as a group, are expected to be the best performers relative to the Value Line universe during the next 6–12 months (100 stocks).
- **Rank 2 (Above Average):**
  These stocks, as a group, are expected to achieve a better-than-average price performance (300 stocks).
- **Rank 3 (Average):**
  These stocks, as a group, are expected to show a price performance in line with the Value Line universe (approximately 900 stocks).
- **Rank 4 (Below Average):**
  These stocks, as a group, are expected to perform below average (approximately 300 stocks).
- **Rank 5 (Lowest):**
  These stocks, as a group, are expected to have the poorest performance (100 stocks).

According to the company's website, changes in the Timeliness ranking can result from either new earnings reports of company forecasts, or changes in the price movement of the stock in comparison with the sample's stocks, or a shift in its relative position versus other stocks.

3.9.2 Evidence of Outperformance

Let us now look at the performance of the ranking, which is shown in Fig. 3.8. The figure, which is extracted from the corporate website,[175] represents the cumulative historical performance of Value Line's Timeliness from April 16, 1965 to June 28, 2013. Note that this graph is based on weekly stock rankings. Value Line provides less frequent ranking re-evaluations, while achieving astonishing expected returns. It is important to recall that the ranking system is based on public information only.

The performance ranges from −99 % to an estimated 40,616 % for June 2013. The Dow Jones Industrial Average Index is used as a benchmark. Let us recall that

[174] http://www.valueline.com/About/Ranking_System.aspx.

[175] http://www.valueline.com/About/Ranking_System.aspx.

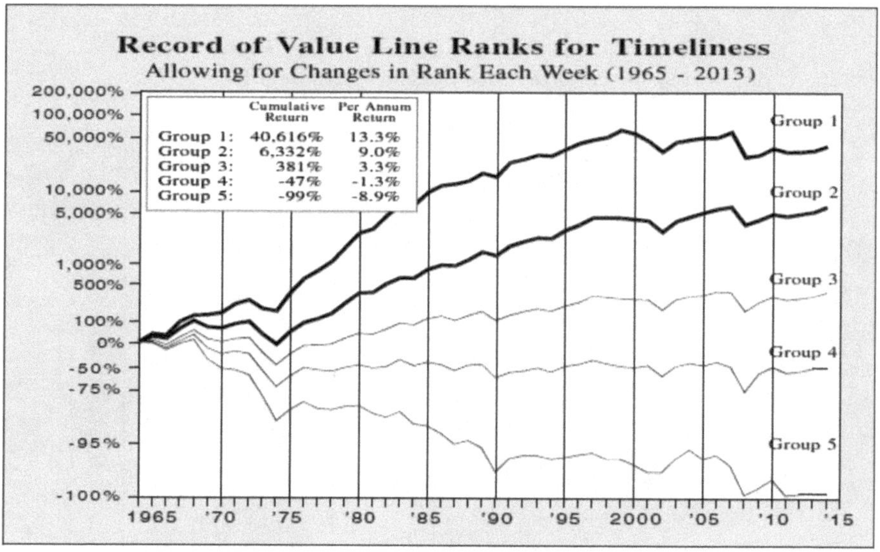

Fig. 3.8 Cumulative performance of Value Line's Timeliness from April 16, 1965 to June 28, 2013. *Source*: Value Line website at http://www.valueline.com/About/Ranking_System.aspx

the DJIA reached 1,000 points on January 18, 1966, and rose to 14,000 points on July 16, 2007, creating a maximum wealth multiple of fourteen.[176] On December 30, 1965, the DJIA stood 969.26 points versus 14909.60 on June 28, 2013, i.e., in this period, an initial investment would have been multiplied by a bit more than fourteen.[177] An investment in the first rank of Value Line would have multiplied by more than three hundred. An investment in the stocks of rank 1 and 2 would have generated a yearly return of 13.3 % over the period.[178] This is far more than the historical DJIA yearly return of 9.4 %.[179]

Unfortunately, only few of the readers of this book could have invested that long ago. Therefore, Fig. 3.9 is extracted from the corporate website.[180] It represents the cumulative historical performance of Timeliness from December 30, 1988 to June 28, 2013.

The performance ranges from 664 to 1,924 % for June 2013, despite two major financial shocks. A second interesting point in this figure is the addition of the S&P

[176] *Source*: Bloomberg (ticker: INDU:IND).

[177] *Source*: http://www.fedprimerate.com/dow-jones-industrial-average-history-djia.htm.

[178] This is the weighting average for rank 1 and 2 stocks, given that there are 100 stocks in rank 1 and 300 stocks in rank 2, we then have to multiple their relative weight by their respective performance.

[179] *Source*: Own, based on historical data. Large differences are still observed even starting in 1900 or in 1982 until 2013.

[180] http://www.valueline.com/About/Ranking_System.aspx.

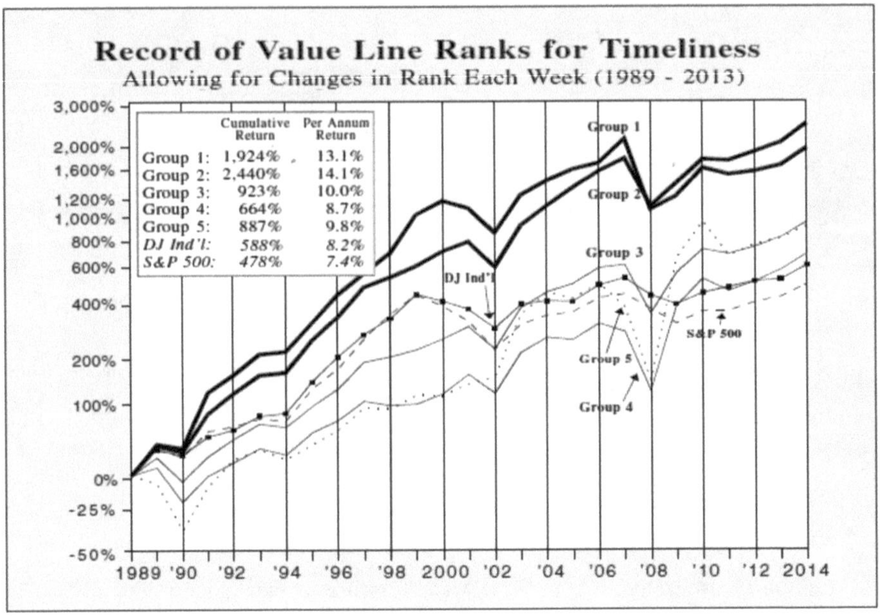

Fig. 3.9 Cumulative performance of Value Line's Timeliness from December 30, 1988 to June 28, 2013. *Source*: Value Line website at http://www.valueline.com/About/Ranking_System.aspx

500 Index and the DJIA for a direct comparison. For the equivalent period, their performance was 478 and 588 %, respectively. In other words, the first rank category of Value Line still presents an outstanding performance, but even the rank 5 stocks outperform compared to a typical benchmark. As a matter of fact, the impressive performance of the rating system led many to refer to it as the *Value Line enigma*.[181] It is very important, however, that the two figures do not include any trading costs, neither for the Value Line stocks nor for the benchmark indices.

In 1973, Fischer Black conducted the first large scale test of the Value Line enigma.[182] In his article, Black admits that he had been a strong believer in the EMH and passive management. He found that the rank 1 portfolio earned annual risk-adjusted returns 20 % greater than the rank 5 portfolio. The system did produce significant excess returns over a 5-year period. According to his report, abnormal returns would have resulted even after taking 2 % out for round trip transactions costs, given that the turnover in the ranked stocks was high.

Copeland and Mayers extended Black's study until 1978.[183] They found equivalent results, with a 7.07 % return differential between rank 1 and rank 5 firms

[181] Porras and Griswold (2000).

[182] Black and Kaplan (1973).

[183] Copeland and Mayers (1982).

for semi-annual portfolios when future benchmarks are computed using a market model. Stickel also found positive risk-adjusted abnormal positive returns using Value Line rankings to form trading strategies, thus challenging the EMH.[184]

Based on strong academic support for its strategy, Value Line decided to launch a mutual fund, the Value Line Centurion fund which invested in rank 1 and rank 2 stocks only (four hundred stocks). However, not only did the real-money fund not keep pace with the paper returns from the top-rated stocks, which continued to outperform, but it hardly outperformed the market during its first years of existence.[185] But over the period from 1979 to 1991, the real Value Line fund had an annualized return of 16.1 %, which is more than the historical long-term average DJIA yearly return of 9.4 %, although it is less than the annualized return of the Value Line paper portfolio of 26.2 %.[186] Value Line continues to be one of the highest ranked newsletters by the Hulbert Financial Digest which does account for costs.[187]

Two questions have to be asked: How is it possible to generate such an abnormal return compared to passively managed indices (i.e., DJIA, S&P 500, etc.)? How come the implementation of the ranking system did not produce the expected results despite a rather conservative approach about trading costs? We will first deal with the question of abnormal returns.

3.9.3 Possible Explanations for Abnormal Returns

There are currently four hypotheses to explain the overperformance, but none is really satisfying:

- Post-announcement effect
- New information provided by Value Line
- Construction scheme of Timeliness
- Economic state factors

3.9.3.1 Post-announcement Drift

The first proposed explanation is related to what is called a post-announcement drift. In other words, the market receives a new positive or negative information, and the price of the given stocks reacts with a time lag. Since Value Line's Timeliness is managed weekly, the outperformance would only be the result of a lag between the information publication and its short-term inclusion in the stock price. Affleck-Graves and Mendenhall addressed the issue whether Value Line was able to predict future stock prices or whether the outperformance of its fund was related

[184] Stickel (1985).

[185] Porras and Griswold (2000, p. 39).

[186] Leinweber (1995, p. 2).

[187] Porras and Griswold (2000, p. 40).

to post-announcement drift.[188] According to their research, the abnormal returns across Value Line Timeliness ranks are no longer significant when post-earnings announcements are deleted. They also add that Timeliness' ranks have no predictive power for firms with small earnings surprises (i.e., unexpected excess earnings compared to consensus expectations). They assume that the Value Line enigma is a manifestation of post-earnings-announcement drift. This is in line with Stickel, who asserts that the price reaction takes place several days later, with a stronger effect on small firms.[189] However, even though this analysis concurs perfectly with the efficient market hypothesis, Choi and Peterson found contrary evidence about the underreaction of agents.[190]

3.9.3.2 New Information Provided by Value Line

The second reason invoked to explain the outstanding paper performance of Value Line's Timeliness is information supply. According to this hypothesis, the company provides additional information that is reflected in the price of its recommended stocks thanks to their ranking. Since the publication of Copeland and Mayers's article, researchers have focused their attention on the market reaction on the release date of Value Line's weekly ranking.[191] In 1995, Peterson wanted to know if abnormal returns are due to pertinent information or post-announcement drift through an examination of the *Stock Highlights* section of Value Line.[192] He concluded that highlighted stocks earned abnormal returns of 2.42 % over a 3-day announcement period.[193] No correlation could be found between abnormal returns and the time lag between the last earnings announcement and the Value Line publication. Choi also analyzed the investment advice implied in the Timeliness rankings from 1965 to 1996. However, to prevent any bias due to earnings surprises, which Peterson did not, he selected a sample which eliminated post-announcement drift. According to Choi, Value Line recommendations exhibit a performance beyond what is predicted by existing models of expected return.[194] In 1999, Graham provided evidence that other investment newsletters view Value Line as the market leader and respond to its recommendations.[195] Apparently investors believe that Value Line does provide additional information. However, this seems inconsistent with the fact that Value Line only uses publicly available information.

---

[188] Affleck-Graves and Mendenhall (1992).

[189] Stickel (1985, p. 121).

[190] Choi (2000) and Peterson (1995).

[191] Copeland and Mayers (1982).

[192] Peterson (1995).

[193] Zhang, Nguyen, and Le (2010) reaches the same conclusion, see p. 372.

[194] Choi (2000).

[195] Porras and Griswold (2000).

3.9.3.3 Construction Scheme of Timeliness

A third hypothesis is related to how Timeliness is constructed. Bad performers are simply not included in the model. In 2000, Porras and Griswold re-examined Copeland and Mayers's study with former and newly found models since the 1980s.[196] Like Copeland and Mayers, they underline that the outperformance is a result of the abnormal negative performance of firms that Value Line ranks poorly, not of the positive performance of firms it recommends. This result appears clearly counter-intuitive. Yet, while a comprehensive explanation should account for each rank of Timeliness, this observation cannot explain the abnormal returns of rank 1.

A second construction-related issue is about firm size. Previous research has shown that small firms on the New York Stock Exchange have performed better with a cumulative return of 20.65 % than large firms with only 1.53 % over 28 years.[197] Given that in an index of 1,700 shares there is a much higher quantity of small firms than, for example, in the S&P 500 Index, an upward bias should be expected. Huberman and Kandel studied the relation between Value Line's successful record in predicting relative stock price movements and the firm size effect.[198] The data suggests little direct relation between the two phenomena. They also remarked that Value Line tends *not* to rank stocks of small firms. Even if they are ranked, they are more likely to receive a low rank than large firms. Stickel further notices that a downgrade from category 1 to category 2 has the most dramatic price effect. This effect is negatively increased in the case of a downgrading of a small-sized firm.[199] As a matter of fact, neither the small firm effect nor the general construction of Timeliness can explain the outperformance.

3.9.3.4 Macroeconomic Factors

A last and fourth explanation is related to macroeconomic factors. According to this hypothesis, Value Line was strongly favored by specific economic developments. For instance, the company could promote firms selling wind turbines when a big rush on renewable energy was foreseeable. In 1990, Huberman and Kandel proposed that the outperformance of the rank 1 portfolio was simply a compensation for holding systematic risk,[200] which contradicts Black and Kaplan (1973).[201] What is more, looking back to Fig. 3.9, rank 1 shares suffered a lot during the subprime crisis in 2008 and the dotcom crisis beginning 2000. It seems unlikely that Timeliness has benefited from particular economic trends. Timeliness also seemed to perform poorly before the bubble burst. Furthermore, the out- or underperformance does not hold for former crises, for example, the oil price shocks in the 1970s. Why did the

[196] Porras and Griswold (2000).

[197] Lustig and Leinbach (1983, p. 46).

[198] Huberman and Kandel (1987).

[199] Stickel (1985, p. 121).

[200] Huberman and Kandel (1990, p. 187).

[201] Black and Kaplan (1973).

rank 1 stocks still perform better? Such an hypothesis, while being plausible, is difficult to prove for a long time period.

From all possible explanations, only the *new information hypothesis* appears plausible and sufficient to explain the Value Line enigma. Value Line does not exactly present new information. However, a plausible explanation would be that it makes available information more visible, based on quantitative and qualitative results, i.e., the ranking process. This definitely plays a role in the information spreading process, but the very existence of Value Line proves a significant lag in the diffusion of information, and as such brings serious harm to the efficient market hypothesis.

After the at least partial explanation of the abnormal returns of the Value Line strategy, we can focus on the implementation shortfall.

3.9.4 Possible Explanations for Implementation Shortfall

The term *implementation shortfall* was coined by Andre Perold who has written extensively on the subject and how to measure it.[202] In the case of the Value Line enigma, two major reasons seem to be plausible for the implementation shortfall:

- Trading costs and cash disposal
- Victim of its own success

3.9.4.1 Trading Costs and Cash Disposal

The first hypothesis for the implementation shortfall is related to trading costs and money disposal. Trading costs reduce the potential benefit that could arise from an information advantage or mispricing. This is very common on stock exchanges which means that strategies that seem to offer an advantage to investors may not work under real world conditions because of transaction costs and other costs.[203] Perold also argued that the larger a portfolio, the harder it is to exploit any informational advantage. Previously, a most probable explanation of the abnormal return of Value Line's Timeliness seemed to be the addition of information. But given that the portfolio contains no less than 1,700 stocks, the value of additional information can be strongly diluted or may be neutralized by costs related to taking advantage of the information, or more simply to obtain this information. Investors must also account for the bid-ask spread, and mutual funds typically have the added burden of not being 100 % invested because of the need to maintain cash reserves. In 2000, Choi, who does recognize model-based abnormal returns, expressed his doubt about the actual realization of these abnormal returns.[204]

---

[202] Leinweber (1995, p. 2).

[203] Perold (1988).

[204] Choi (2000).

3.9.4.2 Victim of Its Own Success

The second hypothesis refers to the outstanding success of Value Line's Timeliness. As previously mentioned, Leinweber computed that, from 1979 to 1991, the Value Line paper portfolio had an annualized return of 26.2 %,[205] whereas the real Value Line fund only earned 16.1 % per year,[206] i.e., the paper returns were not realizable by the mutual fund. The reporter Salomon Jr. states in *Value Line's self-defeating success*: *"Value Line's rankings are a prisoner of their own success: They work so well that too many people try to act on them."*[207] Simply speaking, the anomaly seems to diminish with time passing by as it should in theory. But this observation can hardly be validated for 2012 considering the difficult general economic situation. For 2012 the strategy seems to be successful as shown in Fig. 3.9 although the data are not risk-adjusted. It is also important to note that a court order in the 1960s mandated a delay between the publication of Value Line rating changes and trading in the portfolio.[208] Once again, finding an anomaly and exploiting it are two different things.

The available evidence seems to show that trading costs are the main reason for Value Line's implementation shortfall. This apparently prevents any investor from beating the market. But even assuming trading costs were cut, what is the reason for the persistence of the Value Line enigma?

3.9.5 Persistence

A recent research of Zhang, Nguyen and Le published in 2010 deals with an event concerning the Value Line enigma.[209] Initially, the weekly update of the Timeliness ranking was published on Fridays. However, starting from June 5, 2005, Value Line published it 1 day before: on Thursdays. The publication states that the Value Line effect still exists after more than 50 years of existence. It persists, despite the large academic work about it. The researchers assert, contradicting previous research, that the next-day abnormal return after the announcement has disappeared. They interpret this as an improvement in market efficiency, and as a sign of a possible weakening of the Value Line anomaly. The highest cumulative abnormal return reaches 9.07 % over a 50-day window, and is not achieved by the rank 1 portfolio.[210] Finally, despite the evident mutation of the Value Line enigma, post-earning-announcement drift is once again disproved, thus confirming the new information hypothesis as the explanation for price increases.

---

[205] Leinweber (1995, p. 41).

[206] Leinweber (1995, p. 42).

[207] Salomon Jr. (1998).

[208] Porras and Griswold (2000, p. 40).

[209] Zhang et al. (2010).

[210] Zhang et al. (2010, p. 362).

The Value Line enigma has not been completely solved yet. It seems that the observed abnormal returns are related to the re-release of already public information by Value Line. The performance of Timeliness may be achieved by the exclusion of poorly performing stocks rather than by the aggregation of hot stocks. Value Line Centurion, the mutual fund implementing the strategy, has to face three technical issues. The first one is implementation shortfall, which is common on the stock exchanges. The second is that strategies back-tested by historical data do not necessarily indicate future returns in correlation with past ones. The third one is related to trading costs. The high turnover in each Timeliness rank deteriorates the performance. The fund managers are also tied by regulatory issues: to observe a delay between the publication of the ranking and trading. Therefore, the exploitation of the anomaly by the fund is not optimal. Private investors face even higher difficulties because trading costs are for them higher than for investment institutions. The Value Line anomaly persists, while possibly changing due to recent economic conditions.

3.10 Expiry of IPO Lockups

The expiry of IPO lockups is another interesting structural market anomaly, the last one we will look at. Actually, the anomaly has been widely documented and several explanations have been suggested. But as for many other anomalies, the abnormal returns do not seem to have a single explanation but many possible ones. In this section we describe the anomaly and the pertinent technical terms. Then, evidence and explanations are presented before discussing the persistence of the anomaly.

3.10.1 Description

An IPO is the introduction of a new firm on a stock exchange. It stands for *initial public offering*. Lockups are agreements made by insiders of the stock-issuing firms to abstain from selling shares for a specified period of time after the issue. The lockup period typically ranges from 90 to 180 days, but is not always mandatory. Figure 3.10 plots the number of IPOs and their lockup lengths in days in each year from 1988 to 1997 for 2,529 companies. During this period, 180 days seemed to be the rule. For the subsequent decade, analogous results were found.

Early studies by Reilly and Hatfield (1969) and Stoll and Curley (1970) show a significant difference between the offering price of IPOs (determined by the firm and the underwriter) and the first-day or first-week closing market price.[211] From 1990 to 2001, first-day returns on U.S. IPOs were approximately 25 %.[212] This raises the question whether firms and underwriters misprice IPOs because it appears as if they

[211]Reilly and Hatfield (1969) and Stoll and Curley (1970).

[212]Ritter and Welch (2002, Table 1, p. 4).

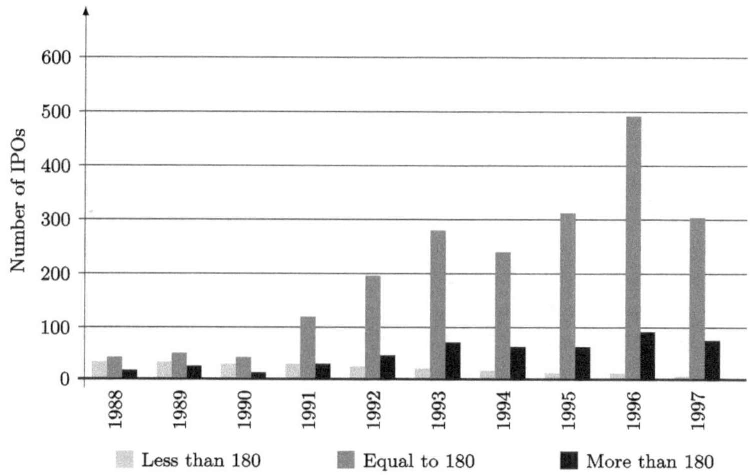

Fig. 3.10 Number of IPOs and lockup length by year from 1988 to 1997. *Source*: Bradley, Jordan, Roten, and Yi (2001, Fig. 1, p. 48)

could have sold the shares at a higher price. The second issue lies in the fact that apparently, even if the final date of the lockup is publicly disclosed, the expiry of IPO lockup effect remains. The anomaly arises from an abnormal estimated loss of −1.5 % in the time immediately surrounding the announcement day.[213] However, on August 16, 1999, Healtheon's stock[214] fell 18.5 % in a single day.[215] More recently in 2012, Facebook[216] shares grew 11 % on the lockup end date, while investors had been nervous about this previously hot stock falling apart for months.[217] A large scale analysis is required to define the general properties of the anomaly if there are any. They would help investors to understand the two issues connected to IPO lockups, and would prevent fear effects that might distort stock prices.

[213]Field and Hanka (2001, p. 472).

[214]Healtheon was a dotcom startup company. Healtheon's business plan was to streamline communication and paperwork in the United States health care system. They developed software that placed their company between physicians, patients, and health care institutions, eliminating unnecessary paperwork and facilitating networking and communication amongst the three.

[215]Field and Hanka (2001, p. 472).

[216]Facebook is a social networking service launched in February 2004. In 2012, Facebook had over one billion active users.

[217]*Source*: Bloomberg (ticker: FB:US).

3.10.2 Evidence

Ritter demonstrated that long-term investors in IPOs, who buy shares immediately after the offering at market prices, realize low returns.[218] Together with Loughran, he also showed that IPOs underperform non-IPO firms by approximately 25 and 50 % on a respectively 3-year and 5-year horizon, respectively.[219] This empirical result is interpreted as evidence for the *overvaluation* of IPOs during the first days of market trading.[220] About at the same time, Aggarwal and Rivoli used a large scale IPO database to study the abnormal return generated by a buy and hold position following the introduction date of the IPO and measured negative performance.[221] This contradicts the popular belief that IPO investing is smart.

Ten years later, Fields and Hanka analyzed 3,217 lockup agreements that prevented insiders from selling shares immediately after the IPO from 1988 to 1997. In the week the lockup period expires, they found a permanent 40 % increase in average trading volume and a cumulative abnormal return of −1.8 % on the studied sample.[222] The abnormal return seems to be stable over the 10 year sample period, yet this is not due to changes in the proportion of trades at the bid price. Technically speaking, there is no mean reversion. The abnormal return is much more pronounced when the firm is venture-financed. Actually, venture funds sell more aggressively than other pre-IPO shareholders.[223]

Figure 3.11 illustrates the cumulative abnormal return from day −5 to day +23 around lockup expiration for the sample of 2,529 firms. Cumulative abnormal returns are calculated using the value-weighted index from 1988 through 1997. The figure, which is taken from the empirical study of Bradley et al. (2001), shows that after a single business week, the price drift reaches −5 % and seems to remain at the same level through the following month.[224] There is, however, a decrease a few days before the actual end date of the lockup period. This might be interpreted as an adaptation of market agents to upcoming future downward price movements.

Figure 3.12 illustrates the impact of the lockup end date on the average daily trading volume for venture-backed and non-venture-backed IPOs around lockup expiration from 1988 through 1997. Day 0 represents the lockup expiration. In the case of a non-venture-backed security, the impact on the volume appears negligible. It is a non-event. However, venture backing proves to have a significant impact on trading volumes. This might be a caused by inside traders who make money from their investment and spread a much larger volume of stock on the stock exchange.

[218]Ritter (1991).

[219]Loughran and Ritter (1995, p. 30).

[220]Loughran and Ritter (1995, p. 49).

[221]Aggarwal and Rivoli (1990).

[222]Field and Hanka (2001, Table IV, p. 482).

[223]Field and Hanka (2001, Table IV, p. 482).

[224]Bradley et al. (2001, p. 14).

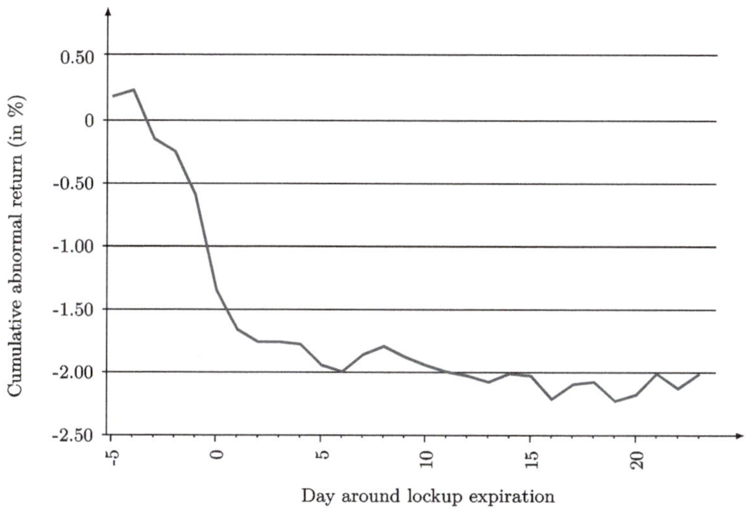

Fig. 3.11 Cumulative abnormal return from 1988 to 1997. *Source*: Bradley et al. (2001, Fig. 2, p. 49)

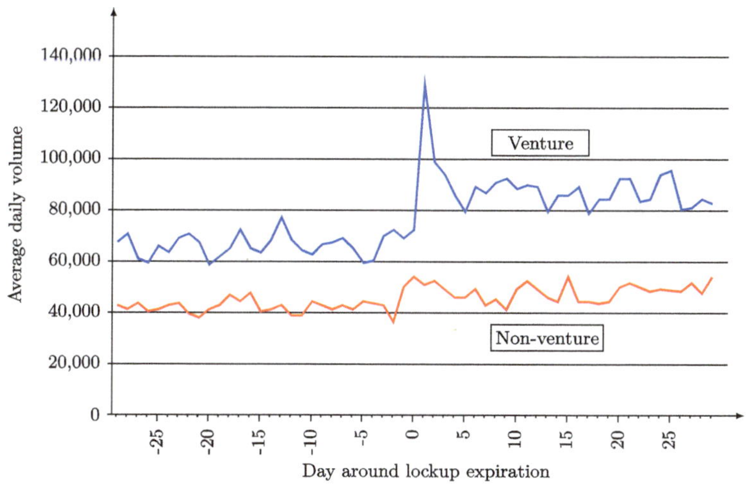

Fig. 3.12 Average daily trading volume for venture-backed and non-venture-backed firms from 1988 to 1997. *Source*: Bradley et al. (2001, Fig. 5, p. 52)

A last characteristic of IPO lockups is related to their agents. Fields and Hanka found that prior to the scheduled expiration date, 6 % of lockup agreements are abrogated by substantial insider share sales.[225] These unrestricted investors liquidate positions prior to the scheduled lockup release, thus possibly generating negative

[225] Field and Hanka (2001, p. 473).

returns. It seems that negative abnormal returns are more robust for firms that are not influenced by SEC Rule 144.[226] It seems logical that insider sales have an impact on price movements. Yet, clear explanations are still needed.

3.10.3 Explanations

Three major hypothesis exist to explain this anomaly:

- Adverse selection
- Stock momentum
- Market inefficiency

3.10.3.1 Adverse Selection

Brau, Lambson, and McQueen (2005) consider IPO lockups as a solution to the *adverse selection problem*[227] which results from information asymmetries at the time of the stock issue.[228] Insiders who stick to their lockup commitments send a positive signal, whereas insiders who sell faster than expected send a bad signal. Therefore, to forgo sales within the lockup period is a costless way for insiders to prevent negative price movements. If the hypothesis holds true, lockups should be shorter when the degree of asymmetric information is small (high-transparency firms) and when the cost of mimicking is high (risky firms). The empirical results for a sample of 4,013 initial public offerings and 3,279 seasoned equity offerings between 1988 and 1999 support these predictions. Therefore, adverse selection may appear as the best explanation. It explains the underpricing of IPOs at least in the short term and takes into account that venture capital investors and inside traders tend to dispose of their shares at an early stage. But it does not explain the long-term effect of negative abnormal returns.

3.10.3.2 Stock Momentum

The second explanation represents another interpretation of the usefulness of the lockup period. Does the lockup period answer to a psychological need to protect bankers and potential investors from a large negative price drift? According to Brav and Gompers, evidence shows that lockups are a bonding solution for a moral hazard

[226] Rule 144 allows the public resale of restricted and controlled securities if a number of conditions are met. For example, holding period, adequate stock information, personal information and trading volume are criteria limiting the resale. The complete rule set is available at the following website: http://www.sec.gov/investor/pubs/rule144.htm. Also see Keasler (2001).

[227] Adverse selection, anti-selection, or negative selection is a term used in economics. It refers to a market process in which undesired results occur when buyers and sellers have asymmetric information (access to different information); the *bad* products or services are more likely to be selected.

[228] Brau et al. (2005).

problem and not a solution for an adverse selection problem.[229] In 2002, Aggarwal, Krigman and Womack developed another theory: managers strategically underprice IPOs to maximize personal wealth from selling shares at lockup expiration.[230] According to this hypothesis, the underpricing generates information momentum by attracting attention to the stock and thereby increasing the demand for the stock. This allows managers to sell shares at lockup expiration at higher prices than they would have obtained otherwise.

A sample of IPOs in the 1990s shows that higher ownership by managers is positively correlated with first-day underpricing. Underpricing is positively correlated with research coverage, and research coverage is positively correlated with stock returns and insider selling at lockup expiration. These results are consistent with the model of Rajesh, Krigman and Womack. The IPO mispricing is done on purpose with the objective to generate a momentum effect.[231] Given that a momentum is the combination of an underreaction, followed by an overreaction, inside traders and bankers can generate abnormal returns at a later date (lockup expiration date) as compared to the introduction date of the new stock. While being very satisfying in many aspects, this explanation does not cover the 6 % of early sales of inside traders.[232]

3.10.3.3 Market Inefficiency

The third hypothesis holds that IPOs are launched at the appropriate price. Markets, however, are then believed to be inefficient in the post-IPO period, as prices do not reflect all available information. If the market were efficient, newly introduced stocks should perform at least as well as their counterparts on the stock exchange. In 2001, Ofek and Richardson investigated the volume and price patterns around the lockup ending period.[233] They showed that, even though the event is totally anticipated, stock prices drop from −1.15 to −3.09 %, and the trading volume increases by 38 %, when the lockup expires.[234] Does the market adapt to upcoming events? The researchers tested bid-ask bounce,[235] liquidity effects,[236] and biased expectations of supply shocks,[237] but found little support for the relevance of these factors. Nevertheless, the evidence points to a *downward sloping demand curve for*

229 Brav and Gompers (2003).

230 Aggarwal, Krigman, and Womack (2002).

231 Aggarwal et al. (2002).

232 Field and Hanka (2001, p. 473).

233 Ofek and Richardson (2000).

234 Ofek and Richardson (2000, p. 2).

235 Closing ask price on the previous day can be taken as today's bid price. While creating no abnormal return, the anomaly would be related to normal market frictions between bought and sold stocks.

236 Investors prefer liquid assets and pay a premium for them. Illiquid assets perform less well.

237 Large demand or supply expectations are a fear factor for investors. In the case of not knowing the appropriate expectation, a small drift toward the correct stock price can be observed.

shares, with the most likely explanation pointing to a permanent, long-run effect.[238] With a positive shift in supply[239] and a downward sloping demand curve, the price would be expected to fall. That is why Ofek and Richardson used two popular explanations for this possible result: price pressure (i.e., temporarily downward sloping demand curve) versus long-term demand effects. The evidence supports the significance of long-term demand effects, but not entirely, as some results are inconsistent with implications from long-term downward sloping demand curves.

Markets seem to have a low demand for a newly listed stock in the years following its IPO. This low demand can only be explained by the unability of the investors to include pertinent information in the price of the share. This hypothesis holds for early stock disposal. Inside traders know more, and thus sell before the stock falls in value. It also holds for the short-term effect. Corporate owners sell their additional shares on the lockup expiration date. It also holds for a price drift due to the inability of market agents to incorporate accurate information into the price. Yet, it does not hold for a 10-year period.

Professionals and researchers have reached a consensus about the undervaluation of IPO prices and the momentum exploitation. The momentum effect has a long persistence as illustrated in the former section about industry portfolio momentum. But how does it affect IPO lockups?

3.10.4 Persistence

Stock momentum and the slow subsequent price drift are long-lasting phenomena. In the case of IPO lockups, however, the time span seems to be exceptionally long, namely more than 10 years.[240] Protection against the price fall on lockup expiration is, unfortunately, nearly impossible. Since arbitrageurs should know in advance about the expected price drop, short selling is the consequence. According to Ofek and Richardson's report, the magnitude of the price drop is related to the stock's underlying volatility. Further research in this direction might bear fruit.

To sum up, newly introduced stocks seem to be undervalued on the IPO date. This undervaluation is bigger for expensively priced stocks, and smaller for less expensive stocks. This mispricing seems to trigger a stock momentum that bankers and corporate owners can exploit at the lockup expiry, typically 180 days following the IPO date.

---

[238]Ofek and Richardson (2000).

[239]The shift in supply can be explained by the fact that on the IPO date, company owners typically sell 15–20 % of their stock. On lockup expiry, more shares can be sold.

[240]Field and Hanka (2001).

3.11 Summary

Through this section, the reader has been able to witness the following market anomalies:

- Weekend effect
- January effect
- Turn-of-the-month effect and holiday effect
- S&P 500 effect
- Trading by insiders
- Momentum of industry portfolio
- Home bias
- Value Line enigma
- Expiry of IPO lockups

Each of these anomalies has been fully presented, described and explained where possible, based on academic research and case studies. Each of them presents a challenge to standard modern portfolio theory. For instance, home bias should not appear if the capital asset pricing model holds fully true. If the efficient market hypothesis holds true, even under its semi-strong form, the Value Line enigma would not be observed. The aim of this chapter is not to disprove the progress and benefits of both theories, and the large set of evidence in favor of market efficiency and optimal portfolio construction are not challenged. However, it seems evident that a blind belief in these models appears perilous. They are not as valid as the $1 + 1 = 2$ equation is true for fundamental arithmetic. A model is an attempt to explain empirical phenomena so that it accounts for most of their effects. Anomalies constitute evidence that these models can and have to be improved. Human behavior, sometimes irrational and sometimes constrained by market-distorting regulations, is a strong driver for further research in the world of market finance.

In the 1980s a large set of market anomalies was discovered. The ones presented above are strictly stock-related and were found on various financial markets. But since these anomalies cannot be exploited by arbitrage, some researchers argue that the efficient market hypothesis holds. Some anomalies are cyclical, i.e., they appear and disappear at periodic times without clear explanations. It is believed that a large variety of anomalies are yet to be discovered or publicly published. Logically, a smart investor would present the anomaly only after he has arbitraged it for a long time. The presented explanations account for most of the frequently provided hypotheses used to explain anomalies. For instance, information asymmetry is a typically spread concept to explain them.

It is important to note that the anomalies presented in this chapter are relatively harmless. They deviate from the efficient market hypothesis and are not easy to exploit. However, the stock market crashes, which have occurred in the market for more than 400 years, are much more severe deviations. We will look at them in the next chapter.

References

Affleck-Graves, J., & Mendenhall, R. R. (1992). The relation between the Value Line enigma and post-earnings-announcement drift. *Journal of Financial Economics, 31*(1), 75–96.

Aggarwal, R. K., Krigman, L., & Womack, K. L. (2002). Strategic IPO underpricing, information momentum, and lockup expiration selling. *Journal of Financial Economics, 66*, 105–137.

Aggarwal, R. K., & Rivoli, P. (1990). Fads in the initial public offering market? *Financial Management, 19*(4), 45–57.

Agrawal, A., & Tandon, K. (1994). Anomalies or illusions? Evidence from stock markets in eighteen countries. *Journal of International Money and Finance, 13*(1), 83–106.

Arbel, A. (1985). Generic stocks: An old product in a new package. *Journal of Portfolio Management, 11*(4), 4–13.

Ariel, R. A. (1987). A monthly effect in stock returns. *Journal of Financial Economics, 18*(1), 161–174.

Ariel, R. A. (1990, December). High stock returns before holidays: Existence and evidence on possible causes. *Journal of Finance, 45*(5), 1611–1626.

Barclay, M. J., & Warner, J. B. (1993). Stealth trading and volatility: Which trades move prices? *Journal of Financial Economics, 34*(3), 281–305.

Basu, S. (1977). Investment performance of common stocks in relation to their price-earnings ratios: A test of the efficient market hypothesis. *Journal of Finance, 32*(3), 663–682.

Benesh, G. A., & Pari, R. A. (1987). Performance of stocks recommended on the basis of insider trading activity. *Financial Review, 22*(1), 145–158.

Bettis, C., Vickrey, D., & Vickrey, D. W. (1997). Mimickers of corporate insiders who make large-volume trades. *Financial Analyst Journal, 53*(5), 57–66.

Bhabra, H. S., Dhillon, U. S., & Ramirez, G. G. (1999). A November effect? Revisiting the tax-loss-selling hypothesis. *Financial Management, 28*(4), 5–15.

Bhardwaj, R. K., & Brooks, L. D. (1992, June). The January anomaly: Effects of low share price, transaction costs, and bid-ask bias. *Journal of Finance, 47*(2), 553–575.

Black, F. S., & Kaplan, R. S. (1973). Yes, Virginia, there is hope: Tests of the Value Line ranking system. *Financial Analysts Journal, 29*(5), 10, 12, 14 and 92.

Blume, M. E., & Edelen, R. M. (2002, April). *On replicating the S&P 500 index*. Working Paper 08-02. Philadelphia: The Rodney L. White Center for Financial Research, Department of Finance, Wharton School, University of Pennsylvania.

Bos, R. (2000, September). *Quantifying the effect of being added to an S&P index*. New York: Standard & Poor's.

Bradley, D. J., Jordan, B. D., Roten, I. C., & Yi, H. C. (2001). Venture capital and IPO lockup expiration: An empirical analysis. *Journal of Financial Research, 24*(4), 465–493.

Brau, J. C., Lambson, V. E., & McQueen, G. (2005). Lockups revisited. *Journal of Financial and Quantitative Analysis, 40*(3), 519–530.

Brav, A., & Gompers, P. A. (2003). The role of lockups in initial public offerings. *Review of Financial Studies, 16*(1), 1–29.

Brockman, P., & Michayluk, D. (1998). The persistent holiday effect: Additional evidence. *Applied Economic Letters, 5*(4), 205–209.

Cadsby, C. B., & Ratner, M. (1992, June). Turn-of-month and pre-holiday effects on stock returns: Some international evidence. *Journal of Banking and Finance, 16*(3), 497–509.

Chan, L. K. C., Jegadeesh, N., & Lakonishok, J. (1999). The profitability of momentum strategies. *Financial Analyst Journal, 55*(6), 80–90.

Chang, E. C., & Pinegar, J. M. (1986). Return seasonality and tax-loss selling in the market for long-term government and corporate bonds. *Journal of Financial Economics, 17*, 391–415.

Chang, S., & Suk, D. Y. (1998). Stock prices and the secondary dissemination of information: The Wall Street Journal's "Insider Trading Spotlight" column. *Financial Review, 33*(3), 115–128.

Chen, H., Noronha, G., & Singal, V. (2003). The price response to S&P 500 index additions and deletions: Evidence of asymmetry and a new explanation. *Journal of Finance, 59*(4), 1901–1930.

Chen, H., & Singal, V. (2003, July/August). A December effect with tax-gain selling. *Financial Analysts Journal, 59*(4), 78–90.

Choi, J. J. (2000). The Value Line enigma: The sum of known parts. *Journal of Financial and Quantitative Analysis, 35*(3), 485–498.

Chui, A. C. W., Titman, S., & Wei, K. C. J. (2003). Intra-industry momentum: The case of REITs. *Journal of Financial Markets, 6*(3), 363–387.

Clarke, R. G., & Tullis, R. M. (1999). How much international exposure is advantageous in a domestic portfolio? *Journal of Portfolio Management, 25*(2), 33–44.

Coën, A. (2001). Home bias and international capital asset pricing model with human capital. *Journal of Multinational Financial Management, 11*(4–5), 497–513.

Copeland, T. E., & Mayers, D. (1982). The Value Line enigma (1965–1978): A case study of performance evaluation issues. *Journal of Financial Economics, 10*, 289–321.

Coval, J. D., & Moskowitz, T. J. (1999). Home bias at home: Local equity preference in domestic portfolios. *Journal of Finance, 54*(6), 2045–2073.

Damodaran, A. (1989). The weekend effect in information releases: A study of earnings and dividend announcements. *Review of Financial Studies, 2*(4), 607–623.

Damodaran, A., & Liu, C. H. (1993). Insider trading as a signal of private information. *Review of Financial Studies, 6*(1), 79–119.

Daniel, K., Hirshleifer, D., & Subrahmanyam, A. (1998). Investor psychology and security market under- and overreactions. *Journal of Finance, 53*(6), 1839–1885.

Dash, S. (2002, July). *Price changes associated with S&P 500 deletions*. Standard & Poor's. http://www.spglobal.com/070902pricechanges.pdf. Accessed 1 July 2013.

Dellva, W. L., DeMaskey, A. L.,& Smith, C. A. (2001). Selectivity and market timing performance of fidelity sector mutual funds. *Financial Review, 36*(1), 39–54.

Denis, D. K., McConnell, J., Ovtchinnikov, A. V., & Yu, Y. (2003). S&P 500 index additions and earnings expectations. *Journal of Finance, 58*(5), 1821–1840.

Eckbo, B. E., & Smith, D. C. (1998). The conditional performance of insider trades. *Journal of Finance, 53*(2), 467–498.

Eleswarapu, V. R., & Reinganum, M. R. (1993). The seasonal behavior of the liquidity premium in asset pricing. *Journal of Financial Economics, 34*, 373–386.

Fama, E. F. (1970). Efficient capital markets: A review of theory and empirical work. *Journal of Finance, 25*(2), 383–417.

Fama, E. F., & French, K. R. (1995). Size and book-to-market factors in earnings and returns. *Journal of Finance, 50*(1), 131–155.

Feder, G., Just, R. E., & Schmitz, A. (1980). Futures markets and the theory of the firm under price uncertainty. *Quarterly Journal of Economics, 94*(2), 317–328.

Ferreira, E. J., & Brooks, L.D. (2000). Re-released information in the Wall Street Journal's "Insider Trading Spotlight" column. *Quarterly Journal of Business and Economics, 39*(1), 22–34.

Field, L. C., & Hanka, G. (2001, April). The expiration of IPO share lockups. *Journal of Finance, 56*(2), 471–500.

French, K. R. (1980). Stock returns and the weekend effect. *Journal of Financial Economics, 8*, 55–69.

French, K. R., & Poterba, J. M. (1991). Investor diversification and international equity markets. *American Economic Review, 81*(2), 222–226.

Friederich, S., Gregory, A., Matatko, J., & Tonks, I. (2002). Short-run returns around the trades of corporate insiders on the London stock exchange. *European Financial Management, 8*(1), 7–30.

Gao, P. (2006, December 18). *Herding, information aggregation and momentum effects*. Working Paper. Chicago, IL: Kellogg School of Management.

Goetzmann, W. N., & Kumar, A. (2001, December). *Equity portfolio diversification*. National Bureau of Economic Research (NBER). Working Paper No. 8686.

Gonzalez, M. (1996, April 5). In investing, timing could be everything. *Chicago Tribune*. http://articles.chicagotribune.com/1996-04-05/business/9604050033_1_birinyi-associates-stock-market-retirement-funds. Accessed 4 Jan 2014.

Grinblatt, M., Titman, S., & Wermers, R. (1995). Momentum investment strategies, portfolio performance and herding: A study of mutual fund behavior. *American Economic Review, 85*(5), 1088–1105.

Grundy, B. D., & Martin, J. S. (2001). Understanding the nature of the risks and the source of the rewards to momentum investing. *Review of Financial Studies, 14*(1), 29–78.

Harris, L., & Gurel, E. (1986). Price and volume effects associated with changes in the S&P 500 list: New evidence for the existence of price pressures. *Journal of Finance, 41*(4), 815–829.

Hasan, I., & Simaan, Y. (2000). A rational explanation for home country bias. *Journal of International Money and Finance, 19*(3), 331–361.

Haugen, R. A., & Jorion, P. (1996). The January effect: Still there after all these years. *Financial Analysts Journal, 52*(1), 27–31.

Hegde, S. P., & McDermott, J. B. (2003). The liquidity effects of revisions to the S&P 500 index: An empirical analysis. *Journal of Financial Markets, 6*(3), 413–459.

Hensel, C. R., & Ziemba, W. T. (1996). Investment results from exploiting turn-of-the-month effects. *Journal of Portfolio Management, 22*(3), 17–23.

Holden, C. W., & Subrahmanyam, A. (2002). News events, information acquisition, and serial correlation. *Journal of Business, 75*(1), 1–32.

Hong, H., & Stein, J. C. (1999, December). A unified theory of underreaction momentum trading, and overreaction in asset markets. *Journal of Finance, 54*(6), 2143–2184.

Huberman, G. (2001). Familiarity breeds investment. *Review of Financial Studies, 14*(3), 659–680.

Huberman, G., & Kandel, S. (1987). Value Line rank and firm size. *Journal of Business, 60*(4), 577–589.

Huberman, G., & Kandel, S. (1990). Market efficiency and Value Line's record. *Journal of Business, 63*(2), 187–216.

International Finance Corporation. (1997). *Emerging stock markets factbook 1997*. Washington, DC: International Finance Corporation, Subsidiary of the World Bank.

Jegadeesh, N., & Titman, S. (1993). Returns to buying winners and selling losers: Implications for stock market efficiency. *Journal of Finance, 48*(1), 65–91.

Jeske, K. (2001). Equity home bias: Can information cost explain the puzzle? *Federal Reserve Bank of Atlanta Economic Review* (Third Quarter), 31–42.

Johnson, T. C. (2002, April). Rational momentum effects. *Journal of Finance, 57*(2), 585–608.

Kahle, K. M. (2000). Insider trading and the long-run performance of new security issues. *Journal of Corporate Finance, 6*, 25–53.

Kamara, A. (1997). New evidence on the Monday seasonal in stock returns. *Journal of Business, 70*(1), 63–84.

Kaul, A., Mehrotra, V., & Morck, R. (2000). Demand curves for stocks do slope down: New evidence from an index weights adjustment. *Journal of Finance, 55*(2), 893–912.

Keasler, T. R. (2001). Underwriter lock-up releases, initial public offerings and after-market performance. *Financial Review, 37*, 1–20.

Keim, D. B. (1983). Size-related anomalies and stock return seasonality: Further empirical evidence. *Journal of Financial Economics, 12*(1), 13–32.

Kim, C. W., & Park, J. (1994, March). Holiday effects and stock returns: Further evidence. *Journal of Financial and Quantitative Analysis, 29*(1), 145–157.

Kim, E. H., & Singal, V. (1997, Fall). Are open markets good for foreign investors and emerging nations? *Journal of Applied Corporate Finance, 10*(3), 18–33.

Kunkel, R. A., & Compton, W. S. (1998). A tax-free exploitation of the tun-of-the-month effect: C.R.E.F., *Financial Services Review, 7*(1), 11–23.

Lakonishok, J., & Lee, I. (2001). Are insider trades informative? *Review of Financial Studies, 14*(1), 79–111.

Lakonishok, J., & Smidt, S. (1988, Winter). Are seasonal anomalies real? A ninety-year perspective. *Review of Financial Studies, 1*(4), 403–425.

Leinweber, D. J. (1995). Using information from trading in trading portfolio management 10 years later. *Journal of Investing, 4*(2), 40–50.

Lesmond, D. A., Schill, M. J., & Zhou, C. (2004). The illusory nature of momentum profits. *Journal of Financial Economics, 71*(2), 349–380.

Lewis, K. K. (1999). Trying to explain home bias in equities and consumption. *Journal of Economic Literature, 37*(2), 571–608.

Lintner, J. (1965). The valuation of risk assets and selection of risky investments in stock portfolios and capital budgets. *Review of Economics and Statistics, 47*(1), 13–37.

Liu, B., & Li, B. (2010). Day-of-the-week effects: Another evidence from top 50 Australian stocks. *European Journal of Economics, Finance and Administrative Sciences, 24*, 78–87.

Loughran, T., & Ritter, J. R. (1995, March). The new issue puzzle. *Journal of Finance, 50*(1), 23–51.

Lustig, I. L., & Leinbach, P. A. (1983). The small firm effect. *Financial Analysts Journal, 39*(3), 46–49.

Maxwell, W. F. (1998, Summer). The January effect in the corporate bond market: A systematic examination. *Financial Management, 27*(2), 18–30.

Menzly, L., & Ozbas, O. (2004, December). *Cross-industry momentum*. Working Paper. Los Angeles, CA: University of Southern California.

Moskowitz, T. J., & Grinblatt, M. (1999). Do industries explain momentum? *Journal of Finance, 54*(4), 1249–1290.

Ofek, E., & Richardson, M. (2000, January). *The IPO lock-up period: Implications for market efficiency and downward sloping demand curves*. Working Paper FIN-99-054. New York, NY: Stern School of Business. https://archive.nyu.edu/handle/2451/27072. Accessed 11 Feb 2014.

O'Neal, E. S. (2000). Industry momentum and sector mutual funds. *Financial Analysts Journal, 56*(4), 37–49.

O'Shaughnessy, J. P. (1998). *What works on Wall Street: A guide to the best performing investment strategies of all time* (rev. ed.). New York: McGraw-Hill.

Patel, J. B. (2012). A reexamination of the effect of daylight saving time changes on U.S. stock returns. *Journal of Academy of Business and Economics, 12*(2), 109–114.

Perold, A. F. (1988). The implementation shortfall: Paper versus reality. *Journal of Portfolio Management, 14*(3), 4–9.

Peterson, D. R. (1995, December). The informative role of the Value Line investment survey: Evidence from stock highlights. *Journal of Financial and Quantitative Analysis, 30*(4), 607–618.

Porras, D., & Griswold, M. (2000). The Value Line enigma revisited. *Quarterly Journal of Business and Economics, 39*(4), 39–50.

Reilly, F. K., & Hatfield, K. (1969). Investor experience with new stock issues. *Financial Analysts Journal, 25*(5), 73–80.

Ritter, J. R. (1991, March). The long-run performance of initial public offerings. *Journal of Finance, 46*(1), 3–27.

Ritter, J. R., & Welch, I. (2002, February). *A review of IPO activity, pricing and allocations*. Working Paper No. 8805. National Bureau of Economic Research (NBER).

Roth, G., & Saporoschenko, A. (1999). The informational effects of large insider stock purchases. *Managerial Finance, 25*(1), 37–48.

Rozeff, M. S., & Kinney, W. R., Jr. (1976). Capital market seasonality: The case of stock returns. *Journal of Financial Economics, 3*, 379–402.

Russel, P. S., & Torbey, V. M. (2002). *The efficient market hypothesis on trial: A survey*. Business Quest. School of Business Administration at Philadelphia University and School of Business at Bond University, Gold Coast, Australia. http://www.westga.edu/~bquest/2002/market.htm. Accessed 29 Dec 2013.

Ryan, C. I., & Siebens, J. (2012, February). *Educational attainment in the United States 2009: Population characteristics*. Washington, DC: U.S. Department of Commerce, Economics and Statistics Administration, U.S. Census Bureau.

Salomon, R. S., Jr. (1998, June 15). *Value Line's self-defeating success*. Forbes. http://www.forbes.com/forbes/1998/0615/6112294a.html. Accessed 4 Jan 2014.
Sarkar, A., & Li, K. (2002). Should U.S. investors hold foreign stocks? *Federal Reserve Bank of New York's Current Issues in Economics and Finance, 8*(3), 1–6.
Schoenmaker, D., & Bosch, T. (2008). Is the home bias in equities and bonds declining in Europe? *Investment Management and Financial Innovations, 5*(4), 90–102.
Schulmerich, M. (2014). *Stock market anomalies - An overview*. State Street Global Advisors. SSgA Capital Insights.
Scott, D. L. (2003). *Wall Street words: An A to Z guide to investment terms for today's investor*. (3rd ed.). Boston: Houghton Mifflin Company.
Seyhun, H. N. (1992). Why does aggregate insider trading predict future stock returns? *Quarterly Journal of Economics, 107*(4), 1303–1031. http://www.jstor.org/discover/10.2307/2118390?uid=3737952&uid=2&uid=4&sid=21103293169113. Accessed 8 Feb 2014.
Shleifer, A. (1986). Do demand curves for stocks slope down? *Journal of Finance, 41*(3), 570–590.
Sias, R. W., & Starks, L. T. (1995). Day-of-the-week anomaly: The role of institutional investors. *Financial Analyst Journal, 51*(3), 58–67.
Singal, V. (2006). *Beyond the random walk*. Oxford: Oxford University Press.
Steeley, J. M. (2001). A note on information seasonality and the disappearance of the weekend effect in the U.K. stock market. *Journal of Banking and Finance, 25*, 1941–1956.
Stickel, S. E. (1985). The effect of Value Line investment survey rank changes on common stock prices. *Journal of Financial Economics, 14*, 121–143.
Stoll, H. R., & Curley, A. J. (1970, September). Small business and the new issues market for equities. *Journal of Financial and Quantitative Analysis, 5*(3), 309–322.
Tesar, L. L., & Werner, I. M. (1995). Home bias and high turnover. *Journal of International Money and Finance, 14*(4), 467–492.
World Bank. (2013, March). *The World Bank: Indicators*. Worldbank Online. http://data.worldbank.org/indicator/CM.MKT.LCAP.CD. Accessed 18 Feb 2014.
Wurgler, J., & Zhuravskaya, E. (2002). Does arbitrage flatten demand curves for stocks? *Journal of Business, 75*(4), 583–608.
Zhang, Y., Nguyen, G. X., & Le, S. V. (2010). Yes, the Value Line enigma is still alive: Evidence from online timeliness rank changes. *Financial Review, 45*(2), 355–373.
Ziemba, W. T. (1991). Japanese security market regularities: Monthly, turn-of-the-month and year, holiday and golden week effects. *Japan and the World Economy, 3*, 119–146.

Stock Market Crashes

4

4.1 Introduction

After having introduced the basics of performance and risk measurement in Chap. 1 and the ideas of modern portfolio theory in Chap. 2, Chap. 3 looked at market anomalies as deviations from the theory. Now, in this chapter, we will look at stock market crashes. Stock market crashes are extreme events that have occurred for more than four centuries. However, the frequency and the magnitude of these crashes seem to have increased in the last few decades. This chapter will present a historical overview, starting from the first very prominent crash (the tulip mania) until the current ones.

Section 4.2 will be dedicated to define a stock market crash and tries to draw a general picture of the environment in which a crash occurs. Before a crash a bubble builds up. This will be closer examined in Sect. 4.3 followed by a general characterization of a crash in Sect. 4.4. Thereafter, this chapter presents key crashes grouped into two categories: crashes with only a regional impact (Sect. 4.5) and crashes with a global impact (Sect. 4.6). In detail, we will examine the following crashes with a regional impact[1]:

- Tulip mania, 1637 (Sect. 4.5.1)
- South Sea bubble, 1720 (Sect. 4.5.2)
- Railway mania, 1846 (Sect. 4.5.3)
- Souk Al-Manakh, 1982 (Sect. 4.5.4)
- Dubai real estate bubble, 2009 (Sect. 4.5.5)
- Greek crisis, 2010 (Sect. 4.5.6)
- Flash crash, May 6, 2010 (Sect. 4.5.7)

[1]A summary of the key bubbles and crashes presented in this section can be found in the white paper *Stock Market Crashes and Financial Bubbles: What History Tells us*, see Schulmerich (2010b).

M. Schulmerich et al., *Applied Asset and Risk Management*,
Management for Professionals, DOI 10.1007/978-3-642-55444-5_4

The last one is a non-typical crash, which went mostly unnoticed by the common public yet over a very short period of time had a severe impact on the stock market. International crashes that will be presented are:

- Great Wall Street Crash, 1929 (Sect. 4.6.1)
- Asian financial crisis, 1997 (Sect. 4.6.2)
- Russian financial crisis, 1998 (Sect. 4.6.3)
- Dotcom bubble, 2001 (Sect. 4.6.4)
- Subprime crisis, 2008 (Sect. 4.6.5)

This chapter concludes with a case study on how to identify a bubble. This is done in Sect. 4.7 where we look at the Chinese stock exchange and the Chinese real estate market, where first alarming signs appeared in July 2013. Many experts believe that a crisis in China could have as severe impacts to the world economy as the bankruptcy of the investment bank Lehman Brothers in September 2008.

Stock market anomalies and crashes cannot be explained by traditional finance theory as presented in Chap. 2. This is the reason why the psychology of investors has to be considered. We will do this in the next chapter where we will present behavioral biases and use them in an attempt to explain stock market crashes. Finally, Chap. 6 will then analyze how the extreme events that happened since 2007 have changed the view of the market investors on risk measurement, risk management and investing.

4.2 What is a Crash?

Crashes are known by the general public as a significant drawdown in the stock market. This *definition* is more a vague point of view than a real definition and does not provide a precise description of what actually a stock market crash is and how it happens.

According to Sandeep Patel[2] and Asani Sarkar,[3] a stock market crash is an abrupt decline in the prices of an index relative to a recent price peak. To be called a crash, a decline from the most recent peak has to be more than 20 % for the developed markets, and more than 35 % for the emerging markets.[4]

A typical example of a crash would be the dotcom bubble, which marked the end of the internet hype which started in the mid of the 1990s. Figure 4.1 shows the

[2]Sandeep Patel was the director and head of portfolio analytics in HFDMG at Merrill Lynch from November 2005 to March 2008 and founder and managing member of Governance Performance Management from August 2004 to October 2005. Dr. Patel has 20 years of experience in capital markets.

[3]Asani Sarkar is a visiting lecturer at Princeton University and research officer with the money and payments function of the Federal Reserve Bank of New York.

[4]Patel and Sarkar (1998, p. 6).

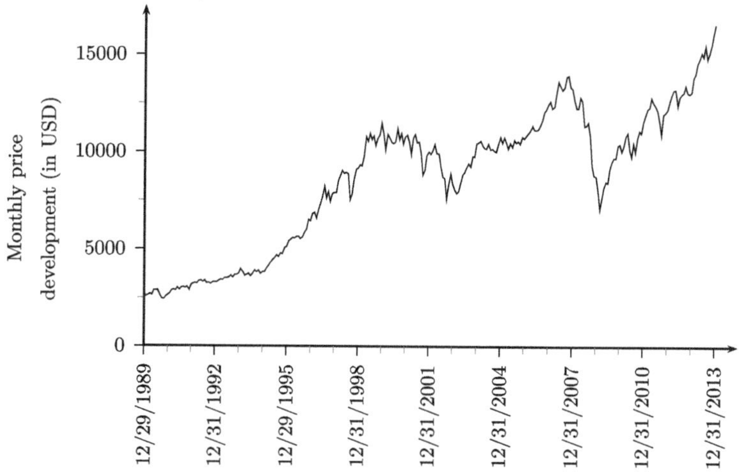

Fig. 4.1 Monthly price development of the Dow Jones Industrial Average Index in USD from January 1993 to December 2013. *Source*: Factset (ticker: DJII)

monthly returns of the Dow Jones Industrial Average (Dow Jones) Index in USD from January 1993 to December 2013. After many months of positive returns the monthly return turned negative in 2001 and 2002. The figure also shows clearly the subprime crisis in 2008 and the recovery in 2009.

The origins of a crash look very different from one to another. However, as will be seen, a crash is always preceded by periods of extreme optimism and closely followed by pessimism. These extreme peaks and troughs are directly related to what is called moral hazard. Moral hazard occurs when a party insulated from risk may behave differently than it would behave if it were fully exposed to the risk.[5] Nobel-prized economist Paul Krugman[6] described moral hazard as *"any situation in which one person makes the decision about how much risk to take, while someone else bears the cost if things go badly."*[7]

The matter is less if practitioners actually are under risky conditions, but rather how they feel risk in the given economic conditions. Suppose the stock markets have yielded a yearly return of 10 % for the past 4 years, then it looks normal, or at least plausible, that the market will continue on this trend. Most recent predictions of a probable economic downturn of a few analysts may not be given any regard.

[5] Ahrens (2008).

[6] Paul R. Krugman was born on February 28, 1953. He is an American economist, columnist and author. He is professor of economics and international affairs at the Woodrow Wilson School of Public and International Affairs at Princeton University, Centenary Professor at the London School of Economics, and columnist for the New York Times. In 2008, Krugman won the Nobel Memorial Prize in Economics for his contributions to new trade theory and new economic geography.

[7] Krugman (2009, p. 63).

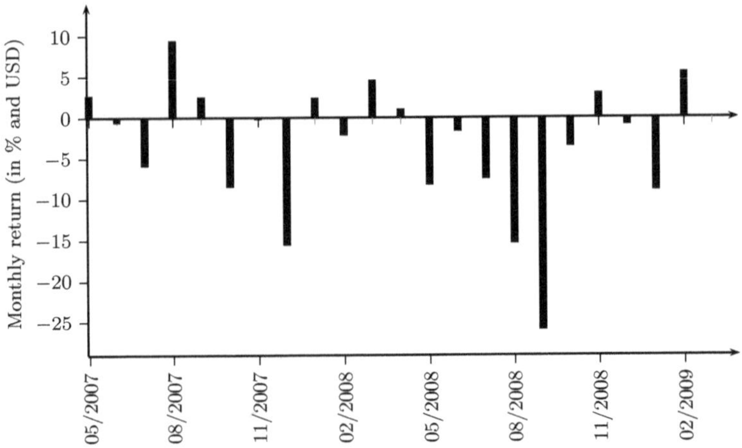

Fig. 4.2 Monthly return of FTSE Strait Times All Share (FSTAS) Index in USD from June 2007 to March 2009. *Source*: Factset (ticker: ASX-FTX)

The beginning of a crash is the time point when the index starts falling. It may fall for days, weeks or even months. For instance, if the FSTAS[8] price dips 25 % below its most recent peak, the time period of this drop will be identified as a crash. Figure 4.2 shows the monthly return of the FSTAS Index during one of the most severe crashes in time, the subprime crisis.

A stock market crash is normally preceded by an abnormal rise in prices. We will look at this phenomenon in the next section.

4.3 Before a Crash: A Bubble

Both movements (rises and crashes) form a speculative stock market bubble. While there exists a clear and precise definition of the word *speculation*,[9] there is no unique and clear definition for market bubbles. Kindleberger and Aliber define a bubble as *"an upward price movement over an extended range that then implodes.*

[8]The FSTAS (Financial Times Stock Exchange Strait Times All Share) is an index of the London Stock Exchange, London, U.K. The FTSE Strait Times All Share Index (FSTAS) is a modified market capitalization weighted index comprising all companies within the top 98 % by full market capitalization of the SGX Mainboard (Singapore Exchange) universe (i.e., large-cap, mid-cap and small-cap indices combined). There are 640 mainboard listings on the SGX.

[9]The Concise McGraw-Hill Dictionary of Modern Economics defines speculation as *"the act of knowingly assuming above-average risks with the hope of gaining above-average returns on a business or financial transaction"*, see Greenwald (1983).

An extended negative bubble is a crash."[10] The Palgrave dictionary[11] describes a bubble as *"any unsound commercial undertaking accompanied by a high degree of speculation."*[12] However, these definitions are rather general and do not allow to precisely identify the starting point of a bubble. They help to recognize a bubble once it has burst, but not when it develops.

A speculative stock market bubble (or speculative bubble in short) starts with an unsustainable increase in prices. The overoptimism of investors leads them to buy on hope and marvelous return promises rather than on genuine, fundamental information about real[13] (or intrinsic)[14] values. The intrinsic value of an asset is the value of the asset which is obtained through a hypothetically complete understanding of the asset's investment characteristics.

To assess the level or *normality* of the prices, valuation ratios are used. One of the most famous and most fundamental valuation ratios for stocks is the price-to-earnings ratio (P/E ratio):

$$P/E\ ratio = \frac{market\ value\ per\ share}{earnings\ per\ share}. \tag{4.1}$$

Earnings per share are the net earnings generated by a company divided by the number of outstanding shares. Net earnings are the gross sales minus taxes, interests, depreciation, preferred dividends and what is called other expenses on a balance sheet. There are multiple variants between basic and diluted earnings per share, ex-post and ex-ante, etc.

If a company's P/E ratio lies above its *normal level*, then the stock is overbought. This normal level is often seen as the historical average over the last years or even decades. The forward P/E (or estimated P/E) uses the estimated net earnings over, for example, the next 12 months and is among the most commonly used indicators. Estimates become less reliable if the macroeconomic environment shows extremely high volatility. This is precisely what happened during the market downturn of 2008, when macroeconomic conditions lead to forecasting challenges. The so-called forecasts were thus very difficult to create.

Let us have a look at Fig. 4.3 which shows the monthly development of the historical P/E ratio[15] of the S&P 500 Index between January 1881 and December

[10]Kindleberger and Aliber (2005, p. 16).

[11]R. H. Inglis Palgrave's original Dictionary of Political Economy (1894–1899), was a landmark in both publishing and economics, a liberal and scholarly overview of the whole sphere of economic thought in its days. Henry Higgs's revised edition, Palgrave's Dictionary of Political Economy (1923–1926), retained the spirit of the original while embracing new concepts in the development of economics as a discipline. An online version is available at http://www.dictionaryofeconomics.com/dictionary.

[12]Eatwell, Milgate, and Newmann (1987, p. 181).

[13]Shiller (2005, p. 2).

[14]Pinto, Henry, Robinson, and Stowe (2007, p. 15).

[15]Historical P/E ratio is also referred to as trailing P/E ratio.

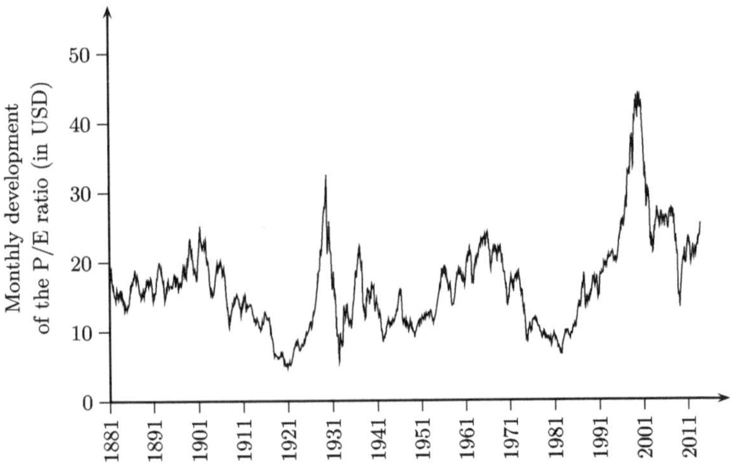

Fig. 4.3 Monthly development of the P/E ratio of the S&P 500 Index between January 1881 and December 2013. *Source*: Factset (ticker: sp50). The P/E ratio is calculated by using the last 12 months (LTM) earnings. Historical data is lagged by 45 days to reflect the amount of information available on the historical dates

2013. It is important to note that the P/E ratio was introduced significantly later than 1881 and was not used at that time. The data in Fig. 4.3 are, therefore, backfilled.

Each index or industry has its own average P/E which stands as a benchmark for the P/E of the companies. During the past 130 years, the P/E of the S&P 500 reached extreme lows and extreme highs as Fig. 4.3 shows. That being said, in general and with all due caution, a P/E is often considered to be quite high when it stands above 25 either in the U.S. or in Europe.[16] Yet, the price-to-earnings ratio can get as high as 80 (for Focus Media[17] in 2007 for instance) and was below one when stock exchanges were introduced decades ago, i.e., the investor bought a stock for less than the realized earnings.

Nevertheless, this ratio only tells us when there might be a bubble, but not when it starts. We thus still face the challenge of distinguishing between a bubble in progress and normal market movements. But a few market indicators derived from technical analysis[18] can be used to evaluate the relative optimism or pessimism of the market.

---

[16]http://www.csinvestor.com/investing-guide/stocks/high-pe-ratio.

[17]Focus Media Holding Limited (ticker: FMCN) is a Chinese company traded on the NASDAQ. The company is China's leading digital media group.

[18]Technical analysis aims at generating profit on stock markets based on historical price chart analysis and trends.

These indicators can be sorted into short-term, middle-term and long-term indicators. The most common and best-known are the VIX,[19] the equity PCR (put-call ratio)[20] and the so-called RYDEX Nova/Ursa ratio.[21] When these ratios reach a certain limit, it is advised to sell or buy. They are valuable since they actually measure the mood of the market by evaluating if the stock price is overbought or oversold. Still, they do not predict precisely whether a bubble is created or not. Worse: sometimes while they show overheating, it is commonly known that they can return to normal level without any bubble burst.

A speculative bubble ends when prices fall back to *normal levels*. This involves a time period of abrupt decline in prices, also called a crash. To date, a smooth return to normal from a mounting bubble has not been documented, but still might be possible.

4.4 Characteristics of a Crash

A speculative bubble bursts with a crash, i.e., usually an abrupt decline of more than 20 % of stock prices in developed markets and 35 % in emerging markets occurs.[22] However, it is important to note that the time span is not an important consideration in the definition of a stock crash. In Japan, we saw a crash that started in the early 1990s and lasted more than 13 years (Japanese asset price bubble), while the subprime crisis lasted around a year.

The fact is that crashes can be neither easily nor intuitively understood. This is partially due to communication failures, but also to the incomprehension of specialists and scholars who keep considering crashes as mere exceptions or extreme financial instabilities that randomly occur and are difficult to study with the available

[19]VIX is the ticker symbol for the Chicago Board Options Exchange Volatility Index, a popular measure of the implied volatility of S&P 500 Index options. A high value corresponds to a more volatile market and therefore more costly options. Often referred to as the fear index, it represents a measure of the market's expectation of volatility over the next 30 day period. The VIX Index was introduced in 1993 in a publication by Professor Robert E. Whaley of Vanderbilt University, see Whaley (1993).

[20]The total put-call ratio is the original indicator introduced by Martin Zweig, a money manager and author who was one of the few traders who escaped the crash of 1987 by buying put options. The put-call ratio is calculated using the following equation: total put options purchased divided by total call options purchased. It measures the mood of investors, see Duarte (2006, p. 163).

[21]Formed in 1993, the RYDEX (Nova + OTC)/Ursa ratio measures the prevailing sentiment among investors, determining whether the majority of mutual fund switchers are generally bullish, bearish or somewhere in between. RYDEX Nova provides 1.50 times the performance of the S&P 500 Index. That is, if the S&P 500 Index rises 1 %, the RYDEX Nova will provide 1.5 % to the investor. RYDEX Ursa provides −1.00 times the performance of the S&P 500 Index. The RYDEX was the first fund family to offer a *short* fund and a *leveraged* mutual fund. See Headley (2002, p. 27).

[22]Patel and Sarkar (1998, p. 6).

set of data.[23] Even worse, we keep believing that stock market returns follow a normal distribution.[24]

As a second issue, when studying stock market crashes, we try to use mathematical tools so as to study the crisis through equivalent conditions. However, adjustments are required. Socio-economical conditions in which a crash happens vary across time and occurrences. As a result, raw data is difficult to use and adjustments, yet imprecise, are made so as to realize a rigorous statistic approach. Therefore normal distribution, which under normal conditions is assumed to closely corresponds to the rate of returns observable on the stock markets, cannot be used to model a crisis.

There was the Great Depression of 1929, Black Monday in 1987, the dotcom bubble in 2000 and the subprime crisis in 2007, although, if we assume a normal distribution, the probability of a daily loss of 22.61 % will on average occur only once every 10^{50} days.[25] But this single number is meaningless, because since the estimated date of the solar system formation, say arbitrarily 6 bn. years ago, *only* $2.2 * 10^{12}$ days have passed.[26] This over-astronomical probability is comparatively as surprising as the frequency and severity of shocks witnessed during the Russian crisis or the dotcom bubble in 2000. Derman[27] describes crashes as follows: *Market crashes are not randomly occurring lightning bolts; they are the consequence of the madness of crowds who are busy avoiding the last mania as they participate in what will turn out to be the current one.*[28]

Being able to forecast these extreme events is thus crucial for central banks and every single investor alike. To do so, a better understanding of their occurrence is required which reaches beyond the general definition stated above and the applied mathematical models.

The use of a portfolio to limit risk, while being very effective under normal conditions, looks inappropriate when confronted with massive drawdowns. The correlation between the various single securities and also and especially between different assets increases significantly. Figure 4.4 shows the correlation amongst European stocks and the correlation amongst global stocks which apparently move in waves and seem to have a positive trend.

On the other hand, using derivatives to protect the portfolio from these drawdowns can become costly very quickly. As a first conclusion, we can underline a

---

[23] Garber (2001, p. 3).

[24] An explanation of normal distribution can be found in Sect. 1.2.2.

[25] On October 19, 1987, the Dow Jones Industrial Average Index lost 22,61 %. See Mandelbrot and Hudson (2004, p. 4).

[26] Mandelbrot and Hudson (2004, p. 4).

[27] Professor Emanuel Derman is the director of the MS program in financial engineering at Columbia University and head of risk management at Prisma Capital Partners. Previously he was the managing director in firm-wide risk at Goldman, Sachs & Co. and a columnist for Risk magazine. He obtained a Ph.D. in Particle Physics at Columbia University.

[28] Derman (2004, p. 10).

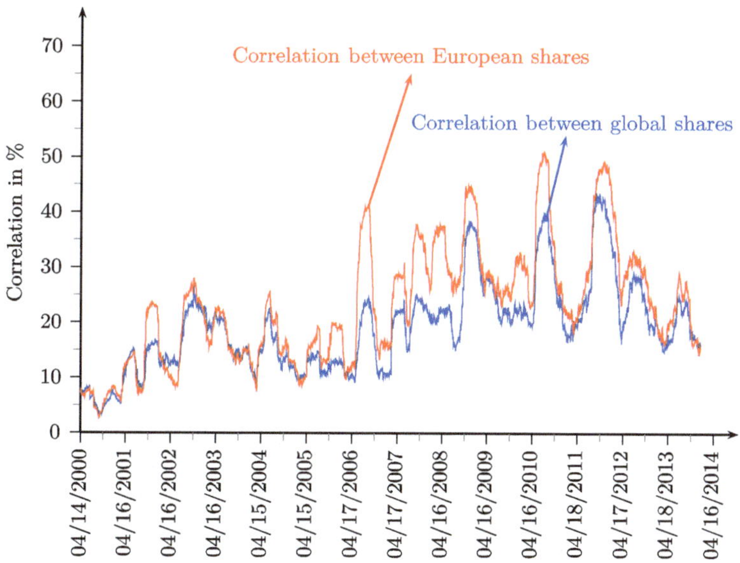

Fig. 4.4 Correlation between European shares and correlation between global shares for time frame of 75 days starting from April 4, 2000 to December 31, 2013. *Source*: Nomura

first misbelief: the mathematical fundamentals in use appear inappropriate during crashes. Yet, another misbelief has to be corrected: the idea that a stock market crash implies a subsequent general country-wide economic downturn.

In 2007, the subprime crisis triggered a banking crisis, which in turn lead to an economic downturn and then an economic crisis.[29] This shortcut is easily made, but it is incorrect. Crashes can be the trigger of poor national economic performances, but this is not necessarily the case as will be shown. Sometimes, because of a few rogue traders, computer trading or some new events, the market goes mad and needs to be regulated if the economy as a whole is to be protected.[30]

This need for regulation is fair and even more understandable if public interest is at stake. This is actually the case in the U.S. with millions of families relying on 401(k) plans. A 401(k) plan is a defined contribution plan offered in the U.S. by a corporation to its employees, which allows employees to set aside tax-deferred income for retirement purposes. In some cases, employers will match their contributions dollar for dollar. Taking a disbursement before a certain specified age, usually the retirement age, will trigger a penalty tax. The name 401(k) comes from

[29]Shiller (2008, pp. 6–9).

[30]Khor and Khor (2001, p. 65).

the IRS[31] section pertaining to the retirement program.[32] Logically, more and more citizens are counting on regulation, which reached a major advance in 2008.[33]

However, as will be seen for the Souk Al-Manakh crisis in Kuwait, regulation is not always an optimal choice, since a crash does not necessarily produce an economic downturn or crisis. Figure 4.5 represents the growth rate in percent of the U.S. GDP[34] from 1950 until 2012. The GDP is commonly used to measure the relative economic health of a country. Now, it will help us to see if stock market crises and economic growth trends are systematically related.

According to Shiller, the beginning of the subprime crisis dates back to 2007.[35] However, the U.S. GDP growth rate decreased before the crash occurred, i.e., in 2004.

The subprime crisis was preceded by this economic downturn: it followed in the wake of 2 years of slowing growth. But the opposite idea, that an economic upswing excludes a financial crash, is equally false: On October 19, 1987, and September 16, 1992, there were the respective crashes of Black Monday and Black Wednesday. Yet in both cases the GDP was still growing higher relative to the previous year!

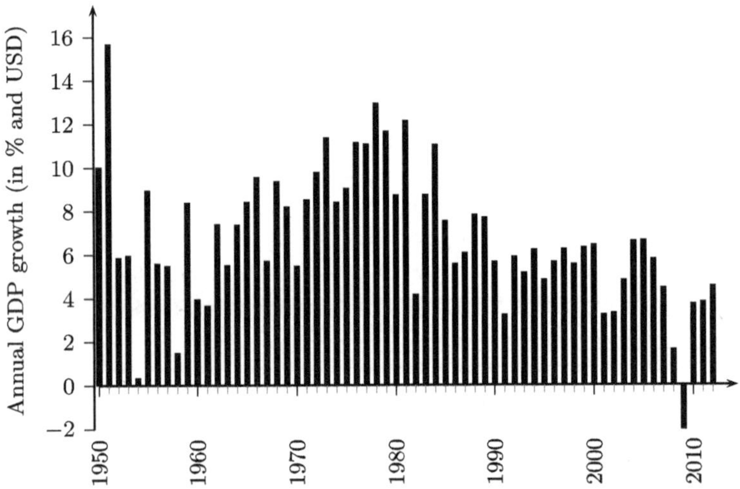

Fig. 4.5 Annual U.S. GDP growth in USD from 1950 to 2012. *Source*: BEA (2014, Sect. 1, Table 1.1.5)

[31] The Internal Revenue Service IRS is the U.S. government agency responsible for tax collection and tax law enforcement.

[32] For more information on the 401(k) plan, an official pedagogic resource is available at http://www.401k.org.

[33] Since 2008, people who have a 401(k) are able to sue in an attempt to recover losses if they think their accounts have been mishandled. See Stout (2008a).

[34] GDP stands for gross domestic product. The total market value of all final goods and services produced in a country in a given year, equal to total consumer, investment and government spending, plus the value of exports, minus the value of imports. See the website of the Organisation of Economic Co-operation and Development at http://stats.oecd.org/glossary/detail.asp?ID=1163.

[35] Shiller (2008, p. 5).

Another example: in 1967, except for the outrageous number of fatal plane crashes,[36] the NYSE did not suffer much while the U.S. GDP growth slowed heavily. In other words, there is no straightforward relationship between the occurrence of stock market crashes and the economic development.

However, Fig. 4.5 shows something even more interesting. The economic development appears to be cyclical. We can see waves of increasing and slowing growth rates which are between 8 and 11 years long.[37] These specific cycles have been studied for the real economy, but not for the stock markets in the same time.

To illustrate this, let us take the Dow Jones Industrial Average Index (DJIA).[38] If we compute the correlation between the growth rates of the DJIA and the U.S. GDP since 1929,[39] we reach a value around 0.11 for yearly data starting in 1929 and −0.02 for quarterly data[40] starting in 1947. In other words, stock markets and real economy are loosely correlated and connected, at least on the rate of return basis.

It thus appears difficult to predict a crash based on the general economic situation. In our example, we used the U.S. GDP as an indicator, but we could also use other indicators with the same results. A crash occurs whatever the economic situation is and the stock market performance is not correlated with the real economy.

Information is another key factor in evaluating asset values. Recall that according to the EMH presented by Fama in 1970, we assume that all information is available at the same time to all investors and that it is reflected in the asset prices.[41] One could imagine that an information issue might be the reason behind crashes: a wave of stunningly bad information might throw the market into dire straits. For example, an earthquake or a nuclear disaster might significantly diminish the performance of a large amount of companies country-wide or an index.

Quite interestingly, *bad news occurrences* do not seem to be the reason of stock market crashes. According to Shiller,[42] from 1871 to 1988, days of great price moves

[36]On July 18, 1966; on December 10, 1966; on April 20, 1967 and on June 4, 1967. See NTSB (1967).

[37]They are also known as the Juglar cycle of 7–11 years of duration and were discovered in 1862. See Greenwald (1983, p. 593).

[38]The Dow Jones Industrial Average is the best-known U.S. index of stocks, a price-weighted average of 30 actively traded, well-established companies' shares. The Dow, as it is called, is a barometer of how shares of the largest U.S. companies perform. It is one of several stock market indices created by Wall Street Journal editor and Dow Jones & Company co-founder Charles Dow.

[39]We took this date because the oldest publicly available data provided by the Bureau of Economic Analysis are from this year.

[40]The Bureau of Economic Analysis provides quarterly data only since 1947, when this method started to be used.

[41]Fama (1970).

[42]Robert J. Shiller is the Arthur M. Okun Professor of Economics at Yale University, professor of finance and fellow at the International Center for Finance of Yale School of Management and, together with Eugene F. Fama and Lars P. Hansen, winner of the Nobel Prize in Economics 2013. He is the author of numerous books in finance among which stand bestsellers like *Irrational Exuberance* etc.

on the DJIA and the S&P 500 were linked to major public information only 10 % of the time.[43] So, even if media sometimes spread information in a not always impartial and neutral way as their profession would require them to do according to the Munich Charter,[44] it appears that they are not the immediate key reason of big price moves.

Big price moves are not due to information delivery.[45] It is rather the mood of investors and how they perceive the value of their information which creates an asymmetry.[46] If there is general optimism, the index will see its value grow, if the mood of investors is rather pessimistic, the market will go down. When the stock market starts to collapse, it appears that there is no *factual information asymmetry*. But when the bubble bursts, a direct consequence is a factual information asymmetry in addition to the asymmetry related to the cognitive evaluation of information by the investors![47]

We know that a crash cannot be forecast based on a given economic situation. Big price movements, crashes being one particular form, do not seem to be caused by the arrival of important information. However, it is interesting to note the frequency of their occurrence. Following below is a non-exhaustive compilation of important crises which occurred in the past century only:

- 1929 : Wall Street crash (4 years)[48]
- 1973 : Crash due to oil price increase[49]
- 1980 : Silver Thursday[50]
- 1982 : Souk Al-Manakh stock market crash[51]
- 1987 : Black Monday[52]
- 1992 : Black Wednesday[53]

---

[43]Shiller (2005, p. 91).

[44]The Charter of Munich is a code of ethics signed by representatives of most of the European and international journalists organizations in 1971. The sixth article mentions that media *"provides a right,* ***balanced and impartial information****; gives the possibility to different points of view to be expressed when about controversial issues; when somebody is implicated (called into question) in an article, allows him to react and to defend himself if he wishes, either in the related article or in an article to be published later."* An online version is available at http://www.mediawise.org.uk/european-union.

[45]Shiller (2005, p. 91).

[46]Edmans, Garcia, and Norli (2007, p. 1971).

[47]Dunmore (1991, p. 1).

[48]Galbraith (1979).

[49]Perron (1989).

[50]Underwood (2009).

[51]Fitch (2010).

[52]Bogle (2008).

[53]Driver and Martell (2002).

- 1997 : Asian crisis[54]
- 1998 : Russian crisis[55]
- 2001 : Dotcom bubble[56]
- 2007 : Chinese correction[57]
- 2007 : Subprime crisis[58]
- 2009 : Dubai crisis[59]
- 2010 : Flash crash of May 6 and start of the Greek crisis[60]

Recently, crises seem to occur more frequently: the Chinese correction, the subprime crisis and the Dubai crisis erupted within the last 6 years. The feeling that such tail risk events are likely to happen more frequently now because of the increased correlations of financial markets is shared by many market participants.

In order to determine how changing perceptions of tail risk have affected the investment strategies of institutional investors, the Economist Intelligence Unit, on behalf of State Street Global Advisors,[61] conducted a survey of over 300 investors from the U.S. and Europe, including institutions, pension funds, family offices, consultants, asset managers, private banks and insurance funds.[62] Figure 4.6 displays the participants' view by looking at two statements:

1. *Tail risk events are likely to happen more frequently now (as of mid-2012) because of the increased correlations of financial markets.*

   As the upper section of Fig. 4.6 displays, three quarters of the participants agree or even strongly agree to this statement.
2. *Tail risk events are likely to be more severe now (as of mid-2012) than they were in the past because of the increased interconnectedness of financial markets.*

   As the lower section of Fig. 4.6 displays, almost three quarters of the participants agree or even strongly agree to this statement as well.

The rather high frequency of crashes is not a new phenomenon as we can see in Table 4.1. The problem is that in 2001, 2007 and 2009, the crises leaked into the

---

[54]Kaminsky and Schmuklerb (1999).

[55] Dungey, Fry, Gonzalez-Hermosillo, and Martin (2002).

[56]Ofek and Richardson (2003).

[57]Arbeter (2007).

[58]Frank, González-Hermosillo, and Hesse (2008).

[59]Hasan (2010).

[60]Krudy (2010).

[61]State Street Global Advisors (SSgA) is, as of December 31, 2013, the second largest institutional asset manager worldwide with roughly \$ 2 tn. as asset under management spanning almost all asset classes.

[62]Economist Intelligence Unit (2012, p. 2).

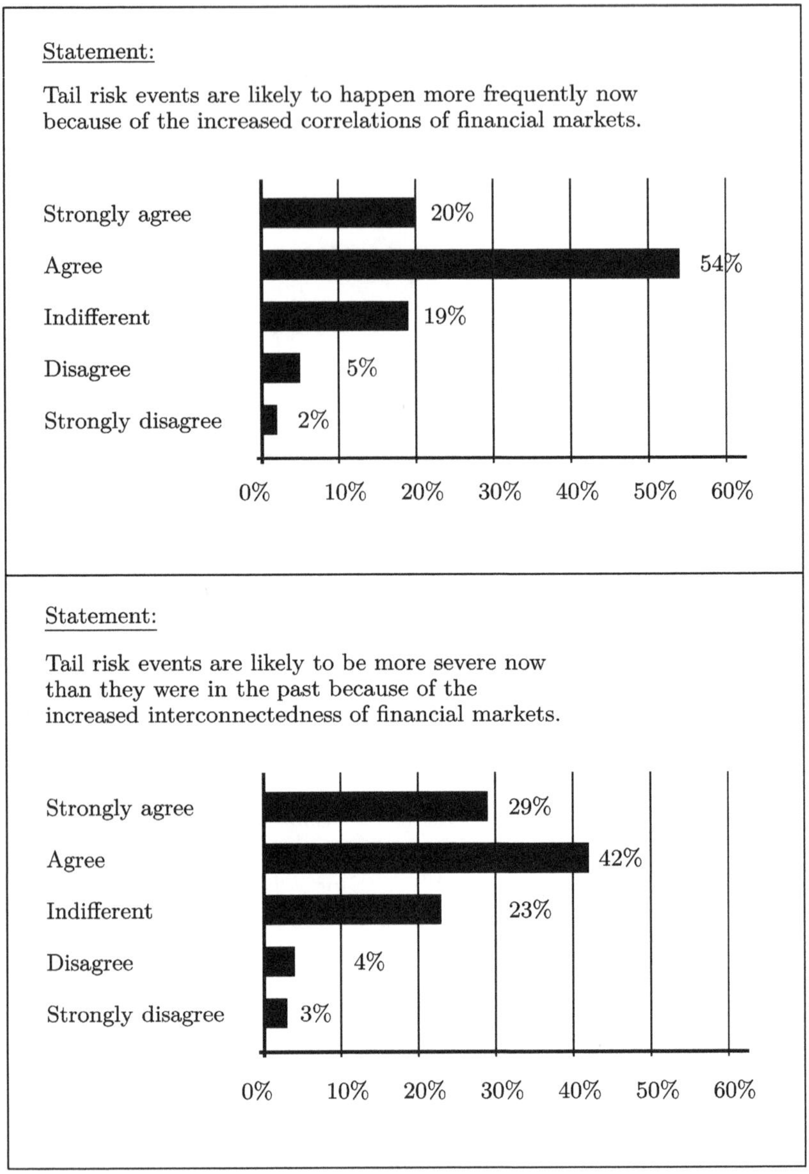

Fig. 4.6 EIU survey on the frequency and severeness of tail risk events nowadays (as of mid-2012) versus the past. *Source*: Economist Intelligence Unit (2012, p. 21)

real economy, with a greater speed and impact than earlier in history. The second problem is the time span required to regain the losses of the crash periods. It appears that a given portfolio takes more time than earlier to retrieve its value from before the bubble burst.

Table 4.1 Largest 1-year real stock price index decreases from 1960 to 1999

Country	Begin	End	Price decrease (in %)	Price change (in %) over subsequent 1-year period
Thailand	08/1997	08/1998	−62.8	71.9
Hong Kong	08/1997	08/1998	−55.5	90.0
Korea	06/1997	06/1998	−61.9	167.0
Indonesia	03/1997	03/1998	−48.1	−45.1
Jamaica	01/1993	01/1994	−73.8	69.6
Pakistan	10/1990	10/1991	−59.5	9.0
Finland	02/1990	02/1991	−47.5	6.3
Norway	12/1989	12/1990	−46.1	68.6
Taiwan	10/1989	10/1990	−74.9	85.1
South Africa	07/1985	07/1986	−62.1	48.9
Hong Kong	12/1981	12/1982	−55.1	7.7
Canada	06/1981	06/1982	−47.9	69.4
Colombia	01/1980	01/1981	−47.1	74.2
Spain	10/1976	10/1977	−54.1	−15.6
Sweden	08/1976	08/1977	−63.6	96.6
Italy	04/1974	04/1975	−46.1	−31.3
Norway	01/1974	01/1975	−53.6	−2.1
U.K.	11/1973	11/1974	−63.3	72.7
Philippines	10/1973	10/1974	−61.9	−14.1
Australia	10/1973	10/1974	−53.0	33.6
France	09/1973	09/1974	−49.0	25.3
Denmark	09/1973	09/1974	−45.8	14.7
Denmark	07/1969	07/1970	−56.0	−15.3
Norway	05/1967	05/1968	−54.2	39.9
India	11/1963	11/1964	−58.4	−18.8

Source: Shiller (2005, p. 134)

To resume: a bubble is hard to notice. As Galbraith found, the common factors leading to a crisis are[63]:

- A prolonged period of rising stock prices and excessive economic optimism
- A market where P/E ratios exceed long-term averages
- Extensive use of margin debt[64] and leverage by market participants

Yet, crashes cannot be forecast based on specific bad news, the general state of the country or the world's economy. Their frequency is far from being low, which

[63]Galbraith (1997).

[64]The dollar value of securities purchased on margin within an account. Margin debt carries an interest rate, and the amount of margin debt will change daily as the value of the underlying securities changes. See http://www.investopedia.com/terms/m/margin_debt.asp.

implies that the assumptions for portfolio diversification are not applicable under bubble/burst conditions. Finally, it is very hard to develop a standard model so as to prevent or at least forecast these events because mathematical models are based on a statistical approach which uses constantly evolving elements. The adjustment of the given factors is critical, yet hard to achieve. The task to prevent crashes is certainly crucial, but it appears more difficult than ever.

4.5 Crashes with Regional Impact

A famous Italian poet Filippo Pananti[65] said 250 years ago: *"History is not useful to read the past but rather to read the future."* This is precisely what we intend to do. True, each stock market crash differs from one to another. However, by knowing what happened in the past, one might protect oneself against crashes if ever a similar set of elements would occur again. On the following pages, we will try to capture the elements that we believe are the key factors of the past crises.

We will first describe important historical stock market crashes with a regional impact, starting with the tulip mania. It is the first documented market crash in history even if crises must have formerly happened in the earlier past. For example, when the Roman emperor Julian[66] suspected the merchants of a large city to speculate on rising food prices, he decided to cut the price of staple food, force local farmers to sell their crops at the stated price and massively import from Egypt to put an end to the famine. The result was starvation for multiple months. Unfortunately, little documentation is left from that era.

The second crisis presented in this section is the South Sea bubble, followed by the railway mania. We believed it necessary to also include the Souk Al-Manakh crisis which is less well known but illustrates that stock market crashes can occur even in Muslim countries and despite of strict regulations. This specific crash occurred in Kuwait in the 1970s. As a follow-up, we present the Dubai crisis and the Greek crisis. Both are tightly linked to a major previous crisis, i.e., the subprime crisis, which will be presented in a later section.

4.5.1 Tulip Mania (1637)

The tulip mania (or tulipomania) occurred in 1637, during the Dutch Golden Age,[67] and is considered as the first recorded speculative bubble. This period is well

[65]Filippo Pananti (1766–1837) is a Tuscan poet. Exiled for his liberal republicanism in 1799, he settled in London, where he published burlesque poetry and prose, including his masterpiece, the mordant verse novel *Il poeta di teatro* (1808). Returning to Italy in 1814, he was captured by pirates, as he describes in *Aventure e osservazioni sopra le coste di Barberia* (1817).

[66]Emperor Julian, also known as Julian the Apostate as well as Julian the Philosopher, was Roman Emperor from 361 to 363 and a noted philosopher and Greek writer.

[67]The Dutch Golden Age which lasted through most of the seventeenth century was brought about in part by religious toleration and successful trade with the rest of Western Europe.

known for the extraordinary price rise of tulips, but also for the following abrupt decline.[68]

Recently introduced in Europe from the Ottoman Empire, tulips became very popular in the Netherlands.[69] Soon, they became a luxury item in Dutch gardens and a status symbol. However, a virus called "mosaic" killed the majority of the flowers and induced the petals to develop contrasting colored stripes or *flames*. The more bizarre the bulb was, the greater was the cost of owning it.

The virus helped to set off speculation on tulip bulbs:

- First, the available stock of tulips progressively shrunk (shortage effect).
- Second, bulb merchants simply tried to predict the most popular style for the coming year (fashion effect).
- Then, they bought an extra large amount of the item anticipating a rise in its price (speculation effect).

Many people swapped their belongings such as land, jewels and furniture for bulbs that would make them wealthier (speculation effect). Tulip bulb prices began to rise extraordinarily and the more expensive the bulbs became, the more people considered them as a valuable and interesting investment (loss of sight of intrinsic value).

In his book, *Extraordinary Popular Delusions and the Madness of Crowds*, Charles Mackay[70] observes that *nobles, citizens, farmers, mechanics, seamen, footmen, maid-servants, even chimney sweeps and old clothes women dabbled in tulips*.[71]

Figure 4.7 shows the progression of a standardized price index for tulip bulb contracts created by Earl A. Thompson.[72] However, the author was unable to find prices between February 8 and May 1, 1637.[73] Thus, the shape of the decline is unknown. Nevertheless, the tulip market is known to have collapsed abruptly in February 1637.

Nowadays, *the tulip mania* is often used as a metaphor to describe speculative bubbles that happened years and even centuries later, such as the Great Depression that will be discussed later. As a side note, the movie *Wall Street: Money Never Sleeps* from 2010 (director Oliver Stone) with Michael Douglas especially alludes

[68]At the peak of the tulip mania in February 1637, tulip bulb contracts sold for more than ten times the annual income of a skilled artisan.

[69]Malkiel (2003, p. 35).

[70]Charles Mackay (1814–1889) was a Scottish song writer, poet and journalist.

[71]Mackay (1995, p. 94).

[72]Earl A. Thompson (1931–1978) was a leading American writer of naturalist prose whose premature death prevented him from achieving the long-standing popularity of other writers.

[73]Thompson (2006, p. 3).

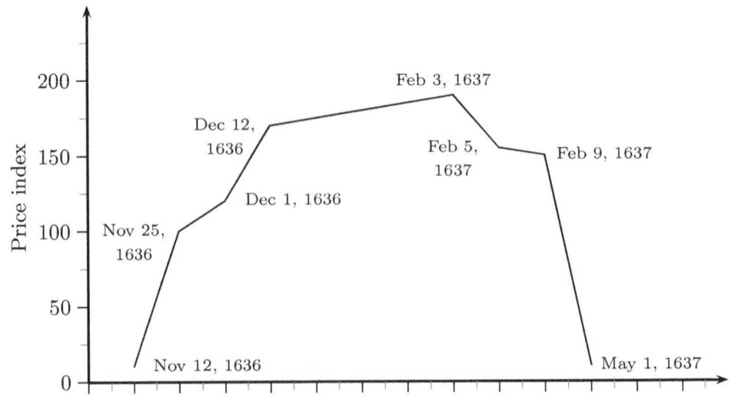

Fig. 4.7 Index of prices recorded in Dutch tulip contracts over the time period 1636–1637. *Source*: Based on Thompson (2006, p. 3)

to the tulip mania as the first important bubble.[74] The next financial crisis would occur on the other side of the Channel around 50 years later: the South Sea bubble.

4.5.2 South Sea Bubble (1720)

In the early eighteenth century, the British Empire prospered, but the government faced difficulties to meet its financial obligations. On the other hand, the number of wealthy British people grew, and many were eager and ready to invest in companies. But to own shares was a privilege: in 1693, only 499 people owned stocks of the British East India Company (BEIC).[75] The BEIC was a British company that traded with the Indian subcontinent and China.

The government took advantage of this situation and founded the South Sea Company (SSC) in 1711 to restore the faith in the government's financial strength.[76] The company was granted the monopoly of slave trade in the South Sea, which comprised most of today's South America, Mexico and the Caribbeans.

The newly founded company took on a government IOU of about £ 10 mn.[77] The initials *IOU* comes from the expression *I owe unto* and the pronunciation of *I owe you* and is an evidence of debt. The government and the company convinced

[74]The movie *Wall Street: Money Never Sleeps* from 2010 is the successor of the blockbuster movie *Wall Street* from 1987 (director Oliver Stone) in which Michael Douglas plays the ruthless investment banker Gordon Gekko. Michael Douglas received his second Oscar award for this role which at the end of the 1980s and in the 1990s inspired many young professionals to enter the finance industry.

[75]Malkiel (2003, p. 39).

[76]Malkiel (2003, p. 39).

[77]Malkiel (2003, p. 40).

the holders of the £ 10 mn. of debt to exchange it with an issue of stock in the company.[78] It is more or less the same process that the government advocated in 2009 to save General Motors in the U.S.

Prospective investors expected a big profit from this conversion and started to become very interested in the stock. The calculation was that a stock could prove to be financially substantially more rewarding than debt because of the much higher rate of return. SSC was also very attractive because the slave trade was a highly lucrative and burgeoning business at this time. In December 1719, the peace contract signed between the British Empire and Spain made the SSC's prospects even more promising. Actually, the South Sea was under Spanish rule, and the peace contract removed all obstacles the British company had to face when trading in an enemy area.

In 1720, the SSC managers decided to further improve the company's image by proposing to fund the entire national debt of £ 31 mn.[79] After the operation was implemented by law by Parliament, the stock price, for the same reasons as given above, rose from £ 130 to £ 300.[80] The stock attracted more and more investors, and even the King of England succumbed by subscribing for this stock with an investment amount of no less than £ 100,000.[81] There could not be any better guarantee for investors. The price was expected to increase and it did, as it went up to £ 340 in a few days only.[82] Afterwards, to ease the public appetite, the SSC directors announced another stock issue at £ 400.[83] The public became voracious and 1 month later, the stock reached £ 550.[84]

On June 15, 1720, after another issue, the payment plan (or dividends) became less binding: 10 % down and no other payment for 1 year. The stock hit £ 800. Half the House of Lords and more than half the House of Commons signed on, providing a feeling of guarantee to late investors. The stock price eventually reached £ 1,000.[85]

The difference between £ 130 and £ 1,000 a share is significant. Expectations were far too high relative to what was humanly and technologically achievable at this time. In August 1720, news about the infeasibility of the SSC's promised projects started to reach the investors. Bad news about the company's performance, but also about the fact that not a single director of SSC had any experience in the South American trade started to hit. Even worse, the SSC directors had taken the cash of the company for their own use.[86] Investors realized that the price of shares

---

[78]Malkiel (2003, pp. 38–39).

[79]Malkiel (2003, p. 41).

[80]Malkiel (2003, p. 41).

[81]Malkiel (2003, p. 41).

[82]Malkiel (2003, p. 42).

[83]Malkiel (2003, p. 42).

[84]Malkiel (2003, p. 40).

[85]Malkiel (2003, pp. 42–44).

[86]Carswell (2002, p. 142).

had no relationship with the real prospects of the company. They sold out and the share price fell back to about £ 100 within a few months.[87]

A set of measures were taken so as to prevent such a crisis to occur again. These measures were so restrictive that they hampered the whole economy. But more than a century passed before the next crisis occurred: the railway mania. Once again, the late consciousness of an unfeasible project would ruin thousands of investors.

4.5.3 Railway Mania (1846)

Again, this crisis occurred in Great Britain. In the 1840s, the introduction of railways granted opportunities for capital investment, and the major success of the railway between Manchester and Liverpool in 1830 gained the attention of investors. The transportation of freight and people proved to be efficient and profitable. But apparently, the railway mania was not caused by the universal success of the railways which were mostly completed when the bubble burst.[88] The then newly founded *Economist* said: *"The new railways are of small extent, and generally branches or extensions of existing lines."*[89] As Andrew Odlyzko notes, by 1845, railways were neither a new phenomenon nor small, as their revenues were already over 1 % of GDP and about 10 % of total government spending.[90]

At the same time, the Bank of England cut interest rates supported by an improving economy. Thanks to the industrial revolution, British investors were also numerous, as a growing middle class population wanted to invest their savings. In 1825, the government put an end to the Bubble Act,[91] which limited partnerships to five investors.[92]

It was also at this time that newspapers were created which, supported by the emerging modern stock markets organization, heavily promoted investments like railways. Everyone could invest in large scale projects, while paying less than 10 % of the price of the share.[93] In exchange, the company had the right to ask for additional cash, when required, for pursuing industrial projects. This payment structure looked particularly efficient for both buyers and sellers. On one side the railway companies had consistent cash flows to continue their efforts and the guarantee to have the necessary resource when required. At this time, investments were made for many years. On the other side buyers kept part of their savings to invest in other assets, while being able to receive the full amount of dividends. As a result, many families, being rich or not, invested all their savings in this investment scheme and investors fought to own railway stocks pushing the prices very high.

[87]Malkiel (2003, p. 44).

[88]Jackman (1916, p. 586).

[89]Odlyzko (2010, p. 74).

[90]Odlyzko (2010, p. 12).

[91]Archives from Hansard, Hansard Database (1825), the official report of debates in Parliament.

[92]This law had been passed in consequence of the crash of the South Sea Company.

[93]Odlyzko (2010, p. 42).

At the same time, the British government did little to regulate the railway industry. Companies only had to submit a bill to Parliament, which would approve the route and give the company the right to buy the required land for the building of the railway.[94] Technically and financially the scheme was extremely efficient and should have worked. However, important problems in this *laissez-faire* system arose. An unlimited number of companies could present projects because the United Kingdom required railway lines in the whole country. Little to no control on the viability of the projects was realized.[95] As a result anyone could create his railway company, mount a project, submit a bill to Parliament, collect the cash on the stock market and flee with the money. This is what happened at the peak of the railway mania, in 1846, when many Members of Parliament were investors in this industry.

The railway projects were very ambitious, especially compared to the success of the first railway between Liverpool and Manchester and the promising intercity lines that the huge British Empire could provide for railway companies.[96] Many *direct* lines were allowed that crossed vast swathes of countryside where the locomotive would barely be able to work. Only a few projects were rejected for their infeasibility.[97] Even stocks in railways that were not only unissued but not yet authorized by the Parliament, were freely exchanged and were purchased even by unwealthy investors in order to sell them rapidly.[98] A shift in the investment philosophy was thus happening.

At the peak of the mania, individual capitalists, in pursuit of private profit, plowed more than twice as much into the construction of the public infrastructure as their nation was spending on the military, one of the biggest in the world at that time.[99]

Once again, we find the same characteristics as for the South Sea bubble. A big number of investors and a sentiment of security because of investments from officials lead to general overoptimism and massive investments. When the unviability of a number of projects became clear, the fraud of a few entrepreneurs was uncovered. For instance, the magnate and MP George Hudson developed routes by amalgamating small railway companies and rationalizing them. He ultimately failed owing to his fraudulent practice of paying dividends from capital.[100]

Coupled to this wave of failures, the Bank of England raised interest rates in late 1845. As banks began to reinvest in bonds, the money began to flow out of railways, undercutting the boom.[101] The share price rise of railways slowed, and when prices eventually began to fall, investment stopped virtually overnight, leaving numerous companies without funding and a lot of investors with no prospect of any return on their investment.

[94]Odlyzko (2010, p. 6).

[95]Parnell (1833, pp. 101–119).

[96]Bright (1837).

[97]Odlyzko (2010, p. 161).

[98]Bright (1837, pp. 140–141).

[99]Odlyzko (2010, p. 4).

[100]Arnold and McCartney (2004, pp. 952–953).

[101]Odlyzko (2010, p. 86).

However, a few firms which had reached a critical size, such as the Great Western Railway and the nascent Midland began to buy up strategic failed lines to expand their network.[102] These lines could be purchased at a fraction of their real value when given the choice between a below-value offer for their shares or the total loss of their investment. Naturally, shareholders preferred a smaller price.[103] This process drastically reduced the number of operating railway companies in the late 1840s and early 1850s, and new lines were only constructed by large companies.[104] Eventually, the economic growth in the 1850s and 1860s was accompanied by smaller booms in railway construction, but these never reached the scale of the railway mania. The reasons include a more thoughtful but still very limited government control, a more cautious attitude of the investors and the fact that the U.K. railway network was approaching maturity, with less "blank canvas" available than in the 1840s.[105]

As for the SSC in the previous example, when investors realized that the plans of the railway companies were overoptimistic, they sold out their stocks and the prices collapsed. Many bona fide[106] investors were involved, and their losses were huge. To conclude, as an investment, it was a disaster as costs were far in excess of estimates, and revenues a fraction of projections. More than a century later, another crash would occur in the Middle East which shares many aspects of the railway mania: the Souk Al-Manakh crisis.

4.5.4 Souk Al-Manakh (1982)

The Souk Al-Manakh[107] is a crash that occurred in 1982 in Kuwait.[108] The Kuwait stock market started operating soon after the country had gained its independence from the British Empire in 1961.[109] In the first oil crisis of 1972 and 1974, the Arab states placed an embargo on crude exports to the U.S. in protest at its support of Israel in the Yom Kippur War. These events drove oil prices to new heights, which created an unprecedented amount of wealth in the oil-rich Persian Gulf countries, with Kuwait receiving a good share of this fortune. Kuwait entered a bull market, and since the U.S. was a bear market, the new rich Kuwaitis turned to the domestic

[102]Odlyzko (2010, pp. 58 and 139).

[103]Odlyzko (2010, p. 131).

[104]Odlyzko (2010, p. 90).

[105]Odlyzko (2010, p. 114).

[106]Bona fide investors are investors with good faith, whose aim is not to buy and sell to make money quickly.

[107]The literal transcription of this name would be *camel market*. Yet, it is highly questionable that it formerly was a real camel market. The name rather refers to the ambiance of these markets in the past.

[108]Rasmaroni (2006).

[109]Hassan, Al-Sultan, and Al-Saleem (2003, p. 7).

stock market to store their wealth.[110] The bull market was enhanced by the scarcity of Kuwaiti stocks, which consisted of *"only a few dozen uninteresting companies [that] were traded on the official exchange."* [111]

These funds prompted a speculation boom in the official stock market that culminated in a small crash in 1976 and 1977.[112] The government responded by bailing out the banks, setting a precedent that survives to this day. In addition to this measure, the Sheik introduced stricter regulations on the listing of new companies on the stock exchange, on forward trading and margin regulations. By 1978, the Kuwait stock market stabilized.[113] The next oil crisis followed in 1979, when Iran's energy production fell after the Iranian revolution. In Kuwait, the new-found wealth flooded the economy suddenly and, in the long run, destructively.

Higher oil revenues in Kuwait and the restrictive stock market led to the establishment of the Souk Al-Manakh by 1981, a market separated from the main Kuwait Stock Exchange. It soon started specializing in the trade of highly speculative non-Kuwaiti companies, such as those incorporated in Bahrain or the United Arab Emirates (U.A.E.), which did not fall under the legal restrictions.[114] Interest in Kuwait's heavily-regulated official market dwindled as the Souk Al-Manakh earned the reputation of being more exciting. Here, it was possible to double one's money within a few months or even a few hours.[115]

Yet, the Royal Sheik was still reluctant to grant the corporate charters necessary for companies to become publicly traded for fear that companies might become vehicles for stock speculation as it had previously happened.[116] In fact, many of the offshore companies whose shares were traded had nothing more than paper assets.[117] The U.A.E. shares were estimated to be traded at 20–30 times their earnings[118] at that time.[119] The IPO market was flourishing. Just as for the coming dotcom bubble in the U.S., people were seeking to take advantage of the roaring stock market, the outstandingly credulous investors and the general public's appetite for shares of all kind. For example, the GulfMedical[120] IPO with a starting price of one dinar was oversubscribed 2,600 times.[121]

---

[110]Financial Times (1995) which published the survey on p. iii of its May 23 edition.

[111]Darwiche (1986, pp. 87–89).

[112]Hassan et al. (2003, p. 7).

[113]Hassan et al. (2003, p. 7).

[114]Darwiche (1986, pp. 20, 86, 87 and 88).

[115]Darwiche (1986, p. 93).

[116]Verenoso's testimony as a consultant in Veneroso (1998).

[117]Financial Times (1995).

[118]This refers to the price-to-earnings ratio mentioned previously in Sect. 4.4.

[119]Darwiche (1986, p. 79).

[120]GulfMedical was initially a hosteling venture, which failed to find the required capital. It then was transformed in a hospital project before being able to go public.

[121]Quinn (2010).

Kuwaiti investors, however, were not concerned with risk. Their collective memory recalled the intervention of the state less than 5 years ago, and they assumed that the government would always be on the sidelines, ready to bail the market out if needed. As a result, Souk Al-Manakh became a vast gambling casino.[122]

Soon wealthy Palestinians, Egyptians and Pakistanis purchased stock through Kuwaiti nominees as only Kuwaitis could trade legally. Non-Kuwaitis were just as eager to invest despite their lack of legal standing. At the boom's peak, in 1980–1981, some stocks were advancing 100 % per month, with some advancing ten times.[123] One stock, The *Kuwait International Investment Company*, even advanced 136-fold.[124] Moreover, many Kuwaitis borrowed from banks to finance their trading which pushed the value of the listed securities to absurd heights. Still, this did not appear to be enough, and traders started to rely on postdated checks.[125] This was decisive for the development of the bubble. The height of the general insanity occurred when eight individuals, known as the Cavaliers, floated a total of $ 55 bn. in postdated checks, seemingly in an attempt to corner the market.[126] The most prolific check writer of the Cavaliers was Jassim Al-Mutawa, a passport office employee in his early 1920s, who managed to pass off $ 14 bn. all by himself. Sloppy accounting permitted Jassim Al-Mutawa and his associates to pile up a staggering $ 3.4 bn. overdraw.[127]

Believing the stock market would continue to soar, the Cavaliers made no payment provisions. The check owners expected a payment on the due date, as a default would violate the cultural tradition of trust and cause a loss of face for the entire family. The rapid inflow of speculative capital caused Kuwaiti stocks to skyrocket, with Souk Al-Manakh shares increasing by nearly 70 % in 1981 compared to its 1977 level.[128] By the early 1980s, the market was the third-largest in the world.[129]

But like all bubbles, the Souk Al-Manakh had to burst. In 1982, a bounced check touched off a cascade of margin calls and losses.[130] Most of Kuwait's banks became insolvent because of the crisis, having lent heavily to traders who could not repay the loans.

In August 1982, a speculator presented one of Jassim Al-Mutawa's postdated checks for payment ahead of its due date, but Al-Mutawa who was responsible for

---

[122]Darwiche (1986, p. 61).

[123]Darwiche (1986, p. 72).

[124]Darwiche (1986, p. 15).

[125]A postdated check is a check issued at present but with a later maturity date so it cannot get encashed at the present date. The check is like a promissory note and no legal action can be taken if the account closes down or there is shortage of funds.

[126]In finance, to *corner the market* refers to getting sufficient control of, for example, a particular stock to allow the stock price to be manipulated.

[127]Quinn (2010).

[128]Darwiche (1986, p. 25).

[129]Fitch (2010).

[130]Fitch (2010).

an astonishing amount of post-dated checks, was unable to pay his debts. The Souk Al-Manakh house of cards imploded, throwing hundreds of speculators into default.

To make matters worse, an overabundance of oil suppressed prices, reducing Kuwait's 1982 oil revenues to only a quarter of those of 1980. The final crush came when Kuwait's new finance minister, Abdelatif Al-Hamad, indicated that he had no intention of supporting the market at the insane levels to which it had risen.[131]

The Souk's decline was so instantaneous and dramatic that it could not even be considered a crash, as *there were simply no bids*.[132] By September 1982, the Ministry of Finance ordered all dubious checks to be turned in for clearance, tallying the value of worthless checks to no less than 27 bn. Kuwaiti dinar KD ($ 92 bn.).[133] Netted out, these checks amounted to KD 7 bn. ($ 23 bn.), more than Kuwait's entire GDP today! Losses were spread across the whole spectrum of the Kuwaiti society and were equivalent to $ 90, 000 for every Kuwaiti man, woman and child.[134] The Kuwaiti government promptly shut down the Souk Al-Manakh, built a new stock exchange and trading could eventually start over again.

Fadwa Ardel Darwiche and Fida Darwiche explain the process of this crash as follows:

> *The Souk Al-Manakh, [...] was based on a working mechanism which consisted of three main components:*
>
> 1. *A wave of exaggerated optimism on the part of investors about the future of share prices, which was reflected in inflated prices.*
> 2. *Almost no connection between information on the shares and the decision to buy.*
> 3. *Easily available finance through payment by postdated cheques, backed up by banking facilities on a huge scale.*
>
> *As a result, Al-Manakh became a market for forward buying on a large scale, where speculation relied on the investors' anticipation that share prices would go up.* [135]

Like the tulipomania and the South Sea bubble before, the Souk Al-Manakh *miracle* was an illusion, an inverted pyramid that relied on the exponential increase of share prices for its continued existence. Echoes of the Souk Al-Manakh crisis are still discernible in Kuwait's speculative and rumor-fed markets as well as in the frequent pleas from investors for government aid during market slumps.

The next crisis to be presented here, the Dubai real estate bubble also took place in the Middle East region. It did not unfold in the same exotic way as the Souk Al-Manakh bubble, but it also had a dramatic impact on the investors. Although it was not exclusively a stock market crisis, it deserves some attention, particularly under the aspect of contagion. First, it seems that it occurred because of a previous crisis: the subprime crisis. Second, it is tightly linked to sovereign and credit issues.

---

[131] Quinn (2010) and Darwiche (1986, p. 99).

[132] Veneroso (1998).

[133] See the 1995 survey in Financial Times (1995) published on May 23, and Darwiche (1986, p. 107).

[134] See the 1995 survey in Financial Times (1995) published on May 23.

[135] Darwiche (1986, p. 87).

4.5.5 Dubai Real Estate Bubble (2009)

In 2007, one of the most disastrous crises since 1929 started: the subprime crisis, which would throw the world into a general economic recession. However, this crisis also affected very different economies in an unpredicted way. It is particularly interesting because of the contagion effect of credit risk on market risk and the impact of the optimism of a leader of the country on investors.

Dubai is part of the United Arab Emirates (U.A.E.).[136] Since 1950, the state diversified its economy into ports, trade, services and finance, mostly successfully.[137] Under the authority of Sheik Mohammed bin Rashid Al Maktoum, the country's wealth expanded dramatically, despite a serious lack of oil in its soils. The investors took confidence in the projects of the monarch, especially after the surprising success of the palm-tree shaped island of Jumeirah.

Surfing on this trend, the sheik expanded his projects and decided to create two other uncommonly gigantic artificial islands: Jebel Ali and Deira. In addition, many city construction projects were started, among them the 828 m high Burj Khalifa Bin Zayed.[138] One of the objectives was to create a resort center for European and Asian holiday guests and to sell beach-side apartments.[139] This objective seemed to have been reached. At the height of Dubai's property boom, one-bedroom apartments were selling for as much as $ 2,776 a square foot on the secondary market, according to Betterhomes, one of Dubai's largest real-estate brokers. One of the most expensive apartments at the time sold for just over $ 3.5 mn.[140]

In his book *My Vision—Challenges in the Race for Excellence*, Sheikh Mohammed bin Rashid Al Maktoum describes his hopes and dreams. He retraces the very ambitious projects and prospects he has for his state and his visions for the U.A.E.

The appetite for grandeur lead the country to borrow heavily. The emirate used financial vehicles specializing in certain industries, like Nakheel, Istithmar and Emaar, to sustain its borrowing capacities. However, Dubai is one of the poorest countries in the U.A.E. and founded its economic model on a borrow-and-build system, requiring a large amount of liquidity while expanding its tourism capacity.[141] The emirate lived on credit for its expansion, and on tourism for its primary source of income.

---

[136]The U.A.E. is a federation of seven sheikdoms, including Abu Dhabi and Dubai, which have separate ruling families and separate budgets, but joint security, immigration and foreign policies. Abu Dhabi is the senior partner in the group and controls 90 % of its vast oil reserves, considered to be the world's fifth largest.

[137]Spencer (2009).

[138]Burj Khalifa Bin Zayed has been the tallest building in the world since August 2010.

[139]Gapper (2009).

[140]Bianchi and Critchlow (2010).

[141]Spencer (2009).

But when the subprime crisis set in, a general economic slowdown was the consequence that put pressure on the oil price. Since Abu Dhabi accounts for 90 % of the oil sources in the area,[142] not Dubai, it was not directly oil that forced the emirate to deal with its large debts. But because of the international confidence crisis which was caused by the subprime crisis and the developing credit crunch, only a few banks were still willing to lend money.[143] And while the diminishing oil price should have increased the number of tourists, the crisis actually made numerous people jobless and created a feeling of economic insecurity among potential travelers. As a result, tourism dropped heavily, limiting the income of the country.[144]

The credit crisis and an international recession which lead to a lack of income from tourism accompanied by a stock market crash plunged the country into liquidity difficulties and left it insolvent with a large amount of unsold properties. On November 25, 2009, Dubai World, a Dubai government-backed conglomerate, asked its bankers for a 6-month stay on debt repayments.[145]

The timing of the declaration was optimal. It was the Eid al-Adha religious holiday in the Middle East and Thanksgiving in the United States. In normal days, the U.S. market would have tried to get information about the Dubai stock exchange movements, and European traders would have followed the U.S. American lead. The markets that were open for business were hit by uncertainty.[146] Moreover, a computer crash happened on this very day at the London Stock Exchange, which is 21 % owned by the Dubai Government, and left dealers unable to trade for three and a half hours.[147]

Yet, markets reacted quickly. European markets fell by 3.2 % on average on Thursday 26, then calmed on Friday to close higher. The FTSE 100 plunged by 171 points to 5,194—its biggest 1-day fall in 8 months in one of the most jittery days in the financial markets since the depths of the banking crisis.[148] Markets in Asia were sharply lower on November 25, with the Hang Seng in Hong Kong falling 4.8 %.[149] The Dow Jones Industrial Average Index tumbled 154.48 points, or 1.48 %, to 10,309.92 on November 25, recovering from a 225-point loss early on. The S&P 500 Index fell 19.14 points, or 1.72 %, to 1,091.49.[150]

Figure 4.8 shows the daily price development of the Dubai Financial Market General Index (DFMGI) in the last quarter of 2009 while Fig. 4.9 shows the corresponding daily returns. Similarly, Fig. 4.10 shows the daily price development of the Financial Times All Share (FSTAS) over the same time period. As we can see, the price declines occur simultaneously, yet not with the same intensity.

[142] Salama (2010).

[143] Anderson and Heather (2007).

[144] Spain (2009).

[145] Greenwood (2009, pp. 1–2).

[146] Hosking and Robertson (2009).

[147] Hosking and Robertson (2009).

[148] Bianchi and Critchlow (2010).

[149] Bianchi and Critchlow (2010).

[150] Bianchi and Critchlow (2010).

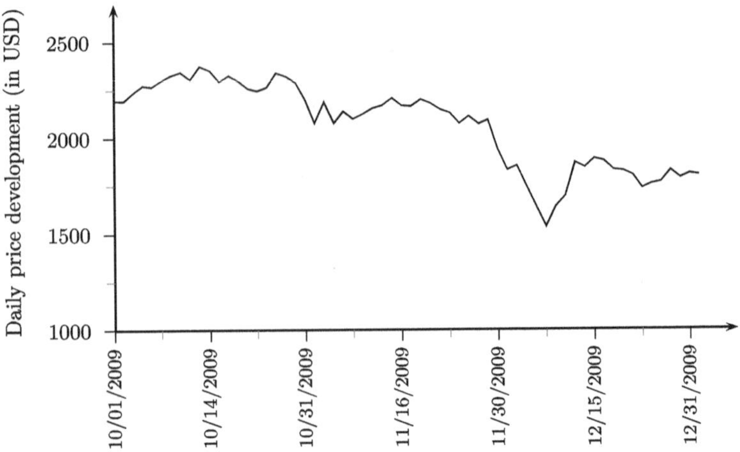

Fig. 4.8 Price development of the Dubai Financial Market General Index (DFMGI) in USD (daily data) from October 1, 2009 to December 31, 2009. *Source*: Factset (ticker: 166097)

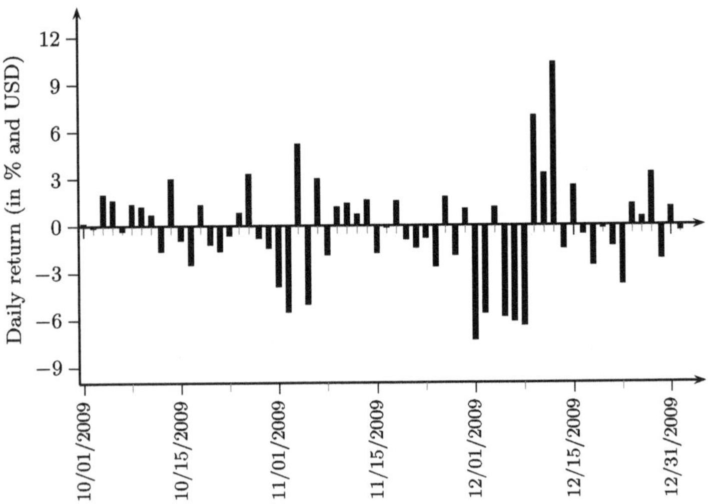

Fig. 4.9 Daily return of the Dubai Financial Market General Index (DFMGI) in USD from October 1, 2009 to December 31, 2009. *Source*: Factset (ticker: 166097)

In 2009, house prices fell 50 % in less than 6 months,[151] leading to a steep stock correction. Emaar's shares, once the most valuable real estate stock in the world, had

[151]Spencer (2009).

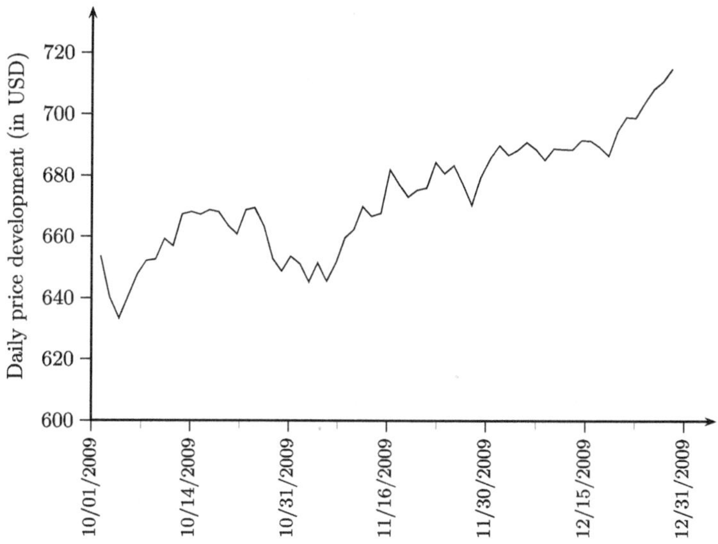

Fig. 4.10 Daily price development of the Financial Times All Share (FSTAS) Index in USD from October 1, 2009 to December 31, 2009. *Source*: Factset (ticker: ASX-FTX)

fallen 86 % since hitting an all-time high of AED 28.7 ($ 7.80)[152] on September 18, 2005, at the peak of Dubai's housing bubble.[153] On November 30, the stock closed 3.4 % lower at AED 4.02 , with investors skeptical of whether the opening of the Burj Dubai was enough to revive its fortunes.[154]

This leads us to answer the following questions:

- Was the Dubai crash connected to the fall of Lehman Brothers 1 year before?
- Why was a group of state-owned company in a relatively small country able to move the stock markets of the whole planet?

At first, Nakheel[155] was known to be in trouble, but investors assumed that as a state-owned company it would not default on its debt. Yet Dubai had to face the major problems stated above. A research note from UBS underlines:

> *Abu Dhabi's support for Dubai might be less generous than the markets have assumed so far. Perhaps Abu Dhabi has forced Dubai to tackle the problem of excessive corporate debt in-house first before extending more financial support.* [156]

---

[152]AED stands for United Arab Emirates dirhams.

[153]Irish (2010).

[154]Bianchi and Critchlow (2010).

[155]Nakheel is one of the world's largest real estate developers. It is one of the financial vehicles used by Dubai to implement its investing strategies in real estate.

[156]Kennedy (2009).

The fact is that the Dubai government refused to issue detailed statements of how it was to handle Dubai World's debt problems, and discredited those who said that the crash had undermined Dubai's development model.[157] Moreover, the debt, initially estimated at $ 60 bn. in November 2009, was reevaluated[158] to $ 80 bn., $ 90 bn. and then $ 110 bn.[159] Therefore, the collapse of Lehman Brothers has very little connection with this matter, even though the company was in the middle of the CDS business, a key element of the global contagion.

The logic of the answer to the second question is fairly simple. Credit default swaps (CDS) which came to prominence during the subprime crisis, were a strongly contagious element. Figure 4.11 shows the 1-year Dubai Sovereign Credit Default Swap (CDS) yield in bps[160] from March 13, 2009 to December 12, 2009.

On November 24, 2009, i.e., the day before the announcement of Dubai World, the 5-year CDS price was 318 bps. Two days later, it jumped to 541 bps which is a 70 % increase and on November 27 it reached 647 bps. This means that to insure a $ 10 mn. transaction on November 27, 2009, the required amount was $ 647, 000, which was more than double the price of a week before.[161]

Figure 4.12 below represents the daily evolution of 10-year Gulf State Sovereign CDS from March 13, 2009 to August 12, 2009. The price of the CDS increased

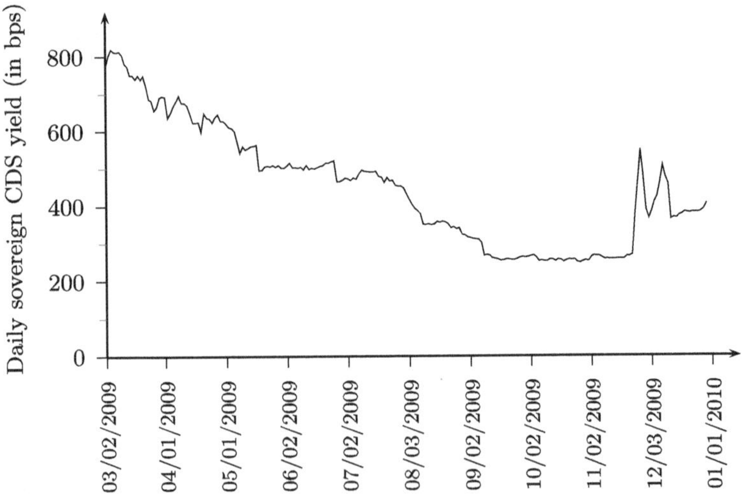

Fig. 4.11 Dubai Sovereign Credit Default Swap (CDS) yield in bps from March 13, 2009 to December 12, 2009. *Source*: Factset (ticker: CDSDU1Y-FDS)

[157] Spencer (2009).

[158] Greenwood (2009, p. 2).

[159] Sakoui and Kerr (2010).

[160] Bps stands for basis point. One basis point is a percent of a percent.

[161] Hernandez (2009).

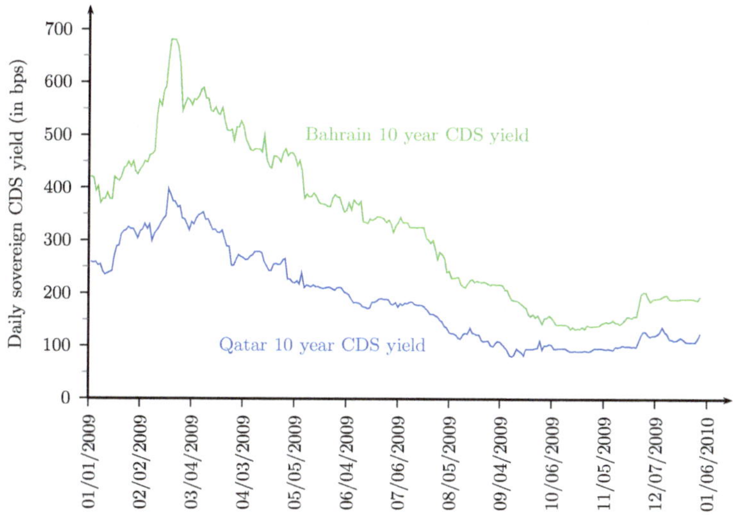

Fig. 4.12 Daily Gulf State Sovereign CDS yield from March 13, 2009 to August 12, 2009. *Source*: Factset (ticker: CDSQA10Y-FDS)

sharply for Dubai, but also for other countries of the U.A.E. This shows that the market treated this debt issue which was originally a corporate one (Nakheel only) as a sovereign debt issue (Dubai and the U.A.E. as a whole), involving the countries of the area, which invested massively in Dubai.

The problem was not limited to the countries of the U.A.E. On November 27, 2009, Indonesia's cost increased by $ 12,000 and China's by $ 6,000 for any CDS insuring a $ 10 mn. transaction. United States Treasury prices rose, as did the dollar, as investors fled equities for the safety of government-backed debt. The 10-year Treasury note rose 18/32 to 101 14/32, and the yield fell to 3.21 %, from 3.27 % on late Wednesday, December 2, 2009.[162]

The resulting uncertainty induced a high volatility. Very few investors had noticed how perilous investments in Dubai, Abu Dhabi and other countries of the U.A.E. were. The oil industry and the strong partnership among members of the U.A.E. seemed to provide sufficient guarantees to any investor. At the same time, only a few were able to realize the threat to European and U.S. banks, although everybody knew that the U.A.E. countries had invested massively abroad. A large debt would force them to give up their U.S. and European investments, which might lead to a major downturn on the stock exchanges of these geographic areas.

On the other side, people feared that the debt issue could become a real investment threat and result in a systemic crisis in emerging markets.[163] The latter

[162]Hernandez (2009).

[163]Systemic risk: the risk of collapse of an entire financial system or entire market, as opposed to risk associated with any one individual entity, group or component of a system. See Working Paper Kaufman (1999, pp. 28–29), also published in Kaufman (2000).

proved to be wrong: as it had happened 10 years before in Malaysia,[164] the systemic risk did not materialize. Nevertheless, the stock exchanges in developed markets reacted negatively to the rise in risk underlined by the CDS, fearing heavy losses and selling U.A.E. related stocks.

So as to calm down the market, Dominique Strauss-Kahn, managing director of the International Monetary Fund, said in an address in London in late November 2009 that *"the storm has passed"* but *"the global economy remains very much in a holding pattern - stable, and getting better, but still highly vulnerable."* He added: *"There is a lot of uncertainty in the air."* [165] In Europe, a research note from Credit Suisse estimated that European banks might be the hardest hit if Dubai World could not meet its obligations, with the total exposure estimated at € 13 bn. ($ 19.5 bn.).[166]

On November 26, 2009, shares in HSBC slumped by 5 %, wiping $ 9.6 bn. from its value. According to the U.A.E. Banks Association, HSBC had $ 17 bn. of loans outstanding to the U.A.E., of which Dubai is one. More than $ 4 bn. was slashed from the value of Barclays, while Lloyds and Royal Bank of Scotland saw their respective values fall by $ 2.6 bn. and $ 2.3 bn.[167] These banks were among the major investors in Dubai's corporate and sovereign debt or real estate projects.

Yet, Dan Alpert,[168] protested on November 27, 2009: *"The fact is the equity markets globally have gotten way ahead of themselves. The stock markets and the bond markets are in violent disagreement, and at some point, it is going to be resolved by a sell-off in the equity markets."* [169] But by July 2010, such a move had not been observed yet.

Analysts were looking for an orderly resolution of the issue to help restore confidence in Dubai's economy and its $ 110 bn. government-related debt.[170] On July 22, 2010, Dubai World invited creditors to a take-it-or-leave-it agreement on a $ 23 bn. debt restructuring. At this date, the creditors that held about 60 % of the $ 14.8 bn. of the holding company's debt—the seven banks on the coordinating committee of lenders—had already agreed on the initial terms of the restructuring proposal, which offered to pay back the principal over 5–8 years.[171] These lenders to Nakheel, which lies at the heart of Dubai World's debt woes, also had met a week before on July 15, 2010. The government had pledged up to $ 9.5 bn. in funding for the government-owned company that had been hit hard by the global financial crisis and Dubai's real estate crash.[172]

[164] Hernandez (2009).

[165] Schuman (2009).

[166] Hernandez (2009).

[167] Hosking and Robertson (2009).

[168] Dan Alpert is managing partner of Westwood Capital. The company provides financial services in mergers & acquisitions, restructuring, financial advisory and litigation support.

[169] Hernandez (2009).

[170] Sakoui and Kerr (2010).

[171] Sakoui and Kerr (2010).

[172] Sakoui and Kerr (2010).

To summarize, Dubai's overoptimistic real estate projects forced the country to its knees. Across the world, the exaggerated expectation for Dubai investments pushed stock and real estate to unsustainable prices. When the missing capacity of Dubai to meet its interest payments became apparent, a wave of fear and pessimism submerged the market. The risk increased steadily, as well as the volatility in the financial markets. Luckily, the announcement of Abu Dhabi to back Dubai may have saved the world from a general panic and a systemic crisis. Otherwise, the European and U.S. markets might have plunged even more dramatically because of the fear of a massive sell-off of U.A.E. investments in these geographic areas. On the other hand, it would have been irrational for Abu Dhabi not to help a member of the union, which might have brought the union to an end. This very situation would soon sound familiar for Greece.

4.5.6 Greek Crisis (2010)

The Greek crisis is another illustration of the emotional excesses in the world of finance. After the subprime crisis, the world economy was weakened and the small economies were heavily hit. Greece was among these countries, even though this was not at first obvious. Despite the unprecedented global reach of the recent crisis, the European Commission believed that some of its causes were similar to previous crises, such as the Asian crisis in the late 1990s or the crisis of the Nordic countries in the early 1990s.[173] The common feature was that all these crises were preceded by *"long periods of rapid credit growth, low-risk premiums, abundant availability of liquidity, strong leveraging, soaring asset prices and the development of bubbles in the real estate sector."* [174]

From 2003 to 2007, the country's growth averaged 4 % a year, with heavy consumer spending and easy access to credit. The $ 32,100 per capita GDP of the country stood just below Japan, Italy, Israel, Taiwan and Korea.[175] Yet, since November 7, 2007, the Greek stock market went through a crisis, dropping from 1053.116 to 239.140 on March 05, 2009.[176] Figure 4.13 shows the development of the MSCI Greek stock market index between January 1, 2008, and December 31, 2013, which illustrates the situation.

Between March 5 and November 6, 2009, the MSCI Greece Index (large- and mid-cap equities) soared by 136 %.[177] Then, a stock market correction occurred beginning November, which seemed normal after the previous strong upward moves. Yet, on November 25, 2009, Dubai declared to reschedule its debt repay-

---

[173] European Economy Series (2009).

[174] European Economy Series (2009, p. 1).

[175] Mucha (2010). Updated data is available through the World Bank online database at http://data.worldbank.org/indicator/NY.GDP.PCAP.CD.

[176] *Source*: FTSE/Athens Stock Exchange Large Cap Index (Ticker: FTASE:IND).

[177] *Source*: Bloomberg, as of September 30, 2013 (Ticker: FTASE:IND).

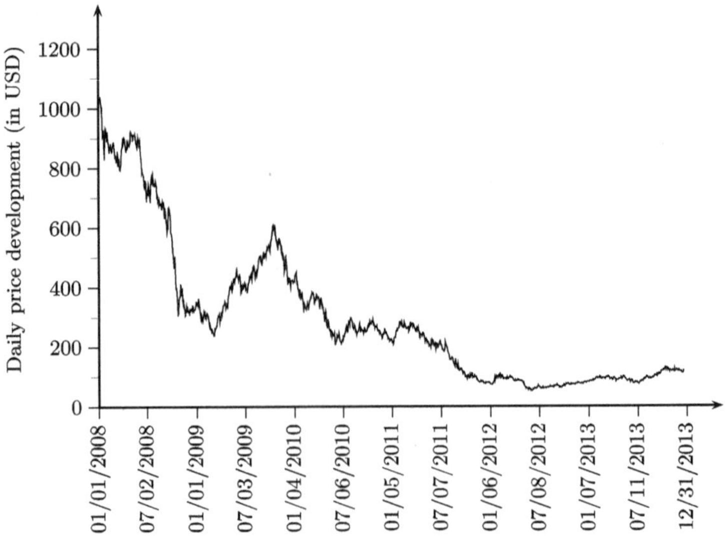

Fig. 4.13 Daily price development of the MSCI Greece Standard Index (large- and mid-cap, net dividends reinvested) in USD from January 1, 2008 to December 31, 2013. *Source*: Factset (ticker: 9300000)

ment, which created an increased volatility on the markets and shook the Greek economy. The Athenian stock market continued to drop. In fact, on October 22, 2009, the first real stroke had already occurred when Fitch Ratings[178] downgraded the Greek sovereign debt from "A" to "A-", with negative perspectives. The next downgrading happened on December 8, 2009, with a new notation of BBB+ for the country of Plato.

This was soon followed by Standard & Poor's and Moody's, who on December 16 and 22, 2009, respectively, downgraded Greek sovereign debt from A1 to A2[179] because of the economic degradation of the country.[180]

Actually, the Greek economy rested on two pillars: tourism and transports. After 2008, these two industries suffered significantly and lost 15 % of their income. Moreover, Greece suffered from a chronic structural deficit averaging 100 % of the

[178]There are three big audit companies, whose role is to asses the quality of investment tools and companies present on the market. Each company has its own grading system. Yet, the grades given by the audit companies are comparable. The companies are Fitch Ratings, Standard & Poor's and Moody's.

[179]These grades are important for credit access conditions. The higher the grade, the lower the yield a country or a company has to pay. An asset, say a sovereign debt of a given country, is less risky when graded "A" than when graded "A-". The downgrading means that a country has to pay more if it wants to seek financial resources on the markets.

[180]Rascouet (2009).

GDP from 2001 to 2010, and did not manage to profit from the booming years 2000–2007 to improve public finances. Moonlighting reached incredible heights, especially in tourism, and amounted to around 20 % of Greece's GDP according to estimates. Even Georges Papandreou deplored rampant corruption and fraudulent statistics before his arrival at the head of the state.[181]

The second blow came after Georges Papandreou was elected Prime Minister, on October 4, 2009. Officially, Greece had a public deficit of 6 %, which was already alarming. According to the European stability pact, a country of the European Union should not have a public deficit above 3 % of the GDP and a public debt above 60 %.[182] But after the subprime crisis, even a big European country like France had a public deficit of 7.9 %.[183] However, the Greek government sharply revised its public deficit estimation on October 4, 2010, when the Prime Minister declared he actually expected it to be 12.9 % for year 2009 and 9.4 % for year 2010.[184] Even worse, this number was later revised by Eurostat[185] to 13.6 %.[186]

The public debt was estimated to be 113 % of the GDP at the end of 2009,[187] i.e., $ 406 bn., against an estimated 120 % for end 2010,[188] but the revised Eurostat number amounted to 115.1 %. This made the country the second highest indebted country in the European Union. The Greek government wanted to rely on the financial markets as long as possible rather than seek the help of the eurozone and the IMF,[189] but finally, after a short delay, it had to submit a request for help.[190] The interest rate gap between Greek and German 10-year bonds[191] widened to a record high of 5.67 % on April 21, 2010.[192] This translated into interest rates above 8 %, a level not long sustainable for the country. On April 27, 2010, the interest rate reached 9.73 %[193] and 2 days later, it topped 11.076 %,[194] an all-time high

[181]Berteloot and Hebert (2010).

[182]See the official and updated elements regarding the Stability and Growth Pact on the website http://ec.europa.eu/economy_finance/economic_governance/sgp.

[183]See the article from the Agence Francaise de Presse (2010) in the newspaper Liberation.

[184]Kennedy and Ross-Thomas (2010).

[185]Eurostat is the European Union's statistics agency in charge of evaluating different economic factors and issues.

[186]Becatoros (2010).

[187]Tupy (2010).

[188]Becatoros (2010).

[189]Becatoros (2010).

[190]France 24 (2010).

[191]Germany is considered a benchmark for stability in the European Union. Sooner in history, the country had seen its national currency heavily devalued after World War II, and since then has tried to keep a very strong hold on the stability of its economy and currency.

[192]Becatoros (2010).

[193]Becatoros (2010).

[194]Daily Telegraph (2010).

for a eurozone member. The Greek 2-year bonds reached 18 %.[195] On April 27, 2010, the Athenian stock market lost 6 % in a single day.[196] Ben May, a European economist for Capital Economics,[197] illustrated this situation: "*Greece is in the midst of another hellish week and now faces no choice but to seek to formally activate the European rescue package.*"[198]

These were seriously bad news, especially when considering that the country needed to borrow no less than $ 71 bn. in 2010 alone, with $ 11.3 bn. maturing on May 19, 2010.[199] Voices were raised against Greece and demanded its exit from the EU, because in order to join the union, the country had rigged its public accounting and again did so in 2004.[200] Marc Faber,[201] the publisher of *The Gloom, Boom & Doom report*,[202] said:

> *The best would be to kick out Greece and the countries that abuse the system. They didn't have the fiscal discipline that was essentially imposed by EU.*[203]

On the other hand, Angela Merkel[204] noticed:

> *They will certainly understand the necessity of a reconsideration of penalization applicable against those who bypass the rules of the stability pact. [...] We have to draw lessons from this crisis so that the euro currency keeps its strength and stability.*[205]

She also proposed to modify the treaty of Lisbon so as to allow the exclusion of countries which do not follow the rules of the European Union.[206] Yet, such a modification was very unlikely to happen at this time, because it required the unanimity of the voters, including Greece.[207] An exit from the EU went against the Greek interests: A Greek currency would be largely devalued and the country could not benefit from EU help and a favorable access to financial resources.

---

[195] Berteloot and Hebert (2010).

[196] Berteloot and Hebert (2010).

[197] Capital Economics is a leading macroeconomic research consultancy.

[198] Becatoros (2010).

[199] Becatoros (2010).

[200] Tupy (2010).

[201] Marc Faber studied economics at the University of Zurich and, at the age of 24, obtained a Ph.D. in economics magna cum laude. He is also the author of several books including *Tomorrow's Gold—Asia's Age of Discovery* which was first published in 2002 and highlights future investment opportunities around the world, see Faber (2002). This book was for several weeks on Amazon's best seller list and is being translated into Japanese, Chinese, Korean, Thai and German. Dr. Faber is also a regular contributor to several leading financial publications around the world.

[202] *The Gloom Boom & Doom Report* highlights unusual investment opportunities.

[203] Kennedy and Ross-Thomas (2010).

[204] Angela Merkel, born in 1954, has been the Chancellor of Germany since 2005.

[205] Agence Francaise de Presse (2010).

[206] Berteloot and Hebert (2010).

[207] Berteloot and Hebert (2010).

Letting Greece default might have been catastrophic for the euro currency and could have started a domino effect. Actually, Greece accounted for less than 3 % of the EU economy.[208] However, when Greek interest rates became more attractive, more investments from other countries flowed into Greece to buy sovereign debt. A default on Greek debt could create default in other countries, if foreign investors who lost their investments in Greece might not be able to pay off their own debt. At least, it would create an unwanted European economic slowdown. A domino effect could occur and seriously harm the weakest economies. That is why the EBRD[209] warned that Bulgaria, Romania and Serbia might be hurt by the crisis in Greece.[210] Nevertheless, Greece was not the only country hammered by a large debt. Table 4.2 below shows the countries called the *PIIGS*[211] by the British press which were facing the biggest sovereign debt and public deficit during the Greek crisis.

The amounts are quite staggering and are thought to be the next challenge of the European countries.[212] Therefore, the President of the European Central Bank at that time, Jean-Claude Trichet, called a Greek default *out of question*.[213] Yet, the increased risk on the sovereign debt of Greece and other countries like Portugal, which was also downgraded,[214] made the euro currency slide from $ 1.50 in January to $ 1.26 on May 6, 2010.[215] Fears of credit contagion were real and could have been disastrous for the euro currency and the eurozone countries as a whole.

Many investors panicked as BNP Paribas observed: *"People worry that if Greece is Bear Stearns, Portugal is Lehman and Spain AIG."*[216] Panic spread across the world. A widespread stock market sell-off was triggered on April 27, 2010,[217] with the downgrading of Greece to BB+, putting its sovereign debt into the category of speculative investments.[218] Moody's Investor Services would soon do the same, with a new A3 grade on June 14, 2010.[219] On April 27, the FTSE 100 fell 1 %, Hong Kong dipped 1.3 %, Wall Street shed 1.9 % overnight, Athens slid 1.93 %, Tokyo fell

---

[208]Pearlstein (2010).

[209]EBRD stands for European Bank for Reconstruction and Development. Founded in 1991, it is the largest financial investor in its region of operations which stretches from central Europe and the Western Balkans to central Asia.

[210]Gaunt (2010).

[211]PIIGS is an acronym which stands for Portugal, Ireland, Italy, Greece, Spain. They held a large amount of debt which went well above the limits fixed by the European Union and the Stability Pact.

[212]Sakoui (2010).

[213]Blackstone (2010).

[214]Berteloot and Hebert (2010).

[215]Pearlstein (2010).

[216]Bear Stearns, Lehman Brothers and AIG are banks that went bankrupt after the subprime crisis. See Gaunt (2010).

[217]Gaunt (2010).

[218]Berteloot and Hebert (2010).

[219]Oosterveld and Carlson (2010).

Table 4.2 Peripheral eurozone total financing needs

	2010	2011	2012	2013	2014
Greece					
Debt maturing	15,772	31,317	31,735	24,878	103,702
Fiscal deficit	21,770	14,341	7,392	5,469	48,972
Total financing need	37,542	45,658	39,127	30,347	152,674
Portugal					
Debt maturing	17,939	15,913	8,569	8,190	50,611
Fiscal deficit	13,467	10,920	7,832	4,915	37,134
Total financing need	31,406	26,833	16,401	13,105	87,745
Spain					
Debt maturing	76,563	84,037	61,235	51,506	273,341
Fiscal deficit	102,641	79,520	57,357	33,361	272,879
Total financing need	179,204	163,557	118,592	84,867	546,220
Italy					
Debt maturing	251,532	192,211	168,208	100,444	712,395
Fiscal deficit	86,804	86,612	89,407	90,020	352,843
Total financing need	338,336	278,823	257,615	190,464	1,065,238
Ireland					
Debt maturing	8,587	4,648	5,955	6,028	25,218
Fiscal deficit	18,636	16,219	12,006	7,947	54,808
Total financial need	27,223	20,867	17,961	13,975	80,026
Total	613,711	535,738	449,696	332,758	1,931,903

Source: Bank of America (2010)
Note: All numbers present in table are expressed in thousands of U.S. dollars

2.6 %, Madrid declined 3 % and Lisbon lost 6 %.[220] An analyst at IG Markets, Ben Potter, declared: *"Any hope that the Greek issue was finally coming under control took a huge blow yesterday with the country's sovereign debt being downgraded to junk."*[221]

A few days later, on May 6,[222] the Dow Jones lost 9 % before recovering 3 % before the end of the day. The MSCI had lost 11 % since mid-April. The price of crude oil had fallen 14 % in the previous 5 days with growing concerns on Chinese growth.[223] Commodities had lost 5 % during the former week while emerging markets sovereign debt yields had blown 55 basis points. A real confidence crisis was about to form again after the credit crunch in the wake of the subprime crisis, while in the meantime, the euro currency was attacked by speculative bets.[224]

[220]Gaunt (2010).

[221]Gaunt (2010).

[222]This day was also referred to as the flash crash as will be explained later. See Gaunt (2010).

[223]This day was also referred to as the flash crash as will be explained later. See Gaunt (2010).

[224]Berteloot and Hebert (2010).

The fear crossed the Atlantic, and the then CEO and Co-CIO of PIMCO, Mohamed El-Erian, predicted: *"Problems with Greek debt are about to spread to other countries and could infect the U.S. unless the nation tackles its own mounting problems."* [225] Practitioners were on the verge of a global panic and stock crashes seemed imminent, not exactly because of a Greek default, but rather because of the overreactivity of the markets at this time. Andrew Bosomworth, head of portfolio management at PIMCO in Germany declared: *"It is very difficult to find a near-term equilibrium from a policy and markets positioning perspective."* [226]

A solution could have been to reschedule debt repayments as in the case of Dubai. However, this could have been disastrous for many European banks holding Greek debt, as Herman Van Rompuy[227] declared. A rescue package looked preferable. Actually, the German and French interests were obvious. French banks were exposed to up to $ 80 bn., while German banks had bought $ 45 bn. of Greek debt.[228] Hypo Real Estate[229] for instance possessed $ 8 bn. of Greek debt.[230]

Yet, the European reaction toward Greece was far from being unanimous. A poll from a left-wing newspaper showed that while 60 % of French people were willing to offer Greece financial support, less than 20 % of the Germans agreed.[231] It has to be noted that during the past 10 years the inflation differential between Greece and Germany had widened to 30–40 %,[232] i.e., Greece had lost its competitiveness even in the eurozone. As a result, a return to equilibrium thanks to financial help looked doubtful. Yet, the European Central Bank had already bought $ 53 bn. of sovereign debt, including $ 32 bn. of Greek debt. A few weeks before, its president Jean-Claude Trichet had said he would not do so,[233] which sheds light on the lack of independence of this institution.[234]

On March 25, 2010, eurozone leaders agreed to put emergency measures in place so as to prevent a debt crisis contagion to other countries.[235] Germany would provide

---

[225] Cox (2010).

[226] Gaunt (2010).

[227] Herman Van Rompuy was the President of the European Council during the Greek crisis. The European Council is the institution of the European Union responsible for defining the general political direction and priorities of the union.

[228] Blackstone (2010).

[229] The Hypo Real Estate Holding AG is a holding company based in Munich, Germany which comprises a number of real estate financing banks. The company's activities span three sectors of the real estate market: commercial property, infrastructure and public finance, and capital markets and asset management. Hypo Real Estate is the second largest commercial property lender in Germany. See Associated Press (2008b) on October 4.

[230] Berteloot and Hebert (2010).

[231] Hureaux (2010).

[232] Hureaux (2010).

[233] Pearlstein (2010).

[234] Spiegel Online (2010).

[235] Mahier and Duval (2010).

$ 8.4 bn. and France would pay $ 7.2 bn. to help the Greek government.[236] In all, the eurozone rescue package amounted to $ 40 bn. of bilateral loans to be provided *if and when needed* and was supplemented by a $ 13.2 bn. rescue package from the IMF[237]—this was more than Greece's officially estimated needs at this stage.[238]

The effect was not immediate and uncertainty kept the markets very volatile. It has to be noted that this help appeared difficult to ensure. First, the rescue package created moral hazard for the future by intrinsically protecting countries with doubtful economic and accounting policies.[239] Second, EU members did not help Poland, Hungary and other countries in difficulty just after the subprime crisis.[240] In the case of Greece, their only legal recourse was the article 122-2 of the Lisbon Treaty, that allows EU members to help a country that faces catastrophic and unprecedented events. Yet, the events that the article refers to, are natural disasters rather than financial crises.[241] A last issue is the rescue package itself. Neither France nor Germany had the required funds. However, by borrowing on the financial markets at 3 % and lending at 5 %, the French state would win $ 200 mn. and the same would happen for Germany.[242] In exchange for the rescue package, Greece engaged itself to fulfill a number of conditions. The most significant ones follow below[243]:

- Tax reform unifying the tax rate (+$ 1,450 mn., 0.5 % of GDP)
- Special tax on profitable businesses and real estate value (+$ 1,390 mn., 0.4 % of GDP)
- Surcharges on tobacco, alcohol, petrol and mobile phones (+$ 1,470 mn., 0.5 % of GDP)
- Reduction of tax evasion (+$ 1,590 mn., 0.5 % of GDP)
- Reduced fraud in social security contributions (+$ 1,590 mn., 0.5 % of GDP)
- Reduced expenditure on military equipment (−$ 605 mn., 0.2 % of GDP)
- Reduced subsidies to pension funds (−$ 715 mn., 0.2 % of GDP)
- Reduced hospital grants (−$ 1,850 mn. , 0.6 % of GDP)
- Ten percent reduction in premiums paid to civil servants (−$ 860 mn., 0.3 % of GDP)
- Deleting one third of the outstanding short-term contracts in the public sector (−$ 160 mn., 0.1 % of GDP)

---

[236] Spiegel Online (2010).

[237] IMF stands for International Monetary Fund. Founded in 1944 after the Bretton Woods agreements, its aim is to help countries in economic distress because of economic and monetary constraints.

[238] Becatoros (2010).

[239] Tupy (2010).

[240] Mahier and Duval (2010).

[241] Mahier and Duval (2010).

[242] Berteloot and Hebert (2010).

[243] The full version of the conditions is available online, see Ministry of France (2010).

- Blocking of recruitment in 2010 (−$ 105 mn., 0.0 % of GDP)
- Four of five retirements not replaced in the public sector
- No wage increases above $ 2,650[244]

Predictably, the Greek people was against these drastic reforms, which could cause much hardship, as it had happened to Latvia before. In 2009, Latvia had required the help of the IMF with the objective to reach a fiscal deficit of 5 % for 2009. But the austerity plan of the IMF led to an economic contraction of 18 %, i.e., three times worse than forecasts![245] On May 5, 2010, under a general unrest fueled by left-wing unions, a nationwide strike closed down airports, government offices, courts and schools.[246] The rescue package for Greece could reduce the purchasing power of its citizens by 10–30 %,[247] according to the French socialist politician Laurent Fabius.[248]

To summarize, the Greek crisis was mainly caused by doubtful accounting by the Greek government and the illusion of wealth maintained by an easy access to credit and a large level of debt per citizen. The subprime crisis affected the income from tourism and transports, which made more people jobless, increased social expenses and put a heavy burden on the public treasure.

With a deepening debt and capital expenditure and shrinking revenues, the Greek state faced difficulties to meet its debt obligations. Rating Agencies like Fitch, Moody's and Standard & Poor's downgraded the country significantly, and Greece faced unsustainable yields if it wanted to borrow on the markets. This endangered the strength and the stability of the euro currency and the volatility and panic on the stock markets reached new highs. A general economic pessimism struck all asset classes and threatened even the U.S. markets. After many appeasement speeches from politicians and officials and the guarantee of a joint rescue package from euro members and the IMF, the market seemed to slowly recover its sanity.

4.5.7 Flash Crash of May 6, 2010

In this section, we do not exactly deal with a crisis or a bubble. But we believe that flash crashes, an interesting kind of market eccentricities, should be presented here. They are not as new as it appears, given that the first one was observed several decades ago. Flash crashes may become much more frequent with the structural changes which have occurred on the stock markets during the past

[244] Mahier and Duval (2010).

[245] Tweed and Eglitis (2010).

[246] Blackstone (2010).

[247] Berteloot and Hebert (2010).

[248] Laurent Fabius served as French prime minister from July 17, 1984 to March 20, 1986. He was 37 years old when he was appointed and has been, so far, the youngest prime minister of the Fifth Republic.

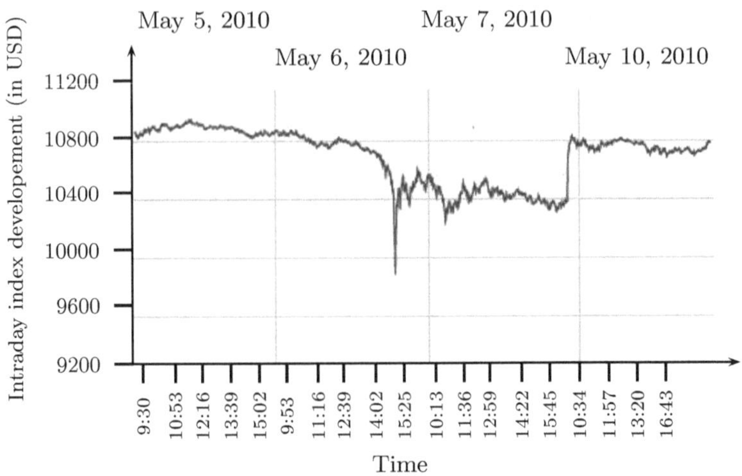

Fig. 4.14 Dow Jones Industrial Average Index on the four trading days between May 5 and 10, 2010. *Source*: Bloomberg (ticker: DJIA INDU)

decades. Information technologies have significantly improved, and approximately a half of the stock exchange trading in the U.S. is computer-based. High-frequency trading plays an increasingly important role with orders passed at an outstanding speed. While the crashes of yesterday lead to losses of 10 % within days or weeks, such brutal market moves can now occur within minutes.

On May 6, 2010, by 2.32 pm an impressive stock market decrease occurred on U.S. American indices. In 8 min, the Dow Jones Index fell by 998.5 points, a decline of 9.2 % equivalent to a loss of between $ 700 bn. and $ 850 bn. as can be seen in Fig. 4.14.[249]

A flash crash does not exactly fit the definition of a financial crash by Patel and Sarkar,[250] i.e., a decrease of 20 % in developed markets. However, it provides further evidence that markets are not as efficient as standard financial theory would suggest. In the following, we present the context in which the dip occurred. In the days after the flash crash, U.S. officials and market participants sought to explain the extreme market volatility. We will present the major reasons which were put forward as well as the countermeasures established by the market authorities.

4.5.7.1 A Description

Starting in early February 2010, U.S. equities performed relatively well, even though signs had begun to appear at the end of the third quarter of 2009 that the economy faced new challenges. The Federal Reserve announced plans to withdraw liquidity from the money markets, and the outlook for corporate profits took a negative

[249] Schulmerich (2010a, p. 1) and Rice (2010, p. 1).

[250] Patel and Sarkar (1998, p. 6).

turn as analysts revised their earnings estimates downwards.[251] Because of these changes in fundamental conditions, investors became increasingly risk-averse as 2010 progressed, and selling pressure grew, very similar to what happened in Russia in 1997, before the crisis.

Moreover, European sovereign debt concerns moved from a healthy state to a disturbing level as sovereign credit default swap (CDS) spreads rallied +175 % from the beginning of the year.[252] Combined with uncertainties related to the upcoming financial regulation reform legislation and a looming criminal suit against a respected global broker which encouraged financial and general profit taking (the market having risen 76 % over the previous 14 months), there were good reasons for investors to be risk-cautious.[253]

At the start of trading on May 6, macroeconomic issues about the European sovereign debt risk sent the U.S. stock markets down by 2–3 %.[254] By late morning, the relative volume of NYSE-listed stocks trading on a downward trend grew to a size that had not been seen since the morning when trading resumed after 9/11. A combination of fundamental and technical factors precipitated the sharp decline that ensued, and *circuit breakers* or so-called *liquidity replenishment points* (LRPs) were triggered in large numbers to mitigate the plunge.[255]

LRPs are activated by excessive movements in stock prices within short periods of time. When triggered, they slow down the market orders, and designated floor-based market makers step in to supplement liquidity in the affected share and to determine whether a trade is correct or not. During this period, which can last from 30 s to 2 min, trading is not automatically executed on the NYSE. Automatic trading only resumes after a manual trade takes place or after the time period for the LRP has expired. This mechanism, which may be set in motion repeatedly if warranted by market conditions, *only applies to the NYSE*, and if trading is slowed down there, stock trading may shift to electronic platforms at other venues.[256] Did the LRPs, present on the NYSE *only*, have an impact on a completely different index, i.e., the DJIA? What were the reasons for the flash crash?

4.5.7.2 The Official Reason for the Flash Crash

While several explanations have been put forward for the events on May 6, it is clear that the flash crash was *not* a speeded-up version of the *quant crisis*.[257] During July and August 2007, many quantitatively managed equity portfolios exhibited extreme volatility, saw their value shrink and then rebounded sharply. This effect was global, occurring at first in the U.S. and then rapidly in the rest of the world. In fact, the

[251] Schulmerich (2010a, p. 2).

[252] Rice (2010, p. 1).

[253] Rice (2010, p. 2).

[254] Schulmerich (2010a, p. 1).

[255] Schulmerich (2010a, p. 2).

[256] Barclays Capital Equity Research (2010, p. 2).

[257] Schulmerich (2010a, p. 2).

crash was primarily an operational issue which lead to renewed calls to address the fragmentation of the U.S. equity market.[258]

The official explanation came less than a week after the crash from Gary Gensler, chairman of the Commodity Futures Trading Commission (CFTC).[259] He told the U.S. House Committee on Financial Services that on May 6 one trader alone accounted for 9 % of the trading volume on the most actively traded stock index derivative contract, the S&P 500 e-mini futures contract.[260]

Because of an algorithm, this one trader only issued sell orders, whereas most of the other 250 traders who were active in the same market on that day both bought and sold securities. As the trader's sell orders went through, the futures index on the Chicago Mercantile Exchange (CME) which is the world's largest futures and options exchange began to plummet, precipitating the DJIA's 9.2 % decline.[261] However, it should be noted that the mutual fund manager did not intend to corner the market.

4.5.7.3 Practitioners Explain the Flash Crash

Dissatisfied by the official explanation, a number of market participants offered their views on the events of May 6. In the following we present two plausible explanations.

Analysts from Barclays Capital[262] distinguish between two different ways of trading: high-frequency algorithmic trading and low-frequency trading by quantitative investors (the more *traditional* investors). Their explanation highlights the significant role high-frequency trading played in the crash.

They incriminate the LRP trigger that followed the shift in trading venues on May 6.[263] When the circuit breakers took effect, orders for the affected NYSE-listed stocks were moved to other markets including NASDAQ and NYSE Arca.[264] At these venues, numerous trades were executed at significantly reduced prices. Thousand of transactions were carried out at levels that cannot be considered *best execution*, i.e., an appropriate price in an appropriate timing. As a result, an equally

[258] Schulmerich (2010a, p. 2).

[259] The CFTC aims to protect market users and the public from fraud, manipulation and abusive practices related to the sale of commodities, financial futures and options, but also to foster open, competitive and financially sound futures and option markets.

[260] Wyatt and Bowley (2010).

[261] Wyatt and Bowley (2010).

[262] Roger Freeman (Brokers/Exchanges Analyst) and Matt Rothman (Quant Analyst) from Barclays Capital. See Barclays Capital Equity Research (2010).

[263] Barclays Capital Equity Research (2010, p. 2).

[264] At the NYSE Euronext stock exchange, NYSE Arca is an all-electronic U.S. trading platform that provides fast execution with open, direct and anonymous market access. According to the analysts at Barclays Capital, around 30 % of the average daily trading volume in the U.S. equity market is executed electronically, i.e., off-exchange, but misunderstandings about the use and users of electronic trading systems are widespread. While quantitative investors and traders account for a large proportion of the order flow, a significant amount can also be attributed to trading by fundamental investors.

vast number of transactions were simply canceled. Even when the NYSE switched back to *normal mode*, trades on NASDAQ and NYSE Arca continued to occur at significantly lower prices than those being displayed on the NYSE's order book. This anomaly was a direct consequence of the inconsistencies of the regulation schemes which were put into place in previous decades in order to implement the law of a single price for a given share throughout the U.S.

According to Larry Leibowitz, chief operating officer of NYSE Euronext, the trades that were sent to electronic networks then fueled the DJIA's decline.[265] The first part of the DJIA's plunge probably reflected normal trading which pushed the index down to around 10,600 from about 10,870 at the opening. However, the second half of the decline to a bottom of 9,869 occurred when orders were sent to venues where liquidity was lacking.[266] The triggering of LRPs on May 6 pushed orders to alternative platforms with few, if any, buyers, hence worsening the decline.

4.5.7.4 Consequences of the Flash Crash

As a direct consequence of the crash, investors lost confidence, and according to the Investment Company Institute (ICI),[267] U.S. stock mutual funds withdrew approximately $ 57 bn. from the market within 4 months,[268] the largest amount during any 4-month period between 2008 and 2011. The apparent market dislocation left participants, onlookers, regulators and investors stunned and perplex.

To prevent further comparable mishaps, the Security Commission of Exchange (SEC)[269] issued four proposals[270]:

- A single stock circuit breaker which would halt trading in a stock that dropped 10 % in 5 min.
- Moving broad market circuit breakers from the Dow Jones Industrial Average Index to a broader index such as the S&P 500 Index as well as tightening the triggering levels.
- Stricter market maker requirements or obligations for automated market makers (high-frequency trading).
- Initiating rules to clarify the process for breaking erroneous trades.

---

[265] Westbrook (2010).

[266] Westbrook and Mehta (2010).

[267] The Investment Company Institute (ICI) is the national association of U.S. investment companies. ICI encourages adherence to high ethical standards, promotes public understanding of funds and investing, and advances the interests of investment funds and their shareholders, directors, and advisers.

[268] Rice (2010, p. 1).

[269] The U.S. Securities and Exchange Commission (SEC) is an agency of the United States federal government. It holds primary responsibility for enforcing the federal securities laws and regulating the securities industry, the nation's stock and options exchanges, and other electronic securities markets in the United States.

[270] Securities and Exchange Commission (2011).

Since then, several initiatives have been undertaken that seek to prevent a repetition of May 6, 2010. For instance, in the second week of June, the SEC approved a pilot program to run until December that would require the U.S. exchanges and the Financial Industry Regulatory Authority (FINRA)[271] to pause trading in any stock for 5 min if its price has moved by 10 % in the preceding 5 min.[272] The SEC also worked with the markets to recalibrate the circuit breakers that were in place across all equity trading venues and futures markets.[273]

4.5.7.5 The Flash Crash from 1962

May 6, 2010, was not the first flash crash in Wall Street history. The *original* flash crash happened on May 28, 1962, when the DJIA Index declined sharply in 20 min, with some shares dropping by over 9 % in less than 12 min.[274] The SEC launched an investigation into the causes for the plunge, but the origin of that crash remains undetermined.

However, there are lessons relevant to the crash of May 6, 2010. In particular, a sharp market distortion is not necessarily due to market fragmentation or electronic trading. Market fragmentation was non-existent in 1962, since trading took place on a physical floor and *high-frequency trading*, which requires a large computational power and microsecond orders was out of question.

The original flash crash occurred as quickly as the crash in May 2010. One of the recommendations from the SEC in 1962 was the *temporary interruption of trading in individual securities under predefined circumstances*.[275] Almost 50 years later, this idea is the basis of the current proposals to introduce harmonized rules on circuit breakers across trading venues.

The 1962 crash revealed a range of complex reasons for investors behavior, including *rational and emotional motivations*.[276] A surprising result of the SEC's investigation of the 1962 crash was the large role played by retail investors. Regarding the interaction of market pressures and personal behavior, the SEC report from 1962 concluded: *"[these motivations] in turn may change the impact of various normal market mechanisms, and thus temporarily impair the market's fair and orderly character."*[277] Today, the debate focuses more on flaws in the market structure than on investor behavior. But while market structures change and investors adapt, basic behavior and reflexes may not.

---

[271] FINRA is the Financial Industry Regulatory Authority. FINRA is dedicated to investor protection and market integrity through effective and efficient regulation of the securities industry.

[272] Ketchum (2014).

[273] Schulmerich (2010a, p. 4).

[274] Schulmerich (2010a, p. 6).

[275] Schulmerich (2010a, p. 6).

[276] Schulmerich (2010a, p. 6).

[277] Schulmerich (2010a, p. 6).

4.5.8 Conclusion

In this section, we have presented many crashes with a regional impact. However, these are not the only ones that happened. The Mississippi bubble in the Americas occurred in 1720 and the U.K. saw an earlier railway bubble before the one described above. In the 1970s, there was a surge in bank loans to Mexico and other developing countries, and between 1985 and 1989, Japan suffered from a real estate bubble which ended in a 13 years long crisis, also called the lost decade. In the same time, Finland, Norway and Sweden saw their stock and real estate skyrocket before a heavy crash. Between 1990 and 1993, a surge in foreign investment in Mexico created another bubble which, again, eventually burst with the *Tequila Crisis*.[278]

However, the easy access to international stock markets and the development of financial engineering opened new horizons to investors. Now, a company based in the suburbs of Chicago can invest in the Malaysian stock market while watching the news of what is going on the Dubai stock exchange. This development favored diversification strategies and was considered to diminish risk. However, the opposite occurred during heavy downturns.

4.6 International Crashes

Since the 1970s, market volatility has reached unexpected highs. Currency crises largely outnumber bank crises (158 vs. 54 from 1975 to 1997)[279] and above all, the importance of the crises is getting stronger and stronger. They also occur at an astonishing speed, compared to former times. A good understanding of the crucial driving forces that operate in the markets is thus vital. Crashes are not mere exceptions or extreme cases that should occur less than once in a generation.

We will begin our presentation of international crashes with the presumably best-known crisis in recent history: the Great Wall Street Crash of 1929. The social, political and psychological impact of this crash is believed to be at least partly responsible for Hitler's rise to power and World War II. We will then deal with the Asian financial crisis and the Russian financial crisis. While less dramatic in appearance, both threw the afflicted countries' economies two decades back into the past. We will continue with the notorious dotcom bubble, which was generated by the invention of the Internet. Very similar to the impact of the railway, this invention also led to many investment mistakes. This section closes with the subprime crisis whose fallout was so enormous that it has been compared to the 1929 crash in its effect on the economy as a whole.

[278]Kindleberger and Aliber (2005, pp. 8 and 108).

[279]International Monetary Fund (1998, p. 77).

4.6.1 The Great Wall Street Crash (1929)

The Great Crash, also called the Wall Street crash of 1929 or the stock market crash of 1929 is another excellent example of overoptimism on a huge scale which developed in a specific economic context. During the so-called *Roaring Twenties*, the economy was growing thanks to technological innovations, like radio, automobiles, aviation, telephone, power grid, etc. Companies like Radio Corporation of America or General Motors saw their stock soar.[280]

The 1920s were years of a general mood of optimism. Investors were extremely attracted by stock returns: the yields of common stocks were high and prices were low. However, to explain the Wall Street crash, one has to consider what happened a few years earlier in Florida.

The first signs of a euphoric mood became visible in the state of Florida as soon as 1920. The American middle class grew and the freshly available capital needed to be invested. Fleeing the overcrowded Manhattan and Chicago for brighter sun and milder weather, many families tried to acquire new houses in Florida.[281] At that time, it was possible to buy housing lots for approximately 10 % of their face value.[282] The multitude of investors led to the belief that prices could only go higher and higher. Credit access for buyers was easier than in the previous years,[283] and collateral was not always required.[284] At the peak of the real estate boom between 1924 and 1925, prices could double in a matter of weeks.[285] It was also at that time that the well-known Charles Ponzi conceived and applied a real estate plan that would later be called the *Ponzi scheme*.[286]

The real estate market finally collapsed in autumn 1926, but not exactly because of a significant change in mood. Two devastating hurricanes destroyed the region, leaving thousands of families homeless. Miami bank clearings were a bit more than $ 1,066 mn. in 1925, against only $ 143 mn. in 1928.[287] Yet, this catastrophe did not dent the general optimism of the nation. The worst was still to come.

In 1926, Calvin Coolidge became U.S. president after the death of Warden G. Harding. He proposed the Coolidge-Mellon plan,[288] which aimed at reducing the

[280] Brooks (1999, p. 65).

[281] Galbraith (1997, p. 72).

[282] Galbraith (1997, p. 72).

[283] Sheeran and Spain (2004, p. 119).

[284] Vickers (1994, p.63).

[285] Galbraith (1997, p. 72).

[286] Galbraith (1997, p. 73). A Ponzi scheme is an investment scam that pays returns to its investors from existing or new funds paid by new investors, rather than from profit earned from an investment made. Operators of Ponzi schemes usually attract new investors by offering higher returns than other investments, in the form of short-term returns that are either abnormally high or unusually consistent. The scheme is named after Charles Ponzi (March 3, 1882–January 18, 1949) who became notorious for using such a scheme in the U.S. in 1920. His scheme ran for over a year before it collapsed, costing his "investors" $ 20 mn.

[287] Galbraith (1997, p. 75).

[288] Greenberg (2006).

taxes of all Americans, and exempted the poorest from any tax payments until 1929. This welcome tax cut increased the optimistic mood, and by early 1928, a strong upward market trend started. Investors had *vision for future, and boundless hope and optimism.*[289] On March 12, 1928, the volume of trading reached 3,875,910 shares at the New York Stock Exchange which was an all-time high.[290] Until November 16, 1928, prices rose once more, resulting in an all-time trading record of 6,641,250 shares.[291] The market rise was associated with an increase in trading on margin.[292] The banks, although forbidden by law to trade on the stock exchange, had little difficulties to circumvent restrictions, and fueled the boom even more.[293]

In addition, in 1927, the governors of European banks asked the Federal Reserve to ease the monetary policy and the Fed agreed to reduce the discount rate from 4 to 3.5 %. The released funds became available for investments in the stock market.

However, there were also drawbacks to this policy. In the early 1920s, brokers' loans—used to finance purchases on margin—averaged $ 1–1.5 bn. By November 1928, the amount reached $ 6 bn.[294] At that time, the interest rates on such loans yielded 12 % to lenders[295] and eventually reached 15 %![296] This led to a flood of gold converging on Wall Street from all over the world to finance the purchase of stocks on margin. Trading on margin made investors buy more stocks and drove the prices higher and higher for almost 1 year.

Seeing the situation was out of control, the Federal Reserve Board declared in spring 1929 that it might tighten interest rates to arrest the boom.[297] In response to this threat, Charles E. Mitchell, at the head of the National City Bank, who wanted to continue to ride on this wave, declared:

> *We feel that we have an obligation which is paramount to any Federal Reserve Warning, or anything else, to avert any dangerous crisis in the money market.*[298]

As a matter of fact, he would then lend money as necessary to offset any restraint by the Federal Reserve. His position was backed by Joseph Stagg Lawrence, a Princeton Economist: *"The consensus of judgment of the millions whose valuations function on that admirable market, the Stock Exchange, is that stocks are not at present overvalued. [. . .] Where is that group of men with the all-embracing wisdom which will entitle them to veto the judgment of this intelligent multitude?"*[299] This

---

[289]Galbraith (1997, p. 1).

[290]Galbraith (1997, p. 43).

[291]Galbraith (1997, p. 46).

[292]Trading on margin consists in resorting to loans to invest in stocks that are expected to rise so as to make profit.

[293]Brooks (1999, p. 101).

[294]Galbraith (1997, p. 49).

[295]Galbraith (1997, p. 49).

[296]Galbraith (1997, p. 77).

[297]Galbraith (1997, p. 75).

[298]Sobel (1999, p. 369).

[299]Shiller (2005, p. 178).

was enough to drive the market even higher. To illustrate, Celanese[300] stock jumped from $ 66 in 1927 to $ 118 in September 1929, while Peabody[301] rose from $ 46 to $ 110 during the same period.[302] But this uptrend would soon lose its strength.

Quite astonishingly, the crash was far from being brutal. Actually, the whole year 1929 was marked by a depressed market, with a few sectors heavily slowed or downturning. For the past 3 years a sort of hidden crash had hit investors, but left the most glamorous and prestigious stocks untouched.[303] On September 3, 1929, when the then all-time high was reached, no one could foresee what was about to happen in the following months. Yet, more and more stocks were underperforming. News of disastrous sales spread, while investment from industrial companies increased. Finally, what had to happen occurred.

On October 21st, a Monday, prices started to fall sharply. The trend continued on Wednesday and the market dipped badly on Thursday. The great bankers of that time, Thomas Lamont of Morgan's, Mitchell of National City and Wiggin of Chase invested massively to cut the downtrend, but in vain.[304] Thursday 24 and Monday 28 were dubbed Black Thursday and Black Monday. Figure 4.15 which represents the monthly returns of the DJIA from December 31, 1919 to December 31, 1929, shows the quickness and strength of the market dips. These consecutive shocks were a major reason for the 10-year long Great Depression in the United States.

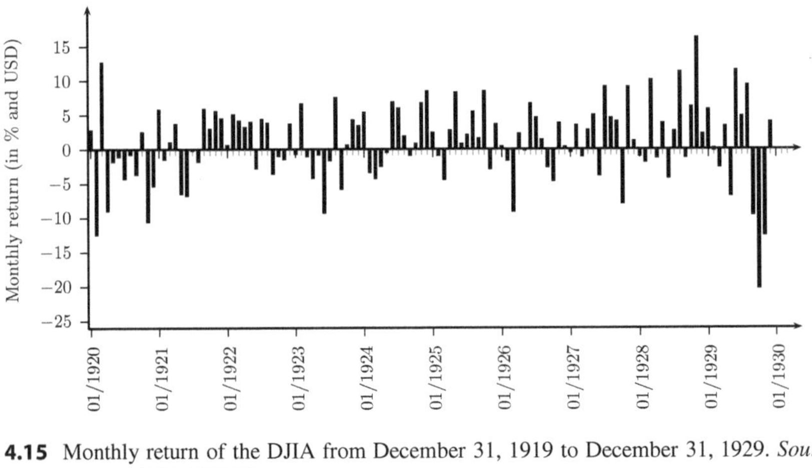

Fig. 4.15 Monthly return of the DJIA from December 31, 1919 to December 31, 1929. *Source*: Bloomberg (ticker: DJIA INDU)

---

[300] Celanese sold and exported a wide variety of chemical products.

[301] Peabody was the largest private sector coal company.

[302] Brooks (1999, p. 110).

[303] Brooks (1999, p. 109).

[304] Galbraith (1997, p. 81).

The trades on margin devastated the whole economy and were linked to an incredible madness which is tellingly obvious in the case of the United Founders Group. Close to bankruptcy, it had been saved by a $ 500 cash infusion in 1921. The company then borrowed money and sold stocks to finance the purchase of other securities whose market value was, before the crash, around $ 1 bn.[305]

The crisis marked the start of the most savage bear market of all times in which share prices fell by 89 % from peak to trough.[306] To compare, the worst U.K. bear market lasted from early 1973 to late 1974, during which shares fell by 73 % *only* on the FTSE Index. The impact of the 1929 crisis which is still vivid in popular culture today, dampened the mood on the stock markets for a quarter century. The following quotation gives the reason: *"Anyone who bought stocks in mid-1929 and held onto them saw most of his or her adult life pass by before getting back to even."* [307]

To summarize, many causes for the October 1929 crash can be cited:

- Stock prices were too high because of speculation.
- Stocks prices were unrelated to fundamentals (overoptimism).
- A real downturn in business activity.
- Excessive buying on margin.
- Panic caused by the press and politicians that often referred to an orgy of speculation.[308]

4.6.2 The Asian Financial Crisis (1997)

In 1997 an unpredictable crisis occurred in Southeast Asia. During the late 1980s and beginning of the 1990s, Asian countries witnessed very fast growth and transformed their agrarian economies into booming industrializing economies with cutting-edge technologies which outranked Europe or America in some sectors.[309] This phenomenon was then dubbed the *Asian economic miracle*[310] and affected the following countries and states:

- Brunei
- China
- Hong Kong
- Indonesia
- Laos
- Malaysia
- Philippines
- Singapore
- South Korea
- Taiwan
- Thailand
- Vietnam

[305]Galbraith (1997, p. 77).

[306]Atherton (2009).

[307]Salsman (2004, p. 16).

[308]Bierman (1998, pp. 16–17).

[309]Krugman (1994, p. 62).

[310]From 1965 to 1990, East Asia had a remarkable record of high and sustained economic growth. 23 economies grew faster than those of all other regions, while the U.S. and Europe were stuck under a growth perspective. See Page (1994, p. 221).

Out of these countries, five particularly important ones were called the Asia-5, namely Indonesia, Malaysia, the Philippines, Thailand and South Korea. They also were the worst hit by the crisis. In less than 6 months, they saw a massive outflow of foreign investments. Starting March 1996, the net private inflows dropped from $ 93 to $ −12.1 bn., a swing of $ 105 bn., out of which $ 77 bn. came from commercial bank lending.[311] While direct investment remained constant, portfolio equity fell by $ 24 bn. and non-bank lending by $ 5 bn.[312] Compared to the pre-shock combined GDP estimated at $ 935 bn., this represents a drawdown equivalent to roughly 11 % of GDP.[313]

This staggering figure is at the origin of what is called the Asian financial crisis which took place in 1997. To understand what happened, we will first present the initial market conditions that allowed the Asian economic miracle. We will then consider the warning signs of the oncoming crisis and continue with the trigger events of the massive investment outflow and its impact on the economy. We conclude with a summary of the regional market conditions in the aftermath of the massive cash withdrawal from foreign investors.

4.6.2.1 Initial Market Conditions

In the first half of the 1990s, worldwide markets conditions had profoundly evolved. Capital markets started to internationalize significantly in industrialized countries. Eurobond[314] lending increased and new bank syndicates emerged as well as new bond and equity mutual funds thanks to deregulation and various financing innovations.[315] Low interest rates in Europe, Japan and the U.S. and the lack of investment opportunities in these regions urged investors to find new sources of growth. Investors found Asian countries to be very interesting for many reasons. Indonesia, Malaysia, Singapore, South Korea and Thailand experienced consistently high growth rates during decades with an average of 6.7 % from 1970 to 1996.[316] Table 4.3 shows the annual growth rate of the five biggest regional economies, the Asia-5, and their GDP per capita in 1996.

Dollar values are based on 1996 U.S. dollars, growth percentages are on a per annum basis. As can be seen, while the disparity of the GDP per capita is large, ranging from 3,060 to 12,410 USD, 1996 value, the average GDP growth rate is very high relative to the U.S. or the eurozone countries, which at best averaged 2–3 % during this period. Moreover, these figures are based on a 26-year period, which explains the relatively high confidence of foreign investors toward these emerging economies.

[311] Krugman (2000, p. 111).

[312] Krugman (2000, p. 111).

[313] Krugman (2000, p. 111).

[314] A eurobond is an international bond that is denominated in a currency not native to the country where it is issued.

[315] Krugman (2000, p. 117).

[316] Karunaratne (1999, p. 20).

Table 4.3 The Asia-5 annual growth rate of GDP and GDP per capita from 1970 to 1995

Country	GDP per capita 1996 (in USD)	Growth percentage of GDP
Indonesia	4,280	6.8
Malaysia	9,703	7.4
Philippines	3,060	3.6
Thailand	8,370	7.5
South Korea	12,410	8.4
Average	7,565	6.7

Source: Karunaratne (1999, p. 20)

Thanks to the Asia Pacific Economic Cooperation (APEC),[317] these countries had committed to a rather ambitious schedule of trade liberalization with the goal of free trade in the region by 2020, without forming an economic bloc.[318] This scheme would have led to low cross-border taxes, a low level of corruption and an increased level of trade among these countries. Prospects appeared very positive.

Furthermore, a wide-ranging financial deregulation allowed domestic banks and corporations to tap into foreign capital to finance their domestic investments. Unfortunately, the deregulation was not followed up by an adequate supervision. These conditions fostered high rates of foreign borrowing, especially in Thailand. The lax supervision allowed banks to take on substantial foreign currency and maturity risks.[319] High interest rates relative to the U.S. or Japan were thus a serious incentive to invest in the Asia-5 countries.[320]

To promote a virtuous circle, and to prevent currency and foreign exchange risk, nominal exchange rates were effectively pegged to the U.S. dollar. On one side, Thailand, Malaysia, Korea and the Philippines managed their currency so that it was moving through a narrow range. If at any point in time, the domestic currency was too strong or too weak, the central bank intervened to correct the drift. On the other hand, countries like Indonesia, while not maintaining a peg, made sure that any currency drift was highly predictable. This relative stability encouraged capital inflows as it increased the confidence of foreign investors by diminishing their perceived risk.[321]

[317]Asia-Pacific Economic Cooperation, or APEC, is the premier forum for facilitating economic growth, cooperation, trade and investment in the Asia-Pacific region. APEC is an intergovernmental grouping that operates on the basis of non-binding commitments, open dialogue and equal respect for the views of all participants. Unlike the WTO or other multilateral trade bodies, APEC has no treaty obligations required of its participants. Decisions made within APEC are reached by consensus and commitments are undertaken on a voluntary basis. A more in-depth presentation of the organization is available at http://www.apec.org/About-Us/About-APEC.aspx.

[318]Park (1996, p. 357).

[319]Maturity risk is the risk that the value of a bond may change from the time it is issued to when it matures. It is specific to bonds and credits.

[320]Krugman (2000, p. 118).

[321]Krugman (2000, p. 118).

Finally, domestic banks and companies were given special incentives to borrow from foreign institutions.[322] For instance, banks operating in Bangkok received special tax breaks. In the Philippines, revenues from foreign exchange loans were subject to a 10 % tax rate versus the regular 35 % corporate income tax rate.[323] The effect was a heavy reliance on foreign resources, which were considered more secure than domestic ones.

4.6.2.2 Alerts and Signals

These conditions were strong incentives to foreign investors to invest in the region. They benefited from tax advantages, little deposit limitations and very few controls from public authorities. Until 1996, very few analysts anticipated a crisis in this region, with the very exception of a researcher called Yung Chul Park, who wrote: *"Financial market opening, which has promoted the investment boom, has made several East Asian countries highly vulnerable to speculative currency attacks and now threatens some of them with financial crisis."* [324] Despite this warning, capital inflow and bank investment remained high. From the end of 1996 to mid-1997, bank lending expanded further to $ 274 bn., i.e., by 10 % at an annual rate.[325]

The risk premium attached to loans to emerging market economies remained low, while it should have risen if the perceived risk by investors increased so as to compensate for the additional risk taken. The opposite actually happened. Cline and Barnes showed that in 1997 bond spreads fell in emerging markets relative to U.S. Treasury bonds, which represents a risk-free asset in financial theory.[326]

Rating agencies did not signal increased risk until after the beginning of the crisis itself, just as for the subprime crisis.[327]

Various other institutions also delivered analyses of market conditions and country risk, but sent no warning signals before the crisis started.[328] Among them was the Euromoney Country Risk Assessment, a well-known independent company which provides ongoing risk analyses and well-regarded assessments.

Investment banks also published regular analyses of industries and geographic sectors. One of the most famous ones, Goldman Sachs, did not only miss to anticipate the crisis, but even worse, after the beginning of the massive foreign cash outflow, they forecast that a rebound would occur in 1997.[329] Unfortunately, such a rebound would come, at best, 2 years later, but the countries' economies would not reach the same level as before 1997.

---

[322] Krugman (2000, p. 118).

[323] Krugman (2000, p. 119).

[324] Park (1996, p. 358).

[325] Krugman (2000, p. 112).

[326] Cline and Barnes (1997, p. 1).

[327] Krugman (2000, p. 119).

[328] Krugman (2000, p. 119).

[329] Krugman (2000, p. 119).

The International Monetary Fund produces similar analyses as a vital support for its ongoing operations. There are two kinds of reports: an overall market forecast contained in the periodic World Economic Outlook and a country assessment that can be found in the report of the Executive Board discussions, in the article IV consultations with member countries. There again, no macroeconomic risk warning was given.[330]

The single real indicator that could have induced a crisis was a serious drop in Thailand's and South Korea's main stock indices, starting January 1996 (−40 %), and continuing until mid-1997(−20 %).[331] However, it is not rare to see such stock fluctuation in emerging markets. According to Patel's definition, a crisis is happening when a stock market in emerging economies plunges by at least −35 % as opposed to developed economies, where −20 % is enough to declare a crisis. As explained above, analysts among the best-regarded banks were counting on a rebound. No ringing bell really alerted the market practitioners of the imminent crash.

4.6.2.3 Triggering Events

At the beginning of 1997, large companies started to show severe difficulties regarding their working capital needs, despite healthy exports. To begin with, in January 1997 a large company called Hanbo Steel,[332] which was burdened by a $ 6 bn. debt, collapsed. It was the first bankruptcy of a chaebol[333] and comparable to the undoing of a whole branch of General Motors, a large car manufacturer in the U.S. In a matter of months, it was the turn of two other South Korean mastodons: Sammi Steel and Kia Motor, a steel producer and a car manufacturer. In Thailand, Samprasong Land Development, a property development company, was 10 months behind in payments on a $ 80 mn. euro-convertible bond in September 1997.[334]

These are two examples of the difficulties that local companies and banks were facing. Panic broke out among investors who started to repatriate their cash flows. Starting mid-1997, Asian economies were hit by a significant investment outflow from their economies toward foreign countries. In 1996, net inflows of commercial bank loans amounted to 5.9 % of GDP versus 2.8 % for the first half of 1997 and −3.6 % in the second half. This represents −9.5 % for the region as a whole, but

---

[330]Krugman (2000, p. 119).

[331]*Source*: Bloomberg, ticker KOSPI:IND and SET:IND for the stock exchanges of South Korea and Thailand.

[332]Hanbo Steel was a steel producing company later purchased in 2004 by Hyundai, a car manufacturer.

[333]A chaebol is a South Korean form of business conglomerate. It is typically a global multinational owning numerous international enterprises, controlled by a chairman who has power over all the operations.

[334]Krugman (2000, p. 119).

−11 % for the Asia-5 countries. This reversal happened too fast to be the result of changing economic fundamentals and bankruptcies started to pile up.[335]

These bankruptcies put many commercial banks under pressure since most of the foreign borrowing of the companies was channeled through commercial banks. During the following 6 months, the Bank of Thailand lent over $ 8 bn.[336] to distressed financial institutions through a financial vehicle designed for this refunding mission: the Financial Institutions Development Fund (FIDF). As concerns began to grow, the Bank of Thailand also used a significant part of its liquid foreign exchange reserves in forward contracts. This was very beneficial for speculators who understood that an exportation slowdown combined with the defense of the currency would ultimately lead the baht, the Thai currency, to be devalued.[337]

In late June 1997, the Thai government stopped to support the largest Thai finance company, Finance One, and announced that domestic and foreign creditors would incur losses. Additionally, the government declared it would be forced to devalue the domestic currency. This declaration was very harmful, since only 6 months before, the government had guaranteed to keep Finance One alive. It had also assured to have enough foreign reserves to defend the peg with the U.S. dollar. The new statement logically created a crisis of confidence in addition to the economic slowdown and also accelerated the withdrawal of foreign funds and prompted the currency depreciation on July 2, 1997.[338] In turn, the Thai baht devaluation triggered the capital outflows from the rest of East Asia, and speculators also attacked the South Korean won and the Hong Kong dollar.

4.6.2.4 Effects of the Withdrawal

A natural consequence of a rapid outflow of capital and liberalization should be a slowdown of foreign borrowing, given a higher expected financial risk. The reason is that most profitable investments opportunities are to be seized early and an overinvestment in non-tradable goods, i.e., real estate, becomes evident. Consequently, a slowdown in exportation and lower capital inflow should be observed. In fact, this was the case, but the quick capital outflow of $ 105 bn. as described above was contrary to all logical expectations.[339]

By the mid-1990s, Thailand, Indonesia and South Korea had large private current account deficits. The maintenance of fixed exchange rates encouraged external borrowing and led to excessive exposure to foreign exchange risk in both the financial and corporate sector. Furthermore, the macroeconomic environment started to change, when the Chinese and Japanese currencies, the renminbi and the yen, were devalued. The results were rising U.S. interest rates which strengthened the U.S. dollar. At the same time a sharp decline in semiconductor prices, an industry

---

[335] Krugman (2000, p. 135).

[336] Krugman (2000, p. 135).

[337] Krugman (2000, p. 135).

[338] Krugman (2000, p. 135).

[339] Krugman (2000, p. 111).

with an important share in domestic GDP, adversely affected economic growth.[340] At first, the effect was limited thanks to the defense of the peg, but the exchange rate ultimately depreciated for all local currencies.

As the U.S. and Europe saw their economies recover from a recession in the early 1990s, the U.S. Federal Reserve Bank under Alan Greenspan began to raise U.S. interest rates to limit inflation while the European countries stopped currency depreciation by starting the stabilization process required for the introduction of the euro currency by 1999. This made the U.S. a more attractive investment destination relative to Southeast Asia, which had been attracting money flows through high short-term interest rates, and raised the value of the U.S. dollar. In Europe, however, the situation still looked grim and domestic investors preferred investing in the U.S., South-American and African economies.

The countries which had currencies pegged to the U.S. dollar (i.e., Thailand, Malaysia, Korea and the Philippines) became less competitive in the global markets as the stronger U.S. dollar made their exports more expensive. At the same time, Southeast Asia's export growth slowed dramatically in the spring of 1996, deteriorating the current account position of the whole region.

To counter this situation, domestic interest rates soared on the withdrawal of foreign credits. Domestic credit conditions tightened, even before the central banks defended their currencies. Since the withdrawal of credits led to a sharp reduction of credit availability, the nominal exchange rate and real exchange rate both depreciated.

The combination of the exchange rate depreciation and the increase in domestic interest rates led to a rapid rise in nonperforming loans in the banking industry of the Asian economies. Especially real estate projects were postponed, then pursued, and finally abandoned. The biggest issue was the borrowing technique. In many cases, real estate developers borrowed in U.S. dollars but did not hedge their position against their domestic currency. The combination of large underperforming loans and direct balance sheet losses due to quick currency depreciation destroyed a large part of the market value and capitalization of banks in Indonesia, Thailand and Korea. Banks started to face difficulties to meet their obligations and the legal ratios of financing, especially in these countries. Logically, with limited foreign investment support and daunting conditions on the domestic market, these banks became illiquid.

4.6.2.5 The International Monetary Fund Intervention

Given the explosive economic situation, the IMF offered its financial help in exchange for the implementation of a Western approach to financial management. The IMF required six elements to be applied[341]:

- Immediate closure of banks with highly problematic accounting
- Quick restoration of minimum capital adequacy standards

[340] Moreno (1998).

[341] Krugman (2000, p. 116).

- Tightening of domestic credit (stricter delivery conditions of credit)
- High interest rates on central bank discount facilities
- Fiscal contraction (less taxes and less governmental expenses)
- Significant structural changes in the non-financial sector

The first three elements had dire effects which led to a stark economic contraction and banking panic, a significant impoverishment of the population, social turmoil, a rise in criminality, etc.[342] Bank lending virtually stopped overnight in Indonesia, South Korea and Thailand. This had the effect to frighten foreign investors who increased their massive inflow reversion. Unfortunately for the countries, a last move would throw their economies 10 years back into the past.

4.6.2.6 The Rating Agencies' Fatal Role in the Crisis

On December 22, 1997, Moody's downgraded the sovereign debt of Indonesia, South Korea and Thailand below investment grade status, relegating them to junk bond status. As a rule of thumb, all domestic companies must have a credit rating lower than the sovereign one, due to risk diversification purposes. The banks, already in a problematic situation, were automatically and immediately downgraded to junk bond status.[343]

As a first consequence, most of the commercial banks in these countries could no longer issue internationally recognized letters of credit for domestic exporters and importers. Given that the banks were believed to be insolvent or highly speculative, their letters of credit would not promise much actual financial security. International trade was severely hampered.

As a second consequence, the downgrade immediately prompted an additional round of debt liquidation, since many portfolio managers are constrained by law to hold investment grade only assets in their portfolio.

A third consequence concerned the derivatives market. The downgrade triggered the exercise of various put options linked to credit ratings, enabling borrowers to call in loans immediately on the downgrade.[344]

As the creditors panicked, a bank run and a sovereign downgrade ensued which threw South Korea, Indonesia and Thailand into partial debt defaults. In Korea, these defaults were handled by an emergency standstill of debt repayment, followed by a concerted rollover of the short-term repayments in longer-term repayments. The debt renegotiation was backed by Korean government guarantees and represented no less than a third of the Korean debt due in 1998. For other countries like Indonesia or Thailand, defaults were unilateral, without governmental guarantees.[345]

[342] Krugman (2000, p. 116).
[343] Krugman (2000, p. 116).
[344] Krugman (2000, p. 115).
[345] Krugman (2000, p. 116).

4.6.2.7 Summary

To sum up, the Asian financial crisis can be understood as a combination of factors that transformed a virtuous into a vicious circle. A bank failure and a bank closure in Thailand and Indonesia helped trigger the exodus of investments and credits from the Asian region.

Second, the corporate failures raised general doubts on the financial health of domestic companies and about the expected profits to be realized in the APEC area. A wave of distrust and fear prompted a massive debt recall and the sell-off or exercise of debt derivatives.

The political uncertainty especially in South Korea, the Philippines and Indonesia hastened credit withdrawal. The governments were not believed to be trustworthy, while new elections were seen as a danger to measures declared by the present governments.

The contagion effect also played an important role. Many creditors perceived the region as a whole and assumed that if Thailand was in trouble, the other countries had similar difficulties. Such a mistake was also made in the more recent subprime crisis concerning the debt of different European countries. Given that tax policies, government types, political sensibilities or even demographic conditions are significantly different in Asia, such a general perception was heavily biased.

The intervention of the IMF was also significant, but not necessarily in a positive sense. By forcing the immediate suspension or even the closure of some commercial banks and other financial institutions, it sent a strong negative signal to potential investors, which may have intensified the withdrawal effect.

Some researchers have proposed the growing exports of China as a contributing factor to the export growth slowdown of the ASEAN nations,[346] though these economists maintain that the main cause of the crisis was a real estate bubble due to excessive speculation.[347] China had begun to compete effectively with other Asian exporters particularly in the 1990s after the implementation of a number of export-oriented reforms. However, such an argument is dubious given that both ASEAN and China experienced simultaneous rapid export growth in the early 1990s.[348]

Many economists believe that the Asian crisis was created not by market psychology or technology, but by policies that distorted incentives within the lender/borrower relationship. The resulting large quantities of credit that became available generated a highly leveraged economic climate, and pushed up asset prices

[346]The Association of Southeast Asian Nations is a geopolitical and economic organization of ten countries located in Southeast Asia, which was formed in 1967. Its aims include accelerating economic growth, social progress and cultural development among its members, protection of regional peace and stability, and providing opportunities for member countries to discuss differences peacefully.

[347]Palma (2000, p. 12).

[348]Eccleston, Dawson, and McNamara (1998, p. 109).

to an unsustainable level.[349] As presented earlier, this hypothesis would not be able to explain the reasons of the rapid and massive cash withdrawal.

Finally, voices were raised, especially by local governments, that speculators had a big hand in the crisis, one of the most famous being George Soros.[350] The combined effect of all speculators, which was undoubtedly present during the bubble burst, does not seem to be large enough to explain the crisis.

4.6.3 The Russian Financial Crisis (1998)

The Russian financial crisis, which is described in this section, happened just a few years before the dotcom bubble. It was less important than the South East Asia crisis, but also disastrous for the country. Fortunately, Russia was able to overcome the crisis thanks to its trade balance surplus. In the following, we examine the initial state of Russia in the wake of the founding of the Russian Federation. We then present the mechanics leading to the dramatic financial crisis and its aftermath and further expand the explanation on how a currency crisis contaminated other economies around the world.

4.6.3.1 Inheritance of the USSR and Optimistic Views: 1992–1997

Following the implosion of the former Soviet Union, Russia went through a series of necessary and deep economic reforms. Privatization and macroeconomic stabilization were achieved with relative success. However, in August 1998, after a first year of positive economic growth since 1991, Russia was forced to default on its debt and devalue its domestic currency, the ruble.

The USSR officially collapsed on December 25, 1991, leaving a host of independent republics as a legacy, and a large amount of unpaid debt. Following the collapse, Russia had to face a tremendous debt burden inherited from the poorly managed Soviet era.

After 6 years of economic reform in Russia, the results seemed to be positive, in particular the trade balance moved toward an equilibrium between exports and imports. Figure 4.16 shows the Russian Merchandise Trade Balance from the first quarter 1994 to the last quarter of 2001.

During this period, exportation, represented by the blue line, constantly remained above importations, and the positive foreign trade balance helped to repay the large debt burden. The figure shows a widening gap following the first quarter of 1998 which seems to develop steadily with the same strength in the following years. This

[349] Edwards (2000, pp. 43–60).

[350] George Soros, born in 1930, is a Hungarian-American investor and the chairman of Soros Fund Management. He is known as *The Man Who Broke the Bank of England* because of his $ 1 bn. in investment profits during the 1992 Black Wednesday U.K. currency crisis. He also invested massively in Asia at that time and researchers investigated his role in the crisis in this region. See Brown, Goetzmann, and Park (1998, p. 2).

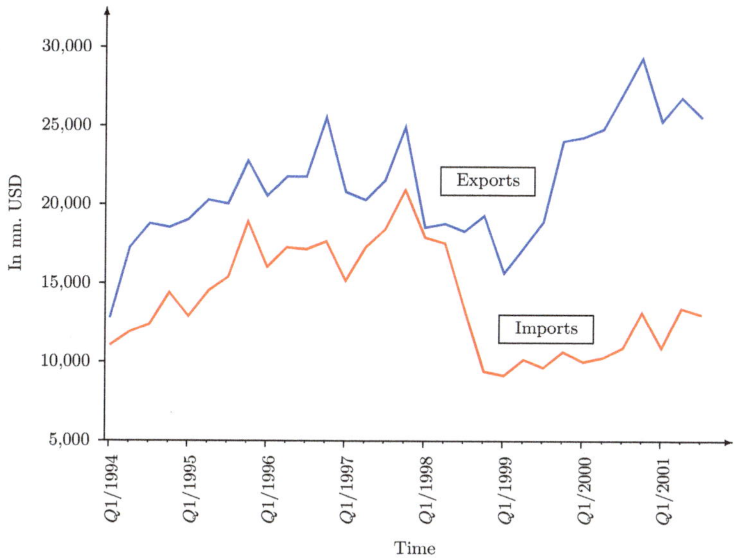

Fig. 4.16 Russian merchandise trade balance. *Source*: Chiodo and Owyang (2002, p. 11)

denotes a healthy foreign trade surplus starting 1999 which was supported by a relatively high oil price at that time of \$ 23 per barrel. Fuels accounted for more than 45 % of Russia's main commodities exportation in 1997.[351]

In addition to a positive trade surplus, the government managed to limit inflation from an average 131 % in 1995 to 22 % in 1996. It reached a mere yet still high 11 % in 1997 as illustrated in Fig. 4.17.

Figure 4.17 represents the consumer price index in Russia from September 1995 to March 2001. There is an impressive hike in June 1998 which fits with the collar observed during the same period in Fig. 4.16. An explanation on this point will be provided in the following.

The financial help provided by the World Bank and the International Monetary Fund in exchange of economic and governmental reforms was promising. The World Bank was even prepared to provide an additional assistance of a \$ 2 – 3 bn.[352] package on an annual basis to support the reconstruction of the country that faced important day-to-day financing difficulties. In the meantime, economic and diplomatic relations with Western countries improved.

Domestic output also showed signs of a recovery after the savage liberalization of the former Soviet economies after 1991 had lead into a debacle.[353] This proved

[351]Chiodo and Owyang (2002, p. 11).

[352]Chiodo and Owyang (2002, p. 9).

[353]Chiodo and Owyang (2002, p. 9).

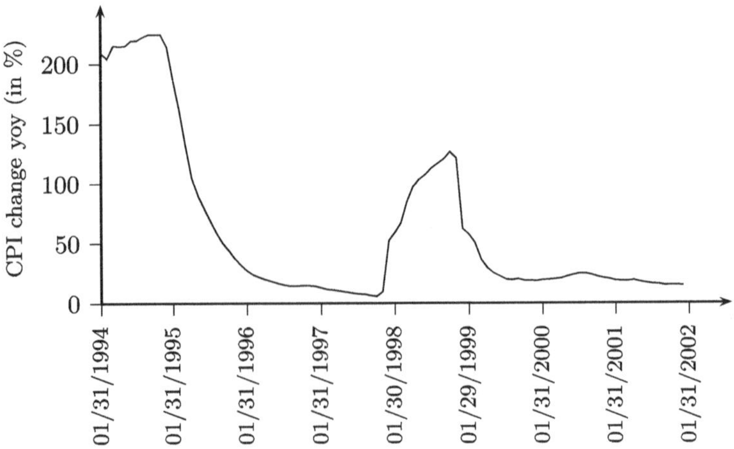

Fig. 4.17 Consumer price index (CPI) development: percent change over previous year from September 1995 to March 2001. *Source*: Chiodo and Owyang (2002, p. 11)

that Russia could start to manage its economy without the continuous monetary infusions from the International Monetary Fund or the World Bank.

The country also managed to control currency fluctuations against major foreign currencies like the U.S. dollar as presented in Fig. 4.18, which shows the exchange rate of the Russian ruble versus the U.S. dollar from September 1995 to March 2001. Until 1998, a peg of around 5 RUB per USD appears relatively stable, thanks to the sound management from the Central Bank of Russia (CBR).

Given its stormy past, the country seemed to perform pretty well. However, significant problems remained. The country had succeeded in renegotiating the Soviet debt for a notional amount of $ 33 bn., but most of its payments were based on trade surplus expectations. For the coming years, the country anticipated growing debt payments when credit granted from the IMF would come due and faced considerable problems regarding interest payments. The government had to decide where and how to increase tax collection, while attempting to decrease domestic borrowing.

Tax collection was problematic. The former administration had worked under a two layer scheme with a regional and a federal layer. The regional administrations encouraged the tax-paying companies to declare falsified incomes to the federal administration,[354] and in exchange, the companies would make transfers to the accommodating regional administration. As a result, the local administrations had revenues shrinking less fast than the federal revenues. However, the federal administration had to face most of the interest payments.

To sum up, the economic landscape of Russia showed signs of improvement with a timid expected 2 % economic growth for 1998. But then, a phenomenon hit

[354]Shleifer and Treisman (2001, p. 91).

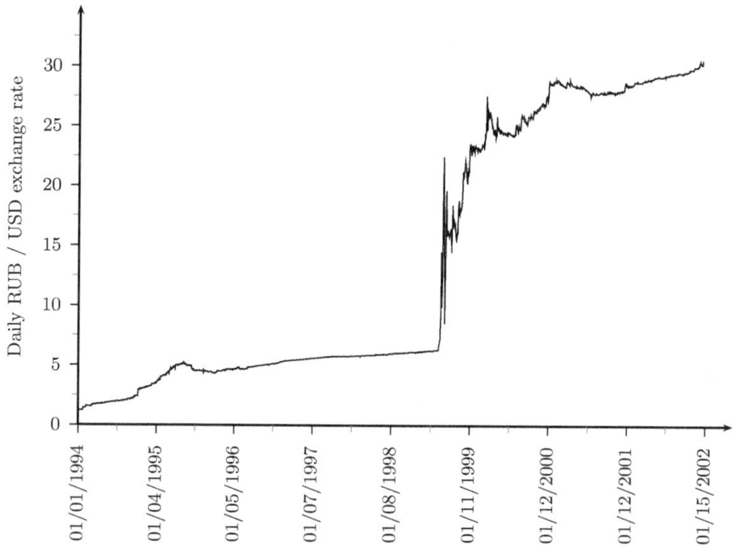

Fig. 4.18 Exchange rate between the Russian ruble and the U.S. dollar from September 1995 to March 2001. *Source*: Chiodo and Owyang (2002, p. 11)

the country that caused the real GDP to decline by 4.9 %: the tsunami wave of the Asian financial crisis from a few months earlier.[355]

4.6.3.2 The Asian Crisis Launches a Currency Crisis

In 1997, as presented in Sect. 4.6.2, the Asian countries faced a currency crisis. Interest rates soared and falling exchange rates depreciated their currencies. In November 1997, after the onset of the Asian financial crisis, also called the East Asian crisis, the ruble started to be attacked by speculators. The reason is fairly straightforward. The energy price is indexed on oil. Production and trade require a lot of oil, notably for transportation purposes. In case of a recession or crisis, the worldwide demand of oil and energy drops and production output shrinks. Given that a significant part of Russian exports are linked to oil and non-ferrous metals, two hard-hit commodities, the Russian balance of trade was dangerously shaken. Unemployment skyrocketed and the country was plunged in chaos and riots, once again.

A slowdown in output combined with a weakening currency is very likely to lead to devaluation. The Central Bank of Russia (CBR), who defended the currency to maintain the narrow exchange rate with the U.S. dollar, lost nearly $ 6 bn. in foreign exchange reserves.[356]

---

[355]Chiodo and Owyang (2002, p. 12).

[356]Chiodo and Owyang (2002, p. 10).

But it is possible to take advantage of this situation by applying standard trading strategies. In anticipation of a possible devaluation, the CBR signed forward contracts with non-resident holders of short-term government bills (called GKOs) to exchange rubles for foreign currency, which enabled them to hedge exchange rate risk. It also significantly increased the rate of return of such contracts to reach 150 %. The basic idea was to prevent what happened in Asia: a massive net outflow of money from Russia to foreign investors countries.[357] It is also important to note that many commercial banks had off-balance sheet liabilities consisting of similar forward contracts with foreign clients. Such contracts were estimated to represent at least $ 6 bn. by the first half of 1998.[358]

To appease the anxiety of the investors, the government submitted a new tax code so as to simplify and guarantee better tax collection in February 1998.[359] But although the code was legally binding, important key aspects were ignored in its implementation. Furthermore, a governmental crisis emerged by March 1998 between the Central Bank of Russia and the executive branch of the government.[360] Such a conflict also emerged in the EU in 2012, when the central bank had to guarantee a stable economy, while some national governments tried to force the central bank to revitalize the domestic market.

The combination of these factors of uncertainty expanded government bond yields by 47 %.[361] With an inflation rate of 11 %, acquiring them could have been reasonable.[362] However, investors became more and more risk-cautious and the government faced difficulties to emit new debt to repay former one. Commercial banks were also facing liquidity difficulties due to gloomy forecasts and the risk aversion of clients. Meanwhile, the additional tax collection diminished the available cash in banks and further reduced their liquidity. The central bank tried to improve the sovereign debt attractiveness with an interest rate increase, but the effect was minimal. Despite a 20 % increase in return, escalating to 50 % by mid-1998, household ruble deposits increased by a mere $ 1.3 bn. in 1998 versus $ 29.8 bn. a year before.[363]

The IMF proposed a rescue plan on July 20, 1998, of $ 11.2 bn., which covered a $ 4 bn. loss due to the further decrease of the oil price (i.e., the barrel price had plunged from $ 23 to $ 11) and included $ 4.8 bn. to be used immediately to meet payments due between May and July 1998.[364] The decline of the oil price is

[357] Desai (2000, p. 51).

[358] Chiodo and Owyang (2002, p. 13).

[359] Chiodo and Owyang (2002, p. 12).

[360] Chiodo and Owyang (2002, p. 12).

[361] Chiodo and Owyang (2002, p. 14).

[362] Chiodo and Owyang (2002, p. 14).

[363] Chiodo and Owyang (2002, p. 13).

[364] Chiodo and Owyang (2002, p. 14).

indirectly shown in Fig. 4.16, as the Russian balance of trade significantly suffered during summer 1998.

To sum up, because of the decisions taken by the Central Bank of Russia interest rates rose to an unsustainable level. Banks ran short of liquidity because of the clients' risk cautiousness and higher tax rates. Part of their liabilities were off-balance sheet, which barely helped them to stick to regulatory ratios, while heavy losses in foreign-denominated liabilities would be registered soon because of forward contracts. A significant part of the population was unemployed and the government suffered from internal fights. Tax collection was low due to the weakening economy and the structural organization of its collection. The interest payments were also 40 % higher than the effectively collected tax.

On August 13, 1998, Russian stock, bond and currency markets weakened as a result of investor fears of devaluation. The markets had to be closed for hours to stem developing panic. On August 17, the Russian government had no choice but to devaluate its currency, default on its domestic debt and declare a moratorium on payments to foreign creditors. On September 2, 1998, the ruble peg to the dollar was abandoned as one can see in Fig. 4.17. The exchange rate jumped from 5 to 30 RUB per USD in less than 3 months, a 600 % change.[365] The year ended with a decrease in real output of 4.9 % and inflation skyrocketed as illustrated in Fig. 4.17.[366] Exports remained high after the ruble devaluation and oil price rose again in 1999, which greatly helped in refinancing the short-term debt. However, this crisis left foreign investors highly dubious about the ability of Russia to transform its economy and the trustworthiness of its government after billions were lost during the crisis.[367]

4.6.3.3 Worldwide Contagion: The Case of Long Term Capital Management

The crisis would not stay regional. Foreign investors who had invested in Russia tried to protect themselves and hedged their investment from currency risk and devaluation by entering into forward contracts in a currency different from the ruble. While being protected from currency risk, they were not safe from counterparty risk or sovereign risk. The former is calculated by the probability of loss if the counterparty defaults multiplied by the invested amount.[368] The latter is the risk that a foreign central bank will alter its foreign exchange regulations thereby significantly reducing or completely nullifying the value of foreign exchange contracts.[369] With Russia defaulting as a whole, investors started to panic and chased for quality assets, in order to escape from counterparty or sovereign risk.

[365]Chiodo and Owyang (2002, p. 10).

[366]Chiodo and Owyang (2002, p. 10)

[367]Chiodo and Owyang (2002, p. 14).

[368]Kaplan Schweser (2012, p. 109).

[369]Kaplan Schweser (2012, p. 109).

The Russian crisis had severe repercussions on the other side of the Pacific Ocean, where a hedge fund called Long Term Capital Management (LTCM)[370] had developed sophisticated quantitative models to calculate the fair price for many assets and to trade on them: Based on the outcomes, LTCM took short and long positions in many securities and in many different asset classes (i.e., equity, fixed income, derivatives, commodities, etc.) and gambled on the convergence of asset prices to create high returns using leverage. The business was highly profitable, generating annual returns of 43 % in 1995 and 41 % in 1996.[371] At the beginning of 1998, LTCM weighted $ 125 mn. with a leverage ratio of 28 for 1. In other words, for one unit of asset in the portfolio they actually held 28 times this position on the financial markets. At some point in time, they had open deals on $ 1 tn.

However, their trading strategies was subject to model risk, i.e., if the underlying hypothesis and input parameters did not hold true, the money generation process would turn into a pitfall. By August 1998, the volatility across the world reached tremendous levels under the combined effect of the Russian and Asian financial crises and Brazil's devaluation of its peso. The required daily marking to market on futures for instance depleted by 44 % the capital of the company in a single month because of margin calls.[372] The general feeling of panic among investors regarding their investments created an unprecedented surge in volatility of any asset class in any country.[373]

The situation was so extreme that the Federal Bank of the United States organized a bailout with fourteen investment institutions to invest $ 3.65 bn. for a 90 % stake in LTCM. The excessive volatility across the world directly linked to the Asian and more notably to the Russian financial crisis had led to a global market turmoil.[374]

4.6.3.4 Summary

The Russian financial crisis of 1998 was linked to a very weak economy, which in the wake of the foundation of the Russian Federation was comparable to an emerging market. The Asian financial crisis had triggered a significant fall in demand for oil and non-ferrous metals, which represented a large part in the balance of trade of the country. Russia relied on its exports to finance at least partly the country's reconstruction. Alas, the attack of speculators combined with a slowdown of their exports pushed the country into disaster. Devaluation and debt defaulting were the necessary consequences of this crisis which then spread worldwide because of LTCM's almost crash.

[370] Long Term Capital Management (LTCM) was a fund founded by John Meriwether, an exceptionally successful bond trader. Two of his consultants (Myron S. Scholes and Robert C. Merton) were Nobel-prized professors who had discovered a closed-form formula to price a European option (Black-Scholes formula).

[371] Kaplan Schweser (2012, p. 109).

[372] Kaplan Schweser (2012, p. 110).

[373] Dungey et al. (2002).

[374] The events around the fall of LTCM are described in detail in Lowenstein (2002).

Surprisingly, Russia bounced back from the August 1998 financial crash starting 1999, notably thanks to a relatively high oil price which created a trade surplus and maintained a healthy economic situation until the next heavy oil price fall in 2008. Another reason for the rebound was that domestic industries, such as food processing, had benefited from the devaluation, which caused a steep increase in the prices of imported goods.[375]

4.6.4 Dotcom Bubble (2001)

In the 1920s, radio companies played a significant role for economic growth. A radio is actually a single transistor. By miniaturization, we can put at least a million transistors into a structure of the size of a hand palm. This is called a microprocessor. This discovery together with the boom of aeronautics and a few other industries lead the stock market to soar again in the 1960s with a flood of IPOs of electronics companies, among which IBM and Texas Instruments.[376] Before going public, a few companies changed their names to finish by *-tron* so as to attract investors.[377]

The microprocessor helped building the first computers, but those were at first not very convenient (expensive, massive, not user-friendly at all). The second evolution came from the software industry. Many attempts were made to create a network between these machines and the first networks were deployed for university and military use. Then Tim Berners-Lee and Robert Cailliau, while working at CERN,[378] invented the World Wide Web in 1990. A few years later, Marc Andreessen released the first web browser, *Mosaic*, which soon became *Netscape*.

Between 1994 and 1996, more and more people were using the Internet, and businessmen, bankers and politicians wanted to build an *information superhighway*.[379] Economic concepts and ideas on how to best use the Internet were flourishing. On the stock exchange, the demand for Internet-related stocks increased significantly.[380] In December 1994, Netcome On-Line Communication Services[381] issued 1.85 mn. shares at $ 13 sold to 41.500 subscribers.[382] Because the company did not pay any dividend and had no recorded profits, a new formula was created which was also applied to the following Internet IPOs.[383] PSINet[384] was the next to go public and

[375]Stiglitz (2003).

[376]Cassidy (2003, p. 70).

[377]Cassidy (2003, p. 70).

[378]CERN is the European Organization for Nuclear Research. The name CERN is derived from the acronym for the French Conseil Européen pour la Recherche Nucléaire.

[379]Cassidy (2003, p. 38).

[380]Cassidy (2003, p. 79).

[381]This company provides Internet access on Sun Microsystems computers running Unix.

[382]Cassidy (2003, p. 79).

[383]Cassidy (2003, p. 79).

[384]PSINet was one of the first internet service providers.

although the company had lost $ 10 mn. the previous year, it still sold 3.8 mn. shares at $ 12. UUNet[385] did even better: 4,725 mn. shares for $ 14. Each of these stocks almost doubled on the first day![386]

Netscape's IPO captured the imagination of investors in an unprecedented manner[387] and for good reasons. The company founder Jim Clark had already founded a former company, which went public in only 5 years. This was quite a feat at this time. With its Internet browser *Netscape Navigator*, the company won 60 % of the browser market share within months.[388] In addition to that, on the very first day of the IPO, the share valued $ 28 soared to $ 71, and turned the Wisconsin farm boy Jim Clark into the head of a $ 663 mn. company.[389] Not bad after a single year of work. Dreams of instant wealth became true: one could become a billionaire in a matter of years and students or entrepreneurs could come out with any idea and would be granted cash in an unprecedented way, as long as it had a link with the World Wide Web. Wall Street's appetite seemed endless and a flood of IPOs soon filled the market, among which stood Yahoo!. The stock rose from its IPO price of $ 1.08 to $ 250 in January 1999.[390] It was worth $ 133 bn., more than Ford and General Motors combined. Between January 1995 and May 1996, the Dow Jones gained 47 % and rose from 3,837.10 to 5,643.20 points, while the NASDAQ increased by 65 %, from 751.31 to 1,243.43 points.[391]

But when the dream came to an end in 1996, companies like CNET had missed to go public, and looked desperately for sources of income.[392] The fact is that many Internet companies did not find a way to make cash out of their business, and needed investors' money just to survive. In the mind of many of its users, Internet was meant to be free. Even *Netscape*, which was not profitable yet, struggled to find its business model. It lived on because it made companies pay for its navigator, but not households or students. Hence, it would soon have to face the competition of the web browser of Microsoft (an improved *Mosaic*). Above all, Internet was meant to be free in the mind of many of its users.

The new valuation formula and the dream to become rich very quickly lead people to invest in a mood of euphoria without first looking at the income and sales of the companies. But investors soon woke up and the first slump occurred during summer 1996.[393] The price of Yahoo! stock dropped from $ 30 to $ 20 within months.[394] The IPO market virtually closed down.

---

[385] UUNET, founded in 1987, was one of the largest Internet service providers and one of the nine tier 1 networks.

[386] Cassidy (2003, p. 80).

[387] Cassidy (2003, p. 85).

[388] Cassidy (2003, p. 81).

[389] Cassidy (2003, p. 85).

[390] Cassidy (2003, p. 295).

[391] Cassidy (2003, p. 85).

[392] Cassidy (2003, p. 85).

[393] Cassidy (2003, p. 102).

[394] Cassidy (2003, p. 102).

Yet the contagious optimism of a few investors was unmitigated. David Shulman,[395] a fierce fighter against what he thought to be a bubble in 1995, wrote a few months later: *The powerful 1995 rally in the stock prices is not a bubble but rather a signal that the valuation paradigm has changed.*[396] His affirmation was backed by the fact that average citizens and investors did not fear inflation anymore and were willing to pay more. In addition to that, the heavy investments in computers and Internet technologies led companies and governments to think that productivity was improving, although the statistics did not show any productivity improvements yet.[397]

These views were echoed by Wall Street. During the first 6 months of 1997, the DJIA Index gained 7 % and reached 6,800 points. Eventually, the positive economic effects of heavy investments on the Internet became visible. During the first quarter of 1997, the U.S. GDP rose at an annualized 5 % rate, the consumer index rose 2.2 % and unemployment fell to the 1973 level of 4.9 %.[398] In the second quarter of 1997, the productivity of the non-farm business sector, i.e., the most important sector of the U.S. economy, reached 2.7 %,[399] a level not reached since *the Glorious Thirties*[400] after it had hovered for more than 25 years around 1–1.5 % since 1972–1974.[401] In a presentation made to the Senate Banking Committee in 1997, Alan Greenspan declared the economic environment in which the U.S. economy was to be exceptional, and predicted that other improvements were to come.[402] Firms would be able to produce just-in-time, meet customer demands easily and reduce unsold inventories and storage costs, as they benefited from large investments in information technology which allowed them to gather all necessary information.[403]

The game could start anew and a bubble mounted again for media companies who delivered content via the Internet. The bubble soon reached the Internet stocks, the media companies understanding that the future of information transmission would be on this canal. However, other problems were to come.

The Russian financial crisis (also called the *ruble crisis*) hit Russia on August 17, 1998.[404] On August 31, 1998, the Dow Jones Industrial Average Index fell 513 points (the second biggest drop in points in its history), while the NASDAQ dropped

---

[395]David Schulman was then an investment strategist at the investment bank Solomon Brothers.

[396]Norris (1995).

[397](Cassidy, 2003, p. 121).

[398]Cassidy (2003, p. 154).

[399]Cassidy (2003, p. 154).

[400]The *Glorious Thirties* are a period between World War II and the 1970s, characterized by a relatively steady growth of the economy and a significant improvement of the living standard in developed countries.

[401]Cassidy (2003, p. 153).

[402]Cassidy (2003, p. 158).

[403]Cassidy (2003, p. 161).

[404]Desai (2000, p. 51).

140.53 points (the biggest all-time dip in points).[405] Excite[406] and Amazon[407] lost 20 %, while Yahoo![408] and America Online[409] fell around 15 % in only 1 day.[410] This event was reflected on the cover of the Time magazine from September 14 with the title *Is The Boom Over?*[411] The world was thought to be on the verge of a global depression with the Asian crisis, the ruble crisis and Latin America weakened by the Mexican crisis.[412]

Because misfortunes never come alone, LTCM and Quantum Fund[413] declared to be on the brink of bankruptcy. Greenspan thus decided to cut interests rates to reboot the economy, after multiple months of pressure to raise them so as to put an end to the irrational rise of the markets.[414]

This move from the chairman of the Fed could be taken as if *the U.S. would stay out of the crisis for at least the coming year*, even though he did not say it this way.[415] Moreover, losses which occurred in Russia, Malaysia and Latin America appeared to be related only to high-risk environments. Investors knew it and could afford the eventual losses.[416] LTCM and Quantum Fund were too unknown to the average Americans to reduce their will to invest. As a result, the general public in the U.S. was far from being disappointed and the final but fatal wave of optimism overcame the ambient pessimism of the market.

The optimism was supported by the expansion of the telecommunication facilities which were necessary to sustain the impressive growth of the Internet and the U.S. economy. Between 1998 and 2000, the amount of buried fiber in the U.S. increased no less than fivefold.[417] By April 1999, this development pushed the stock price of Lucent, a $ 1 bn. sales company which produced the required hardware components, to a 800 % increase in 3 years.[418]

One reason that contributed to this impressive rise was the structure of the market. In fact, most of the Internet stocks were still possessed by the founders of these

---

[405] *Source*: Bloomberg (tickers: DJIA:IND and NASDAQ for the DJIA and the NASDAQ, respectively).

[406] Excite is a web portal company featuring search, content and functionality for personalization.

[407] Amazon is an Internet-based company providing media products, mostly books.

[408] Yahoo! is a web company offering various services on the Internet.

[409] America Online, also known as AOL is an Internet access provider.

[410] Cassidy (2003, p. 186).

[411] Cassidy (2003, p. 186).

[412] See the Financial Times article *This is Unquestionably a Bubble*, Financial Times (1998).

[413] Founded by Georges Soros, this fund had achieved very profitable deals based on event trading. His event-driven strategies lead the fund to take very large positions, which, if failed, could massively endanger economies, even the U.S. economy. See Brown et al. (1998, p. 1).

[414] Cassidy (2003, p. 187).

[415] Cassidy (2003, p. 188).

[416] Cassidy (2003, p. 188).

[417] Cassidy (2003, p. 207).

[418] Cassidy (2003, p. 206).

companies. A little amount of the available stocks combined with a strong demand could only mean increasing prices.[419] Investment banks cared little, because the possibility to generate profit was outstanding and the risk of losing face in front of the clients for helping impressively bad companies to go public was very limited.[420]

As a matter of fact, no less than 300 Internet firms went public during the 1998–2000 period[421] and accounted for a fourth of the IPOs realized at this time in the U.S.[422] A company could go public after only 5 months of existence![423] The general optimism was not limited to the Internet stocks but also reached the U.S. stock exchanges as a whole. For example, the annual average number of underwritings between 1980 and 2009 was 249,[424] but if we exclude the boom years from 1992 to 2000, this number falls to 172![425]

TheGlobe.com[426] and eBay[427] went public with outrageous speculation. TheGlobe.com shares were traded ten times the opening price on the first day.[428] eBay grew so fast that its capitalization was weighted $ 20 bn. by early 2000.[429] Yet, famous analysts still believed the stock undervalued. They recognized the quick progression of the stock prices, but were much more afraid not to profit from the next surge than to lose money through risky assets.[430] Moreover, after the dissolution of the USSR, it became almost unpatriotic to compare the 1990s to the 1920s.[431] People who did not buy stocks were merely laughed at.[432]

There were so many Internet stocks on the market at that time that special indexes were created, like the Dow Jones Internet Composite Index. This helped to keep the illusion of a regular business, despite the huge losses of these firms. Prices were far too high relative to fundamentals. As a result, when Yahoo! bought GeoCities and America Online bought Netscape, it was not with U.S. dollars, but with shares, which were then called Internet dollars.[433]

---

[419] Cassidy (2003, p. 213).

[420] Cassidy (2003, p. 216).

[421] Cassidy (2003, p. 192).

[422] Ritter (2010, Table 11, p. 17).

[423] Cassidy (2003, p. 243).

[424] *Source*: Own, based on Ritter (2010, Table 11, p. 17).

[425] *Source*: Own, based on Ritter (2010, Table 11, p. 17).

[426] TheGlobe.com was an internet start-up founded in 1994 and offered social networking services. The company's stock price collapsed in 1999, and the company retrenched for several years before ceasing operations in 2008.

[427] eBay is an online auction and shopping website where people and businesses buy and sell a broad variety of goods and services worldwide.

[428] Cassidy (2003, p. 196).

[429] Cassidy (2003, p. 212).

[430] Cassidy (2003, p. 213).

[431] Cassidy (2003, p. 69).

[432] Cassidy (2003, p. 192).

[433] Cassidy (2003, p. 203).

The emergence of the Internet and online trading systems helped people to invest personally. As it had happened during the railway mania, they were strongly encouraged by the media.[434] People started to give up their careers so as to become rich quickly.[435] The demand of cash for investments increased and before long people started trading on margin. The amount of margin debt more than doubled between December 1996 and December 2000.[436] All credits taken together (bank loans, credit cards loans, home-equity loans, etc.), the average total household debt stood at a 102 % of personal income.[437] A 20 % increase since 1992![438] In dollar value, the margin debt outstanding amounted to $ 243.5 bn. on January 2000,[439] approximately the level before the crash of 1987![440] Freshly graduated students from top U.S. Universities, like Harvard, MIT or Stanford, strongly considered a career in high-tech start-ups.[441] A whole nation dabbed on the Internet market, just as the Dutch did four centuries ago with tulips.

Index performances beat all records. On March 29, 1999, the Dow Jones closed above 10,000.[442] An analyst, Ralph Acampora,[443] said:

It's no longer a ceiling; it's a floor. As time goes by, we'll be looking down at it.[444]

By December 31, 1999, the Dow Jones climbed to an all-time high: 11,497.12.[445] This represents an annual 25.22 % increase,[446] which is a lot compared to historical data. On November 28, 1998, the NASDAQ closed above 2,000.[447] One year later it reached 3,000,[448] and by end December 4,000.[449] Yet new ideas to support this growth were missing. Mostly already existing solutions were slightly improved.

U.S. officials were concerned by the 107-month growth, the longest ever in the U.S.[450] The *irrational exuberance*[451] had to end. First, invoking the Securities

---

434 Cassidy (2003, p. 286).

435 Cassidy (2003, p. 232).

436 Cassidy (2003, p. 255).

437 Cassidy (2003, p. 256).

438 Cassidy (2003, p. 256).

439 Cassidy (2003, p. 268).

440 Cassidy (2003, p. 268).

441 Cassidy (2003, p. 239).

442 Cassidy (2003, p. 218).

443 Ralph Acampora is a pioneer in technical analysis and worked as the New York Institute of Finance's Director of Technical Analysis Studies.

444 Cassidy (2003, p. 218).

445 Cassidy (2003, p. 261).

446 Cassidy (2003, p. 261).

447 Cassidy (2003, p. 198).

448 Cassidy (2003, p. 244).

449 Cassidy (2003, p. 261).

450 Cassidy (2003, p. 273).

451 The term *irrational exuberance* comes form a speech on December 5, 1996, from the chairman of the Fed, Alan Greenspan. He criticized the behavior of the stock markets' investors as being far from the required rationality on an investment process. See Shiller (2005, p. 231).

Exchange Act of 1934, Alan Greenspan forced investors to deposit at least 50 % of the price of each share bought on margin.[452] Second, in a speech on January 13, 2000, he explained with simple words why the stock market would no longer grow at the pace of the past 5 years and indicated a coming rise in interest rates.[453] Finally, he lifted the rates from 4.5 to 5.75 %, then to 6 %.[454]

There was no immediate panic, with the Dow closing 37.85 points down, i.e., by less than 1 %. However, a first decline started slowly on January 14, 2000.[455] A few stocks, like Value America,[456] once traded at $ 75, fell to $ 5.[457] Amazon announced at this time that for each dollar of goods sold, it lost around three dollars. Even for an Internet-based company, this was too much for Wall Street. The stock lost 30 % on the very day.[458] Investors were hit by another wave of bad news about the income of Internet and Internet-related companies. At the same time, productivity gains started to plummet.[459] While the NASDAQ Index still climbed upwards until March 9, 2000, to 5,046.86 points,[460] more than 80 % of the stocks on the S&P 500 Index fell by 20 %.[461] Investors were increasingly wary of putting money into non-profitable Internet stocks.[462]

The volatility on the market reached levels unseen in 30 years. On March 24, 2000, the journal *Business Week* revealed that Wall Street was peddling high-risk investment to the public,[463] from which professional traders benefited integrally, and accused the media to encourage non-professional investors to gamble with their savings on the stock exchange. A pro-stock exchange analyst from Goldman Sachs, Joseph Cohen, declared he would reduce its share of Internet stocks in his portfolio.[464] Although these statements were stunning, they cannot be the cause for the ensuing crash, because similar statements had also been made previously in May 1999.[465]

As Cassidy underlines: *"Stock Market crashes often happen in a news vacuum. There was no particularly bad news in October 1929 and October 1987 either, but*

---

[452]Cassidy (2003, p. 268).

[453]Cassidy (2003, p. 271).

[454]Cassidy (2003, pp. 273–274).

[455]Cassidy (2003, p. 271).

[456]Value America was a dotcom company. Its business model involved connecting customers on the web directly to manufacturers, with the intent of providing better pricing and faster shipping. See Cassidy (2003, p. 271).

[457]Cassidy (2003, p. 271).

[458]Cassidy (2003, p. 271).

[459]Cassidy (2003, p. 274).

[460]Cassidy (2003, p. 281).

[461]Cassidy (2003, p. 282).

[462]Cassidy (2003, pp. 271–272).

[463]Cassidy (2003, pp. 286–287).

[464]Cassidy (2003, p. 271).

[465]Cassidy (2003, pp. 284–285).

both dates were preceded by periods of increased volatility. Volatility is the stock market equivalent of a nervous rash. It indicates that investors are agitated and unsure what is going to happen next. In such circumstances, any hint of bad news tends to get exaggerated, and stocks can move sharply for no apparent reason."[466]

By April 3, 2000, and on this single day, stocks on the NASDAQ stock exchange fell by more than 10 %,[467] while investors rediscovered Ford and the *Old Economy* stocks. The next day, things went worse. Many investors faced margin calls which, quite often, they could not meet. At noon, the NASDAQ was down 575 points (roughly 12.5 %),[468] while the Dow Jones lost 504 points (approximately 4.5 %).[469] The value of exchanged stocks was outstanding. In only 1 day, 1.5 bn. shares changed hands on the Dow, while 2.88 bn. shares were exchanged on the NASDAQ electronic trading system.[470] To put these numbers into perspective it is worthwhile to remember that on October 19, 1987, when the Dow Jones fell by 22.6 %, there were no more than 600 mn. traded shares, which was already a record.[471]

After 3 days of heavy selling, the volatility, still high, seemed to calm down. Yet, the journal *The Economist* pointed out that most of the shares on the NASDAQ made neither profits nor paid dividends. The trailing price-to-earnings ratio was about 62, while the historic average between 1973 and 1995 never exceeded 21.[472] On April 10, the selling resumed. The NASDAQ lost 5.8 %, and the Dow Jones Internet Composite Index lost 10 %.[473]

Figure 4.19 which shows the monthly return of the NASDAQ Composite Index in USD from January 1995 to December 2003.[474] The movements were driven by the fear of investors, an abnormally high inflation rate (the highest in 5 years)[475] and strong retail sales, which implied further hikes in interest rates. This is precisely what the Fed did, when it increased the interest rates from 6 to 6.5 %.[476]

To resume, the dotcom bubble was a three-act crisis. The first consisted of a wave of unsound IPOs of companies, which encountered difficulties to develop a healthy business model. The second started with a heavy dip because of the Russian crisis. This shock underlined the necessity of a nationwide infrastructure, while not harming too much the American economy. A greater access to trading solutions, an increase in purchases on margin and a wave of successful people in the Internet industry increased significantly the hope to become quickly rich thanks to the stock

[466]Cassidy (2003, p. 287).

[467]Cassidy (2003, p. 287).

[468]Cassidy (2003, p. 288).

[469]Cassidy (2003, p. 288).

[470]Cassidy (2003, p. 289).

[471]Cassidy (2003, pp. 287–289).

[472]Economist Online (2000).

[473]Cassidy (2003, p. 291).

[474]Cassidy (2003, p. 295).

[475]Cassidy (2003, p. 293).

[476]Cassidy (2003, p. 302).

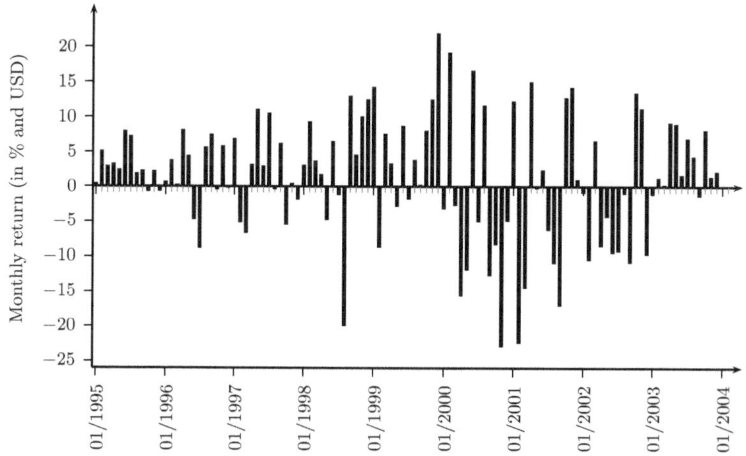

Fig. 4.19 Monthly return of the NASDAQ Composite Index in USD from January 1995 to December 2003. *Source*: Factset (ticker: COMP)

markets. But in the final act, bad fundamentals, rising inflation and the lack of new projects triggered a wave of pessimism which struck the market starting in April 2000. Since its peak on March 10, 2000, with a 170 % rise from fall 1998, the NASDAQ lost 34.2 %.[477] The Dow Jones Internet Composite Index was down 54.6 % from his March 9, 2000, high.[478] In dollar value, Wall Street saw $ 2 tn. evaporate with Microsoft accounting for a bit less than one eighth of this loss.[479]

4.6.5 Subprime Crisis (2007)

The subprime crisis was one of the deepest and most disruptive crises in worldwide financial history and is comparable to the Great Wall Street Crash of 1929. The build-up process of the real estate bubble in the U.S. and its contagion first to the U.S. stock exchanges and then worldwide is complex and was fairly difficult to forecast at that time. It engulfed the whole financial industry and underlined critical failures in the controlling and regulatory processes during nearly all the stages which lead to the crisis.

Everything started with a simple observation. Approximately two Americans out of five could not afford to buy a home.[480] The economic situation of many citizens willing to buy a house did not allow them to benefit from a normal fixed-

[477]Bloomberg under ticker INDU for the DJIA Index, CCMP for the NASDAQ Index and YHOO for Yahoo.

[478]Bloomberg under ticker INDU:IND for the DJIA Index.

[479]Cassidy (2003, p. 295).

[480]Callis and Cavanaugh (2007, p. 1).

rate amortizing loan. The reason lies in the so-called credit score, for example the FICO score.[481] Clients are classified according to the obtained scores which assess their creditworthiness. For a rather low-risk client, the credit file of the client is classified as A-paper, or *Prime*. If riskier, it would be an *Alternative A-paper*, or if even riskier, it would be rated as a *Subprime-paper*.[482] The name of the crisis actually comes from this classification: people were given credit whose risk factors were quite high. Normally, they would have never been able to get access to bank lending.

Almost all presidents from Roosevelt to George W. Bush tried to promote home ownership among American citizens.[483] In 1980, Congress passed the Depository Institutions Deregulation and Monetary Control Act (DIDMCA) and the Alternative Mortgage Transaction Parity Act (AMTPA) in 1982. The former abolished the state usury caps that had limited the interest rates banks could charge on primary mortgages. As a result, people who did not have the appropriate criteria to buy a home were able to get a bank credit in exchange for higher interest payments. It became an incentive to make home loans to citizens with low credit ratings.[484] The AMTPA, on the other side, allowed new types of loans: adjustable-rate mortgages (ARM),[485] balloon payment mortgages,[486] interest-only mortgages[487] and the option-adjustable rate mortgage.[488] It also enabled *non-federally chartered housing creditors* to write adjustable-rate mortgages. In other words, it allowed non-bank organisms, which were not subject to bank regulation, to grant floating-rate loans[489] and gave birth to a *shadow banking system*.[490]

The adjustable-rate mortgages became progressively more popular by the end of the 1980s but did not seem to have much effect on the housing ownership ratio

[481] The FICO score is a score established by the Fair Isaac Corporation, a company founded by an engineer and a mathematician, which sells a credit-scoring system to banks and equivalent credit granting institutions. For more information on the company: http://www.fico.com/en/Company/Pages/about.aspx.

[482] Fabozzi (2002, p. 286).

[483] Office of the Press Secretary (2002).

[484] Birger (2008a).

[485] In ARMs, the interest rate becomes floating after a number of years. See Birger (2008b).

[486] Balloon payment mortgages have an outsized payment when the loan comes due. See Birger (2008b).

[487] Interest-only loans require only the repayment of interest (no principal) during the first few years of the loan, but then hit borrowers with large monthly payment which are reset monthly. See Birger (2008b).

[488] This option allows borrowers to underpay by as much as they want during the first few years. Then, the unpaid accrued interest part has to be paid in addition to the formerly set interest rates. See Birger (2008b).

[489] Birger (2008b).

[490] A shadow banking system is characterized by non-bank originators which grant a credit to borrowers. They are not subject to the same regulations as conventional banks. See Shiller (2008, p. 42).

before 1994.[491] Since then, after 20 years of stagnation, the home ownership rate for all regions, age groups, racial groups and income groups increased[492] from 64 % in 1994 to an all-time high of 69.2 % in 2004.[493] By 2007, ARMs averaged 80 % of outstanding subprime mortgages.[494] Figure 4.20 shows the amount of subprime mortgage quarterly lending in USD from December 31, 1999 to September 30, 2007. The heydays of subprime lending were between 2005 and 2006,[495] when they averaged $ 150 bn. in value, i.e., around 20 years after the laws were enacted.

Therefore, it would be inappropriate to explain this growth only by the effects of deregulation. With the economy booming thanks to the Internet technology, large foreign investments flooded into the United States and granted easy money access to American citizens. This impressive inflow of foreign currency helped fuel a housing market boom and encouraged debt-financed consumption.[496]

Additionally, between 2000 and 2003, the Federal Reserve lowered interest rates from 6.5 to 1.0 %.[497] This was done to soften the effects of the dotcom bubble and

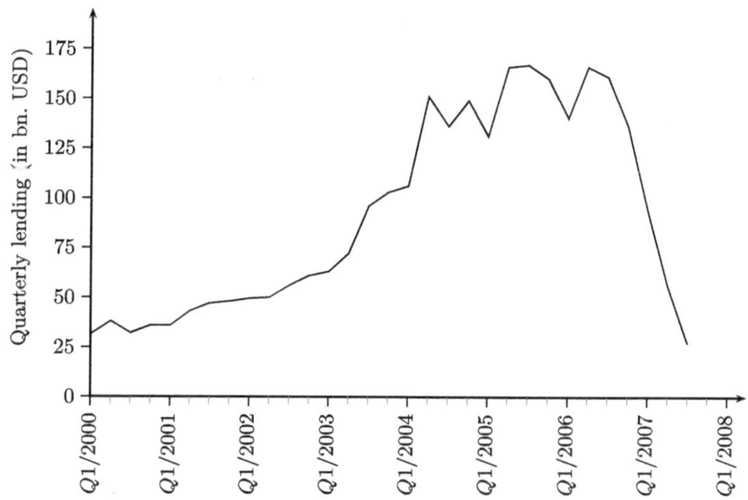

Fig. 4.20 Subprime mortgage quarterly lending from December 31, 1999 to September 30, 2007. *Source*: New York Times (2007)

[491]Shiller (2008, p. 12).

[492]Shiller (2008, p. 12).

[493]Callis and Cavanaugh (2007, Table 4, p. 4).

[494]Uhlfelder (2009, p. 2).

[495]New York Times (2007).

[496]Bush (2008).

[497]Updated information on rates is available on the website of the Board of Governors of the Federal Reserve System, see Federal Reserve Bank (2014a). See also Schulmerich (2003, p. 10).

the September 2001 terrorist attacks and to combat the perceived risk of deflation. With lower interest rates and cheaper credit costs, U.S. citizens started borrowing more to invest in housing. Furthermore, the declining interest rates considerably increased the attractivity of ARMs. Borrowers contracted new mortgages to repay their former mortgages or to get access to loans with lower interest rates payments if they had not contracted an ARM yet, but a fixed-rate mortgage. Logically, property prices started to grow significantly. In a climate of increasing home prices and decreasing interest rates, many started to speculate on housing prices, pushing forward average prices across the country.[498]

Figure 4.21 below displays the price movements in 20 major U.S. cities as represented by the S&P/Case-Shiller Home Price Index,[499] here shown from Q1-2000 to Q3-2013.

On average, housing prices increased 100 % with a close to constant annual growth for almost a whole decade. U.S. real estate was financially very appealing, even for foreign investors who wanted to take advantage of the booming market. But the delivery of loans by U.S-based banks and their non-bank equivalents started to weigh on their balance sheets, and their risk exposure increased. Therefore, in the years leading up to the crisis, the top four U.S. depository banks[500] moved an estimated $ 5.2 tn. in assets and liabilities off-balance sheet into special

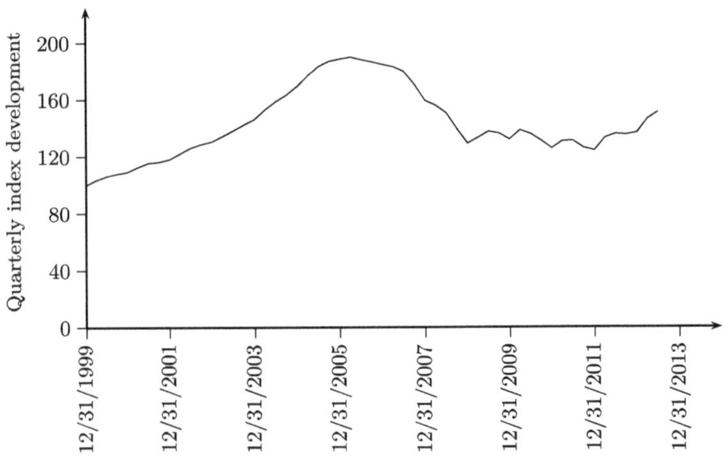

Fig. 4.21 Development of the S&P/Case-Shiller U.S. National Home Price Index from Q1-2000 to Q3-2013 using quarter-end data. *Source*: Factset (ticker: SPCSUSA)

[498]These effects and interactions are nicely described in Schulmerich (2012).

[499]This index is often only called *Case-Shiller Home Price Index*.

[500]The big four are Bank of America Corp., J. P. Morgan Chase & Company, Citigroup, Wells Fargo & Company.

purpose vehicles or other entities to the shadow banking system.[501] These perfectly legal measures enabled them to essentially bypass existing regulations regarding minimum capital ratios, thereby increasing the leverage and profits during the boom but increasing the losses during the crisis.[502] But as this process had its limits, investment banks developed a new way to make additional profits by starting the securitization of mortgages.

Investment banks gathered these loans, often bought from brokers or commercial banks, into a pool and then cut them into different tranches. Each tranche had a specific level of risk and adapted rate of return, and was graded by rating agencies. Because a tranche was far too big for a single investor, it was divided into different securities, namely asset-backed securities (ABS) and mortgage-backed securities (MBS).[503] An ABS could also be restructured to become a collateralized debt obligation (CDO).[504] The investment companies took no risk in this securitization process, but by buying and then selling the assets, they made significant profits thanks to large fees.

On the buy side, MBSs are financially interesting because investors own a share of the whole pool, which limits the default risk from borrowers compared to a conventional credit-granting process. The reason is that this risk is shared among MBS and CDO purchasers. Furthermore, the asset value is guaranteed by the underlying (asset-*backed* securities), so that the level of risk taken by the investor appears to be virtually non-existent.[505]

Finally, an investor, being in the U.S. or not, could participate in the booming housing market by purchasing ABSs or MBSs, which steered cash inflows to U.S. banks from abroad and boosted profits. By this way the subprime crisis was able to export itself abroad.

Yet, as many bankers were aware that risk still existed, a further step in the securitization process was made. Credit default swaps (CDS) were created so that, in exchange of a fixed periodic fee, they protected the CDS buyer in the case of a credit event.[506]

MBS, ABS, CDO and CDS were highly successful, especially when, starting 1995, government-sponsored enterprises (GSEs)[507] like Fannie Mae and Freddie Mac received government tax incentives for purchasing mortgage-backed securities

---

[501] Reserve Bank of India (2010).

[502] Reserve Bank of India (2010).

[503] Mengle (2007, p. 6).

[504] Mengle (2007, p. 6).

[505] Mengle (2007, p. 30).

[506] A credit event occurs when a borrower cannot meet its financial obligations toward a credit and thus is in default. See Mengle (2007, p. 6).

[507] The government-sponsored enterprises (GSEs) are a group of financial services corporations created by the United States Congress. Their function is to enhance the flow of credit to targeted sectors of the economy and to make these segments of the capital market more efficient and transparent.

which included loans to low-income borrowers, i.e., subprime mortgages. Since then, Fannie Mae and Freddie Mac began to get involved with the subprime market.[508] In 1996, the United States Department of Housing and Urban Development (HUD)[509] set a goal for Fannie Mae and Freddie Mac that at least 42 % of the mortgages they purchased be issued to borrowers whose household income was below the median in their area. This target was increased to 50 % in 2000 and 52 % in 2005.[510]

The mounting interest for U.S. mortgages coupled with the mandatory purchases was the cornerstone of the crisis. On one side, the supply of mortgages became scarce, particularly of quality mortgages. On the other side, the mandatory purchases from GSEs created moral hazard. The fact that they were forced to buy mortgages implied that in the case of a massive default from borrowers, the government would help the buyers, namely Fannie Mae and Freddie Mac who together detained more than a half of the outstanding U.S. residential mortgages ($ 5.1 tn.).[511]

In order to feed the foreign investors's appetite for CDOs, CDSs, etc., investment companies became less cautious about the quality of mortgages, if it was controllable at all. Mortgage quality was hard to control because mortgages were mostly bought from commercial banks and brokers. After the Great Wall Street Crash and the Great Depression, the *Glass-Steagall Act* was enacted so as to separate commercial banks and investment banks, in part to avoid potential conflicts of interest between the lending activities of the former and the trading activities of the latter. This forced the investment banks responsible for securitization to call on other financial institutions so as to gather the required mortgages and proceed to the securitization.

When this act was repealed in 1999, it had little effect on the economy. The investment banks had little knowledge of the economic situation of the clients who asked for a credit. But the lenders, brokers and commercial banks cared little about the quality of the mortgages, which in the end would not appear in their balance sheets as they sold and repacked the loans into CDOs, MBSs, etc. They were paid based on the quantity, not quality, of the loans they provided.[512] In a Peabody Award winning video program, NPR[513] correspondents argued that a *"giant pool of money (represented by* $ 70 *tn. in worldwide fixed income investments) sought higher yields than those offered by U.S. Treasury bonds early*

[508] Leonnig (2008).

[509] The United States Department of Housing and Urban Development, also known as HUD, is a cabinet department in the executive branch of the United States federal government.

[510] Wallison (2011, Table 5, p. 61).

[511] Government-Sponsored Enterprises (GSEs) Report from the Fed as of September 17, 2010, in Federal Reserve Report (2010a, Table L.124, line 16), and Federal Reserve Report (2010b, Table L.125, line 2).

[512] Benett (2007).

[513] NPR delivers breaking national and world news and top stories from business, politics, health, science, technology, music, arts and culture. For more information: http://www.npr.org.

in the decade."[514] More alarming: lenders offered more and more loans to higher-risk borrowers,[515] including undocumented immigrants.[516] In addition to wooing higher-risk borrowers, banks and brokers offered increasingly risky loan options and borrowing incentives. In 2005, the median down payment for first-time home buyers was 2 %, with 43 % of those buyers making *no down payment.*[517] Lenders relied on the ever-increasing property prices to recoup their cost in case of a default.

The control of credit files slackened. At first, the *stated income, verified assets* (SIVA) loans were invented which no longer required a proof of income. Borrowers just had to state it and present a proof of cash in a bank account. Then, the *no income, verified assets* (NIVA) loans came out.[518] The lender no longer required a proof of employment, and borrowers just needed to present a bank statement with cash on it. Finally, the *no income, no assets* (NINA) and the *no income, no job and no assets* (NINJA) loan formats were created. NINA and NINJA clients could benefit from official loan products and borrow money without having to prove or even state any owned assets. All that was required for a mortgage was a credit score, for instance a FICO score.[519]

Even the credit rating agencies (CRA),[520] in charge of controlling CDOs, MBSs and the banks failed at evaluating the real level of risk of these asset classes. Critics allege that the rating agencies suffered from conflicts of interest, as they were paid by investment banks and other firms that organized and sold structured securities to investors.[521] Between September 2007 and June 2008, the inaccurate credit ratings granted by CRAs were so staggering that agencies lowered the credit ratings on $ 1.9 tn. in mortgage-backed securities. Financial institutions felt they had to lower the value of their MBSs and acquire additional capital so as to maintain capital ratios. If this involved the sale of new shares of stock, the value of the existing shares was reduced. Thus, rating downgrades lowered the stock prices of many financial firms.[522]

Due to the relaxation in credit restrictions, the volume of subprime mortgages rose from $ 35 bn. in 1994 (5 % of total origination[523]) to $ 160 bn. (13 % of origination) in 1999 and to $ 600 bn. (20 % of origination) in 2006, and fuelled the

---

[514] See Glass (2008).

[515] Kirchhoff and Keen (2007).

[516] Pasha (2005).

[517] Knox (2006).

[518] Pearlstein (2010).

[519] Pearlstein (2010).

[520] A credit rating agency (CRA) is a company that assigns credit ratings for issuers of certain types of debt obligations as well as the debt instruments themselves. In some cases, the services of the underlying debt are also given ratings.

[521] Economist Online (2007).

[522] Birger (2008a).

[523] Origination is the process by which a borrower applies for a new loan, and a lender processes that application.

unrelenting growth in real estate prices.[524] The average difference between subprime and prime mortgage interest rates (the *subprime markup*) declined significantly between 2001 and 2007, according to a study by the Federal Reserve.[525] Clearly, quality was no issue anymore and tricks and teasers were used to attract new clients. An estimated one third of the ARMs which originated between 2004 and 2006 had *teaser* rates below 4 %, which then after a limited period increased significantly, as much as doubling the monthly payment.[526]

During the two former decades ending in 2001, the national median home price ranged from 2.9 to 3.1 times the median household income.[527] This ratio rose to 4.0 in 2004, and 4.6 in 2006.[528] Quite a few homeowners refinanced their homes at lower interest rates, or financed other expenses by taking out second mortgages secured by the price appreciation of their house. U.S. household debt as a percentage of annual disposable personal income was 127 % at the end of 2007 versus 77 % in 1990.[529] It represented $ 14.3 tn. by 2007, i.e., 135 % of disposable personal income.[530] This was far above the $ 705 bn. at year-end 1974, which represented 60 % of disposable personal income! The U.S. home mortgage debt relative to GDP increased from an average of 46 % during the 1990s to 73 % during 2008, reaching $ 10.5 tn.[531] Housing had become tremendously expensive despite easy access to credit and the flood of foreign investors willing to take profit from the rise in housing prices.

Slowly the bubble started to show signs of deflating. Fearing inflation and attempting to stifle the mania for real estate, the Fed progressively raised interest rates from 1 to 5.25 % from July 2004 to July 2006.[532] By 2006, newly built housing properties met the demand for real estate, and the rise of property prices slowed down and then turned into a decrease.

As more borrowers stopped their mortgage payments and went into foreclosure, the effect was first felt by the riskiest MBS and CDO tranches which incurred heavy losses. The mounting number of foreclosures increased the supply of homes for sale. Housing prices dropped, and the value of the borrowers' homes declined. The decrease in mortgage payments also reduced the value of mortgage-backed

---

[524] Bernard (2008).

[525] Bernanke (2008).

[526] Arnold (2007).

[527] Steverman and Bogoslaw (2008).

[528] Steverman and Bogoslaw (2008).

[529] Economist Online (2008).

[530] Please note the difference between the personal disposable income and the debt as a percentage of disposable income. See Balance Sheet of Households and Nonprofit Organizations Report from the Fed as of September 17, 2010, in Federal Reserve Report (2010c, Table B.100, lines 31 and 48, p. 104).

[531] Barr (2009).

[532] Historical Fed rates available at Federal Reserve Bank (2014b).

securities, which eroded the net worth and financial health of banks. This vicious cycle was at the heart of the crisis.[533]

Figure 4.22, which shows the development of the S&P/Case-Shiller Home Price Index over a longer time period (and now using annual data) in comparison to Fig. 4.20, illustrates that by September 2008 U.S. housing prices had on average declined by over 20 % from their mid-2006 peak,[534] on the national level as well as in the ten biggest U.S. cities. This major and unexpected decline in house prices meant that many borrowers had negative equity in their homes, i.e., their homes were worth less than their mortgages. As of March 2008, an estimated 8.8 mn. borrowers—10.8 % of all homeowners—were in this situation, a number that is believed to have risen to 12 mn. by November 2008.[535] These borrowers had an incentive to default on their mortgages as a mortgage is a typical non-recourse debt secured against the property.[536]

Economist Stan Liebowitz argued in the Wall Street Journal that although only 12 % of homes had negative equity, they comprised 47 % of foreclosures during the second half of 2008. He concluded that the extent of equity in the home was the

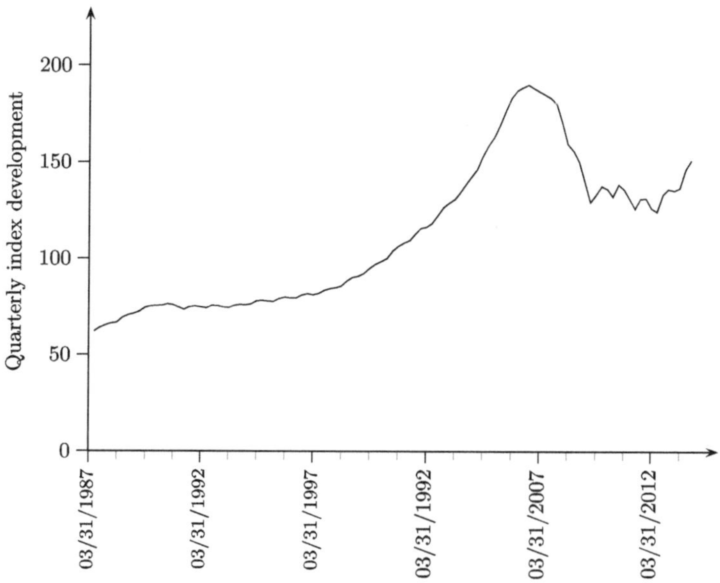

Fig. 4.22 Development of the S&P/Case-Shiller U.S. National Home Price Index from Q1-1987 to Q4-2013 using quarter-end data. *Source*: Factset (ticker: SPCSUSA)

[533]Feldstein (2008).

[534]Blitzer and Guarino (2008).

[535]Blitzer and Guarino (2008).

[536]Blitzer and Guarino (2008).

key factor in foreclosures, rather than the type of loan, the creditworthiness of the borrower, or the ability to pay.[537] Rising foreclosure rates increased the inventory of houses offered for sale. The number of new homes sold in 2007 was 26.4 % less than in the preceding year. By January 2008, the inventory of unsold new homes was 9.8 times the December 2007 sales volume, the highest value of this ratio since 1981.[538] Furthermore, nearly four million existing homes were for sale, of which almost 2.9 mn. were vacant.[539] As prices declined because of this oversupply, more homeowners were at risk of default or foreclosure. As a result, house prices were expected to continue declining until the excess home supply reached *normal* levels.

Because of the rising number of foreclosures which meant unpaid mortgages, MBSs, ABSs and CDOs remained close to worthless in 2007 after 6 years of continuous increase in value as explained earlier. Since firms and banks possessing CDSs were liable to default, the level of unpaid mortgages leveraged through financial engineering caused great damages. Because these financial instruments had been exchanged, repacked and exchanged once again,[540] no one knew exactly what assets they actual possessed in these structured products and how much they were worth.[541]

A confidence crisis built up since only few banks or lending institutions were willing to pursue lending operations.[542] A credit crunch developed, which was particularly harmful for the economy, especially for industries depending on credit access such as the oil industry, building industry, etc.[543]

To solve this issue and to stimulate the stumbling economy, on November 14, 2008, the U.S. government agreed upon a $ 700 bn. rescue package through a strategy called *quantitative easing*.[544] Quantitative easing consists in providing a large amount of money to the financial system so as to ease the pressure on banks and grant them extra capital.[545] It worked at first, but did not prevent the worldwide stock markets to crumble. The exposure to CDOs and CDSs was large, as many financial institutions were relying on the fees based on these products to generate profits.

A worldwide stock market crisis broke out. As we can see in Figs. 4.23, 4.24 and 4.25 showing the U.S., U.K. and Japanese equity markets, respectively, equity markets around the world fell steeply from September 2007 to September 2008.

When the bubble burst, a wide variety of investors had to deal with close-to-worthless assets. Many U.S. universities, among them the famous Massachusetts Institute of Technology (MIT), lost billions during the crisis.[546] On the other side of

[537]Liebowitz (2009).

[538]Associated Press (2008a).

[539]See Coy (2008a) and Coy (2008b).

[540]Stout (2008b).

[541]Yoon (2007).

[542]Anderson and Heather (2007).

[543]Hutchinson (2010).

[544]Stout (2008b).

[545]Elliott (2009).

[546]McLaughlin (2010).

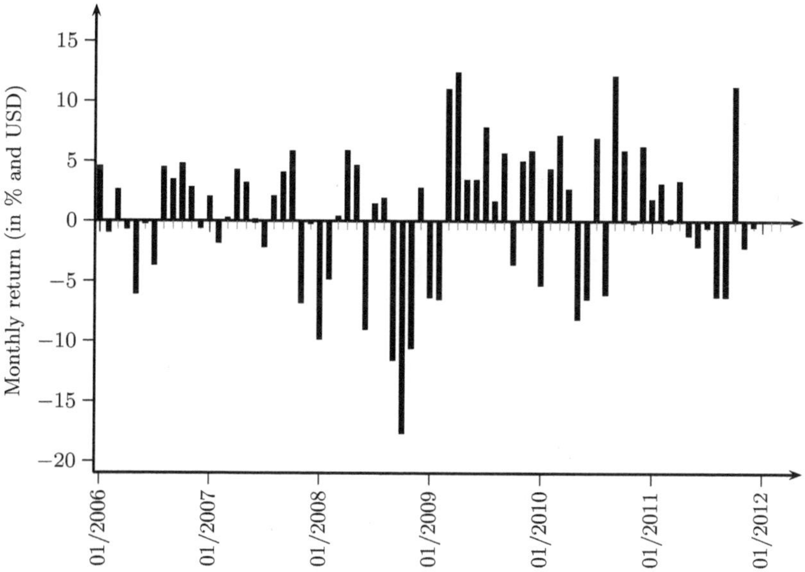

Fig. 4.23 Monthly return of the NASDAQ Composite Index in USD from January 2006 to December 2011. *Source*: Factset (COMP)

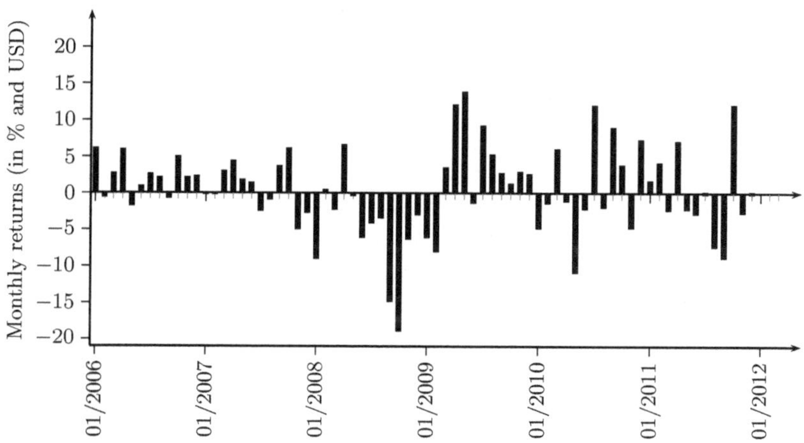

Fig. 4.24 Monthly returns of the FTSE All-Share Index in USD from January 2006 to December 2011. *Source*: Facstset (ticker: ASX:IND)

the ocean, in Norway, a small town called Narvik lost up to 55 % of most of its tax revenues.[547]

[547]The city invested in CDOs and lost 35 mn. kroner ($ 64 mn.) from investments that had fallen to less than 45 % of their original value. See Ivry (2007).

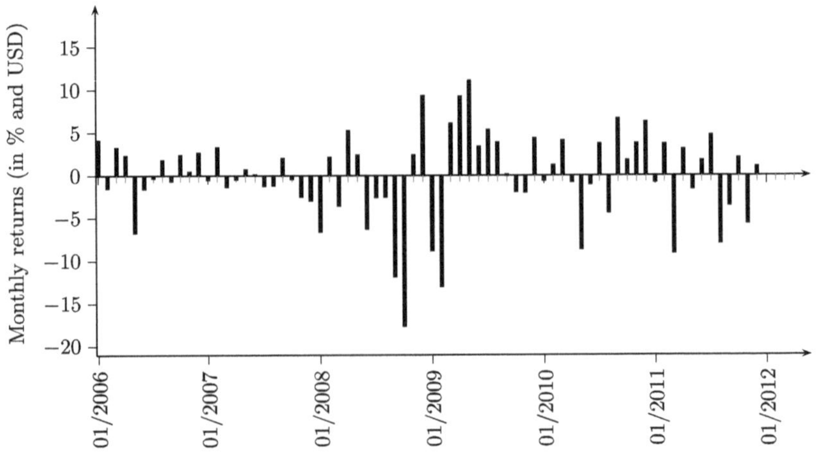

Fig. 4.25 Monthly returns of the Nikkei 225 Index in USD from January 2006 to December 2011. *Source*: Factset (ticker: NKY:IND)

The five largest U.S. investment banks, with combined liabilities or debts of $ 4 tn., either went bankrupt (Lehman Brothers), were taken over by other companies (Bear Stearns and Merrill Lynch), or were bailed out by the U.S. government (Goldman Sachs and Morgan Stanley) during 2008.[548] Fannie Mae and Freddie Mac with a similarly weak capital base, either directly owed or guaranteed nearly $ 5.1 tn. in mortgage obligations, when they were placed into receivership in September 2008.[549]

To put this into perspective, these obligations of seven highly leveraged institutions of $ 9 tn. can be compared to the $ 14 tn. size of the U.S. economy (GDP)[550] or to the total national debt of $ 10 tn. in September 2008.[551]

All this shows the magnitude of this financial crisis, which is only comparable to the Great Depression 80 years before. Therefore, 2008 is also often called the year of the *Great Recession*.[552]

To conclude, let us look at a quotation from the economist Ludwig von Mises[553] who writes in *Human Action: A Treatise on Economics*[554]:

[548]Labaton (2008).

[549]Statement of FHFA Director James B. Lockhart in Russell and Mullin (2008).

[550]Bureau of Economic Analysis database at http://www.bea.gov/index.htm.

[551]Historical and current U.S. debt amounts are available on the Internet, see TreasuryDirect Reports (2014).

[552]The events of the subprime crisis and the fall of Lehman Brothers as the culminating point are described in detail in Paulson (2010).

[553]Ludwig von Mises (1881–1973) was an Austrian School economist.

[554]See von Mises (1966).

There is no means of avoiding the final collapse of a boom brought about by credit (debt) expansion. The alternative is only whether the crisis should come sooner as the result of a voluntary abandonment of further credit (debt) expansion, or later as a final and total catastrophe of the currency system involved. [555]

The real estate bubble infected the domestic and worldwide stock markets and transformed into a global economic slump comparable to the Great Depression in the 1930s. However, the subprime crisis was also the trigger of subsequent crises: the Dubai real estate bubble burst and the Greek crisis, that we have already presented above, and the Irish crisis. The Irish financial crisis is very similar in a way to the Dubai real estate crisis: banks were overexposed to the crumbling property market and, additionally, the large number of foreign companies installed in the country thanks to the low corporate tax rate significantly diminished their investments or withdrew to their home countries.

4.6.6 Conclusion

The year 1929 was a milestone for modern capitalism. Since then, many regional crises have occurred and have hardly spared any developed or developing country. The strength and the width of these crashes seems to have increased significantly, especially during the past two decades. Investors are aware of their occurrence, but their behavior during such dramatic events seems to be as irrational as ever. Massive sell-offs and a flight to quality even occur with a deeper severity than in earlier times.

Five of the biggest crashes in history happened during the past two decades only. The Asian crisis, which followed the Russian crisis, showed how quickly foreign capital can flow out from a whole region.

The dotcom bubble was related to a leap in technology, i.e., the development of the Internet, very similar to the South Sea bubble and the railway mania. The former was brought about by advances in boat engineering and mapping systems. Finally the subprime crisis came which triggered the sovereign debt crisis of Europe. Dubai, Greece and Ireland are a few examples of the subsequent regional crashes because of the transformation of illiquid assets into liquid ones and their cross-border dealing. Would the creation of trade barriers have been of any preventive effect? Certainly not.

For politicians or economists, it is evident that a solution for crises has to be found. The next technological boom might be nano- or pico-technologies or even astrophysical technologies. It could also be a new way of handling energy or resources. Whatever it may be, investments will most certainly reach dramatic levels. If crises will reoccur with the same deeply negative outcomes as we have recently witnessed, only the strongest economies might survive. This is neither the idea of capitalism nor of globalization, which should provide any human with a share of the planet's wealth in relation with his efforts.

---

[555] See von Mises (1966, p. 572).

That is why behavioral factors have to be considered. But as a matter of fact, they are much more related to psychology and sociology than to mathematics and are, as for now, very hard to measure. Yet hopefully the next generation of economists will find a way to cope with this issue and to follow Galileo Galilei's motto *measure what can be measured, and make measurable what cannot be*.[556] Until then, we propose to focus on the available tools to forecast a bubble and a crash which is exactly the topic of the next section. If such a forecast is possible, the dramatic consequences of a bubble or crash could be, at least, diminished in scope and strength.

4.7 Forecasting a Bubble: A Case Study

4.7.1 How to Spot a Bubble?

Up to now, we have looked at past crashes, but it would be interesting to detect a coming crisis or at least the next bubble. By mid-2010, numerous possibilities offered themselves: There were signs of a possible gold bubble, an eventual bubble and excessive speculation on European and U.S. debts, and even a currency crisis between the British pound, the U.S. dollar or the euro. In 2006, investors could also observe a possible mounting bubble in China, followed by a correction.[557] However, there are hints that a further drawdown could occur in the near future. In this section, we will go through a case study and focus on the Chinese stock exchange and real estate so as to see if—as of early 2014—one can predict an upcoming crisis, based on a set of hints. We will also try to see if it is possible to deflate a bubble without quick and dramatic outcomes, i.e., without a crisis.

A few years ago, Alan Greenspan, chairman of the Fed from 1987 to 2006, described the first problem that we meet when dealing with this question: It is difficult to say if and when there is a bubble before a crash occurs.[558] P/E ratio and other indicators may help investors to notice when an asset is over- or undervalued against a trend or against an industry. Yet, when an industry is, as a whole, victim of irrational optimism and exuberance, when many investors think that a new era has come thanks to a new policy, a new technology or just a new way to conceive business, how easy is it to keep control of one's emotions and to stick to traditional paradigms and evaluation processes? Doing so would have greatly undervalued stocks as Amazon or Google[559] at their launch and undermined the performance of the investor's portfolio.

Therefore, we need to develop new tools or have a different approach to what happens on the markets today. Actually, it is not in the best interest of the investor to

---

[556] Kleinert (2009, p. 199).

[557] Arbeter (2007).

[558] Felsenthal (2007).

[559] These two companies, born during the dotcom bubble, survived the crisis and are considered being among the most successful companies of the 2000s decade.

act rationally when there is a bubble.[560] As a result, using P/E ratio, which is based on a relative stability of the expectations of the investors for a given asset, might not be as efficient as one could think.

The second issue raised by the question above is how to identify the methods and levers that governments, companies and investors can use in order to obtain *normal* price levels without creating a massive sell-off. In the following section, focused on China, we will see what makes the world of finance believe that the Chinese market is going through a bubble, what experts and local investors believe and what measures the Chinese government has taken from the subprime crisis to early 2014.

4.7.2 China: A Stock Exchange and Real Estate Case

In 1980, the People's Republic of China drastically changed its way to handle its economy, and the results were visible by the early 1990s. The country is comparable in size to Europe or the United States, but its population accounts for more than Unites States and European Union put together.[561] Three hundred million people are expected to migrate to the cities in the coming decade, who of course will have higher incomes than the vast majority of the still numerous peasants. Moreover, analysts estimated an 8 % annual growth for the coming years versus an average 10 % annual growth during the past decade.[562] In other words, by 2010 the *Chinese Dream* was more than ever valid.

However, starting from 2007 China exhibited the characteristics of great speculative manias on the stock exchange and in real estate. At first glance, one could observe a booming stock market and an exceptionally high growth rate in real estate prices in many cities—more than 15 % a year in certain cities.[563] Yet, these issues are merely the visible part of an iceberg, and a deeper understanding of the various clues that lead us to think that there might be a developing bubble is required, i.e., a review of the economic and political background of the *Chinese Dream*.

At first, we will examine if the achieved growth rate and the growth rate of the Chinese economy as assumed by analysts and experts across the world are sound and logical. Then we will apply our findings to the stock market, and try to see if it is building up a bubble. Finally, we will cover the real estate industry. Both areas were formerly linked in different crises. Given the constant price rises during the past years, it is a possibility that a Chinese real estate bubble develops. Recall that the aim of this section is to find out if it is possible to reduce the size of a bubble so that it vanishes without bursting. Therefore, it is required to discuss the measures

[560]According to Boco, Germain, and Rousseau, an overconfident investor achieves a better performance than a rational investor during bubbles. See Boco, Germain, and Rousseau (2010, p. 2).

[561]Data available on CountryMeters (2010).

[562]Chancellor (2010, p. 3).

[563]Gerrity (2010).

which were implemented from 2008 to 2013 to tackle the mounting risk of a second wave of worldwide economic distress.

The study of the Chinese case is particularly interesting because a slowdown in China would significantly hurt the country and hamper its growth on which most worldwide economies are counting on. If such a crisis were to happen, the European and American economies would have to face enormous difficulties, for example, to repay their debt, which is already very high. We now need to study the various hints that could signal the rise of a bubble.

In the 1980s and 1990s, China shifted from entrepreneurial capitalism towards a more state-directed capitalism, and focused on trophy investment projects[564] (such as the development of the Pudong financial district in Shanghai), foreign direct investment, and exports.[565] This is a big issue because in the years up to the credit crunch, i.e., just after the subprime crisis, China's economic growth largely depended on rising exports.[566] In 2010, China's exports to the West were already at twice the level achieved by Japan in its heyday. This trend was difficult to be maintained, and as a result protectionism measures were quickly undertaken.[567] The threat of an economic slowdown became particularly present since the onset of the Great Recession.[568] A massive sell-off of U.S. dollars during the crisis boosted the value of the Chinese renminbi which lead to a review of the valuation of the Chinese currency.[569]

But the biggest issue for the Chinese government were the state-owned companies (SOEs). For decades, they had been the extended arm of the government. By 2013, their role had declined, but they still received preferential access to credit from state-controlled banks. They were treated favorably compared to private businesses, which were forced to borrow at much higher rates outside of the formal banking system.[570] A study by researchers at the Hong Kong Monetary Authority found that the aggregate profits of the SOE would have been wiped out if they had borrowed at the rates paid by the private sector.[571] The same study found that more than half of

[564]Such projects are not for the economic benefit of a nation, but rather for the prestige they are assumed to generate for the leaders of the projects.

[565]Huang (2008, pp. 74–86).

[566]Chancellor (2010, p. 5).

[567]Across a wide range of sectors, China implemented protectionist measures in order to boost its own domestic firms and lock out foreign competitors. See Moore (2010). Yet the World Trade Organization also warned the United States to stop equivalent measures toward Chinese companies. See Xinhua (2010b).

[568]The Great Recession is the period following the subprime crisis, which covers the general economic downturn across the world and the recession of most industrialized countries. See Chancellor (2010, p. 5).

[569]The Chinese currency is not freely exchangeable and is valued against a basket of currencies since mid-2010, among which the U.S. dollar takes a good share. Before, the Chinese currency was pegged to the U.S. dollar.

[570]Chancellor (2010, p. 5).

[571]Ferri and Liu (2009).

all private businesses had no access to bank credit. This situation was not conducive to healthy growth, because it favored investment not based on productivity but rather on political acquaintances.

Historical and economic studies have shown that directed economies fail to allocate resources efficiently. This might be the case in China as well as will be shown. Yasheng Huang[572] from MIT argues that China's rapid growth was deceptive because the state could invest resources more quickly than the private sector, yet with a lower quality of investment.[573] More problematic: local governments in China raised funds through various vehicles to circumvent regulations that prevent them from borrowing directly. A crackdown on these loans could trigger a *gigantic wave* of bad debts as projects are left without funding, Northwestern University Professor Victor Shih[574] said in March 2010.[575]

The problem with targets imposed by a central authority is that they are liable to being manipulated. Goodhart's Law states that whenever an economic indicator is made a political target, then it loses the information content that would qualify it to play such a role.[576] Unfortunately, this theory has proved to be true in China.

For example, a $ 5 bn. *ghost town* called Ordos has emerged from the desert sands of Inner Mongolia, but 5 years after the construction began, by the end of 2010, this city which was conceived for one million people was still empty.

Another example: it was planned to invest $ 105 bn. into the construction of 18,000 km of high-speed railroads until 2020. This is approximately halfway to travel around the world. Indeed, in a distant future, it may come true what Citigroup analyst Ally Ma predicted: *"There's no doubt that high-speed rail will defeat airlines on all the routes of less than 800 kilometers."* [577] But a transport researcher at the governmental National Development and Reform Commission warned on March 3, 2010, that this project would face problems in recouping the costs and *"might not be able to achieve its minimum passenger loads to break even."* [578] This gigantic type of projects is similar to the Dubai construction projects in 2009 or the Internet fiber optic installations in the U.S. 10 years earlier.

Still, only a few people did care about the performance of the SOEs and the general overoptimism about the so-called unstoppable rise of China. The reason lies in the past, when policy lending to state-owned enterprises had put China's banks under heavy pressure. At the turn of the century, Standard & Poor's estimated

[572] Yasheng Huang is professor in international management at the Massachusetts Institute of Technology. He wrote *Capitalism with Chinese Characteristics: Entrepreneurship and the State*. See Huang (2008).

[573] Chancellor (2010, p. 5).

[574] In 2010, Victor Shih was an assistant professor of political science at Northwestern University specializing in Chinese politics.

[575] Akkermans (2010).

[576] Chrystal and Mizen (2001, pp. 2–3).

[577] Shen (2010).

[578] Chancellor (2010, p. 6).

Chinese nonperforming loans at around 50 % of total outstanding loans.[579] The defaulting banks were then recapitalized and bad loans sold to so-called asset management companies. Acquired at face value, these bad loans were then issued as government-backed notes to the banks. When they arrived at maturity, these junk bonds were simply rolled over.[580] As a result, many people count on the trick to be played once again. The strong belief that this super-government will help the enfeebled SOEs is bolstered by a $ 2.4 tn. foreign currency reserve.[581]

From a conservative perspective, China appears vulnerable notwithstanding its enormous forex reserve. While a strong forex reserve is helpful for issues of currency exchange and international trade, it cannot help a seriously enfeebled banking system and correct bad investment decisions. It is also interesting to notice that the only two countries which have previously accumulated such large foreign exchange reserves relative to global GDP were the United States in 1929 (Black Thursday) and Japan in 1989 (Japanese housing bubble).[582]

In addition to these overoptimistic projects which were poorly managed, according to a Morgan Stanley report, infrastructure investment (private and public sector merged) accounted for two thirds of the 2008 stimulus spending.[583] In 2009, Chinese fixed asset investment climbed by 30 % and contributed to 90 % of the 2009 economic growth.[584] This staggering figure was confirmed by the National Bureau of Statistics of China on February 2, 2010.[585] Figure 4.26 below shows this incredible growth of the fixed income investment share of GDP in China from 1980 to 2009. As one can see, the growth rate of Chinese GDP seems to be artificially maintained by infrastructure investment in general.

Yet, Chinese newspapers report that real estate investment in the country accounts for 20 % of the country's total investment, and that the sector takes up only a small share of the China's GDP which sounds much less alarming.[586]

Therefore, even though one could feel inclined to assume that based on GDP growth and production output the country was working efficiently, this was, in fact, not exactly the case. Shaken by the crisis, the government discouraged layoffs in non-export sectors, and encouraged massive investments in nearly all sectors through very low interest rates, despite low capacity utilization. No less than 10 tn. renminbi (RMB) ($ 1.5 tn.), i.e., the equivalent of a third of the Chinese annual GDP[587] flooded into the economy. Manufacturing industries (private and public companies), which account for just under a third of total fixed asset investment,

---

[579] Studwell (2003, p. 208).

[580] Chancellor (2010, p. 9).

[581] Bloomberg, as of August 2, 2010.

[582] Eichengreen and Mitchener (2003, p. 83).

[583] Chancellor (2010, p. 9).

[584] Chancellor (2010, p. 5).

[585] Xinhua (2010a).

[586] Xinhua (2010c).

[587] National Bureau of Statistics of China (2013).

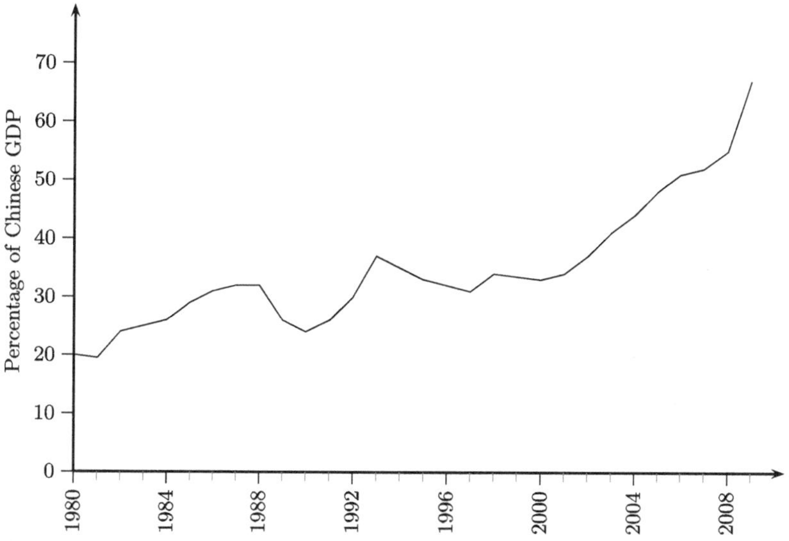

Fig. 4.26 Chinese fixed asset investment share of GDP. *Source*: Chancellor (2010, p. 6)

boosted capital spending at an annualized 27 % rate (from January 2009 to October 2009).[588] As a consequence, the risk of overheating became manifest. For example, a report by the European Chamber in November 2009 identified overcapacities in shipbuilding, flat glass, iron, steel, cement, polycrystalline silicon, and wind power.[589] In the cement industry only, capital expenditures increased by two-thirds despite capacity utilization running at an estimated 78 % of already existing production facilities.[590]

From an economic point of view, investing at a time of overcapacity may have been justified by China's high growth expectations and productivity, but we have shown that these high expectations were hard to maintain. An economic slowdown significantly limited exports and should affect the GDP growth. The SOEs are underperforming entities and political GDP targets lead to poor investment decisions. The capital used by many of these firms hampers the creation of wealth that the country aims for. Their behavior leads to significant moral hazard, increased by an illusion of invincibility thanks to high forex reserves.

To face the crisis, the country flooded the economy with twice as much money as the United States did for its economy. The risk of inflation became very high, and could have destroyed China's GDP growth. Yet, the annual growth objective was artificially maintained thanks to large unrequired investments in numerous industries. One can fairly imagine that such a manipulated growth as it occurred

588 Chancellor (2010, p. 6).

589 European Chamber Report (2010, p. 7).

590 Chancellor (2010, p. 6).

in Spain during the previous decade is perilous for the world's second biggest economy. By the end of 2010, abnormally high rates of return were just not sustainable in the real economy.

Michael Pettis[591] explained on July 29, 2009, that the average growth rate of Chinese GDP should reach a 5–7 % rate in the coming years, contradicting the usual 10–13 % growth rate forecast by the Chinese government.[592] Some analysts, like Marc Faber, thought that economic reports coming out of China were *greatly exaggerated* and he estimated growth at a mere 2 %.[593] Both emphasize that the consumption of American households who had been living on credit for a decade would shrink in the coming years, and grow slower than the GDP.[594] The situation was similar for Europe with its large debt. Expenditures on Chinese products should diminish, and on the other side the productivity of Chinese manufacturers should diminish too, partly because of demographic reasons.

Until 2010, the Chinese economy had lived on cheap labor coming from rural areas.[595] However, while the city population is expected to grow, the Chinese population as a whole is supposed to decrease starting from 2015. The Chinese wages should lose their competitiveness with the passing years, and more investment on more productive machinery (less human labor) is to be expected. In his essay *Myth of Asia's Miracle* from 1994, Nobel-prized economist Paul Krugman suggested that the Asian tiger economies had come to depend on ever-increasing investment inputs.[596]

Such investment, however, was subject to diminishing returns. This point of view was heavily supported by a newly published Federal Reserve report in March 2013 which names several factors that will significantly and adversely affect the Chinese GDP growth by 2030. These factors include an aging population and a declining productivity.[597]

In other words, the implicit rates of return that many investors expected from their investments, for instance in shares, were *overstated compared to achievable goals* and should have been revised starting from 2010. Overinvestment in industries subject to overcapacity, real estate investment on steroids, artificial GDP growth, an aging population and diminishing marginal rates of return for every invested dollar should be sufficient warnings for any investor about the achievable investment goals. The critical point lies in the following fact: who would rely on numbers coming from an emerging country, which is the only one in the world to know its GDP growth rate 3 years in advance?[598] In addition, the People's Republic slipped to 79th place in Transparency International's 2009 Corruption Perceptions Index, just

---

[591]Michael Pettis is an analyst for the Financial Times and specialist for Asian markets.

[592]Pettis (2009).

[593]Harrison (2009).

[594]Pettis (2009).

[595]Chancellor (2010, p. 4).

[596]Krugman (1994, p. 2).

[597]Bull (2013).

[598]Harrison (2009).

below Burkina Faso.[599] The New York Times estimated that half of the sales of luxury goods in China, the fastest growing market in the world in this industry, are given as bribes.[600]

We may assume that the Chinese market has a large potential, but the transition to a developed country or a society of mass consumption is long and extremely hazardous. The Chinese average wage is a tenth of the American average. Up to 2009, the government limited wage increase in order to maintain the competitiveness in the industrial sectors the party chose to focus on. Wage growth was frozen since 2004, even in SOEs,[601] and has stayed below the country's growth rate for the past 10 years. There were more and more riots as citizens demanded better incomes,[602] and feared to lose their purchasing power. Since 2010, wages have rapidly risen in all areas of the country and Chinese leaders were suddenly calling for an increased standard of living.[603] Wages could climb by as much as 30 % from 1 month to another in certain firms.[604] The trend continued even in 2013 with an average expected wage growth of 9 % in the industrial sector[605] which unlike 2008–2009 showed no signs of cooling in the job market.

All this means that China has to take hard decisions which will have huge consequences for its future. The ending of the peg with the U.S. dollar in mid-2010, a steep rise in wages, albeit expected to cool down within the coming years, lead to a high uncertainty about the future growth rate of the country.[606] A lower growth rate may be expected as well as a higher uncertainty about the general economic development of the economy in the next years.

How did the stock markets react to these developments? Figure 4.27 shows the Shanghai Stock Exchange Index from January 2006 to December 2013.

As can be seen in Fig. 4.27, the market rushed through a bubble from beginning 2006 to October 12, 2007, which approximately is the time when the subprime crisis began. The crash, called the *Chinese correction*, lasted approximately 12 months and ended in October 2008. Recall that since China is an emerging market, we speak of a crash if the peak-to-trough movement of the index exceeds 35 %.[607] This was the case since the index slid from 5,903 to 1,729.

However, what happened next is of interest, because the market rebounded from its fall, notably thanks to the rescue package of the People's Republic government.

---

599 Transparency International (2009).

600 Chancellor (2010, p. 7).

601 China Real Estate News (2010).

602 Nguyen (2010).

603 Wong (2010).

604 HSBC Global Research (2010, p. 4).

605 Manek (2013).

606 Wei and Bull (2010) and HSBC Global Research (2010, p. 4).

607 Patel and Sarkar (1998, p. 6).

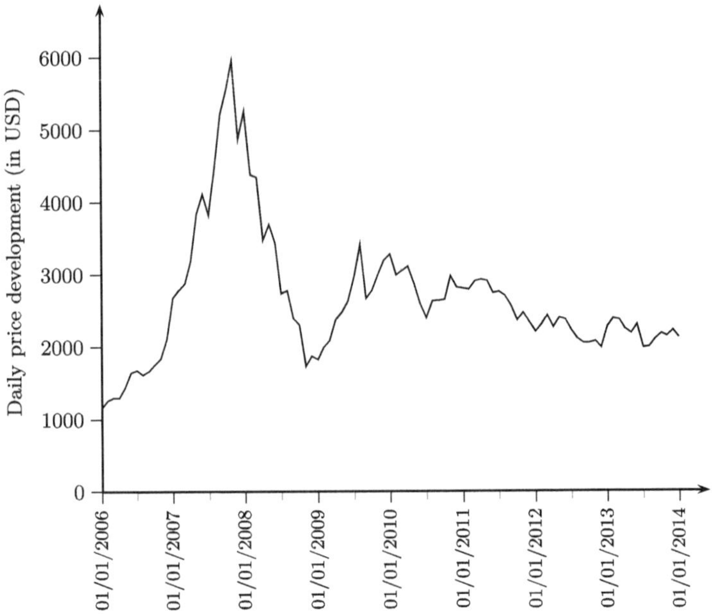

Fig. 4.27 Daily price development of the Shanghai Stock Exchange Composite Index in USD from January 1, 2006 to December 31, 2013. *Source*: Factset (ticker: SHCOMP:IND)

On July 31, 2009, the index reached a new high at 3412.062.[608] While being far below the 2007 peak, it was substantially higher than the long-term trend on this market. Since then, the mania for new IPOs regained strength. For example, when the Agricultural Bank of China was introduced on the stock exchanges of Shanghai and Hong Kong on July 15 and 16, it became the world's largest IPO with a new record of $ 22.1 bn.[609] Yet, the risk of overheating remained strong. In October 2010, many analysts waited for a $ 500 bn. to $ 1 tn. stimulus in the United States. This finally happened beginning November 2010 with a $ 900 bn. stimulus over 2011.[610] Since this American cash inflow would definitely cross the ocean, China being a major commercial partner from the U.S., it justified the rise of the index and bolstered the performance of Chinese large-caps, implying nearly guaranteed turnovers and profit margins thanks to the *quantitative easing II* scheme.[611]

[608] *Source*: Fig. 4.27, Factset.

[609] Anderlini (2010).

[610] Censky (2010).

[611] Quantitative Easing II was a rescue package designed to revitalize the U.S. economy in 2010, the first and third one being launched so as to boost the American economy in 2009 and 2012, respectively. See Censky (2010).

It was possible that the Chinese government would rely on these stimuli to strengthen the economy in the coming year, though this would not be sufficient. The government could also put other industrial titans on the stock exchange to cool down the overheating effect. Finally, in order to achieve this cooling effect and to curb production and limit overcapacity, on June 21, 2010, commercial banks were forced to restrict credit and loans for industries with overcapacity to eliminate backward production capacities, while energy savings and emissions reduction were promoted, according to an official from the China Banking Regulatory Commission (CBRC).[612]

This credit crunch had an effect on real estate, which showed signs of overheating in many cities. In the second half of 2009, and despite higher credit constraints, home prices increased by 7.8 % in 70 large and medium cities in China,[613] compared to the same period in 2008. The top five Chinese cities with the biggest price hikes were Guangzhou (19.9 %), Jinhua (14.9 %), Shenzhen (14.3 %), Haikou (13.4 %) and Beijing (13.2 %).[614] Tangshan, counting 7.1 mn. inhabitants, was the only big city which saw housing prices fall with a modest decline of 2 %.[615] During the same time interval, office building prices increased by an average of 6.4 % compared to December 2008, while business centers increased by 2.7 %.[616] In May 2010, property prices in the country rose 12.4 %, the second-fastest pace on record.[617] Property-market analyst Nicole Wong at brokerage CLSA[618] argued that Hong Kong may inevitably be heading for *another boom and bust* in its real estate sector, due to a combination of tight supply and easy money. *"The answer to the question could there be a property bubble in Hong Kong is yes."*,[619] she said. This trend continued during 2010 with an average 10–13 % growth.[620]

Yet in 2010, the People's Republic government tightened restrictions to credit access made easy thanks to heavy foreign money supply provided during the past years,[621] forbid third home acquisitions,[622] and encouraged the development of

612 Wang (2013).

613 Gerrity (2010).

614 Gerrity (2010).

615 Gerrity (2010).

616 Gerrity (2010).

617 Akkermans (2010).

618 CLSA is Asia's leading independent brokerage and investment group. It provides brokering, investment banking and asset management to corporate and institutional clients around the world.

619 Schuman (2009).

620 Gerrity (2010).

621 Until mid-2010, China pegged its currency to the U.S. dollar, establishing a fixed exchange rate between the two currencies. This was the basis of a large money supply coming from abroad, made even bigger with the declining value of the U.S. dollar on forex markets. This flood of money has granted easy access to money for Chinese citizens for the past years.

622 Akkermans (2010).

the real estate market as the main growth driver of the national GDP.[623] In order to reduce forex reserves and to counter the threat of a rising currency for the Chinese export industry, Wen Jiabao, the People's Republic Prime Minister, signed big investment contracts in Greece[624] and Italy,[625] while Hu Jintao, the Chinese President, did the same in France.[626] In 2013, officials of a southern Chinese province signed a contract with a Ukrainian agricultural firm on 3 mn. hectares of Ukrainian farmland for the next 50 years.[627]

Still, this did not appear to be sufficient. In 2013, China announced it would increase the down-payment requirements and, by 20 %, the taxes on capital gains following the sale of real estate. Individuals were no longer allowed to buy two properties.[628] Real estate consulting firm Knight Frank said that China housing prices rose 9.3 % in 2012, while Hong Kong housing prices rose the most out of all the 55 countries and cities in their Global House Price Index, up 23.6 %.[629] Even worse: According to the National Bureau of Statistics (NBS), real estate prices extended their rally in 62 out of 70 Chinese cities until February 2013,[630] with first-tier cities leading the gains (i.e., Beijing, Hong Kong).[631]

Would the restrictions be enough to contain the bubble? Nothing remains less certain today. The China Banking Regulatory Commission[632] saw growing credit risks in the nation's real estate industry and an increasing pressure from nonperforming loans.[633] This could generate a *chain effect*, endangering the whole industry. The organization also stated that the record 9.59 tn. RMB ($ 1.4 tn.) of loans granted in 2009—almost double of that of the previous year—to combat the effects of the global financial crisis might be causing asset bubbles.[634]

Wall Street analysts did not appear overly concerned by these news, arguing that China's credit system was robust because loans were not securitized as in the U.S.[635] But Chinese banks follow comparable dubious practices. Fitch Ratings

---

623 Gerrity (2010).

624 On October 3, 2010, Prime Minister Wen Jiabao signed 13 investment contracts in the Greek tourism and transportation industry. See Guichard (2010).

625 A € 2.25 bn. agreement ($ 3.07 bn.) was signed on October 7, 2010, concerning solar energy, telecommunications, etc. See Solaro (2010).

626 On November 4–5, commercial agreements amounting for € 20 bn. ($ 27.3 bn.) were signed in aeronautics, nuclear energy, telecommunications, etc. See Lefief and Boucey (2010).

627 Jourdan (2013).

628 Rapoza (2013).

629 Rapoza (2013).

630 The latest available data are as of July 2013. This trend may continue in the subsequent period.

631 Rapoza (2013).

632 The China Banking Regulatory Commission is a regulatory entity meant to enhance security and transparency in the Chinese financial industry by ensuring supervisory and regulatory measures.

633 Akkermans (2010).

634 Akkermans (2010).

635 Chancellor (2010).

calculated that many bank loans were repackaged and sold on to retail investors, other financial institutions and corporations. *"Selling banks"*, wrote Fitch, *"will sometimes enter into a counter agreement to repurchase the loan at a later date. In these instances, the loan may not appear on either the seller's or the buyer's financial statements."* [636]

These off-balance sheet loans allow banks to stay within official lending growth targets.[637] It is thus tricky to evaluate the robustness of the Chinese credit system, notably due to the reluctance of the banks to report on problematic loans. In 2006, Ernst & Young published a report that estimated the total amount of nonperforming loans (NPLs) at $ 900 bn., which at that time already exceeded the total Chinese foreign currency reserves.[638]

To summarize, the outrageous growth in real estate is the symptom of a more problematic situation. China seems to develop what looks to be a speculative bubble in the real estate industry: an index which contains 70 large to medium Chinese cities shows an annual real estate price increase of 7.8 % per annum.[639] This impressive number does not bode well when we take into consideration the small part of the population who is able to buy a property, the rising amount of bad loans, partly encouraged by the Chinese financial stimulus, and the *ghost loans*. However, a bubble might also threaten the stock markets thanks to investors who are overoptimistic about a seemingly manipulated GDP growth and a second quantitative easing plan coming from the United States. Yet, if China's economy is to slow below Beijing's 8 % growth target, complications for the government are liable to arise. Much of the new infrastructure would turn out to be superfluous, excess capacity would burden many industries, the real estate bubble would burst and the banking system would face a serious problem with nonperforming loans and bad quality projects financed for SOEs. Investors who are immersed in this *Chinese Dream* may ignore this scenario, but could face a rude awakening.

4.8 Summary

Stock market crashes are dramatic events for an economy. Among the significant historical crashes presented in this chapter, many had only a regional impact, but the most severe ones had global effects. Bubbles and crashes seem to have happened with an increased frequency during the past three decades. The size of the crashes is related to the development of the global economy. The more wealth is created, the more people want to achieve their goals and fulfill their dreams. However, quite frequently, people are mislead and forget to look at fundamentals like in the times of the dotcom bubble and the subprime crisis. This is obviously a

[636]Chu and Wen (2009).

[637]Chancellor (2010).

[638]E&Y Research Report (2006).

[639]Gerrity (2010).

deviation from modern portfolio theory, which was presented in Chap. 2. In modern portfolio theory, the market participants are assumed to be rational investors (*Homo economicus*) that make their investment decisions using objective quantitative methods. The data would be the fundamentals of firms that lead, for example, to estimations of the expected returns needed for the Markowitz efficient frontier approach when constructing a portfolio.

However, one of the most important lessons of speculative bubbles is that people make investment decisions that are *not* based on fundamental analyses. Often, their investment decisions are driven by human psychology which, in turn, will likely then affect market prices. The dotcom bubble was a typical example of such a behavior.

In consequence, MPT does not help us to explain the behavior of investors in many time periods. Therefore, new explanations have to be found for what happens during such turbulent time periods which seem to get more and longer.

A promising avenue is to take account of psychological factors when looking into the decision-making process of investors. This is the field of behavioral finance which argues that human psychology leads investors to make the same mistakes over and over again, creating the possibility of recurrent and predictable patterns in security prices.[640] We will look closer at this in the next chapter.

Knowing all this, the ultimate question for an investor is: how can I capitalize on and how can I protect myself from heavy drawdowns like we saw in 2001–2002 and 2008? This calls for considering another research area which became prominent after the dotcom crisis: risk management. Therefore, the final chapter, Chap. 6, focuses on the investors and analyzes how they have changed their behavior over the past 10 years, especially regarding the used risk measures and, consequently, the chosen investment strategies.

References

Agence Francaise de Presse. (2010, May 1). *Crise grecque: Sarkozy pour l'action, Merkel pour la sanction*. Liberation. http://www.liberation.fr/economie/2010/05/01/crise-grecque-sarkozy-pour-l-action-merkel-pour-la-sanction_623736. Accessed 29 Dec 2013.

Ahrens, F. (2008, March 19). Moral hazard: Why risk is good. *Washington Post*. http://www.washingtonpost.com/wp-dyn/content/article/2008/03/18/AR2008031802873.html. Accessed 29 Dec 2013.

Akkermans, J. (2010, June 15). *China's bank regulator sees growing real estate risks*. Bloomberg. http://www.bloomberg.com/news/2010-06-15/china-s-bank-regulator-warns-risks-growing-from-real-estate-chain-effect-.html. Accessed 4 Jan 2014.

Anderlini, J. (2010, August 13). AgBank IPO officially the world's biggest. *Financial Times*. http://www.ft.com/intl/cms/s/0/ff7d528c-a6bc-11df-8d1e-00144feabdc0.html?siteedition=intl#axzz2se5bVrDG. Accessed 7 Feb 2014.

Anderson, J., & Heather, T. (2007, August 31). Why a U.S. subprime mortgage crisis is felt around the world. *New York Times*. http://www.nytimes.com/2007/08/31/business/worldbusiness/31derivatives.html?pagewanted=all&_r=0. Accessed 29 Dec 2013.

[640] Chung, Rosenberg, and Tomeo (2004, p. 3).

Arbeter, M. (2007, January 8). Stocks: The Chinese correction? *Bloomberg Businessweek*. http://www.businessweek.com/stories/2007-01-08/stocks-the-chinese-correction-businessweek-business-news-stock-market-and-financial-advice. Accessed 29 Dec 2013.

Arnold, A. J. & McCartney, S. (2004). *George Hudson: The rise and fall of the railway king*. London: Hambledon & London.

Arnold, C. (2007, August 7). Economists brace for worsening subprime crisis. *National Public Radio*. http://www.npr.org/templates/story/story.php?storyId=12561184. Accessed 29 Dec 2013.

Associated Press. (2008a, January 28). New home sales fell by record amount in 2007: Sale prices posted weakest showing in 16 years. *NBC News*. http://www.nbcnews.com/id/22880294/#.UsCcabS24-M. Accessed 29 Dec 2013.

Associated Press. (2008b, October 5). Deal to save Germany's hypo real estate fails. *Jakarta Post*. http://www.thejakartapost.com/news/2008/10/05/deal-save-germany039s-hypo-real-estate-fails.html. Accessed 20 Jan 2014.

Atherton, M. (2009, May 8). The beginner's guide to stock markets. *Times Online*. http://archive.is/xCBNv. Accessed 29 Dec 2013.

Barclays Capital Equity Research. (2010, May 10). *Brokers, asset managers & exchanges: Market commentary/strategy*. New York: Barclays.

Barr, C. (2009, May 27). The $4 trillion housing headache: House prices have returned to 2002 levels, but mortgage debt hasn't deflated from its bubbly highs. *CNN Money*. http://money.cnn.com/2009/05/27/news/mortgage.overhang.fortune/. Accessed 29 Dec 2013.

BEA. (2014, January). *Gross domestic product*. U.S. Bureau of Economic Analysis (BEA). http://www.bea.gov/iTable/iTable.cfm?ReqID=9&step=1#reqid=9&step=3&isuri=1&903=5. Accessed 20 Jan 2014.

Becatoros, E. (2010, April 22). Greek debt crisis deepens; ratings downgraded. *NBC News*. http://www.nbcnews.com/id/36712807/ns/business-world_business#.UsB1fuKET0c. Accessed 29 Dec 2013.

Benett, W. F. (2007, May 6). *Mortgage brokers get fatter payoff for selling riskier loans*. U-T San Diego. http://www.utsandiego.com/news/2007/May/06/mortgage-brokers-get-fatter-payoff-for-selling/. Accessed 29 Dec 2013.

Bernanke, B. (2008, March 14). *Speech from chairman Ben S. Bernanke at the National Community Reinvestment Coalition annual meeting, Washington, DC*. Board of Governors of the Federal Reserve System. http://www.federalreserve.gov/newsevents/speech/bernanke20080314a.htm. Accessed 7 Feb 2014.

Bernard, S. (2008, October 10). Lehman debt auction gives clue to potential losses. *USA Today*. http://usatoday30.usatoday.com/money/economy/2008-10-10-909442932_x.htm. Accessed 29 Dec 2013.

Berteloot, T., & Hebert, D. (2010, April 29). Onze questions-réponses sur la crise grecque. *Le Nouvel Observateur*. http://tempsreel.nouvelobs.com/economie/20100429.OBS3199/onze-questions-reponses-sur-la-crise-grecque.html. Accessed 29 Dec 2013.

Bianchi, S., & Critchlow, A. (2010, January 5). World's tallest skyscraper opens in Dubai. *Wall Street Journal*. http://online.wsj.com/news/articles/SB10001424052748703580904574638111667658806. Accessed 29 Dec 2013.

Bierman, H. (1998). *The causes of the 1929 stock market crash*. Westport: Greenwood Press.

Birger, J. (2008a, August 6). The woman who called Wall Street's meltdown. *CNN Money*. http://money.cnn.com/2008/08/04/magazines/fortune/whitney_feature.fortune. Accessed 29 Dec 2013.

Birger, J. (2008b, January 31). How congress helped create the subprime mess. *CNN Money*. http://money.cnn.com/2008/01/30/real_estate/congress_subprime.fortune. Accessed December 2013.

Blackstone, B. (2010, April 28). Greek crisis poses ECB risk. *Wall Street Journal Online*. http://online.wsj.com/news/articles/SB10001424052748703832204575210334213031028. Accessed 29 Dec 13.

Blitzer, D., & Guarino, D. (2008, November 25). *National trend of home price declines continues through the third quarter of 2008 according to the S&P/Case-Shiller home price indices*. New York: Standard & Poor's.

Boco, H., Germain, L., & Rousseau, F. (2010, April). *When overconfident traders meet feedback traders*. Working Paper, ESC Toulouse.

Bogle, J. C. (2008). Black Monday and black swans. *Financial Analysts Journal, 64*(2), 30–40.

Bright, J. F. (1837). *A history of England*. Longmans: Green and Company.

Brooks, J. (1999). *Once in Golconda: A true drama of Wall Street 1920–1938*. New York: Wiley.

Brown, S. J., Goetzmann, W. N., & Park, J. M. (1998, February). *Hedge funds and the Asian currency crisis of 1997* . Working Paper No. W6427, NYU Stern School of Business, Yale School of Management (International Center for Finance), National Bureau of Economic Research (NBER) and Paradigm Capital Management.

Bull, A. (2013, March 26). Fed study says China's growth could slow sharply by 2030. *Reuters*. http://www.reuters.com/article/2013/03/26/us-usa-fed-china-idUSBRE92P14T20130326. Accessed 29 Dec 2013.

Bush, G. W. (2008, September 24). President Bush's speech to the nation on the economic crisis (transcript). *New Tork Times*. http://www.nytimes.com/2008/09/24/business/economy/24text-bush.html?pagewanted=all&_r=0. Accessed 20 Jan 2014.

Callis, R. R., & Cavanaugh, L. B. (2007, April 27). *Census bureau reports on residential vacancies and homeownership*. Washington, DC: U.S. Department of Commerce News, CB07-56.

Carswell, J. (2002). *The South Sea bubble*. Stroud: Sutton Pub Ltd.

Cassidy, J. (2003). *Dot.con*. New York: Harper Perennial.

Censky, A. (2010, November 3). QE2: Fed pulls the trigger. *CNN Money*. http://money.cnn.com/2010/11/03/news/economy/fed_decision/. Accessed 29 Dec 2013.

Chancellor, E. (2010, March). *China's red flags*. GMO White Paper.

China Real Estate News. (2010, June 17). *Seven years, most working-class salary increase pressure slowly climbing prices*. China Real Estate News.

Chiodo, A. J., & Owyang, M. (2002, November/December). *A case study of a currency crisis: The Russian default of 1998* (Vol. 84, pp. 7–18). The Federal Reserve Bank of St. Louis.

Chrystal, K. A., & Mizen, P. D. (2001, November 12). *Goodhart's law: Its origins, meaning and implications for monetary policy*. Festschrift in honour of Charles Goodhart to be held on 15–16 November 2001 at the Bank of England.

Chu, C., & Wen, C. (2009, December 17). *Chinese banks - Annual review and outlook*. Fitch Ratings. http://fr.scribd.com/doc/56045526/Chinese-Banks-Annual-Review-and-Outlook-1. Accessed 29 Dec 2013.

Chung, S., Rosenberg, M., & Tomeo, J. F. (2004, Summer). Hedge fund of fund allocations using a convergent and divergent strategy approach. *Journal of Alternative Investments, 7*(1), 44–53.

Cline, W. R., & Barnes, K. J. S. (1997). *Spreads and risk in emerging market lending*. Institute of International Finance, Inc. (IIF Research Paper No. 97-1, Washington, DC).

CountryMeters. (2010, October 6). China. CountryMeters Online. http://countrymeters.info/en/China/. Accessed 18 Feb 2014.

Cox, J. (2010, May 6). Greek debt crisis on verge of "Going Global": Pimco's El-Erian. *CNBC*. http://www.cnbc.com/id/36992469/Greek_Debt_Crisis_On_Verge_Of_Going_Global_Pimco_s_El_Erian. Accessed 3 Jan 2014.

Coy, P. (2008a, January 30). Housing meltdown. *Bloomberg Businessweek*. http://www.businessweek.com/stories/2008-01-30/housing-meltdown. Accessed 3 Jan 2014.

Coy, P. (2008b, January 31). Housing meltdown: Why home prices could drop 25% more on average before the market finally hits bottom. *Bloomberg Businessweek*. http://www.businessweek.com/stories/2008-01-30/housing-meltdown-why-home-prices-could-drop-25-percent-more-on-average-before-the-market-finally-hits-bottom. Accessed 20 Jan 2014.

Daily Telegraph. (2010, April 28). Greek debt crisis spreading "like Ebola" and Europe must act now, OECD warns. *Daily Telegraph*. http://www.telegraph.co.uk/finance/financialcrisis/7644709/Greek-debt-crisis-spreading-like-Ebola-and-Europe-must-act-now-OECD-warns.html. Accessed 20 Jan 2014.

Darwiche, F. A. (1986). *The gulf stock exchange crash: The rise and fall of the Souq Al-Manakh*. London: Croom Helm Ltd.

Derman, E. (2004). *My life as a quant*. Hoboken: Wiley.

Desai, P. (2000). Why did the ruble collapse in august 1998? *American Economic Review, 90*(2), 48–52.

Driver, S., & Martell, L. (2002). *Blair's britain* (1st ed.). Malden: Blackwell Publishing.

Duarte, J. (2006). *Futures and options for dummies*. Hoboken: Wiley Publishing.

Dungey, M., Fry, R., Gonzalez-Hermosillo, B., & Martin, V. (2002, April). *International contagion effects from the Russian crisis and the LTCM near-collapse*. International Monetary Fund Working Paper, IMF Institute.

Dunmore, P. V. (1991, March). *A test of the effects of changing information asymmetry in a capital market*. Working Paper, Working Paper Series 3/91, Graduate School of Business and Government Management, Faculty of Commerce and Administration, Victoria University of Wellington.

Eatwell, J., Milgate, M., & Newmann, P. (1987). *The new palgrave - A dictionnary of political economics* (1st ed.). Basingstoke: Palgrave Macmillan. http://www.palgraveconnect.com/pc/doifinder/10.1057/9780230279803. Accessed 10 Feb 2014.

Eccleston, B., Dawson, M., & McNamara, D. J. (1998). *The Asia-Pacific profile* (1st ed.). Routledge: Chapman & Hall.

Economist Intelligence Unit. (2012). *Managing investments in volatile markets*. State Street Global Advisors. http://www.ssgainsight.com/reports/Economist_Managing_Tail_Risk_Report_pr.pdf. Accessed 8 Feb 2014.

Economist Online. (2000, April 6). Monopoly money. *The Economist*. http://www.economist.com/node/300621. Accessed 4 Jan 2014.

Economist Online. (2007, September 6). Credit and blame. *The Economist*. http://www.economist.com/node/9769471. Accessed 4 Jan 2014.

Economist Online. (2008, November 20). The end of the affair. *The Economist*. http://www.economist.com/node/12637090. Accessed 3 Jan 2014.

Edmans, A., Garcia, D., & Norli, Ø. (2007). Sports sentiment and stock returns. *Journal of Finance, 62*(4), 1967–1998.

Edwards, S. (2000). *Capital flows and the emerging economies: Theory, evidence, and controversies* (1st ed.). Chicago: University of Chicago Press.

Eichengreen, B., & Mitchener, K. (2003, September). *The great depression as a credit boom gone wrong*. Bank of International Settlement (BIS), Working Papers, No. 137, Monetary and Economic Department. http://citeseerx.ist.psu.edu/viewdoc/summary?doi=10.1.1.384.3306. Accessed 11 Feb 2014.

Elliott, L. (2009, January 8). Quantitative easing. *The Guardian*.

European Chamber Report. (2010, March 29). *Overcapacity in China: Causes, impacts and recommendations*. European Chamber.

European Economy Series. (2009, September 14). *Economic crisis in Europe: Causes, consequences and responses*. European Economy Series 7/2009, ISSN 0379-0991. http://ec.europa.eu/economy_finance/publications/publication15887_en.pdf. Accessed 18 Feb 2014.

E&Y Research Report. (2006, May 3). *Global nonperforming loan report*. Singapore: Ernst & Young.

Faber, M. (2002). *Tomorrow's gold - Asia's age of discovery*. Hong Kong: CLSA Ltd.

Fabozzi, F. J. (2002). *Fixed income securities* (2nd ed.). New York: Wiley.

Fama, E. F. (1970). Efficient capital markets: A review of theory and empirical work. *Journal of Finance, 25*(2), 383–417.

Federal Reserve Bank. (2014a, February). *Open market operations archive*. Board of Governors of the Federal Reserve System. http://www.federalreserve.gov/monetarypolicy/openmarket_archive.htm. Accessed 24 Aug 2014.

Federal Reserve Bank. (2014b, February). *Selected interest rate (daily) - H.15*. Board of Governors of the Federal Reserve System. http://www.federalreserve.gov/releases/h15/data.htm. Accessed 18 Feb 2014.

Federal Reserve Report. (2010a, September 17). *Z.1 - Historical tables: L.124 - agency- and GSE (Government-Sponsored Enterprises)-backed mortgage pools 1*. Federal Reserve Bank. http://www.federalreserve.gov/releases/z1/current/accessible/l124.htm. Accessed 10 Feb 2014.

Federal Reserve Report. (2010b, September 17). *Z.1 - Historical tables: L.125 - issuers of Asset-Backed Securities ABS*. Federal Reserve Bank. http://www.federalreserve.gov/releases/z1/current/accessible/l125.htm. Accessed 10 Feb 2014.

Federal Reserve Report. (2010c, September 17). *Z.1 Historical tables: B.100 balance sheet of households and nonprofit organizations*. Federal Reserve Bank. http://www.federalreserve.gov/releases/z1/current/acces-sible/b100.htm. Accessed 10 Feb 2014.

Feldstein, M. (2008, November 18). How to help people whose home values are underwater: The economic spiral will get worse unless we do something about negative equity. *Wall Street Journal*. http://online.wsj.com/news/articles/SB122697004441035727. Accessed 10 Feb 2014.

Felsenthal, M. (2007, September 13). Greenspan says didn't see subprime storm brewing. *Reuters*. http://www.reuters.com/article/2007/09/13/us-usa-economy-greenspan-idUSWBT00756820070913. Accessed 3 Jan 2014.

Ferri, G., & Liu, L. -G. (2009, April). *Honor thy creditors beforan thy shareholders: Are the profits of Chinese state-owned enterprises real?*. Hong Kong Institute for Monetary Research, HKIMR Working Paper No.16/2009, Hong Kong Monetary Authority.

Financial Times. (1995). *Survey* Kuwait: Financial Times Ltd.

Financial Times. (1998, April 22). This is unquestionably a bubble. *Financial Times*.

Fitch, A. (2010, March 13). The Souk al Manakh: Kuwait's first big bubble. *The National*. http://www.thenational.ae/business/the-souk-al-manakh-kuwaits-first-big-bubble. Accessed 4 Jan 2014.

France 24. (2010, May 10). The Greek crisis explained. *France 24*. http://m.france24.com/en/20100504-greek-crisis-eu-imf-bailout-debt-economy-germany-france-unions-jobs-euro-markets. Accessed 20 Jan 2014.

Frank, N., González-Hermosillo, B., & Hesse, H. (2008, August). *Transmission of liquidity shocks: Evidence from the 2007 subprime crisis*. IMF Working Paper, WP/08/200, International Monetary Fund.

Galbraith, J. K. (1979). The great crash: Keeping the memory green "is a useful service." *Journal of Portfolio Management, 6*(1), 60–62.

Galbraith, J. K. (1997). *The great crash 1929* (1st ed.). Boston: Mariner Books.

Gapper, J. (2009, November 26). Dubai's financial crash mirrors that of Florida. *Financial Times*. http://blogs.ft.com/businessblog/2009/11/dubais-financial-crash-mirrors-that-of-florida/. Accessed 4 Jan 2014.

Garber, P. M. (2001). *Famous first bubbles: The fundamentals of early manias*. Cambridge: MIT Press.

Gaunt, J. (2010, May 7). Global markets weekahead: Greek crisis goes viral. *Reuters*. http://www.reuters.com/article/2010/05/07/markets-weekahead-idUSLDE6461MM20100507. Accessed 4 Jan 2014.

Gerrity, M. (2010, March 8). Despite growth, 85% of Chinese households cannot afford to buy real estate. *World Property Channel WPC*. http://www.worldpropertychannel.com/international-markets/residential-real-estate/real-estate-news-china-real-estate-sales-optim-consult-china-real-estate-bubble-chinese-housing-market-russian-real-estate-investors-2146.php. Accessed 20 Jan 2014.

Glass, I. (2008, May 9). Giant pool of money. *Chicago Public Media - This American Life Broadcast*. http://www.thisamericanlife.org/radio-archives/episode/355/the-giant-pool-of-money. Accessed 29 Dec 2013.

Greenberg, D. (2006). *Calvin coolidge* (1st ed.). New York: Times Books.

Greenwald, D. (1983). *The concise McGraw-Hill dictionary of modern economics* (3rd ed.). New York: McGraw-Hill.

Greenwood, J. (2009, November 30). *Market commentary: Dubai's debt crisis*. Kansas City: Invesco.

Guichard, G. (2010, October 3). Regain d'intérêt de la Chine pour la Grèce. *Le Figaro*. http://www.lefigaro.fr/conjoncture/2010/10/03/04016-20101003ARTFIG00077-regain-d-interet-de-la-chine-pour-la-grece.php. Accessed 4 Jan 2014.

Hansard Database. (1825, June 2). *Repeal of the Bubble Act, HC Deb* (Vol. 13, pp. 1018–1023). http://hansard.millbanksystems.com/commons/1825/jun/02/repeal-of-the-bubble-act.

Harrison, E. (2009, July 30). Marc Faber: China's numbers are fake. *Credit Writedowns*. http://www.creditwritedowns.com/2009/07/marc-faber-chinas-numbers-are-fake.html. Accessed 4 Jan 2014.

Hasan, Z. (2010, July–September). Dubai financial crisis: Causes, bailout and after - a case study. *Journal of Islamic Banking & Finance, 27*(3), 47–55. http://mpra.ub.uni-muenchen.de/26397/. Accessed 9 Feb 2010.

Hassan, K. M., Al-Sultan, W. S., & Al-Saleem, J. A. (2003). Stock market efficiency in the gulf cooperation council countries (GCC): The case of Kuwait stock exchange. *Scientific Journal of Administrative Development, 1*(1), 1–21.

Headley, P. (2002). *Big trends in trading: Strategies to master major market moves*. New York: Wiley.

Hernandez, J. C. (2009, November 27). Dubai's investment fund crisis unnerves investors for a second day. *New York Times*. http://www.nytimes.com/2009/11/28/business/28markets.html?_r=1&. Accessed 4 Jan 2014.

Hosking, P., & Robertson, D. (2009, November 27). Dubai in deep water as ripples from debt crisis spread. *The Times*. http://www.thetimes.co.uk/tto/business/markets/gulf/article1892198.ece. Accessed 4 Jan 2014.

HSBC Global Research. (2010, June 7). *China economic spotlight*. London: HSBC.

Huang, Y. (2008). *Capitalism with Chinese characteristics*. Cambridge: Cambridge University Press.

Hureaux, R. (2010, May 1). *Crise grecque: Le grand concours de mensonges européen*. Marianne. http://www.marianne.net/Crise-grecque-le-grand-concours-de-mensonges-europeen_a192302.html. Accessed 4 Jan 2014.

Hutchinson, M. (2010, February 10). The unequal credit crunch. *Asia Times Online*. http://atimes.com/atimes/Global_Economy/LB10Dj01.html. Accessed 4 Jan 2014.

International Monetary Fund. (1998). Section IV - financial crises: Characteristics and indicators of vulnerability. In *IMF world economic outlook May 1998* (pp. 74–97). Washington, DC: International Monetary Fund (IMF).

Irish, J. (2010, January 4). Dubai's Emaar to shift focus after tower opening. *Reuters*. http://in.reuters.com/article/2010/01/04/dubai-emaar-idINLDE6031NX20100104. Accessed 4 Jan 2014.

Ivry, J. B. (2007, December 31). Savannah cries over bicycle lost after subprime reset (Correct). *Bloomberg*. http://www.bloomberg.com/apps/news?pid=newsarchive&sid=asc7NT7OKecE. Accessed 4 Jan 2014.

Jackman, W. T. (1916). *The development of transportation in modern England*. Cambridge: The University Press.

Jourdan, A. (2013, September 22). China to invest in 3 mln hectares of Ukrainian farmland. *Reuters*. http://www.reuters.com/article/2013/09/22/china-ukraine-idUSL3N0HI04620130922. Accessed 4 Jan 2014.

Kaminsky, G. L., & Schmuklerb, S. L. (1999). What triggers market jitters? A chronicle of the Asian crisis. *Journal of International Money and Finance, 18*(4), 537–560.

Kaplan Schweser. (2012). *FRM part I book I: Foundations of risk management* (1st ed.). La Crosse: Kaplan Schweser.

Karunaratne, N. D. (1999, January/February). The Asian miracle and crisis: Rival theories, the IMF bailout and policy lessons. *Intereconomics, 34*(1), 19–26.

Kaufman, G. G. (1999, November 15). *Banking and currency crises and systemic risk: A taxonomy and review*. Working Paper, Loyola University Chicago and Federal Reserve Bank of Chicago.

Kaufman, G. G. (2000, May). Banking and currency crises and systemic risk: A taxonomy and review. *Financial Markets, Institutions & Instruments, 9*(2), 69–131.

Kennedy, S. (2009, November 27). UBS: Support for Dubai may be less than assumed. *Wall Street Journal*. http://www.marketwatch.com/story/ubs-support-for-dubai-may-be-less-than-assumed-2009-11-27. Accessed 4 Jan 2014.

Kennedy, S., & Ross-Thomas, E. (2010, April 28). Greek debt contagion presses EU to broaden bailout. *Bloomberg*. http://www.bloomberg.com/news/2010-04-27/greece-s-junk-contagion-pressures-eu-to-broaden-bailout-after-market-rout.html. Accessed 4 Jan 2014.

Ketchum, R. G. (2014, February 6). *Speech transcription from Richard G. Ketchum, chairman and chief executive officer of the financial industry regulatory authority (FINRA)*. FINRA online, speech for National Association of College and University Business Officers. http://www.finra.org/Newsroom/Speeches/Ketchum/P443489. Accessed 18 Feb 2014.

Khor, K. P., & Khor, M. (2001). *Rethinking globalization: Critical issues and policy choices*. London: Zed Books.

Kindleberger, C. P., & Aliber, R. Z. (2005). *Maniacs, panics and crashes* (5th ed.). Basingstoke: Palgrave MacMillan.

Kirchhoff, S., & Keen, J. (2007, April 25). Minorities hit hard by rising costs of subprime loans. *USA Today*. http://usatoday30.usatoday.com/money/economy/housing/2007-04-25-subprime-minorities-usat_N.htm. Accessed 4 Jan 2014.

Kleinert, A. (2009). Der messende Luchs. Zwei verbreitete Fehler in der Galilei-Literatur. *N.T.M. Zeitschrift für Geschichte der Wissenschaften, Technik und Medizin, 17*, 199–206.

Knox, N. (2006, January 17). 43% of first-time home buyers put no money down. *USA Today*. http://usatoday30.usatoday.com/money/perfi/housing/2006-01-17-real-estate-usat_x.htm. Accessed 4 Jan 2014.

Krudy, E. (2010, June 11). Is the May 6 "flash crash" in U.S. markets the new normal? *Reuters*. http://in.reuters.com/article/2010/06/10/idINIndia-49214020100610. Accessed 4 Jan 2014.

Krugman, P. R. (1994). The myth of Asia's miracle. *Foreign Affairs, 73*(6), 62–78.

Krugman, P. R. (2000). *Currency crisis* (1st ed.). Chicago: University of Chicago Press.

Krugman, P. R. (2009). *The return of depression economics and the crisis of 2008* (2nd ed.). New York: W.W. Norton & Company

Labaton, S. (2008, October 2). Agency's '04 rule let banks pile up new debt. *New York Times*. http://www.nytimes.com/2008/10/03/business/03sec.html?pagewanted=all&_r=0. Accessed 4 Jan 2014.

Lefief, J. -P., & Boucey, B. (2010, November 6). Après la France, le Portugal courtise Hu Jintao. *L'Express*. http://www.lexpress.fr/actualites/2/apres-la-france-le-portugal-courtise-hu-jintao_934495.html. Accessed 4 Jan 2014.

Leonnig, C. D. (2008, June 10). How HUD mortgage policy fed the crisis. *Washington Post*. http://www.washingtonpost.com/wp-dyn/content/article/2008/06/09/AR2008060902626.html. Accessed 4 Jan 2014.

Liebowitz, S. (2009, July 3). New evidence on the foreclosure crisis. *Wall Street Journal*. http://online.wsj.com/news/articles/SB124657539489189043. 4 Jan 2014.

Lowenstein, R. (2002). *When genius failed: The rise and fall of long-term capital management*. London: Fourth Estate.

Mackay, C. (1995). *Extraordinary popular delusions and the madness of crowds*. Ware: Wordsworth Editions Limited.

Mahier, C., & Duval, G. (2010, March 25). Grèce: Crise et conséquences. *Alternative Economique*. http://www.alternatives-economiques.fr/grece---crise-et-consequences_fr_art_633_48622.html. Accessed 8 Feb 2014.

Malkiel, B. G. (2003). *A random walk down Wall Street* (8th ed.). New York: W.W. Norton & Company.

Mandelbrot, B. B., & Hudson, R.L. (2004). *The (mis)behavior of markets*. London: Profile Books.

Manek, N. (2013, March 14). China: Factory wages to rise 9%, survey. *Financial Times*. http://blogs.ft.com/beyond-brics/2013/03/14/china-factory-wages-to-rise-9-survey/#axzz2pLoOFMQX. Accessed 4 Jan 2014.

McLaughlin, T. (2010, January 4). MIT on track despite $1.8B investment loss. *Boston Business Journal*. http://www.bizjournals.com/boston/stories/2010/01/04/daily4.html?page=all. Accessed 4 Jan 2014.

Mengle, D. (2007). Credit derivatives: An overview. *Federal Reserve Bank of Atlanta, Economic Review* (Fourth Quarter), 1–24.

Ministry of France. (2010, January 15). *Update of the hellenic stability and growth programme*. Ministry of France, Ref. Ares (2010) 23198. http://ec.europa.eu/economy_finance/sgp/pdf/20_scps/2009-10/01_programme/el_2010-01-15_sp_en.pdf. Accessed 18 Feb 2014.

Moore, M. (2010, March 15). Britain warns China against protectionism. *Daily Telegraph*. http://www.telegraph.co.uk/finance/china-business/7448633/Britain-warns-China-against-protectionism.html. Accessed 4 Jan 2014.

Moreno, R. (1998, August 7). *What caused East Asia's financial crisis?* Federal Reserve Bank of San Francisco, FRBSF Economic Letter 1998-24. http://www.frbsf.org/economic-research/publications/economic-letter/1998/august/what-caused-east-asia-financial-crisis/. Accessed 8 Feb 2014.

Mucha, T. (2010, May 5, updated May 30). Greek debt crisis: The revolt begins? *Global Post*. http://www.globalpost.com/dispatch/commerce/100505/greek-debt-crisis-unrest. Accessed 3 Jan 2014.

National Bureau of Statistics of China. (2013, February). Annual statistical data. *National Bureau of Statistics of China Online*. http://www.stats.gov.cn/english/statisticaldata/AnnualData/. Accessed 18 Feb 2014.

New York Times. (2007, December 17). The tightening credit market: A chronology of warnings ignored. *New York Times*. http://www.nytimes.com/imagepages/2007/12/17/business/20071218_SUBPRIME.html. Accessed 4 Jan 2014.

Nguyen, A. (2010, June 24). Ford plans $450 million thai plant after anti-government riots. *Bloomberg*. http://www.bloomberg.com/news/2010-06-24/ford-plans-450-million-factory-in-thailand-after-anti-government-riots.html. Accessed 3 Jan 2014.

Norris, F. (1995, December 5). Market place; last big Wall Street bear defects to optimists. *New York Times*. http://www.nytimes.com/1995/12/05/business/market-place-last-big-wall-st-bear-defects-to-optimists.html. Accessed 3 Jan 14.

NTSB. (1967). *Aviation: Accident synopsis by month*. National Transportation Safety Board. http://www.ntsb.gov/aviationquery/Month.aspx. Accessed 29 Dec 2013.

Odlyzko, A. (2010, January 15). *Collective hallucinations and inefficient markets: The British railway mania of the 1840s*. School of Mathematics and Digital Technology Center, University of Minnesota, Preliminary Version.

Ofek, E., & Richardson, M. (2003). DotCom mania: The rise and fall of internet stock prices. *Journal of Finance, 58*(3), 1113–1138.

Office of the Press Secretary. (2002, October 15). President hosts conference on minority homeownership. *U.S. Census Bureau News - Office of the Press Secretary*. http://georgewbush-whitehouse.archives.gov/news/releases/2002/10/20021015-7.html. Accessed 10 Feb 2014.

Oosterveld, B., & Carlson, S. (2010, June 14). Moody's downgrades Greece to Ba1 from A3, stable outlook. *Moody's*. https://www.moodys.com/research/Moodys-downgrades-Greece-to-Ba1-from-A3-stable-outlook--PR_200910. Accessed 4 Jan 2014.

Page, J. (1994). The East Asian miracle: Four lessons for development policy. In S. Fischer & J. J. Rotemberg (Eds.), *National bureau of economic research (NBER) macroeconomics annual* (Vol. 9, pp. 219–282) Cambridge: MIT Press.

Palma, G. (2000, November). *The three routes to financial crises: The need for capital controls*. CEPA Working Paper Series III, Working Paper number 18, Center for Economic Policy Analysis, Cambridge University.

Park, Y. C. (1996). *East Asian liberalization, bubbles, and the challenge from China*. Brookings Papers on Economic Activity (Vol. 2, pp. 357–371).

Parnell, H. B. (1833). *A treatise on roads*. Harlow: Longman, Rees, Orme, Brown, Green and Longmans. Reprint by ULAN Press.

Pasha, S. (2005, August 8). Banking on illegal immigrants. *CNN Money*. http://money.cnn.com/2005/08/08/news/economy/illegal_immigrants/. Accessed 3 Jan Feb 14.

Patel, S., & Sarkar, A. (1998, April). *Stock market crises in developed and emerging markets*. Research Papers Series, No. 9809. New York: Federal Reserve Bank of New York.

Paulson, H. M., Jr. (2010). *On the brink*. New York: Business Plus.

Pearlstein, S. (2010, May 7). Greek crisis exposes cracks in Europe's foudation. *Washington Post*. http://www.washingtonpost.com/wp-dyn/content/article/2010/05/06/AR2010050606375.html. Accessed 3 Jan 2014.

Perron, P. (1989). The great crash, the oil price shock and the unit root hypothesis. *Econometrica, 57*(6), 1361–1401.

Pettis, M. (2009, July 29). Get ready for lower Chinese growth. *Financial Times*. http://www.ft.com/intl/cms/s/0/ef4bce88-7d38-11de-b8ee-00144feabdc0.html#axzz2pGD3bLGO. Accessed 4 Jan 2014.

Pinto, J. E., Henry, E., Robinson, T. R., & Stowe, J.D. (2007). *Equity asset valuation*, Hoboken: Wiley.

Quinn, J. (2010, January 22). Souk al-Manakh's omens for U.S. financial crisis. *Rapid City Journal*. http://rapidcityjournal.com/business/columnists/john-quinn-souk-al-manakh-s-omens-for-u-s/article_61e42d70-06df-11df-8545-001cc4c002e0.html. Accessed 3 Jan 2014.

Rapoza, K. (2013, March 30). China takes another stab at housing bubble. *Forbes*. http://www.forbes.com/sites/kenrapoza/2013/03/30/china-takes-another-stab-at-housing-bubble/. Accessed 3 Jan 2014.

Rascouet, A. (2009, December 22). Greece's credit rating cut to A2 by moody's on debt (update 5). *Bloomberg*. http://www.bloomberg.com/apps/news?pid=newsarchive&sid=ak5z0mcuutZ8. Accessed 3 Jan 2014.

Rasmaroni, W. K. (2006, May 29). Capital markets are about more than market-cap. *Daily News Egypt*. http://www.masress.com/en/dailynews/105232. Accessed 3 Jan 2014.

Reserve Bank of India. (2010, July). Genesis and the nature of the crisis. *Reserve bank of India report on currency and finance* (pp. 7–58).

Rice, C. T. (2010). *Lessons from the "Flash Crash"*. State Street Global Advisors.

Ritter, J. R. (2010, February 6). *IPO data*. University of Florida. http://bear.warrington.ufl.edu/ritter/ipodata.htm. Accessed 3 Jan 2014.

Russell, C., & Mullin, S. (2008, November 11). *Statement of FHFA director James B. Lockhart*. Federal Housing Finance Agency. http://www.fhfa.gov/webfiles/23/FHFAStatement9708final.pdf. Accessed 3 Jan 2014.

Sakoui, A. (2010, May 17). Fears for sovereign debt refinancing. *Financial Times*. http://www.ft.com/intl/cms/s/0/0aa3eee6-61da-11df-998c-00144feab49a.html#axzz2pGD3bLGO. Accessed 4 Jan 2014.

Sakoui, A., & Kerr, S. (2010, July 11). Dubai world seeks creditors' support. *Financial Times*. http://www.ft.com/intl/cms/s/0/dea93ef8-8d0e-11df-bad7-00144feab49a.html#axzz2pGD3bLGO. Accessed 4 Jan 2014.

Salama, V. (2010, July 25). Abu Dhabi feels Dubai chill as Emirate accepts money is scarcer. *Bloomberg*. http://www.bloomberg.com/news/2010-07-25/abu-dhabi-feels-dubai-desert-chill-as-emirate-accepts-money-is-now-scarcer.html. Accessed 4 Jan 2014.

Salsman, R. M. (2004, June). The cause and consequences of the great depression, part 1: What made the roaring '20s roar. *The Intellectual Activist*. http://www.worldcat.org/search?fq=x0:jrnl&q=n2:0730-2355. Accessed 3 Jan 2014.

Schulmerich, M. (2003). Einsatz und Pricing von Realoptionen: Einführung in grundlegende Bewertungsansätze. In U. Hommel, M. Scholich, & P. N. Baecker (Eds.), *Reale optionen* (pp. 63–96). Berlin: Springer.

Schulmerich, M. (2010a). *The eight-minute, $700 billion "Flash Crash": Explanations and implications*. State Street Global Advisors. SSgA Capital Insights.

Schulmerich, M. (2010b). *Stock market crashes and financial bubbles: What history tells us*. State Street Global Advisors. SSgA Capital Insights.

Schulmerich, M. (2012). Creating corporate value with the exposure to financial innovations: The case of interest rates. In U. Hommel, M. Fabich, E. Schellenberg, & L. Firnkorn (Eds.), *The strategic CFO* (pp. 295–315). Berlin: Springer.

Schuman, M. (2009, November 30). The lesson of Dubai: The crisis is not over. *Time*. http://content.time.com/time/business/article/0,8599,1943392,00.html. Accessed 3 Jan 2014.

Securities and Exchange Commission. (2011, September 27). *SEC to publish for public comment updated market-wide circuit breaker proposals to address extraordinary Market volatility*, U.S. Securities and Exchange Commission, online release 2011-190. https://www.sec.gov/news/press/2011/2011-190.htm.

Sheeran, P., & Spain, A. (2004). *The international political economy of investment bubbles*. Burlington: Ashgate Publishing.

Shen, I. (2010, February 10). 'Invincible' high-speed trains steal China southern's customers. *Bloomberg*. http://www.bloomberg.com/apps/news?pid=newsarchive&sid=aH_iVz7ir970&pos=7. Accessed 3 Jan 2014.

Shiller, R. J. (2005). *Irrational exuberance* (2nd ed.). Princeton: Princeton University Press.

Shiller, R. J. (2008). *The subprime solution*. Princeton: Princeton University Press.

Shleifer, A., & Treisman, D. (2001). *Without a map: Political tactics and economic reform in Russia*. (1st ed.). Cambridge: MIT Press

Sobel, R. (1999). *Panic on Wall Street: A history of America's financial disasters*, Washington, DC: Beard Books.

Solaro, A. (2010, October 7). Accord commercial de plus de 2 milliards d'euros entre la Chine et l'Italie. *Le Monde & AFP*. http://www.lavieeco.com/depeches/l-italie-et-la-chine-signent-pour-3-2-milliards-de-dollars-d-accords-commerciaux-newsmlmmd.6e8069c39c8696b270bf7ec61f56f130.5a1.xml.html. Accessed 20 Jan 2014.

Spain, W. (2009, November 30). Dubai tourism already defaulting: Economic downturn, overcapacity bite into once-hot travel destination. *Wall Street Journal*. http://www.marketwatch.com/story/dubai-tourism-already-defaulting-2009-11-30. Accessed 9 Feb 2014.

Spencer, R. (2009, November 27). Dubai's financial crisis: A Q&A. *U.K. Telegraph*. http://www.telegraph.co.uk/finance/financialcrisis/6668281/Dubais-financial-crisis-a-QandA.html. Accessed 3 Jan 2014.

Spiegel Online. (2010, May 29). EZB kauft griechische Staatsanleihen: Bundesbanker vermuten französisches Komplott. *Der Spiegel*. http://www.spiegel.de/wirtschaft/soziales/ezb-kauft-griechische-staatsanleihen-bundesbanker-vermuten-franzoesisches-komplott-a-697489.html. Accessed January 2014.

Steverman, B., & Bogoslaw, D. (2008, October 18). The financial crisis blame game. *Bloomberg Businessweek*. http://www.businessweek.com/stories/2008-10-18/the-financial-crisis-blame-gamebusinessweek-business-news-stock-market-and-financial-advice. Accessed 4 Jan 2014.

Stiglitz, J. (2003, April 9). The ruin of Russia. *The Guardian*. http://www.theguardian.com/world/2003/apr/09/russia.artsandhumanities. Accessed January 2014.

Stout, D. (2008a, February 20). Justices say 401(k) participants can Sue. *New York Times*. http://www.nytimes.com/2008/02/20/washington/20cnd-bizcourt.html?_r=2&oref=slogin&. Accessed 4 Jan 2014.

Stout, D. (2008b, September 20). The Wall Street bailout plan, explained. *New York Times*. http://www.nytimes.com/2008/09/21/business/21qanda.html?adxnnl=1&adxnnlx=1388772534-0B1f06xCrLbwsVkMEdhduw. Accessed 4 Jan 2014.

Studwell, J. (2003). *The China dream: The elusive quest for the last great untapped market on Earth*. New York: Grove Press.

Thompson, E. (2006). The tulipmania: Fact or artifact? *Public Choice, 130*(1–2), 99–114.

Transparency International. (2009). Corruptions perception index 2009. *Transparency International Online*. http://www.transparency.org/research/cpi/cpi_2009. Accessed 4 Jan 2014.

TreasuryDirect Reports. (2014). The debt to the penny and who holds it. *TreasuryDirect*. http://www.treasurydirect.gov/NP/debt/current. Accessed 18 Feb 2014.

Tupy, B. (2010, February 13). Opinion: Greek debt crisis a hazard for EU. *Global Post*. http://www.globalpost.com/dispatch/worldview/100213/opinion-greek-crisis-paves-way-more-regulation. Accessed 3 Jan 2014.

Tweed, D., & Eglitis, A. (2010, August 4). Dombrovskis says Latvia's austerity program may ease as economy recovers. *Bloomberg*. http://www.bloomberg.com/news/2010-08-04/dombrovskis-says-latvia-s-austerity-program-may-ease-as-economy-recovers.html. Accessed 3 Jan 2014.

Uhlfelder, E. (2009, February 10). *Son of sub-prime: Another wave of defaults*. Advisor Perspective. http://www.advisorperspectives.com/newsletters09/pdfs/Son_of_Sub-Prime_Another_Wave_of_Defaults.pdf. Accessed 4 Jan 2014.

Underwood, H. (2009, March 25). Silver Thursday and the fall of the bunker hunt, the world's richest man. *The Week*. http://www.theweek.co.uk/politics/24173/silver-thursday-and-fall-bunker-hunt-worlds-richest-man. Accessed 4 Jan 2014.

Veneroso, F. (1998, June 24). Memories of the Souk al Manakh. *The Gold Newsletter*. http://www.24hgold.com/english/contributor.aspx?article=657875950G10020&contributor=Panics,+Manias+and+crashes. Accessed 4 Jan 2014.

Vickers, R. B. (1994). *Panic in paradise: Florida's banking crash of 1926*. Tuscaloosa: University of Alabama Press.

von Mises, L. (1966). *Human action: A treatise on economics*. Chicago: Regnery.

Wallison, P. J. (2011). *Dissent from the majority report of the financial crisis inquiry commission*. Washington, DC: American Enterprise Institute.

Wang, Y. (2013, May 11). China determined to cut industrial overcapacity. *CCTV*. http://english.cntv.cn/program/bizasia/20131105/103971.shtml. Accessed 4 Jan 2014.

Wei, M., & Bull, A. (2010, June 19). Peg is dead as China vows yuan flexibility before G20. *Reuters*. http://www.reuters.com/article/2010/06/19/us-china-yuan-idUSTRE65I11B20100619. Accessed 3 Jan 2014.

Westbrook, J. (2010, May 8). SEC said to consider new rules after market plunge. *Bloomberg*. http://www.bloomberg.com/news/2010-05-08/sec-said-to-consider-new-rules-to-slow-trading-as-it-investigates-plunge.html. Accessed 4 Jan 2014.

Westbrook, J., & Mehta, N. (2010, May 11). Exchanges to report back to SEC on plan for uniform circuit breakers. *Bloomberg*. http://www.bloomberg.com/news/2010-05-11/u-s-exchanges-to-report-back-to-sec-on-plan-for-uniform-circuit-breakers.html. Accessed 1 Apr 2014.

Whaley, R. E. (1993). Derivatives on market volatility hedging tools long overdue. *Journal of Derivatives, 1*(1), 71–84, Fall.

Wong, E. (2010, June 24). China's export economy begins turning inward. *New York Times*. http://www.nytimes.com/2010/06/25/world/asia/25china.html?_r=0. Accessed 4 Jan 2014.

Wyatt, E., & Bowley, G. (2010, May 12). Market inquiry focuses on one trader. *New York Times*. http://query.nytimes.com/gst/fullpage.html?res=9F0DE4DC133BF931A25756C0A9669D8B63. Accessed 1 Mar 2014.

Xinhua. (2010a, February 2). Investment contributes over 90% to China's GDP growth. *China.org.cn*. http://www.china.org.cn/business/2010-02/02/content_19356083.htm. Accessed 29 Dec 2013.

Xinhua. (2010b, September 15). Statistics chief hopes for 10% growth, 3% inflation in 2010. *China Daily*. http://www.chinadaily.com.cn/business/2010-09/15/content_11307604.htm. Accessed 29 Dec 2013.

Xinhua. (2010c, September 30). WTO warns against U.S. possible trade protectionism measures. *China Daily*. http://www.chinadaily.com.cn/world/2010-09/30/content_11367736.htm. Accessed 29 Dec 2013.

Yoon, A. (2007, May 1). Subprime CDO pricing getting perilous, managers say. *Reuters*. http://www.reuters.com/article/2007/05/01/us-usa-subprime-cdo-idUSN3045916020070501. Accessed 3 Jan 2014.

Explaining Stock Market Crashes: A Behavioral Finance Approach

5

5.1 Introduction

Over 60 years ago, Markowitz's ideas of how to create a portfolio started to shape the world of modern investing. In the 1970s, Eugene Fama, arguing that financial markets are efficient, founded the efficient market hypothesis (EMH). He defined an efficient market as

> *a market in which prices always fully reflect available information* [1]

and named three forms of market efficiency as already introduced briefly in Sect. 3.1: the weak form, the semi-strong form and the strong form efficiency[2]:

- **Weak form:**
  Only historical price data is reflected in today's stock price. Hence, technical analysis[3] *cannot consistently generate excess returns (outperformance, overperformance, alpha).*
- **Semi-strong form:**
  All relevant public information is already reflected in today's price. Both fundamental[4] and technical analysis *cannot consistently generate excess returns.* When speaking about *efficient markets*, this refers to the semi-strong form of market efficiency.

---

[1] Sewell (2011, p. 6).

[2] Fama (1970).

[3] Technical analysis consists in studying graphs and their global shapes and properties to determine future price directions.

[4] Fundamental analysis is primarily based on studying the accounting information of a company to assess the fair value of its stock and then to compare it to the market price. If the estimated fair value is above or below the market price, a *buy* or *sell* advice is given.

M. Schulmerich et al., *Applied Asset and Risk Management*,
Management for Professionals, DOI 10.1007/978-3-642-55444-5_5

- **Strong form:**
  All relevant public and private information is reflected in today's price. Even the use of insider information[5] *cannot consistently generate excess returns.*

Since the early 1970s, markets were believed to be extremely efficient and until today, every academic lecture on investment management relies heavily on the three concepts of the efficient frontier, the capital asset pricing model (CAPM) and the EMH, despite their shortcomings as discussed in Chap. 2. According to this view, when information arises, the news spreads very quickly and is incorporated into the prices of securities without delay. The basic argument is that the price of an asset is based on the accounting value and on updated news about the asset.

The Markowitz efficient frontier approach and the CAPM formula serve to create an optimal portfolio, based on the expected rate of return of each asset constituting the portfolio and on the standard deviation of each asset. Asset prices will change under the influence of fresh news, but according to the EMH, only one price is rationally acceptable for a given asset all across the world which is known as the law of one price. Therefore, asset mispricings should be easily spotted and corrected (known as *no arbitrage*) and active portfolios should not, in the long run, be able to generate excess returns above the market.

Historically, however, some active asset managers have consistently succeeded in outperforming market indices although MPT assumes that an index represents an optimal portfolio and that it is optimal for an investor to passively invest in the market.

The market efficiency concept gave rise to many questions ever since it was presented. First, doubts arose when a television program called *Wall Street Week* with Louis Rukeyser was aired in 1970.[6] Stocks recommended on this program beat the market by 4 % per year. Another example is the *Pros vs. Darts* contest of The Wall Street Journal.[7] In this contest, four professional securities analysts select a single stock and four Wall Street Journal staffers throw a dart at a sheet of different stocks. The recommendations of the analysts have beaten the S&P 500 and the DJIA as well by on average 5 % a year.

These examples may be seen as isolated and short-term incidents. But there is other evidence of market inefficiency which was described in Chap. 3 of this book. In the 1980s, numerous anomalies which are recurrent under- or overperformances related to a set of factors were discovered. Among the most famous ones is the January Effect, which was first noticed by Donald Keim, who was then a graduate student at the University of Chicago.[8] He observed that stock prices increase during the month of January in almost every year and that this increase cannot be traced back to fundamentals. Another example contradicting the EMH is the weekend

[5] Insider information is information that has not been publicly revealed yet.

[6] Shefrin (2002, p. 69).

[7] Shefrin (2002, p. 70, 74).

[8] Siegel (2008, pp. 302–303).

effect which has persisted over very long periods and on many international markets.[9] It refers to the difference in returns between Mondays and the other days of the week. The returns on Mondays are significantly negative, in contrast to every other day of the week.

Although financial literature also uses the term Monday effect, it actually is a weekend effect since the bulk of the negative returns occurs between the Friday close and the Monday opening of the markets. Research has shown that there is no piece of information that is responsible for the prices decrease during the weekend.[10] Numerous other anomalies were also discovered, like the small firm effect,[11] the P/E ratio effect[12] or the S&P Index effect[13] among others.

Moreover, stock market crises as presented in the previous chapter represent another deviation from EMH which can persist over a very long time. A typical example is the build-up of the subprime crisis between 2004 and 2008. Then, in mid-2008, the market suddenly fell sharply across a wide range of asset classes. Obviously, a sudden fall or rise in prices can occur without the publication of major new information which means that the EMH fails to explain stock market crashes.[14]

In order to find the reasons for bubbles and crashes, a psychological approach like behavioral finance should be considered. In this chapter, we will first define behavioral finance (Sect. 5.2), and give a brief introduction into this theory (Sect. 5.2.1). Thereafter, Sect. 5.2.2 provides a historical overview of its development since its origins. The key of behavioral finance are behavioral biases which each represent different aspects of human behavior concerning investing. However, this chapter will not provide a complete overview of all these behavioral biases. It will focus on market crashes as deviations from MPT and on how these deviations can be explained.

Section 5.3 will introduce the key behavioral biases which are linked to stock market crashes:

1. Availability bias (5.3.1)
2. Representativeness bias (5.3.2)
3. Herding bias (5.3.3)
4. Overoptimism bias (5.3.4)
5. Overconfidence bias (5.3.5)
6. Anchoring bias (5.3.6)
7. Prospect theory (5.3.7)

Each of these biases and theories will be explained and illustrated by numerous examples. Thus, in Sect. 5.4, we will deal with the very prominent crash of October

---

[9]Damodaran (2002, pp. 141–142).

[10]Damodaran (2002, pp. 141–142).

[11]Banz (1981, p. 3).

[12]Basu (1977, p. 663).

[13]Harris and Gurel (1986, p. 815) and Shleifer (1986, p. 579).

[14]Shiller (2005, p. 91), and Sornette (2003).

1987. It is described in Sect. 5.4.1, and in Sect. 5.4.2, the so-called *positive feedback trading model* which can be nicely linked to behavioral biases is presented in order to provide further explanations. This model is one of the major advances in behavioral finance. The objective of this chapter is to present recurrent behavioral anomalies and to compare them with the behavior of a perfectly rational person. Since these behavioral biases are shared by a large amount of practitioners, they can affect stock market prices and may even cause stock market crashes. A brief summary of the introduction to behavioral finance including its application in asset management can be found in Schulmerich (2014) or Schulmerich (2011).

5.2 What Is Behavioral Finance?

In all speculative bubbles that we have previously described, the psychology of investors had an influence on the abnormal upward and downward price movements. Table 5.1 sums up the factors that set up the dotcom bubble according to Shiller and include psychological factors.[15]

Therefore, our next goal is to analyze how the psychology of investors influences abrupt downward price movements. Behavioral finance has developed the tools to do so during the past decades.

Behavioral finance is the study of the influence of psychology on the behavior of financial practitioners and its effect on market prices.[16] It has burgeoned when the advances made by social psychologists came to the attention of economists[17] and became a hot topic when the dotcom bubble burst,[18] since it tries to explain the

Table 5.1 Factors explaining past stock market crashes until 2008

Structural factors	Cultural factors	Psychological factors
New technology	Media	Behavioral finance
Expansion of reporting business news	New Economy	Factors that are common to all speculative bubbles, i.e., overoptimism over an item such as tulip bulbs, shares of railway or IT-companies
Growth of mutual funds	Public attention to the market	
Expansion of the volume of trades: brokers, day traders, 24 h trading		

Source: Own, based on Shiller (2008, pp. 31–173)

[15] Shiller (2005).

[16] Sewell (2011).

[17] Shefrin (2002, p. 69).

[18] Pompian (2006, p. 3).

inefficiency of markets and the occurrence of stock market crashes. The underlying questions are: how can it be explained that markets can be more or less inefficient and that stock market crashes happen like the ones described above?

5.2.1 A Short Introduction to Behavioral Finance

Behavioral finance is the application of psychology to finance. It is a part of finance which supplements the standard theories of finance by integrating behavioral aspects into the decision-making process for investing.[19] The main concern is to help to explain why and how markets might be inefficient.[20] The area of study can overlap with other research fields, namely, behavioral science, investor psychology, cognitive psychology, behavioral economics, experimental economics, and cognitive science.

At university, most students start their Finance 101 lecture with the presentation of present value and future value. Soon after, the capital asset pricing model, explained in the second chapter of this book, is introduced. According to the model, asset prices are calculated based on the expected return and the variance of each asset. Then, advanced lectures with more quantitative contents are given. Students can discover how to price a vanilla option with a single formula, the *Black and Scholes Formula*, and in some cases with more exotic ones. For the happy few who understand computer simulation or stochastic calculation, the mathematical difficulty rises with the accuracy and the complexity of the means and methods used for asset pricing. These calculations are based on stock prices, strike prices,[21] time, variance, the risk-free rate, etc. According to the EMH,[22] any news is immediately included in asset prices which are assumed to be priced by perfectly rigorous and formal means. Meir Statman[23] explains:

> *Standard finance is the body of knowledge built on the pillars of the arbitrage principles of Miller and Modigliani, the portfolio principles of Markowitz, the capital asset pricing theory of Sharpe, Lintner, and Black, and the option-pricing theory of Black, Scholes, and Merton.*[24]

Standard finance theory is designed to provide useful mathematical explanations for financial situations. Many scholars and professional investors believe in this

---

[19] Sewell (2011, p. 1).

[20] Sewell (2010, p. 1).

[21] The strike price is the price at which a financial option can be exercised. This strike price is fixed in advance and the differential between the strike price and the stock price at a later date determines the payoff.

[22] Fama (1970, p. 383).

[23] Meir Statman is the Glenn Klimek Professor of Finance at the Leavey School of Business, Santa Clara University and visiting professor at Tilburg University in the Netherlands. His research focuses on behavioral finance. He attempts to understand how investors and managers make financial decisions and how these decisions are reflected in financial markets.

[24] Statman (1995, p. 1).

theory. They are convinced that human emotions and cognitive errors in financial decisions have a negligible effect if any on asset prices or market efficiency and they think that cognitive and emotional interactions in decision-making processes do not merit a unique category of study. Behavioral finance adherents, however, are convinced that the awareness of at least a limited amount of psychological biases is crucial to assess the fair price of an asset.[25] However, they assume that reality is perceived in an imprecise and sometimes unrealistic or otherwise oversimplified way.[26]

One of these simplifications is the concept of *Homo economicus* and its underlying hypotheses, namely[27]:

- Perfect rationality
- Perfect self-interest
- Perfect information

According to many psychologists, a human being is more likely to follow emotional impulses (for example, fear, love, hate, pleasure, and pain) and instinct than his intellect. The latter is presumed to be used so as to achieve or to avoid these emotional decisions (i.e., achieve pleasure and avoid pain). Second, numerous studies demonstrate that people are not perfectly self-interested. If they were, then teamwork would be against human nature, compassion would be non-existent and military service would be utterly unattractive in the absence of other stimuli, for example, fame or any socially promoted qualities. Self-destructive behavior (for example, suicide, alcoholism, etc.) would also be highly unlikely. Finally, perfect information seems hard to achieve, since a limited number of people, if any, have full access to the complete information about any given subject.[28] As Pompian explains:

> *Many economic decisions are made in the absence of perfect information. For instance, some economic theories assume that people adjust their buying habits based on the Federal Reserve's monetary policy. Naturally, some people know exactly where to find the Fed data, how to interpret it, and how to apply it; but many people don't know or care who or what the Federal Reserve is.*[29]

The major question that behavior finance tries to answer is not how investors *ideally should behave* (which is the objective of standard finance), but how they actually *behave in practice*. Based on empirical studies, this kind of research seeks to discover psychological phenomena at work in investment decision-making under various conditions. There is no universal theory about investment behavior, but a broad array of evidence has been found which points to the ineffectiveness and irrationality of markets and human decision-making.[30]

---

[25]Pompian (2006, p. 3).

[26]Pompian (2006, p. 10).

[27]Aktipis and Kurzban (2004, p. 2).

[28]Pompian (2006, pp. 15–16).

[29]Pompian (2006, p. 16).

[30]Pompian (2006, pp. 15–16).

There are two subcategories of research[31]:

- Behavioral finance macro detects and describes anomalies in the EMH,[32] then attempts to explain them according to behavioral models.
- Behavioral finance micro focuses on psychological biases and behaviors of individual investors that distinguish them from the rational actors envisioned by the proponents of classical economic theory.

Behavioral finance macro can further be detailed into the research of fundamental analysis, technical analysis and calendar anomalies. Fundamental analysis deals with fundamental factors of stocks or portfolios (for example, in order to identify growth stocks[33] and value stocks).[34] An extensive amount of evidence shows that investors consistently tend to overestimate the prospects of growth companies (for example, of Apple in 2011) and underestimate the value of less-in-favor companies (for example, European banks as BNP Paribas or Deutsche Bank during the same period).[35] Technical analysis is concerned with the shapes and forms of the stock price graphics that reflect the relative strength and the moving averages of a stock as well as supports and resistances. The basic idea is to forecast future returns based on past performance.

On the other hand, behavioral finance micro relates to behavioral biases observed in the investment process which contradict the notion of the *Homo economicus*. Behavioral biases are defined, abstractly, in the same way as systematic errors in judgment. Researchers developed a well-furnished list of more than fifty specific biases, each of them applied to individual investor behavior.[36] Considering the biases which were already identified by cognitive scientists, but not applied to practical behavior yet and the future ones to be discovered, the extent of the research work still to be done is astonishing. It is possible to classify these biases by using heuristic methods.

Heuristics refer to experience-based techniques for problem solving, learning, and discovery. They help people make decisions by speeding up the process of finding a satisfactory solution,[37] where an exhaustive search is impractical or

[31]Pompian (2012, p. 14).

[32]Sewell (2011, p. 6).

[33]A growth stock usually pays a small dividend and is likely to be highly valued (i.e., high price-to-earnings ratio). Yet it is believed to have a good price appreciation potential because of its above-average rate of return relative to the market.

[34]A value stock is a stock that tends to trade at a lower price relative to its fundamentals. Common characteristics of such stocks include a high dividend yield, a low price-to-book ratio and/or a low price-to-earnings ratio. In theory, the value stock is the safer investment because of its underlying value.

[35]Chan and Lakonishok (2004).

[36]Pompian (2006, p. 49).

[37]Kahneman, Slovic, and Tversky (1982, pp. 3–4).

inefficient. Examples of this method include using a rule of thumb, an educated guess, an intuitive judgment, or common sense.

In August 2001, David Hirshleifer proposed a convincing scheme in his article *Investor Psychology and Asset Pricing* to classify behavioral biases. This classification is shown in Fig. 5.1. Fortunately, a short selection of these biases is sufficient to explain some types of market or investment anomalies since they are in particular present during stock market crashes and financial distress periods:

1. Availability bias
2. Representativeness bias
3. Herding bias
4. Overoptimism bias
5. Overconfidence bias
6. Anchoring bias
7. Prospect theory

Additional biases like overreactivity or underreactivity can as well be of relevance. However, these biases are more likely attached to markets than to investors. Once combined, these market-related biases become individual-related biases as described by cognitive psychologists.

As a summary, behavioral finance supplements the standard theories of finance by integrating behavioral aspects into the decision-making process. They can be sub-categorized into two categories, i.e., behavioral finance macro and micro and overlap with research in behavioral science, investor psychology, cognitive psychology, behavioral economics, experimental economics, and cognitive science. Behavioral finance macro focuses on market anomalies while behavioral finance micro identifies biases at work in investment decision-making processes. These biases can be sorted into at least four various categories, namely self-deception, heuristics simplification, emotion and social biases. Only a limited amount of these biases are required to explain at least partly stock market crashes.

In the following part, we will describe the emergence of behavioral finance as an independent scientific subject. We will present its pioneers and the most important research in this field.

5.2.2 Historical Overview

When the concept of behavioral finance was first introduced in 1956, most of the investment community did not believe that human behavior influences security prices.[38] But behavioral finance gained in influence among investment experts after stock market anomalies were discovered that did not fit the assumption of a rational

[38] Schulmerich (2011, p. 1).

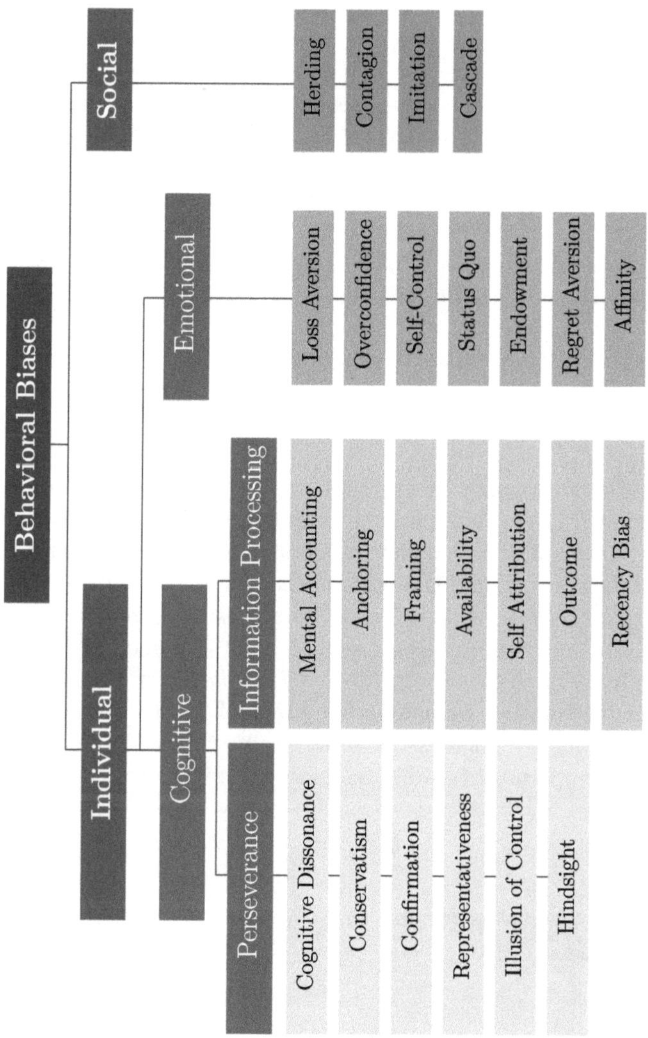

Fig. 5.1 Overview of behavioral biases. *Source:* Based on Hirshleifer (2001)

Homo economicus. In order to understand this field of research and where it stands today, an outline of its history will be given in this section.

The history of behavioral finance is a matter of debate. The irrationality of investors can at least be traced back to the collapse of the tulip mania covered in Chap. 4, and the roots of behavioral finance may date back to as far as 1841, when Charles Mackay[39] published his book *Extraordinary Popular Delusions and the Madness of Crowds*.[40] It deals with phenomena that show the madness of crowds during stock market bubbles.

In 1896, Gustave Le Bon wrote *The Crowd: A Study of the Popular Mind*, one of the greatest and most influential books of social psychology ever written.[41] It would serve as a cornerstone both in social psychology and other psychology-related fields, like behavioral finance.

In 1912, Selden, the author who is known as the first to have addressed the topic of investors' behavior, wrote *Psychology of the Stock Market*.[42] He based his book *"upon the belief that the movements of prices are dependent on the mental attitude of the investing and trading public."*[43] In his book, the author explains:

- How financial markets are driven by emotions such as fear, greed, and panic.
- The role that psychology plays in the movement of the market and of individual stocks.

The human side of economic decision-making also became a theme of classical economics when it emerged in the mid-eighteenth century. At that time, economic utility functions were linked with human psychology so as to assess a given level of satisfaction and the concept of *Homo economicus* was invented to comply with and to support the theory. *Homo economicus* is framed as a perfectly rational human, seeking to reach a maximum level of consumption of goods and services at a minimum cost. In 1759, Adam Smith[44] wrote *The Theory of Moral Sentiments*,[45] which describes the impact of emotional and mental factors both on human and economic interactions and focuses on pride, shame, insecurity, etc. For Jeremy Bentham,[46] a contemporary of Smith, happiness is the ultimate human concern.

[39]Charles Mackay (1814–1889) was a Scottish poet, song writer and journalist.

[40]Mackay (2009) or Mackay (1995).

[41]Sewell (2010).

[42]Selden (1912).

[43]Selden (1912, preface p. 6).

[44]Adam Smith (1723–1790) is the author of *The Theory of Moral Sentiments*. He is widely cited as the father of modern economics and capitalism.

[45]Smith (2011).

[46]Jeremy Bentham (1748–1832), a leading thinker of his time, was an English lawyer, philosopher, and legal and social reformer. His ideas influenced the development of welfarism. He is well known for his advocacy of utilitarianism and animal rights.

Since this is a subjective concept, any mathematical or economic calculation without including an *emotion variable* is impossible.[47]

The belief that human psychology has an important role to play in economics also was a pillar of neo-classical theories and allowed the introduction of the law of demand and supply.[48] Because of the issue of subjectivity, neoclassical economists distanced themselves from psychology and tried to reframe economics in a more quantitative way, giving more concern to the nature of economic agents than to individuals and their beliefs. Individual preferences were seen as irrelevant compared to the effect of the mass and thus economic agents should have a bigger impact. Nevertheless, the neoclassical view emphasizes that humans do not automatically and simultaneously adjust to new economic conditions, while in classical economics they are assumed do so. One reason for this are resource constraints. Let us imagine that we have just invented cars, but everybody uses horses. In terms of finance and logistics it is impossible to provide cars to everyone in an eyeblink, and if this invention is realized in a deserted place without means of communication, the spread of the technology could even take more time

That is why Herbert Simon proposed the concept of *bounded rationality*[49] to counter the weaknesses of *Homo economicus*. Bounded rationality assumes that people first search products or services which promise optimal satisfaction and then may acquire them. These services or products are not given and taken for granted.[50] Put simply, this hypothesis underlines that the choices of individuals are rational but subject to the limitations of knowledge and cognitive capacity.

Moreover, *Homo economicus* does not take into consideration the inner conflicts that most humans face, for example, how to prioritize short- against long-term interests like consuming and saving or social and personal interests. These are important aspects of the human decision-making process and can lead to *irrational* behavior.[51] Another setback to the theory of *Homo economicus* came from experimental economics, a new area of research

Rather than focusing on large scale data like classical and neo-classical macro economics, experimental economists try to study decision-making on the individual scale. Their observations contradicted the assumptions about the behavior of *Homo economicus*. Moreover, these observations were persistent among the people subjected to behavioral experiences. One of the most famous anomalies was discovered by Maurice Allais,[52] and has been named the *Allais paradox*.

[47]Pompian (2006, p. 22).

[48]Pompian (2006, p. 20).

[49]Gigerenzer and Selten (2001, p. 13).

[50]Gigerenzer and Selten (2001, p. 14).

[51]Pompian (2006, p. 24).

[52]Maurice Allais (1911–2010) was a French economist and the 1988 winner of the Nobel Memorial Prize in Economics *"for his pioneering contributions to the theory of markets and efficient utilization of resources."* See Allais (1953, p. 3).

As explained earlier, *Homo economicus* seeks to optimize his satisfaction at a minimum cost (i.e., maximization of the expected utility). But Allais showed that this theory, which had been accepted for decades, does not apply to certain empirically realistic decisions under risk and uncertainty.[53] When crises occur on stock markets, risk and uncertainty are much higher than in normal times which can lead to irrational behavior as will be described below.

Allais carried out a test of two parts where participants were asked to make two hypothetical choices. The first choice was between the two following alternatives[54]:

1. Certainty of receiving 100 million (French francs).[55]
2. Combined probability of 10 % to receive 500 million, of 89 % to receive 100 million and of 1 % to receive zero.

Then the participants were asked to choose between the two following possibilities:

1. Combined probability of 11 % to earn 100 million and probability of 89 % to earn zero.
2. Combined probability 10 % to earn 500 million and probability of 90 % to earn zero.

An expected utility maximizer would opt for the first choice in both cases, since he would focus on the higher probability only and not on the amount to be obtained.[56] However, Allais observed that while the first choice was preferred in the first test, a majority of participants chose the second option in the second test. That is to say, in exchange for losing one percentage point, they could have access to 400 additional million francs. Hence, the decision-making was not as rigorously rational as the *Homo economicus* model requires. This experiment may appear improbable or even trivial, but the phenomenon that Allais reported has subsequently recurred in experiments offering real, yet much smaller quantities of money.[57]

Other experiments also showed evident irrational behavior. A famous example is the prisoner's dilemma which consists of a small game where two people have been arrested and face the choice of cooperating or denouncing the other one. If the two betray each other, each of them gets 5 years of jail. If only one does, then the betrayed one will go to jail for 10 years and the traitor is set free. If the two cooperate then both will be sentenced to only 6 months in prison. In most cases, the game players will denounce each other, not knowing if the other becomes a traitor and at least hoping to gain freedom.

[53] Roth (1993, p. 195).

[54] Allais (1953, p. 527).

[55] The French franc was the former French currency before the introduction of the euro.

[56] Note that the optimal expected gain is not obtained from the first choice.

[57] Pompian (2006, p. 28).

In the mid-1950s and 1960s, experimental economics expanded and numerous anomalies were discovered which highlighted new aspects of human economic decision-making. At the same time, cognitive psychology and decision theory emerged as additional fields of research which are closely linked to experimental economics.

Cognitive psychology is the scientific study of cognition or the mental processes that are believed to drive human behavior. It investigates how people perceive, remember, think, speak, and solve problems.[58] Thanks to the latest models that were developed in this field, researchers were able to frame generally accepted guidelines[59] for making decisions which require a bit more effort than basic everyday choices (for example, the decision about how many sugar cubes to put in a cup of coffee):

1. Take an inventory of the available viable options for obtaining information, for experimentation, and for action.
2. List the events that may occur.
3. Arrange pertinent information and choices/assumptions.
4. Rank the consequences resulting from the various courses of action.
5. Determine the probability of uncertain events occurring.

Under time and resource constraints, high risk and uncertainty, these steps cannot be fully completed. As a result, most people will use shortcuts and rather subjective approaches to decide which course of action should be taken.

In a financial crisis, the conditions are as follows:

- The level of uncertainty is high.
- Time lacks to build an inventory of available information sources.
- Foreseeing the possible future events is difficult.
- Pertinent information cannot be easily classified.
- Due to the crisis the consequences are hard to classify according to the expected severity.
- Assessing the probability of uncertain occurrences is difficult.

It appears evident that under these circumstances irrational decisions are more likely to be made. But decision-making under uncertainty was precisely the research topic that some cognitive psychologists focused on. The first to tie rational choices to a behavioral model was Herbert Simon in 1955.[60] Relatively soon after, in 1968, Howard Raiffa[61] wrote an important work on decision analysis named *Introductory*

[58]Feist and Rosenberg (2009, p. 17).

[59]Pompian (2006, p. 29).

[60]Simon (1955).

[61]Howard Raiffa (born in 1924), the Frank P. Ramsey Professor (Emeritus) of Managerial Economics at Harvard University, is an influential Bayesian decision theorist and pioneer in the

Lectures on Choices under Uncertainty.[62] He normalized and set up three decision approaches so as to provide a more accurate view of the thought process of real people:

- Normative analysis focuses on a rational solution to the problem at hand, i.e., the output of the solution.
- Descriptive analysis focuses on the manner in which real people actually make decisions, i.e., the means.
- Prescriptive analysis focuses on practical advice and tools that may help the subject to achieve results. The idea is to reach a faster and a better solution to the problem at hand.

Raiffa's work was not directly tied to behavioral finance. However, it laid the foundation for many subsequent works in this fast moving research field. Actually, it was Nobel Prize laureate Vernon L. Smith[63] who first introduced the concept of behavioral finance in 1956.[64] In 1961, John Muth[65] wrote *Rational Expectations and the Theory of Price Movements*. This was the first piece of research to study the behavior of market participants.[66]

In 1972, Paul Slovic[67] published *Psychological Study of Human Judgment: Implications for Investment Decision Making* which focuses on the interpretation and selection of information during the investment process.[68] This article is considered to be the cornerstone of behavioral finance. Slovic is referred to as the father of behavioral finance, although he published his article after Vernon Smith.

---

field of decision analysis, with works in statistical decision theory, game theory, behavioral decision theory, risk analysis, and negotiation analysis. See Fienberg (2008).

[62]Raiffa (1968).

[63]Vernon L. Smith (born January 1, 1927, in Wichita/Kansas, USA), professor of economics at Chapman University's Argyros School of Business and Economics and School of Law in Orange, California, is a research scholar at the George Mason University Interdisciplinary Center for Economic Science and a fellow of the Mercatus Center, all in Arlington, Virginia. He shared the 2002 Nobel Memorial Prize in Economic Sciences with Daniel Kahneman, and is the founder and president of the International Foundation for Research in Experimental Economics and an adjunct scholar of the Cato Institute in Washington, DC.

[64]Schulmerich (2011, p. 2).

[65]John Muth (1930–2005) was an American economist, known as *the father of the rational expectations revolution in economics*.

[66]Muth (1961).

[67]Paul Slovic (born 1938 in Chicago, USA), a professor of psychology at the University of Oregon and president of the Decision Research group, was the first to describe the *affect heuristic*. Slovic, who earned his Ph.D. in Psychology at the University of Michigan in 1964, has studied psychological heuristics and published frequently with co-authors like Daniel Kahneman and Amos Tversky. He is considered as one of the leading theorists and researchers in the risk perception field.

[68]Slovic (1972).

1972 is, therefore, considered the birth year of behavioral finance, a view shared[69] by Hersh Shefrin,[70] another preeminent proponent of behavioral finance.

At that time, the investment community did not believe that human behavior influences security prices. However, three cognitive psychologists, i.e., Amos Tversky,[71] Paul Slovic and Daniel Kahneman,[72] continued to analyze the behavioral biases, as they called them, of the investors. They played a central role in the development of behavioral finance.[73]

As Pompian underlines, Tversky and Kahneman brought to light the incidences, causes and effects of human errors in economic reasoning.[74] Building on their growing success they wrote:

- *On the Psychology of Prediction* (1973)[75]:
  This paper explores the rules of intuitive predictions for both category predictions and numerical predictions. Instead of using statistics, people rely on a limited number of heuristics and make predictions based on the degree to which the prediction represents the given evidence. Often there is no correlation with the likelihood or reliability of evidence. Initial studies show that people predict by similarity not by statistical likelihood.
- *Judgment under Uncertainty: Heuristics and Biases* (1974)[76]:
  This article shows that people rely on a limited number of heuristic principles which reduce complex tasks of assessing probabilities and predicting values to

[69]Shefrin (2001, p. xiv).

[70]Hersh Shefrin (born in Winnipeg, Canada), the Mario L. Belotti Professor of Finance at Santa Clara University, is one of the pioneers in the behavioral approach to economics and finance. A 2003 article in the American Economic Review lists him as one of the top 15 economic theorists to have influenced empirical work. In 2009, his book *Beyond Greed and Fear* (Shefrin 2002) was recognized by J.P. Morgan as one of the top ten books published since 2000. Shefrin has long standing collaborations with Richard Thaler and Meir Statman on behavioral finance, being one of the first economists to incorporate ideas from psychologists like Amos Tversky and Daniel Kahneman into working theories.

[71]Amos N. Tversky (1937–1996) was an American cognitive and mathematical psychologist. He was a pioneer of cognitive science and a longtime collaborator of Daniel Kahneman who is a key figure in the discovery of systematic human cognitive bias and in the research of the handling of risk.

[72]Daniel Kahneman, born March 5, 1934, an Israeli-American psychologist and winner of the 2002 Nobel Memorial Prize in Economic Sciences for his work in prospect theory, is renowned for his work on the psychology of judgment and decision-making, behavioral economics and hedonic psychology. He is professor emeritus of psychology and public affairs at Princeton University's Woodrow Wilson School. With Amos Tversky and others, Kahneman laid a cognitive basis for common human errors which arise from heuristics and biases. In 2011, the magazine Foreign Policy named him to its list of top global thinkers.

[73]Schulmerich (2011, p. 1).

[74]Pompian (2006, p. 31).

[75]Kahneman and Tversky (1973).

[76]Kahneman and Tversky (1974).

simpler judgmental operations. Sometimes useful, this can, however, lead to systematic and dramatic errors. The authors compiled their findings in a book under the same name in 1982 in collaboration with Paul Slovic.[77]

- *Prospect Theory: An Analysis of Decision under Risk* (1979)[78]:
  This document criticizes the expected utility theory and introduces a new model which will be extensively discussed in Sect. 5.4. It assumes that people overconsider loss and underconsider gain. Probable outcomes are less considered than certain ones. This paper is now regarded as the seminal work in behavioral finance.

In addition to these works, a paper published by Daniel Kahneman and Mark Riepe in 1998 has to be mentioned: *Aspects of Investor Psychology: Beliefs, Preferences, and Biases Investment Advisors Should Know About*[79] used the research results of Howard Raiffa and for the first time tied decision theory to financial advising. It promotes the necessity of taking into account cognitive and emotional weaknesses while investment decisions are to be taken. It also allowed to categorize behavioral finance under three aspects:

- Biases of judgment which include overconfidence, optimism, hindsight, and overreaction to chance events.
- Errors of preference which include nonlinear weighting of probabilities; the tendency of people to value changes, not states; the value of gains and losses as a function; the shape and attractiveness of gambles; the use of the purchase price as a reference point; narrow framing; tendencies related to repeated gambles and risk policies; and the adoption of short versus long views.
- Biases associated with living with the consequences of decisions which give rise to regrets of omission and commission, and have implications regarding the relationship between regret and risk taking.[80]

This research document is important for many reasons among which stands the practical application perspective. It was the first scholarly work to really challenge financial advisors to examine their practice from a behavioral perspective thanks to a *Checklist for Financial Advisors*.

These discoveries presented serious threats to conventional economic hypotheses, notably to the conservative *Homo economicus* model. Still, behavioral finance only started to flourish when the advances made by psychologists came to the attention of economists during the 1980s.[81] From this time onwards, the idea of market anomalies and inefficiencies became predominant. In July 1985, the Journal

[77]Kahneman et al. (1982).

[78]Kahneman and Tversky (1979).

[79]Kahneman and Riepe (1998).

[80]Kahneman and Riepe (1998).

[81]Shefrin (2002, pp. 7–8).

of Finance published a paper written by De Bondt[82] & Thaler,[83] where a Tversky and Kahneman's representativeness bias was applied to market pricing.[84] Representativeness, as will be described in Sect. 5.3.2, is a rule used by people to judge the probability of an event by considering how much it resembles already available data.[85] Finally, Jack Hirshleifer[86] wrote an article named *Investor Psychology and Asset Pricing*, in which he challenges the capital asset pricing model by including risk and misevaluation in the calculation of the expected returns of assets.[87] He also proposes a framework for understanding decision biases, and provides evidence for the impact of psychology on security prices.

Behavioral issues linked to economic decisions have been discussed by scholars since 1841 at least, when Charles Mackay wrote *Extraordinary Popular Delusions and the Madness of Crowds*[88] where he described the *madness of crowds* in events such as speculative stock market bubbles. The described topic remained a matter of concern for decades, also for classical and neo-classical economists. But human behavior and the quest for happiness were thought to be too subjective, and it was assumed that they could be neglected if the scope of research were only large enough. However, after a series of empirical anomalies were uncovered, the reliability of the *Homo economicus* model was put into question. Three cognitive psychologists, i.e., Amos Tversky, Paul Slovic and Daniel Kahneman, played a central role in the development of behavioral finance and in the 1980s, their research led to a review of basic hypotheses shared by today's economics and modern finance.

Below follows a list of the major research papers in the field of behavioral finance. The list is not exhaustive, but it retraces the major advances in the past 30 years of this scientific field:

- 1912: *Psychology of the Stock Market*[89]
- 1961: *Rational Expectations and the Theory of Price Movements*[90]

---

[82]Werner F.M. De Bondt is Director of the Richard H. Driehaus Center for Behavioral Finance at DePaul University in Chicago. He studies the psychology of investors and financial markets and has examined key concepts of bounded rationality. He received his Ph.D. in Business Administration from Cornell University in 1985.

[83]Richard H. Thaler (born September 12, 1945, in East Orange, New Jersey) studies behavioral economics and finance as well as the psychology of decision-making which lies in the gap between economics and psychology. He investigates the implications of relaxing the standard economic assumption that everyone in the economy is rational and selfish, instead entertaining the possibility that some of the agents in the economy are sometimes philanthropic and subject to error. Thaler is Director of the Center for Decision Research. He received a Ph.D in 1974 from the University of Rochester. He joined the Chicago Booth faculty in 1995.

[84]De Bondt and Thaler (1985).

[85]Kahneman and Tversky (1979, pp. 3–4).

[86]Jack Hirshleifer (born in 1925) is a UCLA economist. He is one of the pioneers of the application of the theories of uncertainty and information in economics.

[87]Hirshleifer (2001, p. 1533).

[88]Mackay (2009).

[89]Selden (1912).

[90]Muth (1961).

- 1972: *Psychological Study of Human Judgment: Implications for Investment Decision Making* [91]
- 1973: *On the Psychology of Prediction* [92]
- 1974: *Judgment under Uncertainty: Heuristics and Biases.* [93]
- 1979: *Prospect Theory: An Analysis of Decision under Risk* [94]
- 1985: *Does the Stock Market Overreact?* [95]
- 2001: *Investor Psychology and Asset Pricing* [96]

Starting from the 1990s, numerous researchers have supplied an abundant amount of evidence for stock market anomalies and irrational behavior. They also provided valuable explanations of their discoveries.

To sum up, behavioral finance answers the need to understand the decision process and the impact of emotions on it. The roots of behavioral finance can be traced back to Charles Mackay and his *Extraordinary Popular Delusions and the Madness of Crowds*. Later on, economists tried to capture the decision-making process under rather optimistic conditions. Bounded rationality was introduced, but it did not account for many everyday life issues when decisions are taken, for example, the conflict between personal and social interests. Pioneers of behavioral finance then emerged, starting in 1956 with Vernon Smith followed by Daniel Kahneman and Amos Tversky. Their masterpiece *Judgment under Uncertainty: Heuristics and Biases* enlightened the investment community with convincing empirical and quantitative results. Still, their influence remained limited until advances made by psychologists came to the attention of economists during the 1980s.

At the same time, further stock market anomalies were continuously discovered. Behavioral finance proposed solutions and empirical results to explain them.

Since the aim of this book is not to discuss in-depth the pace and the importance of these major advances, we will now focus on seven behavioral biases that the investment community assumes to be at least partly responsible for stock market crashes.

5.3 Behavioral Biases and Stock Market Crashes

Today, more than 20 behavioral biases can be distinguished. We have selected seven major biases that are relevant in financial crises and economic distress:

1. Availability bias
2. Representativeness bias

[91] Slovic (1972).

[92] Kahneman and Tversky (1973).

[93] Kahneman and Tversky (1974).

[94] Kahneman and Tversky (1979).

[95] De Bondt and Thaler (1985).

[96] Hirshleifer (2001).

3. Herding bias
4. Overoptimism bias
5. Overconfidence bias
6. Anchoring bias
7. Prospect theory

Each bias will be defined and illustrated by a set of examples from everyday life and within an investment context. We will then present where it can be observed in a situation of financial distress and high uncertainty, i.e., during stock market crashes. For each bias, we also present countermeasures to prevent being adversely affected.

5.3.1 Availability Bias

5.3.1.1 Definition

The availability bias is a mental shortcut that leads people to assess the probability of an event based on how familiar it appears. It is a heuristic that may result in misjudging the probability of events, since events may be considered as less likely to occur if they are harder to imagine.[97]

For instance, most people guess that shark attacks cause more fatalities than airplane parts falling from the sky. In fact, the latter event is 30 times more probable.[98] Casualties resulting from shark attacks are assumed to be more prevalent because sharks provoke a greater fear and receive a far bigger attention from the media. As a consequence, dying from a shark attack is easier to imagine than dying by falling airplane parts.[99]

5.3.1.2 The Subcategories of Availability Bias

Pompian lists four types of availability bias that apply to investors[100]:

1. Retrievability
2. Categorization
3. Narrow range of experience
4. Resonance.

We now present each of the subcategories and its implications in the world of finance.

---

[97]Kahneman and Tversky (1974, p. 1124).

[98]Pompian (2006, p. 94).

[99]Pompian (2006, p. 94).

[100]Pompian (2006, p. 95).

Retrievability

Ideas which are retrieved more easily seem to be more credible, though in reality this is not necessarily the case. In an experiment designed by Kahneman, Slovic and Tversky, a list of names was read, and the participants were asked whether there were more female or male names. The majority of the names were female, but the subset of male names contained a higher frequency of references to celebrities. In accordance with the availability theory, most subjects produced biased estimates and declared that more male than female names were cited.[101]

In the financial sphere, if investors are asked to identify the *best* mutual fund company, they are more likely to mention a firm that heavily advertises and promotes its funds. This shows retrievability bias, since the best-performing funds of today do not use much advertising.[102]

Categorization

When people try to categorize information that matches a certain reference their brains will first generate a set of search terms related to the task or objective they are dealing with. This allows them to efficiently use their classification capabilities and to locate the data they need. However, different references require different search sets (for example, cars vs. trucks).

Let us imagine a French wine connoisseur, who stayed and lived his whole life in his home country, tries to produce two lists of vineyards: a list of high-quality U.S. vineyards and another of high-quality French vineyards. The former list is likely to be more difficult to create for him, if we assume that he is more accustomed to French wines. Consequently, French vineyards are well known to him, and he might think that high-quality vineyards are less numerous in the U.S. than in France, although this might not be true.[103]

When it comes to investing, a majority of U.S. Americans will choose the United States as the country that offers the best investment prospects. This choice is biased because over half of the worldwide equity market capitalization is outside the United States and shows what is called home bias.[104] This example shows that most U.S. investors are affected by the categorization bias.[105]

Narrow Range of Experience

This bias occurs when the frame of reference from which one person can derive estimates is too limited.[106]

Suppose, a successful college basketball player is drafted by a National Basketball Association (NBA) team, where he enjoys several successful seasons. Since

[101] Kahneman et al. (1982, p. 175).

[102] Pompian (2006, pp. 96–97).

[103] Pompian (2006, pp. 95–96).

[104] French and Poterba (1991, p. 222).

[105] Pompian (2006, p. 97).

[106] Pompian (2006, p. 96).

he encounters several other former college basketball players on a daily basis, he is likely to overestimate the relative proportion of successful college basketball players that play professionally in an NBA team and to underestimate the relative number of the college basketball players who fail to join the NBA. In reality, only a small percentage of college basketball players will graduate to the NBA.[107]

As another example of this bias, let us assume an employee who works for a fast-growing high-tech company, say in biotechnology, and very often comes across successful high-tech employees of the same sector. When he is asked which industry has generated the most successful returns over the past 5 years his narrow range of experience may lead him to overestimate the proportion of corporate successes that are high-tech based and to intuitively assume that high-tech companies in biotechnology have been the most profitable over the past 5 years.[108]

Resonance

This bias stems from personal situations that influence the judgment of individuals. In fact, fans of classical music will likely overestimate the portion of the population that listens to classical music and underestimate the portion of those who do not.[109]

In finance, availability-biased investors are likely to favor investments that correspond to their personality. An investor that is attracted by bargains and discount shops will favor value stocks, since the notion of value is highly present in his mind unlike the notion of growth.[110]

In the financial markets, availability bias is often at work. Figure 5.2 shows order imbalance as a *function of news intensity*[111]: the more a stock is advertized, the more it is likely to be bought and vice versa. Research on this bias has provided a large amount of evidence that out of the myriad of more than seven thousand listed stocks investors like to select stocks they find remarkable. for example, Odean and Barber explain that investors tend to pick stocks that recently caught their attention by their trading volume, returns and the news related to it.[112] This is clearly an evidence of availability bias, especially among individual investors who cannot analyze as many stocks as institutional investors.

Availability bias has also been noticed during stock market bubbles alongside heuristics. In fact, Gerding explains that individuals resort to mental shortcuts known as *heuristics* to analyze information and make decisions.[113] These biases are responsible for misjudgments, for example, that stock prices will continue to rise during market booms and to fall during market crashes. The reason is simple:

[107]Pompian (2006, p. 96).

[108]Pompian (2006, p. 97).

[109]Pompian (2006, p. 96).

[110]Pompian (2006, p. 97).

[111]Barber and Odean (2002, p. 1).

[112]Barber and Odean (2002, p. 14).

[113]Gerding (2007, p. 996).

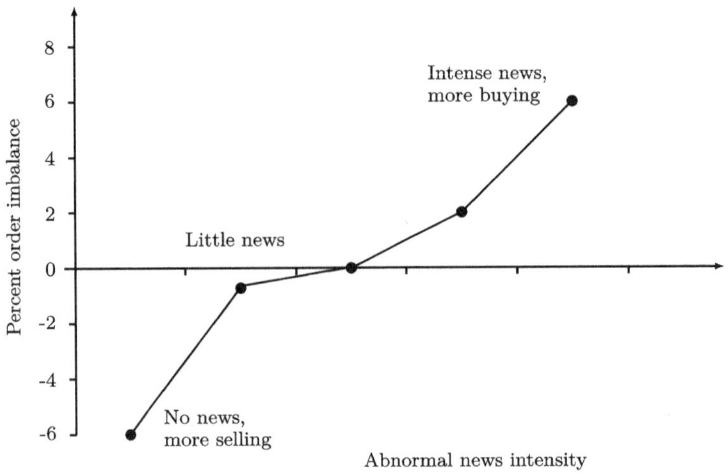

Fig. 5.2 Order imbalance as a function of news intensity. *Source*: Pompian (2006, p. 100), or Barber and Odean (2002, p. 14)

investors tend to hold on their small basket of assets which on average follow a globally upward or downward direction. As an example, let us look at the dotcom bubble. When it built up in 1998–2000, investors were euphoric about the *New Economy* and underestimated risk. Once the market crashed, the same investors lost their confidence and remained focused on the negative results they had experienced.

Moreover, availability bias makes people attribute disproportionate credibility to information which is frequently inaccurate, outdated, partial and sometimes only a patchwork of opinions in various media.

5.3.1.3 Protection Method

Availability bias is categorized as a cognitive bias. It can be easily corrected if the investor focuses on long-term results rather than trying to follow short-term trends, and if he keeps informed and frequently updates his knowledge of the industries he wishes to invest in.[114]

5.3.2 Representativeness Bias

5.3.2.1 Definition

Representativeness, a behavioral bias classified as a heuristic, was first documented in finance by Kahneman, Slovic and Tversky in 1974.[115] It is the tendency to evaluate a situation by comparing it to generalities, stereotypes, repetitions or

[114]Pompian (2006, pp. 102–103).

[115]Kahneman et al. (1982, pp. 3–4).

convenient cases rather than using probabilities.[116] The objective is to gain time through approximation, but this bias can prevent a person from carrying out a thorough analysis. Newly received information is classified in anterior built-in mental categories or new concepts are created for the occasion.

The classification is rather a best-fit method than a rigorous classification approach. Sometimes, incorrect results will be produced when two topics are considered as similar while being drastically different. To illustrate, we will describe an exercise.[117]

5.3.2.2 Example

Linda is 31 years old and a single, outspoken and very bright woman who has majored in philosophy. As a student she was deeply concerned with issues of equality and discrimination. Now, is it more likely that Linda is:

- A bank clerk?
- Or a bank clerk and active in a feminist movement?

Most people answer that it is more likely that Linda is a bank clerk and active in the feminist movement.[118] But this is irrational because:

- The probability of an intersection of two sets of events can never be greater than the probability of one of them.
- There are more bank clerks than bank clerks in a feminist movement. Therefore the first option is the correct answer.

Figure 5.3 schematically reproduces the statements cited above. The intersection is clearly smaller than the two circles. But data for this test show that people tend to make probability assessments according to how closely a given option resembles something else (in this case Linda's past) instead of calculating probabilities.[119]

Two subcategories of representativeness can be distinguished: base-rate neglect and sample-size neglect. Recall that representativeness is the tendency to evaluate a situation by comparing it to generalities. When judging the probability of an event, two types of information might be at hand:

1. Generic information about the frequency of an event such as the prevalence of a disease.
2. Specific information about the case in question such as information about a patient suffering from a disease.

[116] Montier (2002, pp. 5–12).

[117] Montier (2002, p. 9).

[118] Montier (2002, p. 7).

[119] Montier (2002, p. 10).

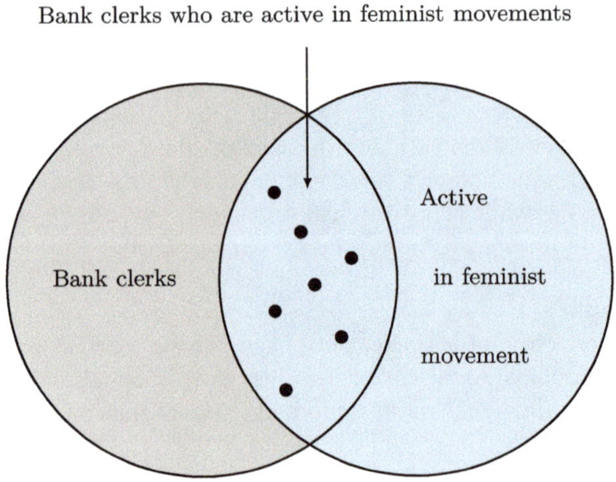

Fig. 5.3 Scheme of crossed probability of a bank clerk to be a feminine activist. *Source*: Montier (2002, p. 36)

Depending on the type of information at hand, base-rate neglect bias and sample-size neglect bias may occur.

5.3.2.3 Base-Rate Neglect Bias

Base-rate neglect is the tendency of people to mistakenly judge the likelihood of a situation by not taking into account all relevant data. Let us imagine a person is facing a situation for which he holds both specific and generic information. Generic information is called the *base-rate* information.[120] If the person leaves aside any specific information about the situation at hand, he fails under the bias called base-rate neglect. This may happen since people who have generic information tend to use it to judge probabilities and to evaluate specific cases. Likewise, some investors tend to rely on stereotypes when making investment decisions. When both generic and specific information is available, it might seem reasonable to give priority to the specific information. But, there is always some possibility that an observation or test may be wrong, and the probability that it is wrong is affected by the base-rate.[121]

Example

In finance, base-rate neglect is displayed by people who think that IPOs offer good long-term investment opportunities. Actually, research has shown that this is true only for a very low percentage of IPOs. There are stocks of investors who made a lot of money in the days following an IPO, but a generalization is not statistically

[120]Kahneman et al. (1982, pp. 153–154).

[121]Kahneman et al. (1982, pp. 155–157).

proven. By ignoring the probabilistic approach, many people fall prey to base-rate neglect.[122]

5.3.2.4 Sample-Size Neglect Bias

The other subcategory of representativeness bias is *sample-size neglect*.[123] In this case investors often fail to accurately consider the sample size of the data on which they base their judgments, and they incorrectly assume that small samples are representative of populations. Some researchers call this phenomenon the *law of small numbers*.[124] When people do not initially comprehend a phenomenon reflected in a series of data, they will quickly make assumptions relying on only a few of the available data points. Individuals prone to sample-size neglect are quick to treat properties reflected in such small samples as properties that accurately describe universal pools of data, although the small sample that is examined may not be representative of the real data.

Example

Let us consider Fig. 5.4 with two sequences of coin toss results, and assume that a fair coin has been used. Which of the pictured sequences do you think is more likely: Sequence 1 or Sequence 2?

Most people choose sequence 1 because it seems more random. In reality, both outcomes have the same probability to occur. Those who pick sequence 2 are affected by the sample-size neglect representativeness bias, also known as the gambler's fallacy or the *Law of Small Numbers*.[125] In fact, if six tosses of a fair coin are heads, the probability that the next toss shows tails is still 50 %. Nevertheless,

Fig. 5.4 Sample-size neglect illustration: which coin toss sequence appears likelier? *Source*: Pompian (2006, p. 72)

[122] Pompian (2006, pp. 64–65).
[123] Kahneman et al. (1982, p. 22).
[124] Kahneman et al. (1982, p. 22).
[125] Pompian (2006, pp. 69 and 72).

we generally still think that in coin tossing, an even ratio of heads to tails should occur, and if one side of the coin appears too often, the other is overdue.[126]

Sample-size neglect representativeness bias can easily be detected in the investment world. For instance, when an investor looks at a money manager's track record, he will mostly judge his performance on the basis of very limited statistical data (commonly for a few quarters), and draw conclusions about the skills of the manager.[127]

Research has also documented cases of representativeness bias in the financial markets with two surveys providing evidence of this bias:

- Case and Shiller surveyed house prices in 1988. They noticed that home buyers in cities, where house prices had risen rapidly in the past, anticipated much greater future price appreciation than home buyers in cities where prices had been stagnant or had fallen.[128]
- Froot and Frankel surveyed the evolution of the U.S. dollar exchange rate in the 1980s.[129] In fact, at that time, the U.S. dollar was rising without any relationship with macroeconomic fundamentals.[130]

Moreover, David Aronson pointed out a representativeness fallacy in the statements of market analysts during the crash of 1987. They forecast a deflationary depression after the 1987 market crash due to very noticeable similarities with the stock market behavior in 1929. But, an examination of the market crashes over the last 100 years shows that a price crash does not automatically precede a deflationary depression.[131]

5.3.2.5 Protection Method

In order to avoid representativeness bias, Pompian recommends to allocate assets so as to increase long-term returns, to invest in a diversified portfolio and to ask oneself four simple questions[132]:

1. How does the fund that you are considering perform relative to similarly sized and similarly styled funds?
2. What is the tenure of the managers and advisers at the fund?
3. Are the managers well known and/or highly regarded?
4. Do the fund's 3-, 5-, and 10-year returns all exceed market averages?

[126]Pompian (2006, pp. 69 and 72).
[127]Pompian (2006, p. 67).
[128]Shleifer (2000, p. 155).
[129]Froot and Frankel (1989).
[130]Shleifer (2000, p. 155).
[131]Aronson (2006, p. 40).
[132]Pompian (2006, p. 74).

But we definitely have to remember that every year produces new top managers and new laggards.[133]

5.3.3 Herding Bias

5.3.3.1 Definition

Herding is a fundamental aspect of human society, and herding bias describes the fact that people who regularly communicate with one another make similar decisions. In order to understand the origin of this kind of behavior, Solomon Asch,[134] an acclaimed social psychologist, conducted an experiment called *conformity study* in 1952 that showed the immense power of social pressure on individual judgment.[135] The results of his research are still cited today.

5.3.3.2 The Asch Experiment

In his famous experiment, Asch placed a participant into a group of seven to nine other people who were unknown to him. He neither knew that these people (called the *confederates*) had been coached by Asch before the experiment started. The whole group was asked to answer a sequence of 12 questions. The questions were related to the length of the lines in Fig. 5.5. One sample question would be: *Is line A, B or C of the same length as the left line in Fig. 5.5?*

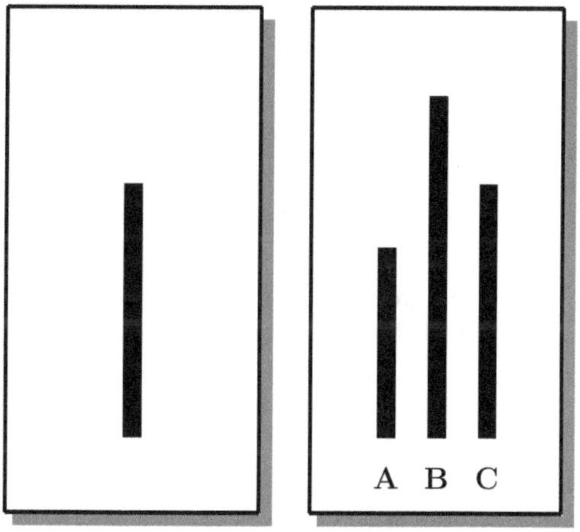

Fig. 5.5 Parallel lines of different length in the Asch experiment. *Source*: Shiller (2005, p. 158).

[133]Pompian (2006, p. 67).

[134]Solomon E. Asch (1907–1996) was a pioneer of social psychology. Born in Warsaw, Poland, he came to the United States in 1920 and received a Ph.D. from Columbia University in 1932. He studied relation-oriented approaches to perception, association, learning, thinking, and metaphor. He became famous in the 1950s with the *Solomon Asch Experiment* which showed that social pressure can make a person say something that is obviously incorrect.

[135]Asch (1952).

The participant would first hear most of the answers given by the confederates, before giving his own answer. But the answers of the confederates to seven of the twelve questions were deliberately wrong. The participant, in one third of the cases, gave the same wrong answers as had been given by the confederates. Furthermore, the participant often showed signs of anxiety or distress, suggesting that the fear of being seen as different or foolish before the group had swayed his judgment.[136]

Herding in financial markets explains speculative stock market bubbles.

- When investors are bullish, they are ready to pay nearly any price. While sellers only want to sell at a higher price, this combination pushes prices up.
- If all market participants are bearish, the opposite effect occurs and drives prices down.

In the history of financial markets, there are many examples where people who did not follow the mainstream were seen as foolish even if they had sound arguments.[137] Herding has a major impact on at least three common theories[138]:

- The assumption that all economic agents are independent of each other in taking their decisions is false.
- The law of supply and demand, where higher prices are supposed to attract more sellers and deter buyers, does not have an immediate effect.
- The idea that asset prices only provide information about fundamentals is incorrect. They reflect a mixture of hard information and soft information, like the crowd's mood, which is difficult to untangle.

According to Didier Sornette, herding during a stock market crash manifests itself in different ways[139]:

- *Information cascades* occur when investors choose to ignore their private information and imitate the actions of other investors who acted previously. They happen when the information of a single investor is not strong enough to reverse the decision of the crowd.
- *Reputational herding* works like information cascades: the investors react like the others in order to protect or improve their reputation by acting in line with a more powerful group.
- *Empirical herding*: There is evidence of herding and clustering among institutional investors when they engage in buying or selling.
- *Other reasons* include the influence of networks, the lack of information, etc.

---

136 Shiller (2005, p. 158).

137 Brunnermeier (2001, p. 165).

138 Schulmerich (2011, p. 7).

139 Sornette (2003, p. 165).

5.3.3.3 Protection Method

Herding bias has to be taken into consideration because it significantly challenges standard economics and finance. As Pompian notices: *"Needless to say, the demise of the Internet-stock bubble demonstrated that even the most massive herd can stampede in the wrong direction."* [140] Herding can be easily countered by justifying an investment with long-term credentials and quantitative statements. It should thus be possible to resist the pressure of selected information and herding.

5.3.4 Overoptimism Bias

5.3.4.1 Definition

Overoptimism is the tendency of people to overrate their abilities, like driving a car or managing companies, and to consider themselves above average. It is linked to the illusion of control or the illusion of knowledge. They either believe they control the outcome of an event or think they know more than the others, while in fact, they neither control anything, nor have superior knowledge.[141] Let us illustrate this bias with several participative tests, and please feel free to answer the questions too, according to the propositions. Four cases will be presented which deal with driving skills, the academic success of students, the performance of companies and the skills of fund managers.

5.3.4.2 Example 1

In his book *Beyond Greed and Fear*, Hersh Shefrin recounts an experiment realized in the United States and in Sweden with the aim to assess the number of drivers who considered themselves above and below average.[142] Participants were asked the following question: *"How good a driver are you relative to the drivers you encounter on the road?"* Possible answers were:

1. Above average
2. Average
3. Below average

Note that the formulation does not imply any emotional aspect. Assuming the sample to be large enough and assuming also that people think rationally—as the *Homo economicus* hypothesis would expect—the central limit theorem (CLT) should apply.[143] In other words, we should reach an approximately symmetric

[140] Pompian (2006, p. 231).

[141] Montier (2007, p. 82).

[142] Shefrin (2002, p. 41).

[143] In probability theory, the central limit theorem states that, given certain conditions, the arithmetic mean of a sufficiently large number of iterates of independent random variables, each with a well-defined expected value and well-defined variance, will be approximately normally distributed.

distribution with 50 % of the drivers above and below the average. In fact, the tests have shown the following results:

- In the test that Myron Scholes mentions, between 65 and 80 % of the participants rated themselves *above average*.
- In Sweden, the same test was conducted. Here, the outcome was even more extreme: 90 % of the drivers rated themselves *above average*.

In both tests, drivers were *overoptimistic* about their driving skills and suffered from an illusion of knowledge: they thought they knew more than the others. Another famous example shows the illusion of knowledge among students.

5.3.4.3 Example 2

During class, students were asked if they thought they would finish in the top 50 % of the class. Like in the previous experiment, we expect a symmetric distribution of *yes* and *no* answers of around 50 %. But the results showed that around 80 % of the students thought they would finish among the top 50 %. Of course, around 30 % of them were disappointed at the end of the course.[144]

5.3.4.4 Example 3

Figure 5.6 present the results from the Duke survey on CFO optimism. CFOs were asked how they evaluated the likelihood of their company to overperform the economy. Assuming that CFOs are rational, we should have on average an equal amount of optimism about the economy and the companies. However, Fig. 5.6 shows a completely different result. In the long run, the CFOs seem to be more

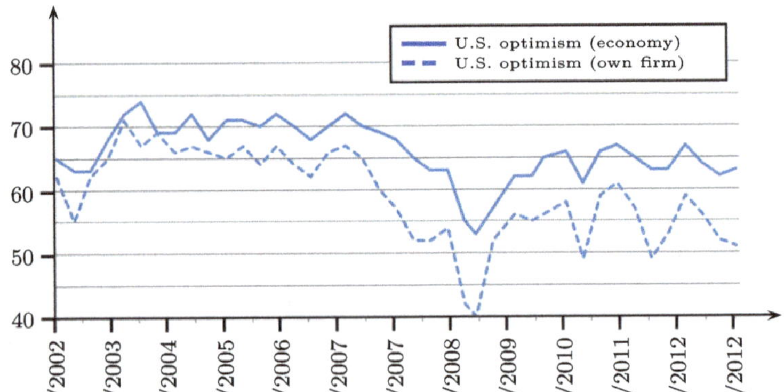

Fig. 5.6 U.S. CFO optimism about their own firm from the second quarter of 2002 to the third quarter of 2013 in percent. *Source*: http://www.cfosurvey.org

144 Montier (2002, p. 3).

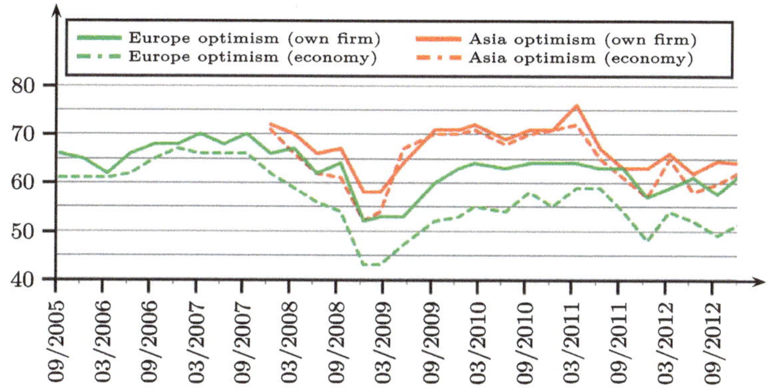

Fig. 5.7 CFO optimism in Europe and Asia about their own firm from the second quarter of 2005 to the forth quarter of 2012 in percent. *Source*: http://www.cfosurvey.org

optimistic about their own firm. This figure can be presented as a proof of the illusion of control.[145] We get a similar picture when we look at Europe and Asia as displayed in Fig. 5.7.

5.3.4.5 Example 4

The same test that was used to assess the overoptimism of drivers and students was given to 200 fund managers.[146] Figure 5.8 shows their self-assessment of their investment skills. Considering the large sample size and the required standards for this profession, once again, we should theoretically see a relatively symmetric distribution around the mean.

But according to the empirical results, 75 % of the managers believe they have superior capabilities compared to their peers. This is not consistent with the *Homo economicus* hypothesis and an evident proof of overoptimism.

During stock market crashes, there are investors who, according to Robert Shiller, expect a rebound to occur. In 1987 for instance, 37 % of institutional investors who expected a rebound answered that such a decline had never occurred without a rebound.[147] Others tried to justify such a belief and cited low prices as a reason to expect a rebound. In this case, investors fell victim to the overoptimism bias thinking they knew more than the others. Even during crashes, investors tend to think that they can rely on their expertise. They behave aggressively with certainty

[145]The Duke survey covers over 500 Chief Financial Officers (CFOs) of major United States companies from all industries.

[146]Montier (2007, pp. 82–83).

[147]Shiller (2005, p. 152).

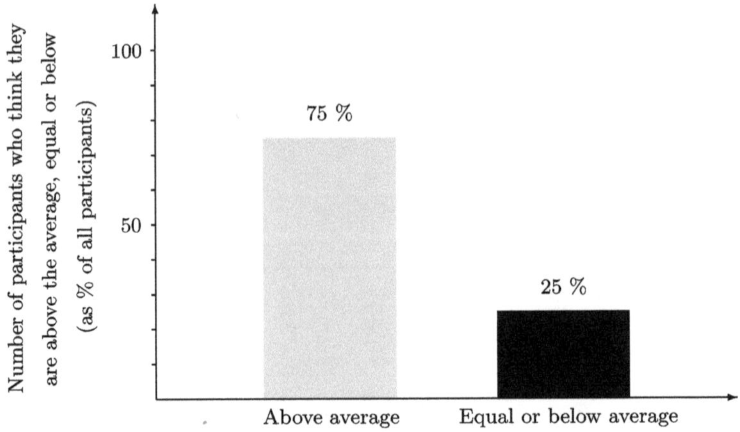

Fig. 5.8 Self-assessment of professional fund managers of their investment skills relative to the average. *Source*: Montier (2007, pp. 82–83)

and are optimistic about their forecasts. According to Shiller, investors showed signs of overoptimism during the strong bull market preceding the 1987 crash.[148]

5.3.4.6 Protection Method

In order to overcome the harmful effects of overoptimism, one should, according to wealth management manager Michael M. Pompian[149]:

1. Save and invest responsibly when opportunities arise, since overoptimism endangers savings.
2. Ensure a sound asset allocation for a well-managed portfolio. Overoptimism can lead us to overweight an asset class. A wise investor always builds a portfolio with a balanced asset allocation.
3. Invest for the long term. Compounding contributes significantly to long-term financial success. Therefore, the long-run accumulation and compounding of assets is reasonable.

Very similar to overoptimism, overconfidence bias can also lead to significant asset misallocation and mistakes.

5.3.5 Overconfidence Bias

5.3.5.1 Definition

People with overconfidence believe that their own judgments are more reliable than objective information. They believe that bad events will not occur to them. The

[148]Shiller (1992, pp. 390–391).

[149]Pompian (2006, p. 170).

concept is derived from psychological experiments that have shown that people tend to think they have better information than they really do. While overoptimism is self-oriented and refers to the qualities of a person, overconfidence concerns the quality of information at disposal. In other words, overconfidence leads a person to overrate exogenous parameters, i.e., elements which have nothing to do with him, whereas overoptimism results in overestimating endogenous parameters, i.e., elements which are directly linked to a person, for example, personal skills and characteristics.

5.3.5.2 Example 1

A given investor has just received information that the sales of Lenovo, a computer company, performed better than expected during the last semester, and assumes this information to be not public. He believes that Lenovo is a hot stock and that he has to invest in it to make a quick profit. He then invests all his money in this company, assuming Lenovo will perform better than Dell, a competitor, without checking Dell's performance. He also forgets to quantitatively and qualitatively check the other companies of the industry. Obviously, he shows overconfidence in the information he just received about Lenovo and overoptimism in his own investment skills, because he assumes he knows better than any other investor where and when to invest.

5.3.5.3 Example 2

Another example is about a job search. Michael, a freshly degreed Ph.D. student from the MIT, has heard that Citibank, a worldwide bank, is hiring a lot of mathematicians for its new branch in Budapest. In the following 3 months, he applies only to this company. At last, he gets an interview, but does not prepare for it, assuming he has the required knowledge. On one side, he proves overconfident. *A lot* of mathematicians is not a number. Will they hire ten or hundred mathematicians? Competitors might hire more in the same location and offer positions which are of interest to him. He also proves to be overoptimistic about the outcome of the interview. He may be well trained, but 3 months have passed since Michael last saw a mathematical formula.

5.3.5.4 Types of Overconfidences

Two types of overconfidence have been documented to date[150]:

- Prediction overconfidence leads people to make estimations with too narrow confidence intervals.
- Certainty overconfidence makes people blind to arguments against their opinions.

We will illustrate both types of overconfidence with examples. The first is a general knowledge game, the second deals with the diagnosis of psychologists. The last one is about hypothetical financial calculation of a famous U.S. index.

[150]Pompian (2006, pp. 51–52).

5.3.5.5 Example 3

Table 5.2 presents a series of questions found in the book by Russo and Schoemaker.[151] Please state a low and high value for each question so that you are 90 % confident that the correct answer is included in your interval:

Now, let us compare for each of the ten questions your answers with the correct answers displayed in Table 5.3. If the true answer is not within your stated interval (i.e., if the true answer is below your low estimate or above your high estimate), then the correct answer is outside of your given 90 % confidence interval. Since you were 90 % certain on each of your given answers, you would expect this to happen only in one out of the ten questions. If it happens more than once, i.e., if you have in less than nine questions the true answer within your interval, you tend to be overconfident.

Let's now compute your accuracy. To do so, for each question, add-up both low and high values of your answers, and then divide the sum by two. For example, if for the first question your low answer happened to be 30 and your high answer happened to be 40, the average is 35. Then, divide this number by the exact number displayed in Table 5.3 to obtain your actual accuracy (in this case 0.8974) for this question. To compute your overall accuracy for the test, just calculate the average of the ten single accuracies you have computed for each question. The test results of Russo and Schoemaker are presented in Fig. 5.9.

Although the level of confidence was 90 %, the participants only reached an accuracy rate of 1 %. This test is an evident case of what is called predictive

Table 5.2 Series of questions to illustrate overconfidence bias

	90 % Confidence	
Question	Low	High
Martin Luther King's age at death		
Length of the Nile River (in miles)		
Number of countries that are members of OPEC		
Number of books in the Old Testament		
Diameter of the moon (in miles)		
Weight of an empty Boeing 747 (in pounds)		
Year in which Wolfgang Amadeus Mozart was born		
Gestation period of an Asian elephant (in days)		
Air distance from London to Tokyo (in miles)		
Deepest (known) point in the ocean (in feet)		

Source: Russo and Schoemaker (1990, p. 71)

[151]Russo and Schoemaker (1990, p. 71).

Table 5.3 Answers to the series of questions to illustrate overconfidence bias

Question	Answers
Martin Luther King's age at death	39 years
Length of the Nile River (in miles)	4,187 miles
Number of countries that are members of OPEC	13 countries
Number of books in the Old Testament	39 books
Diameter of the moon (in miles)	2,160 miles
Weight of an empty Boeing 747 (in pounds)	390,000 pounds
Year in which Wolfgang Amadeus Mozart was born	1756
Gestation period of an Asian elephant (in days)	645 days
Air distance from London to Tokyo (in miles)	5,959 miles
Deepest (known) point in the ocean (in feet)	36,198 feet

Source: Russo and Schoemaker (1990, p. 71)

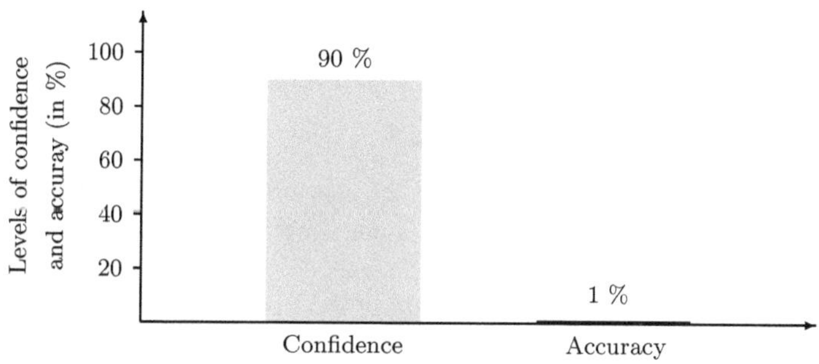

Fig. 5.9 Histogram comparing the level of confidence and the accuracy of the provided answers. *Source*: Russo and Schoemaker (1990, p. 71)

overconfidence. The level of accuracy does not match the level of confidence because of a *too narrow confidence interval* chosen by the participants.

5.3.5.6 Example 4

In the following experiment, psychologists were asked to answer questions regarding a patient's behavioral pattern.[152] The experiment was conducted in four stages. In the first stage, the participants received a short description of the patient, and got more and more information about him in the next stages.

In each stage, the psychologists had to answer the same questions about the patient's behavior pattern, attitudes, interests and typical reactions to real life events. The accuracy was measured in percent by comparing their answers with the correct answers. At the end of each stage, the psychologists had to state how confident in percent they were about the correctness of their answers.

[152] Kahneman et al. (1982, pp. 287–288).

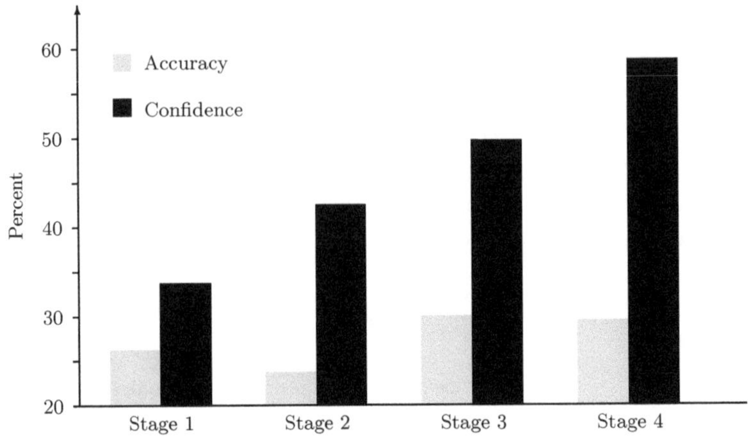

Fig. 5.10 Histogram comparing the level of confidence and the accuracy of the answers for each of the four stages. *Source*: Kahneman et al. (1982, pp. 287–288)

The provided information was as follows:

1. The patient seeks clinical help. He is a 29 year old, white, unmarried veteran of World War II. He works as a business assistant in a floral decorating studio and is a college graduate.
2. One and a half pages of information about the patient's childhood to the age of 12 are added.
3. Two pages of information about the patient's high school and college experience are added.
4. Additional information on the patient's army service and later life up to the present day are given.

Figure 5.10 shows the four steps of the experiment. For each step, the confidence interval is placed next to the accuracy of the prediction. Also pay attention to the vertical axis which starts at 20 % and not zero.

This example illustrates *certainty overconfidence*, also called *experts' overconfidence*. The more information the psychologists get, the more confident they are in their judgments. But the accuracy of the judgments does not increase with the level of confidence. The psychologists—and by extension all experts—are overconfident and underestimate the relevance of numerous prospects and possibilities.[153] They make forecasting mistakes notwithstanding the fact that they are experts in their field.

[153]Kahneman et al. (1982, pp. 287–288).

5.3.5.7 Example 5

The next experiment, designed by researcher Hersh Shefrin, is about the Dow Jones Industrial Average Index, a price index which does not include reinvested dividends. The test participants were told that the DJIA closed the year 1998 at 9,181 points and were asked: *"If the DJIA were redefined to reflect the reinvestment of all dividends since May 1896, when it had a value of 40, what would its value have been at the end of 1998?"* [154]

The participants who were asked this question as part of a survey had to give their best guess, and also to write down a low guess and a high guess, so that they felt 90 % confident that the true answer would lie between the two values. The correct answer is that the Dow Jones Industrial Average Index would have crossed 652,230 at the end of 1998.[155]

None of the participants' answers was reasonably close to the true answer. Like in the first example, the participants set overly narrow confidence bands.

Institutional investors are considered experts in finance. But research has shown that they are overconfident, like the psychologists in the example above.[156] They trade more than private investors because they think that they possess special knowledge. Through his survey on the crash of 1987, Shiller found out that 47.9 % of institutional investors thought that the market would rebound during the day of the crash.[157]

Overconfidence can lead to investment mistakes and to the overestimation of stocks.[158] Investors who are affected by this bias are prone to multiply investments that they consider appropriate. This transforms into high turnover and trading activity and leads to lower returns because of transaction costs and tax issues. Another result is the underdiversification of portfolios, given that one cannot consistently follow every outstanding growth stock. An overconfident investor takes more risk which is sometimes underestimated while the overall portfolio is underperforming.

5.3.5.8 Protection Method

Three rules should be followed in order to avoid overconfidence bias:

1. Build balanced asset allocations so as to prevent overconfidence about a given stock or industry.
2. Use discipline in the assessment of stock performance and in investment.
3. Invest on a long-term horizon or use protective financial strategies, for example, options. Transaction costs will be diminished and the portfolio will be kept well balanced.

---

[154] Shefrin (2002, pp. 18–19).

[155] Clarke and Statman (2000, p. 89).

[156] Shiller (2005, p. 152).

[157] Shiller (2005, p. 152).

[158] Pompian (2006, p. 54).

5.3.6 Anchoring Bias

5.3.6.1 Definition

Anchoring occurs when a person who wants to make a decision and is about to form a quantitative assessment, is influenced by irrelevant and meaningless data. It occurs frequently when a decision is needed, but cannot be easily made.[159] According to Shiller, there are two kind of anchors[160]:

- Quantitative anchors are numbers or quantitative variables used to make estimates, such as past stock prices.
- Moral anchors are found in storytelling or justifications. Much of human thinking is not quantitative, thus resulting in storytelling or justifications. Moral anchoring may tie the market down as people hold their investments against their perceived need to consume the wealth that the investments represent.

The following examples illustrate these anchoring biases.

5.3.6.2 Example 1

Amos Tversky and Daniel Kahneman demonstrated quantitative anchoring in an experiment with a wheel of fortune.[161] On the wheel, which stopped randomly, were numbers from 1 to 100 on it (similar to wheels used in TV game shows).

First, the participants were asked questions, for example, about the percentage of African nations in the United Nations (UN). Before they answered, the wheel was turned and the test persons had to declare whether their answer was above or below the number on the wheel. Finally, the participants were asked their exact answer.

Normally, given that the wheel and its numbers have nothing to do with the number of African UN members, and assuming a high probability that test persons had no clue about the answer, we should reach a substantially equivalent answer among participants, whatever the wheel's result. However, the participants did not know that the wheel was tricked and could only stop at either 10 or 65. The answers of the participants were as follows[162]:

- If the wheel stopped at 10, the median guess of the percentage of African nations in the UN was 25 %.
- If the wheel stopped at 65, the median guess was 45 %.

Figure 5.11 graphically presents the impact of the fortune wheel on the guess of the test participants.

---

[159] Montier (2002, p. 5).

[160] Shiller (2005, p. 148).

[161] Kahneman and Tversky (1979).

[162] Shiller (2005, pp. 147–148).

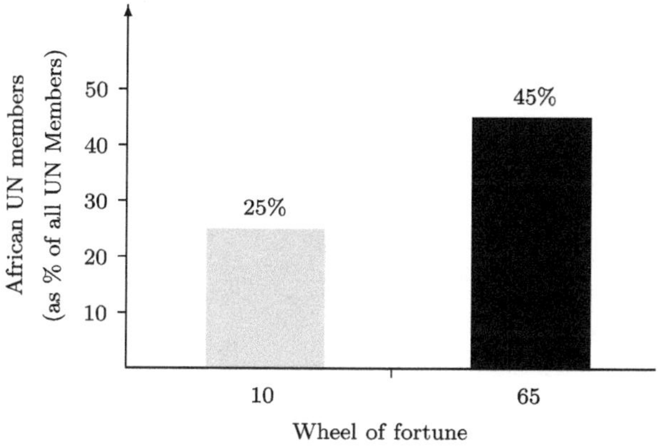

Fig. 5.11 Impact of the fortune wheel on the guess of the proportion of African UN members. *Source*: Shiller (2005, pp. 147–148)

Obviously, the random number on the wheel served as a quantitative anchor and influenced the answers which were given after the wheel number was shown. This experiment exemplifies quantitative anchoring.

5.3.6.3 Example 2

A group of judges was given the same information about the same case. Then, independently, each of them had to roll a pair of dices, the result being either 3 or 9. The judges were told that the result of the dice roll represented the sentencing request from the prosecution. Although the judges knew that the input was totally irrelevant, the result was the following[163]:

- If the dice roll was a 3, the average sentence declared by the judges were 5.3 months.
- If the rolled a 9, they gave a sentence of 7.8 months.

Figure 5.12 graphically presents the impact of the dice game on the sentence. This experiment is another example of quantitative anchoring. Clearly, when the judges determined the sentence for the case they were confronted with, they were influenced by the outcome of the dice roll.

5.3.6.4 Example 3

The following experiment involved experienced real estate evaluators who were asked to determine an achievable price for a house on sale.[164] They were split into two groups and had 20 min to visit the house. They also received an information package including information about the house itself and its surrounding area. The

[163]Englich, Mussweiler, and Strack (2006).

[164]Calverley (2004, pp. 127–128).

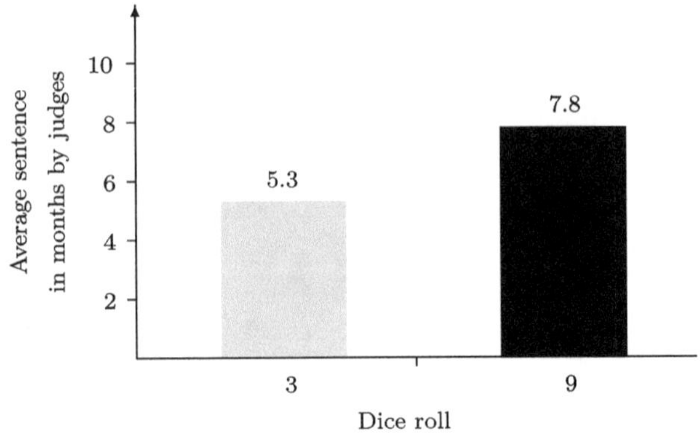

Fig. 5.12 Impact of the dice game on the sentence. *Source*: Englich et al. (2006, p. 188)

information was the same for both groups, but the asking price of the property was different. The outcome of the experiment was as follows[165]:

- The evaluators who were given an asking price of \$119,900 on average suggested that \$111,454 should be achievable for the house.
- The group who were given an asking price of \$149,000 thought on average, that \$127,318 could be achieved.

Here, the initially asked price, which was at hand before a decision was made, was the quantitative anchor that influenced the price proposal of the experts by the realty experts.

5.3.6.5 Example 4

Another quantitative anchoring experiment was performed with a group of fund managers.[166] The idea behind this test was to see if the participants would use their phone numbers as an input when trying to estimate the number of doctors in London. The first part of the participants had phone numbers above the value of 7,000, the second part had numbers below the value of 3,000.

Because phone numbers are absolutely irrelevant while evaluating the number of doctors in London, they should have no impact. Yet, as displayed in Fig. 5.13, they had a significant influence.

We clearly see the correlation between the answers and the values of the phone numbers. This is another evidence of quantitative anchoring.

Quantitative anchoring also occurs in the investment business. When faced with investment decisions people seem to be mostly anchored in past price changes. They use the average price of assets to which they add the compounded average price changes of a previous period. So assuming a house is valued \$100,000 with an

[165] Calverley (2004, p. 128).

[166] Montier (2007, p. 87).

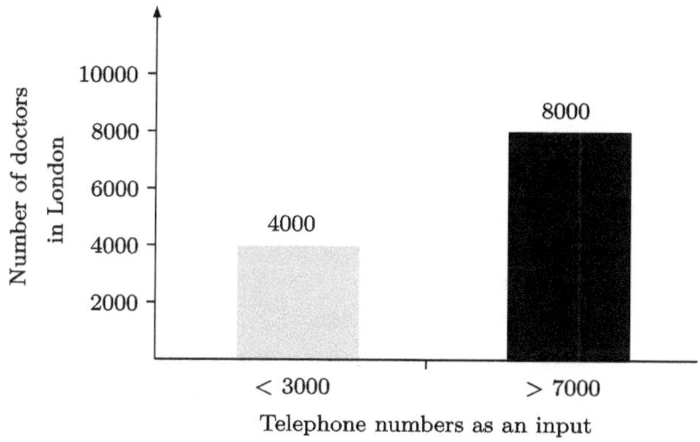

Fig. 5.13 Impact of the phone number on the evaluation of the number of doctors in London. *Source*: Montier (2007, p. 87)

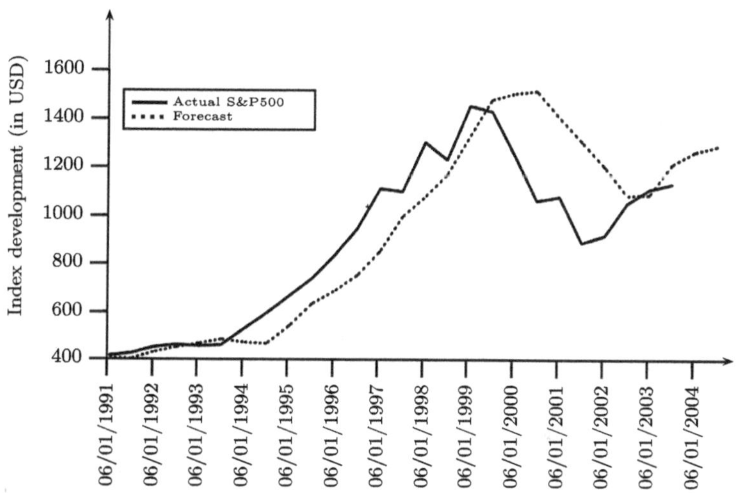

Fig. 5.14 S&P 500 index and forecasts of analysts. *Source*: Montier (2005, p. 11)

approximate price growth rate of 10 % each year for three consecutive past years, many would evaluate the price of the house next year without further research at $121,000.[167]

Analysts also fall prey to quantitative anchoring when forecasting. Figure 5.14 presents the S&P 500 Index and compares it to the average forecasts by analysts. We can see that forecasts of the S&P 500 from June 1991 to June 2004 were highly correlated to the past prices of the index. We will now give examples of moral anchoring.

[167]Shefrin (2005, p. 51).

5.3.6.6 Example 6

Eldar Shafir,[168] Itamar Simonson[169] and Amos Tversky conducted an experiment that showed decision biases which were obviously caused by the desire to find simple reasons for decisions. This bias is called moral anchoring.[170]

The researchers asked their subjects who were divided into two groups to decide on the custody of a child of separated parents. They had to award sole custody to one of the parents, and their options were:

- *Parent A* with an average income, average health, average working hours, a reasonable rapport with the child and a relatively stable social life. This option is called *impoverished.*
- *Parent B* with an above-average income, a very close relationship with the child, an extremely active social life, lots of work-related travel, and minor health problems. This option is called *enriched.*

It is evident that the first option includes no strikingly positive or negative features, while the second one displays both. This information was the same for both groups, but the question they were asked was not.

- The first group was asked:
  To which parent will you award custody, parent A (Option 1) or parent B (Option 2)?
- The second group was asked:
  To which parent will you deny custody?

Figure 5.15 shows the results of each group. When answering the questions, the majority of the participants chose parent B, but the results of the two groups were different and depended on how the questions were asked. This is an example of moral anchoring.

---

[168] Eldar Shafir graduated from Brown University and received his Ph.D. in Cognitive Science in 1988 from the Massachusetts Institute of Technology. He is an American psychologist and the William Stewart Tod Professor of Psychology and Public Affairs at the Princeton University Department of Psychology and the Woodrow Wilson School of Public and International Affairs. He is a faculty associate at the Institute for Quantitative Social Science at Harvard University. He is co-founder and scientific director at ideas42, a social science R&D lab.

[169] Itamar Simonson is a professor of consumer psychology, decision-making, market research, and marketing management. He is known for his work on the factors that determine the choices of buyers. He is the Sebastian S. Kresge Professor of Marketing in the Graduate School of Business, Stanford University. He received his B.A. in Economics and Political Science from the Hebrew University in 1976, his MBA from the UCLA School of Management in 1978, and his Ph.D. in Marketing from Duke University in 1987.

[170] Shiller (2005, pp. 151–152).

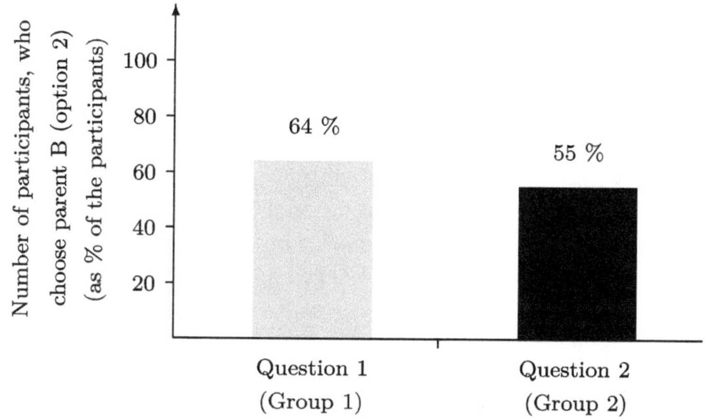

Fig. 5.15 Effect of moral anchoring on choice. *Source*: Shiller (2005, pp. 151–152)

Moral anchoring can also be found in the investment business.[171] In fact, reasons to hold stocks or other investments can take on ethical as well as practical dimensions. Our culture may supply reasons to hold stocks and other savings vehicles that are related to our sense of identity of being responsible, good and level-headed. Let us recall the book *The Millionaire Next Door*,[172] which was a best-seller during the stock market boom of the 1990s. The book affirmed that most millionaires in the U.S. are not exceptional income earners but merely frugal savers. for example, people who are not interested in buying a new car every year for instance are more likely to become rich. The book's enticing story about investment millionaires who do not cash out and consume their wealth was just the kind of moral anchor which helped to sustain an unusual bull market.[173]

Let us now focus on how investors display anchoring during market distress. We have previously seen that investors use two methods to estimate future stock prices: technical or fundamental analysis. The technical analysis is a method that consists of interpreting stock price charts in order to predict the future course of the stock price. To do so, so-called chartists study the past movement of stock prices and the volume of trading.[174]

Fundamental analysis considers the market to be mostly logical and driven by fundamental data. It cares little about the particular pattern of past price movements. Fundamental analysts value stocks with respect to growth rates, dividend payout, cash flow ratios, leverage ratio, the firm's business plan, the seniority of executive management, etc. They visit companies to assess the quality of management through

[171] Shiller (2005, pp. 151–152).

[172] Stanley and Danko (1998).

[173] Shiller (2005, pp. 151–152).

[174] Achelis (1995).

interviews or by direct inspections. When investors assess these factors, they rely on the approach of fundamental analysis. Fundamentalists believe that the market reflects the real worth of a security.[175]

Conclusions from both methods represent anchors for investors. In his survey about the 1987 crash, Robert Shiller asked investors whether they were influenced by price dropping through a 200-day moving average or another long-term trend line. A third of both individual and institutional investors answered *yes*.[176] Seventeen percentage of investors—8 out of 47 who answered questions about other indicators—confirmed that they were influenced by these indicators such as support and resistance levels.

Fundamental indicators also have an impact. They often influence long-term investors who think that corporate earnings or interest rates indicate the long-term trends of stock markets. Therefore, fundamentals can trigger stock market crashes, since heightened interest rates or disappointing corporate earnings (i.e., earnings below the forecast ones) urge many investors to sell, irrespective of their long-term investment focus. Prices may fall abruptly, and a stock market crash may be the consequence.[177]

5.3.6.7 Protection Method

In investments, awareness and rationality are required against anchoring. When forecasting price levels and trends, estimates should be rational and not be based on the performance statistics of previous periods.[178]

Until now we have covered the biases of availability, representativeness, herding, overoptimism, overconfidence and anchoring. A rational investor who is aware of them is less likely to fall under their charms. The discoveries of Kahneman and Tsversky were a great achievement, since they were able to prove empirically that some characteristics of the *Homo economicus* cannot be observed and are even contradicted in practice. Moreover, as these characteristics are shared by most humans, the discovered effect does not disappear under the law of big numbers as it was previously assumed, even if the observed population is significantly increased.

5.3.7 Prospect Theory

5.3.7.1 A First Approach

In 1979, Kahneman and Tversky introduced the *prospect theory*.[179] This masterpiece of research is fairly well pictured in Fig. 5.16.

---

[175] Malkiel (2003, pp. 126–127).

[176] Shiller (1992, pp. 376).

[177] Shiller (1992, pp. 394–395).

[178] Pompian (2006, pp. 81–82).

[179] Kahneman and Tversky (1979, pp. 263–292).

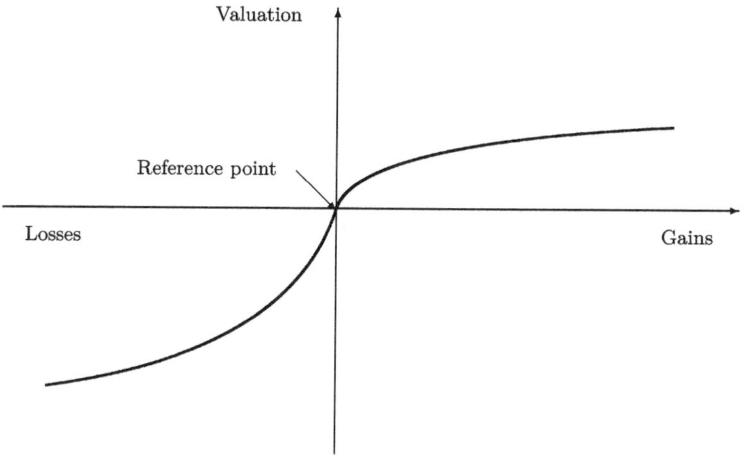

Fig. 5.16 Prospect theory. *Source*: Own based on Gärdenfors and Sahlin (1988, p. 188)

As simple as it may appear, this graphic shows a significant amount of information. The horizontal axis, the gain-loss axis, shows how much money a person has won or lost in a given game with respect to a point of reference. The vertical axis measures how people value their gain or loss. If people were perfectly rational as expected from the classical view and its *Homo economicus* model, the gain or loss of an equal amount of money should produce the same positive and negative value. Logically, the line in the graphic should be straight, but obviously, this is not the case. We observe a value function which is:

- Concave for gains (representing risk aversion)
- Convex for losses (representing risk seeking)
- Generally steeper for losses than for gains (representing loss aversion)

In mathematics, we call a function convex if its shape is outward-oriented, i.e., toward the top. Concave or non-convex functions show a shape which is rather inward-oriented, i.e., toward the bottom. If we divide a perfect circle into a north and a south part, the north part is concave and the south part is convex. The amazing and apparently irrational line in Fig. 5.16 was obtained through several experiments conducted by Kahneman and Tversky. In the following, we will try to show the underlying reasons behind the curvature of the expected straight line and explain the reference point. Finally, we will present the prospect theory and discuss practical observations.

5.3.7.2 The Curves in the Line

In a first test, a group of people was proposed a choice between two possibilities. They could either immediately receive $100, or play a game with an unlimited number of trials where they had a 50 % chance to gain either \$160 or \$60.[180] The vast majority of the tested group chose the first option. While being attracted by \$160, they feared to gain only \$60, despite the fact they could play endlessly. Loss aversion is thus a key aspect of prospect theory.

Numerous subsequent studies have empirically shown a rule of thumb: *"The possibility of a loss is on average twice as powerful a motivator as the possibility of making a gain of equal magnitude."*[181] This logic holds true in terms of percentage. Assume a stock bought at \$10 has lost 50 % of its value and is now valued \$5. To regain its initial value, its has to grow by 100 %. The ratio between a loss of 50 % and a gain of 100 % is 2. This is why risk-averse people will generally ask for twice as much gain in terms of absolute value compared to equiprobable loss, where people tend to think in terms of relative value, i.e., in terms of ratio or percentage. Prospect theory states: *"People underweight outcomes that are probable, i.e., subject to hazard, in comparison with outcomes that are obtained with certainty."*[182]

5.3.7.3 The Reference Point

In a second test, participants were asked to compare two cars, one Porsche and one Skoda. Then they were asked whether they preferred the blue one or the red one. Surprisingly, the subjects were ready to base their decision on a color factor. A typical answer was *I prefer the red one*, despite the fact this was the model of the less well-known car manufacturer. The answer seldom was *I prefer the Porsche* or *I prefer the Skoda*. Here, a key aspect in decision-making processes becomes evident: when comparing two elements, for example, cars, the two models are believed to regroup the same properties. The choice is then made between on what is evidently differing between the two models, for example, color. Prospect theory states: *"People generally discard components that are shared by all prospects under consideration. Therefore, the way the question is asked also influences the answer, i.e., the reference point."*[183]

5.3.7.4 The Model of Prospect Theory

Prospect theory does not explicitly name concrete or relative preferences (for example, the preference of avoiding a loss compared to the realization of a gain).[184] According to Kahneman and Tversky, people gauge all potential gains and losses in relation to a benchmark reference point.[185] Therefore this point is the center of the graph in Fig. 5.16.

---

[180] Note that the expected gains are higher in the probability game than if the players get the money directly. Namely (\$60 + \$160) $*$ 0, 5 = \$110.

[181] Pompian (2006, p. 208).

[182] Asgary and Levy (2009, p. 383).

[183] Kahneman and Tversky (1979, p. 263).

[184] Pompian (2006, p.209).

[185] This point refers to the belief of people following the process of discarding components that are shared by all prospects under consideration. If the subject has equivalent conditions, relative to his

Prospect theory is tightly linked to the concept of loss aversion: the pain that people feel from a loss is stronger than the satisfaction from a gain of equal value. This fits with the observation that people generally feel a stronger impulse to avoid losses than to acquire gains, whereas the perfectly rational *Homo economicus* is assumed to consider one unit of loss as equal to one unit of gain.

Loss aversion as understood in prospect theory also encompasses the *disposition effect*, i.e., the desire of investors to hold losing investments too long (risk-seeking behavior) and to sell winning investments too quickly (risk-avoidance behavior).[186] Let us assume a risk-averse investor buys stocks which decline soon after. Empirically, he tends to stick to his acquisition and believes that money is not really lost as long as a sell order is not passed, waiting for a never coming rebound to reach breakeven. On the other hand, when a gain is realized, the same risk-averse investor is eager to sell very quickly so as to lock-in the profit, fearing a market downturn.

5.3.7.5 The Model in Practice

In July 1985, The Quarterly Journal of Economics published a paper by Barberis, Huang and Santos, which incorporates prospect theory into a model of asset prices.[187] This triggered a vast campaign of research on the topic. Scholars began to discover a host of empirical results that were not consistent with the capital asset pricing model and the efficient market theory. From this time period onwards, the idea of market anomalies and inefficiencies became predominant.

In the twentieth century, equity risk-adjusted returns consistently exceeded fixed income risk-adjusted returns over time. In 1985, Rajnish Mehra and Edward C. Prescott wrote *The Equity Premium: A Puzzle*, which showed that the equity premium[188] had steadily averaged 6.18 % over the twentieth century.[189] Ten years later, a research paper observed that, from 1900 to 2002, U.S. stocks earned a 5.3 % annual premium over Treasury bills.[190] If perfectly rational, investors should have rejected low-paying bonds and purchased equities only.

In 1995, Shlomo Benartzi[191] and Richard Thaler proposed an original solution to this market anomaly thanks to two behavioral finance concepts: loss aversion and

---

peers in the studied situation, this set of conditions will work as the benchmark of the experience and thus constitutes the origin on the figure. See Kahneman and Tversky (1979, pp. 263–292).

186 Shefrin and Statman (1985).

187 Barberis, Huang, and Santos (2001).

188 The equity premium is the difference between the average rate of returns on stocks and the lower average rate of return on riskless Treasury bills.

189 Mehra and Prescott (1985, p.145).

190 Benartzi and Thaler (1995, p. 73).

191 Professor Shlomo Benartzi is a leading authority on behavioral finance with a special interest in personal finance and participant behavior in defined contribution plans. He received his Ph.D. from Cornell University's Johnson Graduate School of Management, and is currently co-chair of the Behavioral Decision-Making Group at The Anderson School at UCLA.

myopia.[192] As explained earlier, *loss-averse* investors value a loss twice as much as a gain, while *myopia* means shortsightedness. When combined together, investors who evaluate their portfolios very frequently are also the most likely to experience losses and to suffer from loss aversion. That is why myopic, loss-averse investors would invest in bonds. According to the simulation of the researchers, whose results are consistent with the realized equity premium, the time horizon of this investor class averages 1 year.[193]

In summary, prospect theory is characterized by a non-linear relationship between the actual and the felt value of a gain versus a loss. People seem to value a loss twice as much compared to a gain. The reference point which marks the shift between loss and gain is based on the discarding of components that are shared by all prospects under consideration, though these could be more relevant than the observed differences. Prospect theory appears to be able to explain major reasons behind stock market anomalies. Yet, while having been confirmed through stock market crashes, it is nevertheless only a part of another theory: the positive feedback trading theory.

5.4 The Positive Feedback Trading Theory

The positive feedback trading (PFT) theory has been put forward by Robert Shiller in his book *Irrational Exuberance*[194] shortly before the dotcom burst. He proposes a qualitative approach to explain stock market crashes and heavy upward and downward price moves.

The idea behind PFT is to focus on endogenous factors and to view extreme market movements as a self-organizing process through two channels, namely a price-to-price channel (investors react to price changes) and a social-psychological channel (investors react to each other).[195] The model holds that previous price moves have an effect on subsequent stock prices, despite the possibility that there may be no fundamental reasons for these movements. Other researchers have tested the feedback trading model and found it relevant for long time periods, much longer than a crisis. They also found that the feedback model is a positive one, i.e., it implies that *"investors buy stocks when prices rise and sell them when prices go down."*[196] The theory was further improved by Jacobs who proposed to include stop-loss in the model, thus enhancing selling orders at critical values.[197]

We first describe the context in which Shiller developed his theory, then explain how the theory can be applied. Despite the fact that it does not explicitly use the

[192] Benartzi and Thaler (1995, p. 73).

[193] Benartzi and Thaler (1995, p. 73).

[194] Shiller (2005).

[195] Sornette (2003, p. 4).

[196] Danielsson and Love (2006).

[197] Jacobs (1999, p. 91).

terms of behavioral finance, the model actually takes into account the most relevant biases which were previously covered.

5.4.1 A Case Study

From the market opening on October 14, 1987, to the market close on October 19, major stock market indices in the U.S. declined by almost 30 %. As shown in Fig. 5.17 which represents the daily return on the DJIA from May 1987 to December 1987, on Monday, October 19, 1987, called Black Monday, the index fell 508 points. A plunge of 22.6 % in 1 day![198]

All major world markets declined sharply from October 14 to 19, 1987. Hang Seng,[199] which lost −45.8 %, is only one example for stock market declines which were greater than 20 % in 19 major industrial countries, indicating a global stock market crash.[200] Several reasons were given for this crash[201]:

- Severe unsustainable price rises. In the U.S., stock prices advanced 31.4 % during the 9 months before the crash. A common explanation of the crash was that stocks were overvalued due to a speculative bubble on high-risk bonds (called *junk bonds*) in the period before.[202]
- Emergence of portfolio insurances (also called CPPI for constant proportion portfolio insurance)[203]

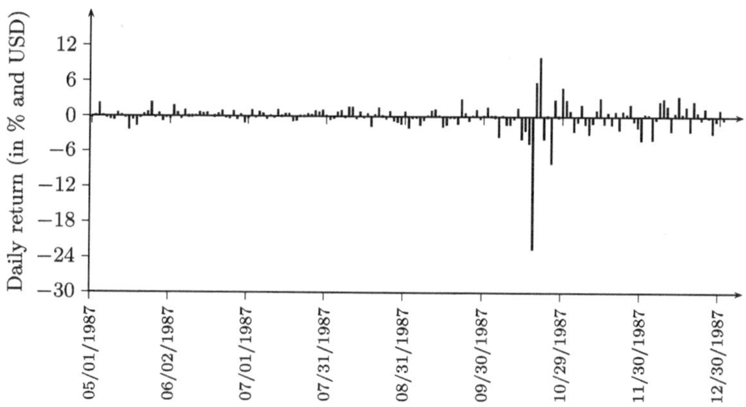

Fig. 5.17 Daily return of the Dow Jones Industrial Average Index from May 1, 1987 to December 31, 1987. *Source*: Bloomberg

198 According to Bloomberg data, as of October 20, 2013 (Ticker: INDU:IND).

199 Hang Seng is the stock index of Hong Kong. Ticker: HSI.

200 Shiller (1992, pp. 386–387).

201 Shiller (1992, pp. 386–387).

202 Galbraith (1993, p. 95).

203 Portfolio insurance (PI) is a strategy for limiting losses that was invented by Hayne Leland and Mark Rubinstein who also marketed it to many institutional investors in the 1980s. PI is a plan

- Computer-assisted trading techniques
- Increasing speculation on stock index futures
- Frequent computer breakdowns etc.

Despite many research efforts, no simple reason was found to explain the crash completely. However, one of the best-known research papers about the 1987 crash, written by Robert Shiller right in its aftermath, involves institutional investors.[204] Many of them agreed that stock prices were overvalued in September 1987, 1 month before the crash. But only a few of them linked the crash to a genuine news story (such as the Prechter[205] sell signal or the U.S. attack on an Iranian oil station) which suggests that the abrupt selling was not caused by fresh information.[206]

Robert Shiller also assessed the relevance of portfolio insurance (PI) for the crash, and asked the investors if they used it to limit losses.[207] While only 5.5 % of the institutional investors answered *yes*; around 10 % of the individual and institutional investors said they had a policy of limiting losses. However, it is possible to limit the losses without an explicit PI scheme, for example, simply by selling. Liquidating a position, on either the cash or the futures markets by stop-loss orders or puts, can be decided and programmed in advance. However, investors can also react without a computer through their brokers, and eventually, stock markets which did not use computer trading also crashed. Therefore, linking the crash to massive computer sell orders due to PI is doubtful. Index futures and derivative securities have also been made responsible for increasing the variability, risk and uncertainty of the U.S. stock markets. But an increase in risk does not imply a massive sell-off.

As a result, none of these reasons can really explain *Black Monday*.

5.4.2 The Positive Feedback Trading Model

Robert Shiller suggested to analyze the market crash differently,[208] i.e., as a self-organizing process.[209] To do so, he focused on two channels by which price declines could drive further prices declines:

---

for selling stocks when the prices start to go down. It specifies trading in index futures that causes index futures contracts to be sold continually as stock prices go down to hedge the portfolio against further losses. See Shiller (2005, p. 101).

[204] Shiller (1992).

[205] Robert R. Prechter (born in 1949), an American author and stock market analyst, is known for his financial forecasts using the Elliott wave principle. The Elliott wave principle is the basis for the Prechter Sell Signal, a form of technical analysis that aims at forecasting market trends by identifying investors' psychological trends through stock price behavior. See Elsner (1987).

[206] Shiller (1992, p. 386).

[207] Shiller (1992, p. 395).

[208] Shiller (1992, pp. 374–375).

[209] Sornette (2003, p. 4).

- A *price-to-price channel* where investors react to price changes
- A *social-psychological channel* where investors react to each other

Why are these channels relevant? The information that Robert Shiller collected on the frequency with which investors checked prices and talked to each other, provided clear evidence that both channels were operating fast enough among the broad masses of investors to play a role in the hour-to-hour movements of the market, especially during the day of the crash:

- Prices were checked with great frequency.
- Communications proceeded very quickly, especially among net buyers and sellers, but *separately*.[210]

The fact that prices were checked with a great frequency is a significant reason for analyzing the feedback between prices and their lagged values to understand the crash and the behavior of the investors. The communication and interaction between the two groups of investors with opposite opinions is also of interest, because it is a basic component of the overall behavioral pattern of investors.

Robert Shiller explained that simple feedback trading models—in which the stock prices depend on the changes of its lagged values, amplifying price changes in a vicious circle—could have explained the price dynamics during the crash of October 1987.[211]

But Shiller was not the only researcher who used a feedback trading model. Jon Danielsson[212] and Ryan Love[213] wrote an article in the International Journal of Finance and Economics about feedback trading where they declare that it also exists in its positive form in the spot USD/EUR currency market. Moreover, they documented that the price impact of order flow is higher when contemporaneous positive feedback trading is included in the model they used. In other words, PFT has more impact when the data is sampled at 1 or 5 min frequency than with an hourly

[210]Shiller (1992, pp. 374–375).

[211]Shiller (1992, p. 374).

[212]Jon Danielsson (born on October 17, 1963) is an Icelandic economist currently teaching finance at the London School of Economics and a member of the Financial Markets Group. His research areas include financial risk, hedge funds, regulation of financial markets, market volatility, liquidity, models of extreme market movements, and the microstructure of foreign exchange markets. He has been a visiting professor at universities in the U.S., the Netherlands, Germany, Iceland, and Spain, and has published in a range of academic and practitioners' journals. He has written extensively on the post-crash situation in Iceland. He received his Ph.D. in Economics of Financial Markets from Duke University in 1991.

[213]Ryan Love has worked in the financial markets since 1996 in various senior roles at international investment banks and financial advisory firms, before establishing Apex Partners in October 2008. He holds a Bachelor of Business from the University of Technology in Sydney, a postgraduate Diploma in Financial Services (Financial Planning), and meets the competency requirements under ASIC's Regulatory Guide RG 146.

or even daily frequency.[214] Other research publications confirm the PFT model for other asset classes such as stock markets in 1993 and U.S. treasury markets in 2001.[215]

Below we will deal with the factors which contribute to PFT and their behavioral finance equivalents:

- Extrapolative expectations about prices: representativeness bias
- Use of technical analysis: anchoring bias
- Trend chasing: herding bias
- Stop-loss orders: loss aversion bias

Extrapolative expectations is another name for what is called representativeness bias in behavioral finance. It is the tendency to evaluate a situation by comparing it to generalities, stereotypes, repetitions or convenient cases. This bias can be seen as a mental mechanism which prevents the decider from analyzing a given process completely and repeatedly. Two surveys have proven extrapolative expectations and illustrated this bias. In the first, Case[216] and Shiller surveyed house prices in 1988.[217] They noticed that home buyers in cities where house prices had risen rapidly in the past anticipated much greater future price appreciation than home buyers in cities where prices had been stagnant or fallen. In the second, written in 1989, Froot and Frankel researched the U.S. dollar exchange rate in the 1980s.[218] At the time of both studies, the U.S. dollar was rising without any relationship with macroeconomic fundamentals.

To which stock market phenomena do these studies correspond? The answer is the upward movement of the prices during speculative bubbles and the price plunge in stock market crashes. During a crash, most of investors sell, and prices go down. Since investors are prone to believe that recent prices movements constitute a new trend that will persist, they keep on selling and the downwards trend continues. Consequently, the bubble bursts. In a stock market boom, the opposite happens.

To predict the future course of stock prices and the appropriate time to buy or sell, investors usually rely on two methods: technical analysis and fundamental

---

[214] Danielsson and Love (2006, p. 35).

[215] Cohen and Shin (2003, p. 1).

[216] Karl E. Case is professor emeritus of economics at Wellesley College where he taught for over 30 years. He is also a founding partner in the real estate research firm of Fiserv Case Shiller Weiss, Inc., and serves as a member of the Boards of Directors of the Mortgage Guaranty Insurance Corporation (MGIC) and the Depositors Insurance Fund of Massachusetts. He is a member of the Standard and Poors Index Advisory Committee, the Academic Advisory Board of the Federal Reserve Bank of Boston and the Board of Advisors of the Rapport Institute for Greater Boston at Harvard University. Professor Case received his Ph.D. in Economics from Harvard University in 1976.

[217] Case and Shiller (1988).

[218] Froot and Frankel (1989).

analysis.[219] Both use past data or anchor values and can create a bias which in behavioral finance is called anchoring. Anchoring is the tendency of decision makers to use fallacious values or beliefs as the basis of their assessment. When making a decision, they will establish a judgment on inappropriate or irrelevant experience, while their views can be influenced by suggestions.

As previously mentioned, investors were influenced in the crash of 1987 by technical indicators, for example, support and resistance, as well as by fundamental indicators, for example, corporate earnings or interest rates.[220] Therefore, both technical indicators and fundamentals can trigger stock market crashes. Prices may fall abruptly, and a stock market crash may be the consequence.[221] The fact that investors use the same indicators, whether technical or fundamental, is one reason for their uniform behavior. But an even more accurate explanation is provided by the behavioral bias of herding.

Recall that according to Shiller's method, the analysis of a stock market crash has to consider a price-to-price channel and a socio-psychological channel. We have shown that the first one can be related to *anchoring*. The social-psychological channel is called *herding* in behavioral finance and refers to the interaction between investors. In his survey about the 1987 crash, Shiller found that the two groups of investors, buyers and sellers, communicated and interacted with each other intensively.

In the financial markets, herding became a central notion since concepts from the social sciences caught the attention of researchers. They started to reject the idea that investors are rational, a key assumption in the standard theories in finance. Herding can be linked to many activities[222]:

- Investment recommendations
- Price behavior of IPOs
- Earnings forecasts
- Corporate conservatism
- Portfolio management etc.

Information cascades occur when investors choose to ignore their private information and imitate the actions of other investors. Once a cascade has developed, the information of an individual investor may not be strong enough to reverse the decision of the crowd. Reputational herding, which works like information cascades, takes place when investors act in line with a more powerful group in order to protect or improve their reputation. There is ample evidence of herding and clustering among institutional investors when they engage in buying or selling. There are

[219]Malkiel (2003, pp. 126–127).
[220]Shiller (1992).
[221]Shiller (1992, pp. 394–395).
[222]Shiller (1992, p. 398).

also other reasons with less impact, like the influence of networks or the lack of information.

So far we have covered the price-to-price channel and the socio-psychological channel introduced by Shiller. Following up on his research, Jacobs proposed to include stop-loss orders as a component of the positive feedback trading theory.[223] Stop-loss orders aim to protect the profits or to cut the losses and instruct the brokerage to sell the stock at a predetermined price that is specified by various methods. for example, in CPPI, the method depends on complex mathematical models. CPPI is a PFT component because it aims at selling stocks when their price decreases. A basic stop-loss order works this way:

- Let us assume an investor buys a stock at $20 a share.
- At the time he buys the stock, a stop-loss is placed at $18 a share.
- The brokerage will automatically sell the stock if the stock price reaches $18 and the loss is limited at $20 minus $18, which equals $2, i.e., −10 % is the maximum loss accepted by the investor.

Stop-loss orders are linked to the previously covered human characteristic of loss aversion which is a pervasive characteristic of investor behavior. Therefore, during stock market crashes, the majority of investors sell to avoid losses. However, there are also buyers, even during crashes. Who are they? Robert Shiller found that in 1987, these investors thought they knew when a rebound would occur.[224] Actually, 37 % of institutional investors who expected a rebound answered that such a decline had never occurred without a recovery. Others cited low prices as a reason to expect a rebound.

In this case, the investors fell victim to overconfidence and overoptimism biases. Even during crashes, investors are overconfident thinking that they can rely on their expertise. As all other professionals, they are certain and optimistic about their forecasts, and this makes them act aggressively. Investors were overconfident during the strong bull market[225] before the crash of 1987, but also *during* the crash as Robert Shiller's survey has demonstrated. Yet, it also shows that in the stock market, but also in real estate investment, investors or prospective investors lose confidence after a stock market crash.

Robert Shiller asked a sample of high-income U.S. Americans if they agreed with the statement: *"The stock market is the best investment for long-term holders, who can just buy and hold through the ups and downs of the market."*[226] In Table 5.4, which relates the results, one can notice the loss of confidence in the stock market after the 2,000 crash, which followed the dot-com bubble.

---

[223] Jacobs (1999, p. 91).

[224] Shiller (1992, pp. 390–391).

[225] Strong bull markets are the main characteristic of speculative stock market bubbles.

[226] Shiller (2005, p. 57).

Table 5.4 Results of the poll among U.S. Americans about the statement: *"The stock market is the best investment for long-term holders, who can just buy and hold through the ups and downs of the market"* (in percent)

	1996	1999	2000	2001	2002	2003	2004
Strongly agree	69	66	67	61	46	42	42
Agree somewhat	25	30	30	31	41	42	41
Neutral	2	2	2	2	3	6	9
Disagree somewhat	2	1	1	3	8	6	4
Strongly disagree	1	1	0	2	2	4	4

Source: Shiller (2005, p. 60)

Table 5.5 Results of the poll about the statement that after a crash the stock market will surely rebound (in percent)

	1996	1999	2000	2001	2002	2003	2004
Strongly agree	38	42	42	38	29	26	21
Agree somewhat	44	44	40	43	44	44	46
Neutral	10	5	10	10	10	12	18
Disagree somewhat	5	7	9	7	12	13	11
Strongly disagree	2	1	0	3	5	5	4

Source: Shiller (2005, p. 62)

Robert Shiller also asked whether the respondents agreed with the statement: *"If there is another crash like October 19, 1987, the market will surely be back up to its former levels in a couple years or so."* [227] Table 5.5 shows that the belief in a stock market recovery after a similar crash like 1987 has been considerably weakened after the 2,000 market crash.

To sum up, the positive feedback trading model holds that investors buy stocks when prices are rising and sell them when prices are declining. In an explanatory model, the price impact of an order flow is higher when the impact of positive feedback trading is taken into account.[228] PFT focuses on endogenous factors and considers extreme market movements as a self-organizing process.[229] Stock prices evolve via two channels, by which price declines can drive further prices declines. These two channels are the price-to-price channel where investors react to price changes and the socio-psychological channel where investors react to each other.

Robert Shiller uses feedback trading models to explain the price dynamics of the stock market crash of October 1987.[230] From the information he collected on the frequency with which investors checked prices and talked to each other during

[227] Shiller (2005, p. 61).

[228] Stock prices are now updated almost every second. This strengthens the effect PFT can have on the markets compared to decades ago when prices where updated less frequently.

[229] Sornette (2003, p. 4).

[230] Shiller (1992, p. 374).

that period, he concluded that both channels can operate quickly enough among the broad mass of investors to influence the hour-to-hour movements of the market.[231] Of course, the game repeats itself again and again, like a Ponzi scheme, a fraudulent investment operation that pays returns to investors from their own money or from money paid by subsequent investors, rather than from any actual profit earned.[232] The late investors pay the returns of the early comers. As a result, one can easily imagine that investors who make decisions based on sensational media news are seldom real winners.

5.5 Summary

Behavioral finance is the part of finance which supplements the standard theories of finance by integrating behavioral aspects into the decision-making process for investing. Its history can be dated back as far as 1841, when Charles Mackay published *Extraordinary Popular Delusions and the Madness of Crowds*, but either Vernon Smith (1956) or Paul Slovic (1972) are credited for its foundation as a separate and distinct area of scientific research. The pioneers and promoters of behavioral finance are Amos Tversky, Daniel Kahneman and Paul Slovic. Their research in heuristics in the financial context provided viable models, among them, the prospect theory. They also deciphered the biases of availability, representativeness, herding, overconfidence, overoptimism, anchoring and loss aversion. These biases play a central role for investors under risky and high uncertainty.

These conditions are found during stock market bubbles and crashes. In order to use behavioral finance to describe stock market crashes, this chapter first briefly presented the history of behavioral finance and then introduced in detail the relevant behavioral biases. These biases were then linked to a model called positive feedback trading which was developed by Robert Shiller. At the end of this chapter, this model was used to explain significant upward and downward price moves, which can result in stock market bubbles and crashes. According to Shiller's approach, stock market crashes are mainly caused by endogenous factors, and are to be analyzed as self-organizing processes. Exogenous factors only play the role of a trigger. Positive feedback trading shows that investors are more likely to buy stocks when prices rise and to sell them when they fall.

The positive feedback trading model is important because it does not tie big price moves to relevant information, but to the redundancy of the phenomenon which is observed and self-alimented by investors. When significant upward or downward price moves occur, loss aversion gains more ground versus risk-seeking. According to Shiller investors react to big unfavorable price movements by becoming underconfident and pessimistic. This in turn leads to a market with lower volatility, lower expected returns and maybe sounder pricing and fundamental data.

[231] Shiller (1992, pp. 374–375).

[232] Shiller (1992, pp. 74–76).

But it is evident that market moods are hard to deal with and not necessarily beneficial if we look, for example, at the consequences of herding bias. It can lead investors to follow a trend of unsustainable prices, in the hope of making higher profits. An example are the events of mid-2008 until early 2009 when the global stock markets took a huge dip then sharply rebounded starting mid-March 2009. Investors betting on momentum in early Q1 2009 incurred significant losses when the stock markets mid-March 2009 started a rally never been seen.

What can investors learn from all this and what can they do in order to increase their return/risk relationship? What *did* they actually do in the year past the Lehman Brothers bankruptcy in terms of measuring the risk involved in their investments and, as a consequence, what investment strategies have they chosen? The final chapter of this book is devoted to answer these questions. In order to do so we cannot simply speak about *the investor* but need to distinguish different types. And, equally important, the analysis has to be broken down in regions as some are traditionally more risk-averse than others.

References

Achelis, S. B. (1995). *Technical analysis from A to Z* (2nd ed.). New York: McGraw-Hill.

Aktipis, C. A., & Kurzban, R. O. (2004). *Is homo economicus extinct? Vernon Smith, Daniel Kahneman and the Evolutionary Perspective*, January 1. Working Paper, Department of Psychology, University of Pennsylvania, Philadelphia, USA.

Allais, M. (1953). Le Comportement de L'Homme Rationnel Devant le Risque: Critique des Postulats et Axiomes de L'Ecole Americaine. *Econometrica, 21*(4), 503–546.

Aronson, D. (2006). *Evidence-based technical analysis: Applying the scientific method and statistical inference to trading signals*. Hoboken: Wiley.

Asch, S. E. (1952). *Social psychology* (1955th ed.). New York: Prentice Hall.

Asgary, A., & Levy, J. (2009). A review of the implications of prospect theory for natural hazards and disaster planning. *International Journal of Environmental Research, 3*(3), 379–394.

Banz, R. (1981). The relationship between returns and market value of common stock. *Journal of Financial Economics, 9*, 3–18.

Barber, B. M., & Odean, T. (2002). *All that glitters: The effect of attention and news on the buying behavior of individual and institutional investors*. Working Paper, University of California at Berkeley.

Barberis, N., Huang, M., & Santos, T. (2001). Prospect theory and asset prices. *Quarterly Journal of Economics, 116*(1), 1–53.

Basu, S. (1977). Investment performance of common stocks in relation to their price-earnings ratios: A test of the efficient market hypothesis. *Journal of Finance, 32*(3), 663–682.

Benartzi, S. & Thaler, R. H. (1995). Myopic loss aversion and the equity premium puzzle. *Quarterly Journal of Economics, 110*(1), 73–92.

Brunnermeier, M. K. (2001). *Asset pricing under asymmetric information: Bubbles, crashes, technical analysis, and herding*. New York: Oxford University Press.

Calverley, J. (2004). *Bubbles and how to survive them* (1st ed.). Boston: Nicholas Brealey Publishing.

Case, K. E. & Shiller, R. J. (1988, November/December). The behavior of home buyers in boom and post-boom markets. *New England Economic Review*, 29–46.

Chan, L. K. C., & Lakonishok, J. (2004). Value and growth investing: Review and update. *Financial Analysts Journal, 60*(1), 71–86.

Clarke, R. G., & Statman, M. (2000, Winter). The DJIA crossed 652,230. *Journal of Portfolio Management, 26*(2), 89–92.

Cohen, B. H., & Shin, H. S. (2003). *Positive feedback trading under stress: Evidence from the U.S. Treasury securities market*, January. BIS Working Papers No. 122, Monetary and Economic Department.

Damodaran, A. (2002). *Investment valuation* (2nd ed.). New York: Wiley.

Danielsson, J., & Love, R. (2006). Feedback trading. *International Journal of Finance and Economics, 11*(1), 35–53.

De Bondt, W. F. M., & Thaler, R. H. (1985). Does the stock market overreact? *Journal of Finance, 40*(3), 793–805.

Elsner, D. (1987, October 23). *When Prechter 'speaks', stock market listens*, Chicago Post. http://articles.chicagotribune.com/1987-10-23/business/8703190908_1_elliott-wave-theorist-stock-market-market-forecasts. Accessed 18 Feb 2014.

Englich, B., Mussweiler, T., & Strack, F. (2006). Playing dice with criminal sentences: The influence of irrelevant anchors on experts judicial decision making. *Personality and Social Psychology Bulletin, 32*(2), 188–200.

Fama, E. F. (1970). Efficient capital markets: A review of theory and empirical work. *Journal of Finance, 25*(2), 383–417.

Feist, G., & Rosenberg, E. (2009). *Psychology: Making connections*. Boston: McGraw-Hill.

Fienberg, S. E. (2008). The early statistical years: 1947–1967. A conversation with Howard Raiffa. *Statistical Science, 23*(1), 136–149.

French, K. R., & Poterba, J. M. (1991). Investor diversification and international equity markets. *American Economic Review, 81*(2), 222–226.

Froot, K. A., & Frankel, J. A. (1989). Forward discount bias: Is it an exchange risk premium? *Quarterly Journal of Economics, 104*(1), 139–161.

Galbraith, J. K. (1993). *A short history of financial euphoria*. New York: Penguin Books.

Gärdenfors, P., & Sahlin, N. E. (1988). *Decision, probability, and utility*. New York: Cambridge University Press.

Gerding, E. F. (2007). Laws against bubbles: An experimental-asset-market approach to analyzing financial regulation. *Wisconsin Law Review, 5*, 977–1039.

Gigerenzer, G., & Selten, R. (2001). *Bounded rationality: The adaptive toolbox*. London: MIT Press.

Harris, L., & Gurel, E. (1986). Price and volume effects associated with changes in the S&P 500 list: New evidence for the existence of price pressures. *Journal of Finance, 41*(4), 815–829.

Hirshleifer, D. (2001). Investor psychology and asset pricing. *Journal of Finance, 56*(4), 1533–1597.

Jacobs, B. I. (1999). *Capital ideas and market realities*. Malden: Blackwell Publishers.

Kahneman, D., & Riepe, M. R. (1998, Summer). Aspects of investor psychology. *Journal of Portfolio Management, 24*, 52–65.

Kahneman, D., & Tversky, A. (1973, July). On the psychology of prediction. *Psychological Review, 80*(4), 237–251.

Kahneman, D., & Tversky, A. (1974). Judgment under uncertainty: Heuristics and biases. *Science, 185*(4157), 1124–1131.

Kahneman, D., & Tversky, A. (1979). Prospect theory: An analysis of decision under risk. *Econometrica, 47*(2), 263–292.

Kahneman, D., Slovic, P., & Tversky, A. (1982). *Judgment under uncertainty: Heuristics and biases*. New York: Cambridge University Press.

Mackay, C. (1995). *Extraordinary popular delusions and the madness of crowds*. Ware: Wordsworth Editions Limited.

Mackay, C. (2009). *Extraordinary popular delusions and the madness of crowds*. Radford: Wilder Publications.

Malkiel, B. G. (2003). *A random walk down Wall Street* (8th ed.). New York: W.W. Norton & Company.

Mehra, R., & Prescott, E. C. (1985). The equity premium: A puzzle. *Journal of Monetary Economics, 15*, 145–161.
Montier, J. (2002). *Behavioural finance*. New York: Wiley Finance.
Montier, J. (2005, November 18). *Seven sins of fund management: A behavioural critique*. Dresdner Kleinwort Wasserstein Macro Research.
Montier, J. (2007). *Behavioural investing*. Hoboken: Wiley Finance.
Muth, J. F. (1961, July). Rational expectations and the theory of price movements. *Econometrica, 29*(3), 315–335.
Pompian, M. M. (2006). *Behavioral finance and wealth management*. Hoboken: Wiley Finance.
Pompian, M. M. (2012). *Behavioral finance and investor types*. Hoboken: Wiley Finance.
Raiffa, H. (1968). *Decision analysis: Introductory lectures on choices under uncertainty*. Reading: Addison-Wesley.
Roth, A. E. (1993). The early history of experimental economics. *Journal of the History of Economic Thought, 15*(2), 184–209.
Russo, J. E., & Schoemaker, P. J. H. (1990). *Decision traps*. New York: Fireside/Simon & Schuster Inc.
Schulmerich, M. (2011). *Behavioral Finance in Asset Management: A Primer*. State Street Global Advisors. SSgA Capital Insights.
Schulmerich, M. (2014, January/February). *Behavioral Finance in Asset Management: A Primer*. Investments & Wealth Monitor (pp. 42–47).
Selden, G. C. (1912). *The psychology of the stock market*. New York: Ticker Publishing Company.
Sewell, M. (2010). *Behavioural finance*, 2010 (revised). Working Paper, University of Cambridge, Cambridge, U.K.
Sewell, M. (2011). *History of the efficient market hypothesis*, August (amended September). Working Paper, University College London.
Shefrin, H. (2001). *Behavioral finance*. Northampton: Edward Elgar Publishing Ltd.
Shefrin, H. (2002). *Beyond greed and fear*. New York: Oxford University Press.
Shefrin, H. (2005). *A behavioral approach to asset pricing*. Boston: Elsevier Academic Press.
Shefrin, H., & Statman, M. (1985). The disposition to sell winners too early and ride losers too long: Theory and evidence. *Journal of Finance, 40*(3), 777–790.
Shiller, R. J. (1992). *Market volatility*. Cambridge: MIT Press.
Shiller, R. J. (2005). *Irrational exuberance* (2nd ed.). Princeton: Princeton University Press.
Shiller, R. J. (2008). *The subprime solution*. Princeton: Princeton University Press.
Shleifer, A. (1986). Do demand curves for stocks slope down? *Journal of Finance, 41*(3), 570–590.
Shleifer, A. (2000). *Inefficient markets*. New York: Oxford University Press.
Siegel, J. (2008). *Stocks for the long Run* (4th ed.). New York: McGraw-Hill.
Simon, H. A. (1955). A behavioral model of rational choice. *Quarterly Journal of Economics, 69*(1), 99–118.
Slovic, P. (1972). Psychological study of human judgment: Implications for investment decision making. *Journal of Finance, 27*(4), 779–799.
Smith, A. (2011). *The theory of moral sentiments* (7th ed.). New York: Gutenberg Publisher.
Sornette, D. (2003). *Why stock markets crash*. Princeton: Princeton University Press.
Stanley, T. J., & Danko, W. D. (1998). *The millionaire next door* (reprint ed.). New York: Gallery Books.
Statman, M. (1995). *Behavioral finance versus standard finance*, Leavey School of Business, Santa Clara University. http://www.aiinfinance.com/Statman.pdf. Accessed 3 Jan 2014.

Investor Risk Perceptions and Investments: Recent Developments

6

6.1 Introduction

In the preceding chapters we have answered the first key question of this book: Why do crashes happen when in theory they should not? This was done in five steps. Chapter 1 provided the mathematical prerequisites to measure risk and return, i.e., the key parameters needed in modern portfolio theory. Thereafter, Chap. 2 presented MPT by introducing the CAPM and the Fama–French three-factor model. This chapter also dealt in detail with the respective empirical tests and looked at the validity of the models which seems doubtful in light of these tests and the market reality. Apparently, traditional finance theory has its weaknesses. This can be seen especially in the prevalence of stock market anomalies which were described in Chap. 3. Chapter 4 looked at the most important stock market crashes in history. As these anomalies and crashes cannot be properly explained by traditional finance theory, the development and the upheavals in the equity markets call for a different approach known as behavioral finance. Chapter 5 explained the ideas of behavioral finance and elaborated in detail on the behavioral biases which are critical to understand stock market anomalies and crashes. On this basis, in the end of Chap. 5, behavioral biases were used to explain the crash of 1987.

Now, the final chapter focuses on the investor, be it an institutional investor, retail investor, sovereign wealth fund or central bank. How did the experience of the investors over the past 10 years change their way of measuring and managing risk? This is the second key question this books aims to answer. We will do so by *taking the perspective of the investors* and breaking their reactions down by investor types and regions within EMEA.[1]

We will also deal in-depth with the implications for the preferred investment strategies. The opinion of the retail end user will be captured by looking at what

[1] The abbreviation EMEA stands for the region covering Europe, Middle East and Africa.

M. Schulmerich et al., *Applied Asset and Risk Management*, Management for Professionals, DOI 10.1007/978-3-642-55444-5_6 is known as *third party* investors which are investment platforms that distribute investment solutions to the retail end user. The third part investors' view reflects, therefore, what retail clients think about investing and what they do actually invest in.

But first we will provide a general overview based on a survey of over 300 investors from the U.S. and Europe, including institutions, pension funds, family offices, consultants, asset managers, private banks and insurance funds. The survey was conducted by the Economist Intelligence Unit (EIU) on behalf of State Street Global Advisors.[2] The past 10 years were characterized by often steep ups and downs in the equity market with the last decline being the most severe crisis after the Great Depression almost 90 years before. Consequently, investors today are more likely to consider the downside of their investment than before 2008. The survey looks at tail risk management as a result of the extreme market movements from various angles. Thereafter, a detailed analysis by investor type and region will be provided. This is done each time by breaking down the past years into four time periods:

1. **Before the crisis:**
   This period covers the years directly preceeding 2007.
2. **During the crisis:**
   Here, we look at the time period from 2008 until the second quarter of 2009.
3. **The years directly after the crisis:**
   This period covers the time mid-2009 until the end of 2011.
4. **Early 2014:**
   This period looks at where investors stand early 2014 (when this book was completed) and how their views have developed based on the 2012 and 2013 experience.

Regardless of investor type and region, the answers are similar and show a clear trend towards the use of a wider range of risk and risk-adjusted performance measures with a primary focus on downside risk measurement, especially directly after the breakout of the subprime crisis and when the euro crisis unfolded. In many cases, this still holds true early 2014.

We also give a brief overview on the attitude of central banks. Usually, central banks primarily care about reputational risk, disaster risk and drawdown risk. This has always been the case. But since the crisis, the focus on drawdown risk has been even stronger.

---

[2]State Street Global Advisors (SSgA) is, as of December 31, 2013, the second largest institutional asset manager worldwide with roughly \$2 tn. as asset under management spanning almost all asset classes.

6.2 A General Overview

6.2.1 Frequency of Financial Shock Events and the Awareness of Investors

In recent years, extreme movements of financial markets have occurred more frequently than in the past. Since the global financial crisis which unfolded 2008, downward movements in stock markets followed by upward movements and above average market volatility can be found during most market periods since 2008. In volatile markets, extreme events can shock financial markets and may have a tremendous impact on investors' returns and how they try to achieve them. Such shocks to financial markets are called *tail risk events* which, mathematically, are often defined as an investment move of more than three standard deviations away from the average investment return.[3]

The most recent tail risk events were the subprime crisis, the eurozone crisis, the tsunami catastrophe in Japan, and the unrest in the Middle East, among others. Clearly, such events and the associated uncertainty have a significant influence on investor behavior.[4] Investors have begun to realize the necessity of protection strategies in order to avoid drawdown periods like 2008.

To get a deeper understanding of this topic, State Street Global Advisors, in summer 2012, commissioned the Economist Intelligence Unit to survey institutional investors and to investigate their views concerning tail risk events and their protection strategies. The survey includes the views of 310 investors from the U.S. and Western Europe (U.K., France, Germany, Italy, Switzerland, and Benelux), including institutional investors, family offices, consultants, and private banks.[5]

At the time of the survey, the financial markets also passed through a negative development in the context of the eurozone crisis which started 2010. Rating agencies predicted in mid-2012 that Greece would leave the eurozone and that Spain would seek a full bailout. And, clearly, a financial shock comes with a lot of cost.[6] Therefore, it is understandable that market agents have begun to rethink their investment strategies. Figure 6.1 on p. 418 provides evidence about investor expectations for the occurrence of future tail risk events and their severeness versus the past[7]:

1. *Tail risk events are likely to happen more frequently now (as of mid-2012) because of the increased correlations of financial markets.*
   The upper section of Fig. 6.1 shows that most investors agree or strongly agree to the statement that tail risk events are likely to happen more frequently now.

---

[3]Economist Intelligence Unit (2012, p. 5).

[4]Economist Intelligence Unit (2012, p. 2).

[5]Economist Intelligence Unit (2012, pp. 4–5).

[6]Economist Intelligence Unit (2012, p. 5).

[7]Please note that Fig. 6.1 is identical to Fig. 4.6 which was shown on p. 258 in Chap. 4 to illustrate the increased frequency and severeness of stock market crashes.

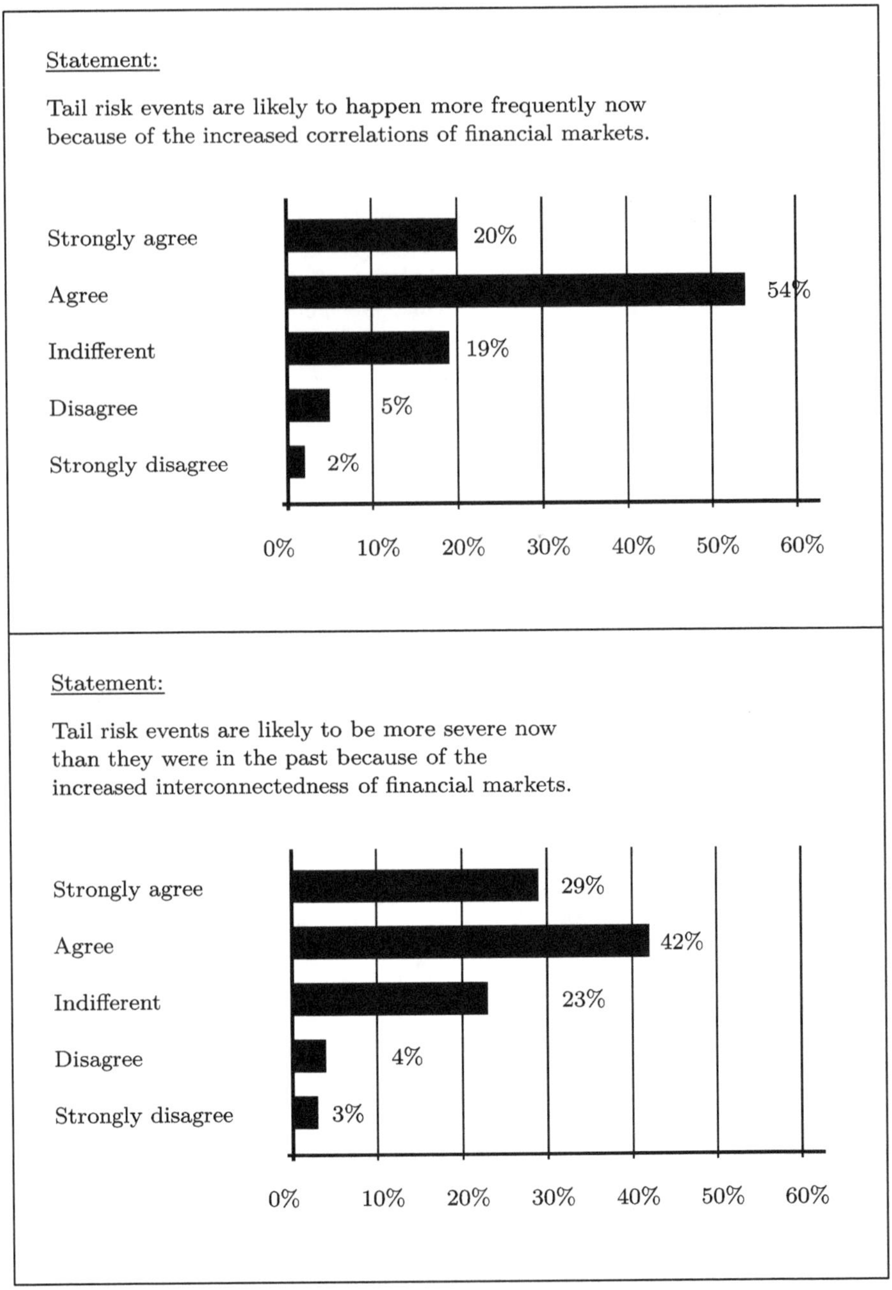

Fig. 6.1 Investor expectations about the frequency and severeness of tail risk events nowadays (as of mid-2012) versus the past. *Source*: Economist Intelligence Unit (2012, p. 21)

Twenty percent strongly agree that tail risk events might occur more frequently and 54 % agree. Nineteen percent are indifferent in their answer while 5 % disagree and 2 % strongly disagree to that point. In aggregate, three quarters of the participants agree or even strongly agree to the statement.

2. *Tail risk events are likely to be more severe now (as of mid-2012) than they were in the past because of the increased interconnectedness of financial markets.* As the lower section of Fig. 6.1 displays, most investors agree or strongly agree to the statement that tail risk events are likely to be more severe now. Twenty-nine percent strongly agree that tail risk events might occur more frequently and 42 % agree. Twenty-three percent are indifferent in their answer while 4 % disagree and 3 % strongly disagree to that point. In aggregate, almost three quarters of the participants agree or even strongly agree to this statement as well.

So clearly, mid-2012—which was already almost 4 years after the Lehman Brothers bankruptcy—was characterized by the fear of three quarters of the investors that a crash was likely to come back with more severeness than observed in past crises. The EIU survey also focused on the awareness of such risk events and the understanding of their impact on financial markets. Figure 6.2 includes the opinion of investors about the appropriate understanding of other market agents of the frequency and severity of tail risk events. It shows that most of the participants believe in an accurate understanding of other markets agents of financial shock events. Fourteen percent of the participants believe that investors have a good understanding of those events, whereas 72 % credit them with some understanding in tail risk events and only 9 % believe that there is no understanding.

It is now interesting to consider the consequences of this market evaluation. This is done in Fig. 6.3 which looks at the question if tail risk management should be an integral part of any comprehensive investment plan. As can be seen, almost 80 % agree or strongly agree that hedging tail risk should be integrated in an investment plan. Only 4 % disagree or strongly disagree. Obviously, in mid-2012, investors had

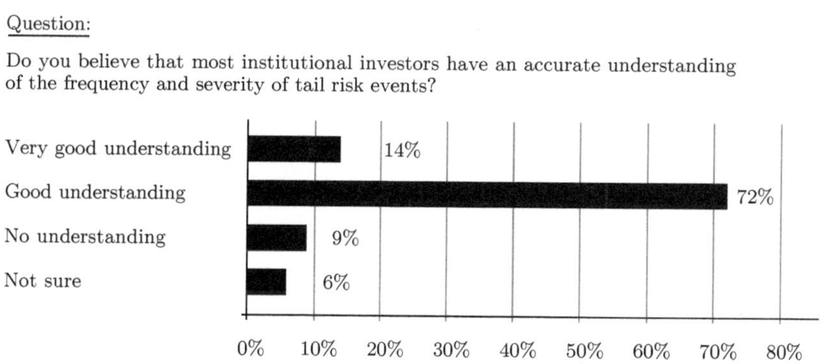

Fig. 6.2 Answers to the question if investors believe that most investors have an accurate understanding of the frequency and severity of tail risk events. *Source*: Own, based on Economist Intelligence Unit (2012, p. 25)

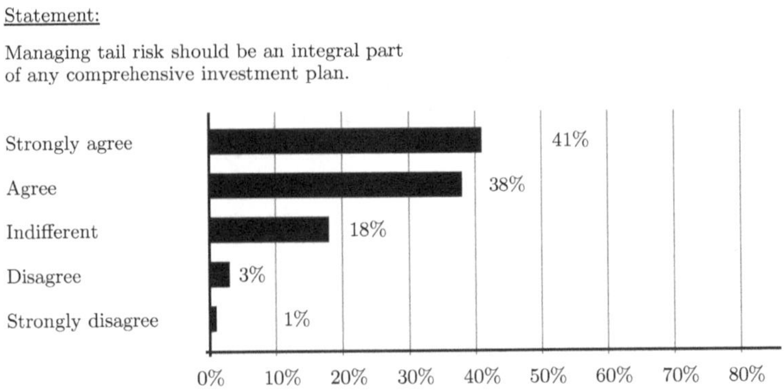

Fig. 6.3 Answers to the question if managing tail risk should be an integral part of any comprehensive investment plan. *Source*: Own, based on Economist Intelligence Unit (2012, p. 21)

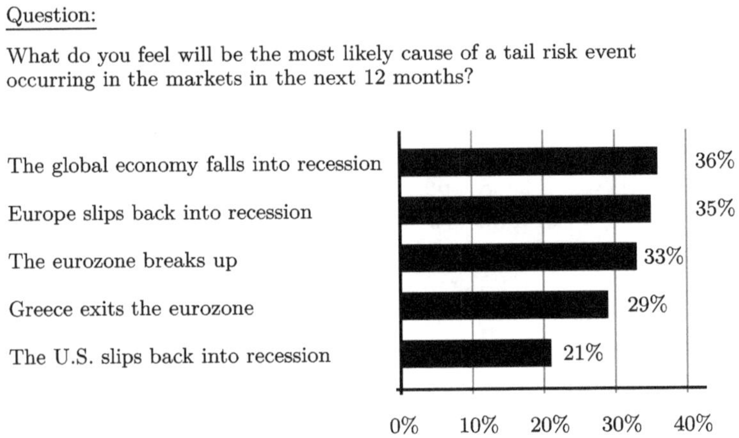

Fig. 6.4 Answers to the question what investors feel to be the most likely cause of a tail risk event occurring in the markets in the next 12 months (i.e., between mid-2012 and mid-2013). *Source*: Own, based on Economist Intelligence Unit (2012, p. 6)

learned their lessons from 4 years of market turmoil with extreme ups like in 2009 but also disastrous downs as experienced in 2008.

6.2.2 Expectations of Tail Risk Events

In this section we will present the results of the survey concerning investor expectations about the cause of future tail risk events. The survey participants were asked for their expectations about the most likely causes of tail risk events within the next 12 month, i.e., between mid-2012 and mid-2013. Figure 6.4 provides an overview about the top five answers.

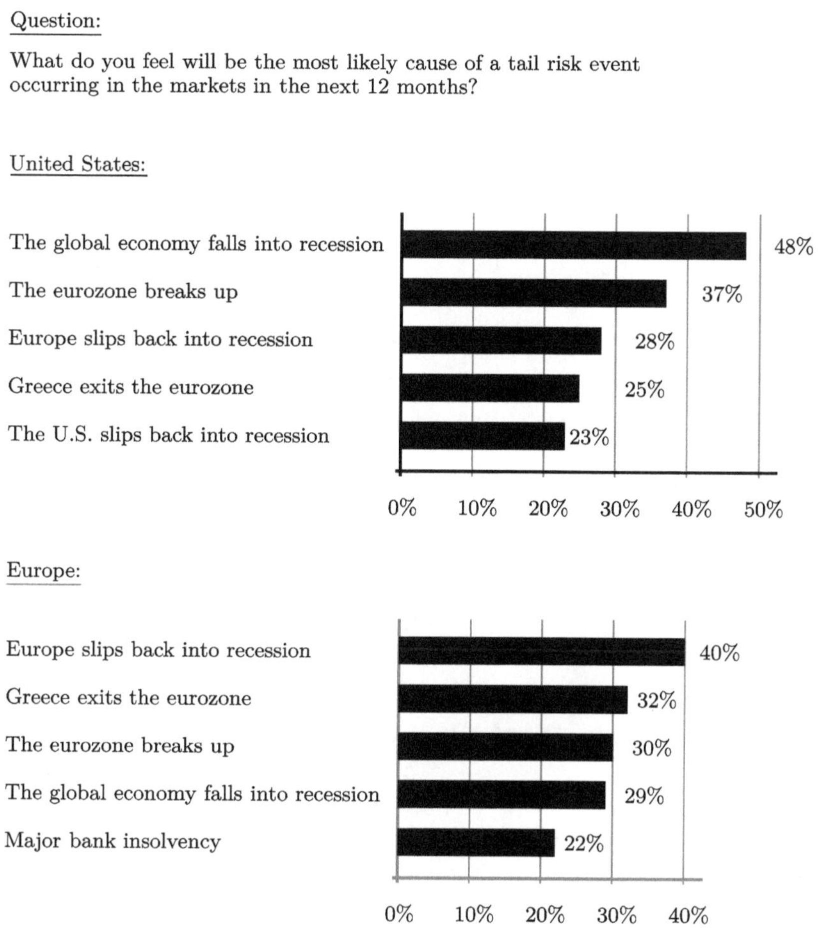

Fig. 6.5 Answers from U.S./European investors about the most likely cause of a tail risk event occurring in the markets in the next 12 months (i.e., between mid-2012 and mid-2013). *Source*: Own, based on Economist Intelligence Unit (2012, p. 7)

Figure 6.5 separates the answers provided by U.S and European investors. While European investors, in mid-2012, considered an European recession and a eurozone crisis as the most likely causes for tail risk events, U.S. investors also mentioned a global recession as one of the most likely reasons. This naturally leads to the next question: how do you, as an investor, prevent such tail risk events?

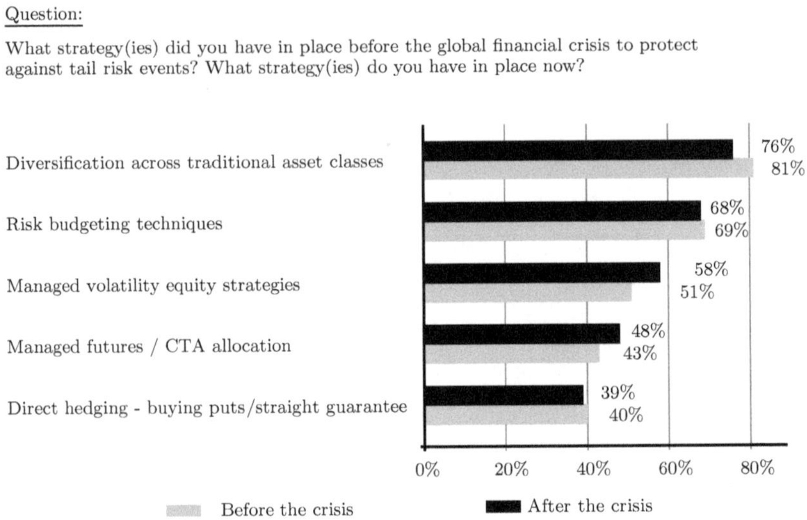

Fig. 6.6 Answers to the question what strategy(ies) did investors have in place before the global financial crisis to protect against tail risk events and what strategy(ies) they do have now. *Source*: Own, based on Economist Intelligence Unit (2012, p. 10)

6.2.3 Change Drivers in Risk Strategy

As tail risk events have more frequently occurred since the global financial crisis, investment behavior has changed, especially concerning protection strategies against shock events. This can be seen in Fig. 6.6.

After the crisis, diversification across different asset classes, risk budgeting and direct hedging decreased, while managed futures and managed volatility equity strategies increased. A managed futures strategy, also known as CTA, is a hedge fund strategy[8] that seeks to generate (absolute) returns through long and short positions in various future contracts and across various markets. Managed futures strategies are in most cases executed by *commodity trading advisors* (CTA) and *commodity pool operators* (CPO) who apply their own individual trading methods and money management techniques to manage their investors' money. A managed volatility equity strategy is a long-only equity strategy that aims to achieve an

[8] For a general introduction to hedge funds see Schulmerich (2014b).

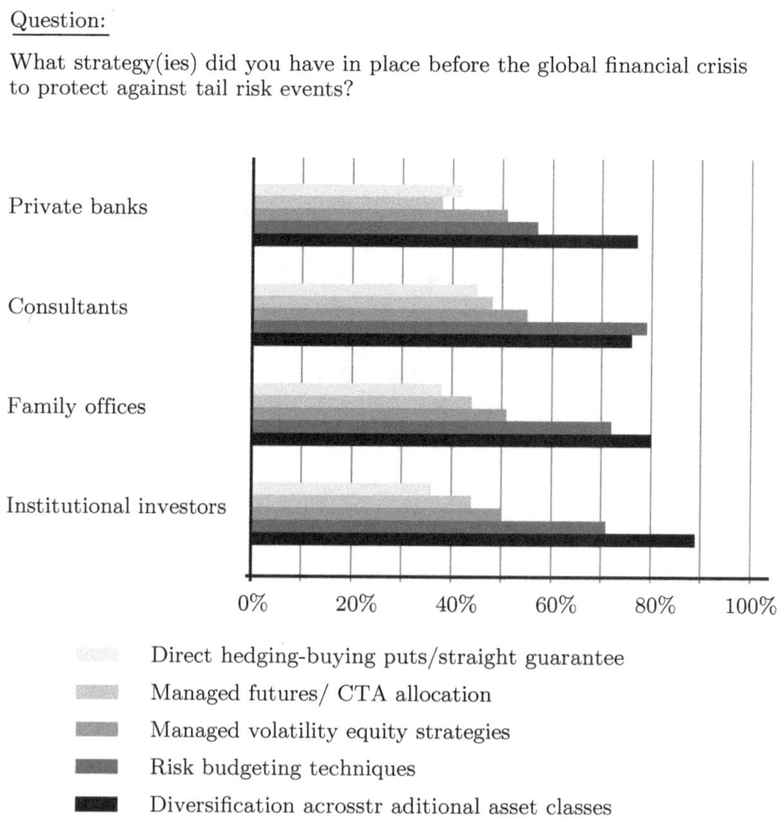

Fig. 6.7 Answers from different types of investors to the question what strategy(ies) they had in place before the global financial crisis to protect against tail risk events. *Source*: Own, based on Economist Intelligence Unit (2012, pp. 12–14)

above-market return with below-market volatility.[9] This is achieved through the focus on low-beta stocks which exhibit a below-market volatility.

What are the answers if we break them down by specific investor type, i.e., private banks, consultants, family offices and institutional investors? This is shown in Figs. 6.7 and 6.8 for strategies used before and after the crisis, respectively.

In the time before the crisis, all investors mentioned diversification across traditional asset classes as the main strategy to protect against tail risk. Also, after the crisis, most investors, except for the consultants, also mentioned diversification between asset classes as the main protection strategy. However, most interesting

[9]For more information see, for example, Schulmerich (2010b).

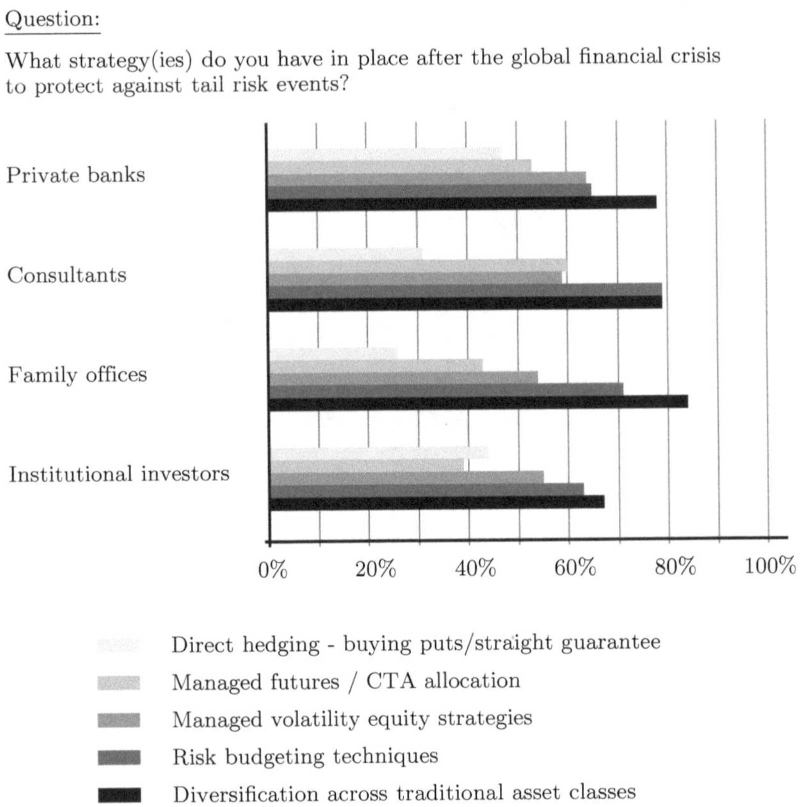

Fig. 6.8 Answers form different types of investors to the question what strategy(ies) they have in place after the global financial crisis to protect against tail risk events. *Source*: Own, based on Economist Intelligence Unit (2012, pp. 12–14)

is that for institutional investors the importance of diversification decreased significantly by 20 % points. This clearly reflects the experience that institutional investors made during the crisis, i.e., that their own diversification did not properly protect them against losses: asset classes which are uncorrelated during normal times often become correlated in crisis times coupled with a spike in volatility.

Figures 6.9 and 6.10 provide a direct comparison of *before* versus *after* the crisis for the various tail hedge strategies broken down by investor type. This is done in order to display a direct comparison which allows to see the differences clearly and easier than by comparing Figs. 6.7 and 6.8.

In the next sections we will look at the various regions in EMEA and break down our results by investor type. These results stem from the investment and market experience of Marcus Schulmerich, the global portfolio strategist for active quant equity portfolios and hedge funds covering all offices in EMEA at State Street

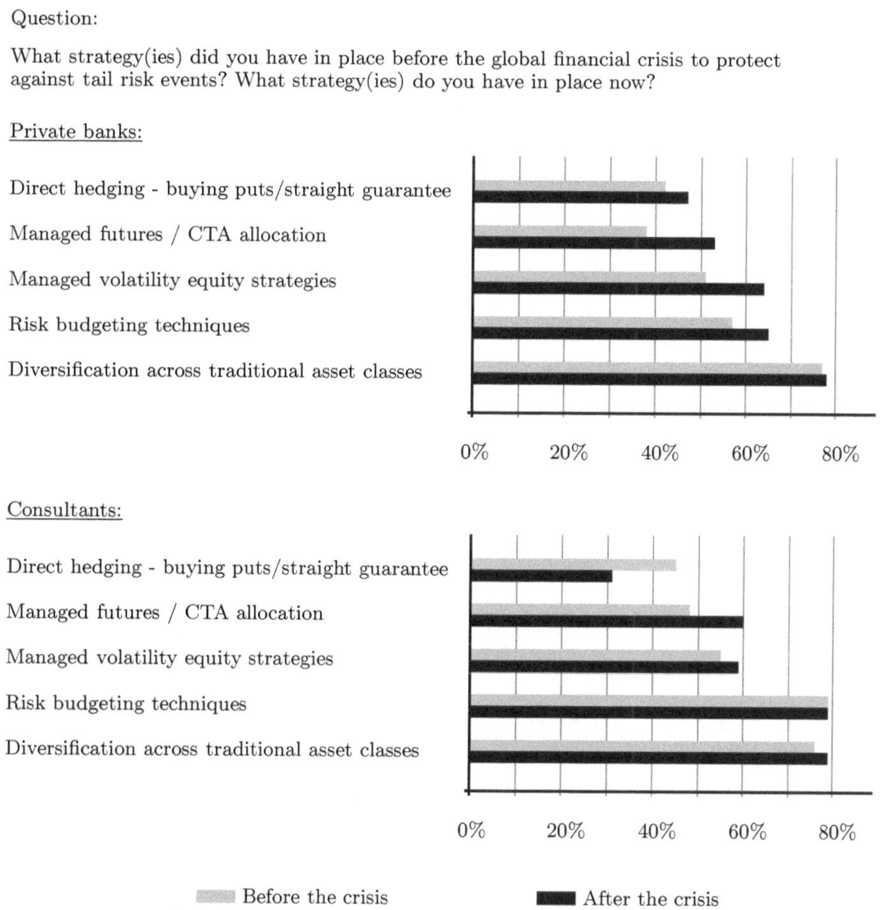

Fig. 6.9 Answers from different types of investors to the question what strategy(ies) they had in place before the global financial crisis to protect against tail risk events and what strategy(ies) they do have now. *Source*: Own, based on Economist Intelligence Unit (2012, pp. 12–14)

Global Advisors (SSgA). He works closely with pension funds, family offices, consultants, asset managers, private banks and insurance funds. The field study results presented in Sects. 6.3–6.12 are based on his asset and risk management experience and on Schulmerich (2011) which analyzes the subprime crisis and how the experience of the years 2008–2010 has changed the EMEA investors' view of risk as well as on Schulmerich (2014a). The latter is an update of the 2011 publication and looks at the most recent developments from 2011 up to early 2014. Important input came from his SSgA sales and client relationship colleagues whom he works with on a daily basis.

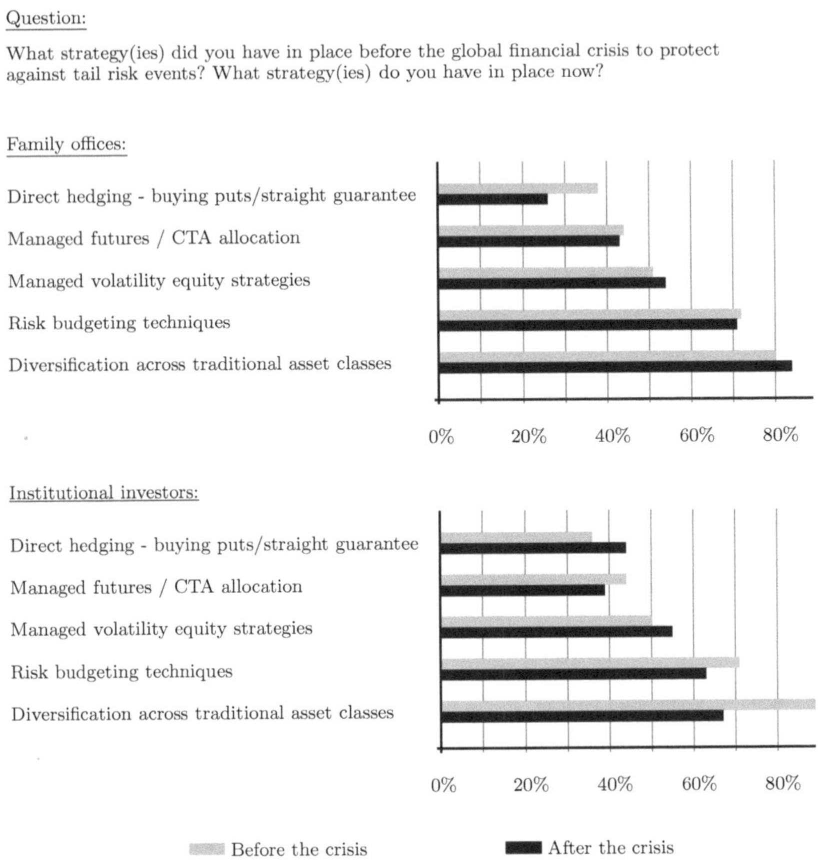

Fig. 6.10 Answers from different types of investors to the question what strategy(ies) they have in place after the global financial crisis to protect against tail risk events and what strategy(ies) they do have now. *Source*: Own, based on Economist Intelligence Unit (2012, pp. 12–14)

6.3 Germany and Austria

Let us first see what asset managers in Germany encounter in their current work environment today. This is summarized in an interesting survey conducted by Isarvest in cooperation with Ketchum Pleon and published in September 2013.[10] This survey includes the views of 124 asset managers in Germany and provides a good overview of the broad asset management topics in Germany end of 2013.

The key part of their survey was to assess the lasting influence of the financial crisis up to now on asset managers and their investment targets. In detail, they asked

[10]Jordan, Rüther, and Winkelhaus (2013).

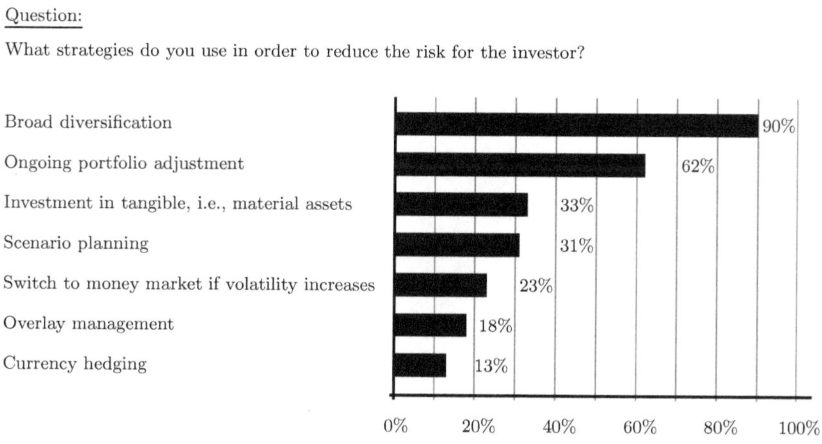

Fig. 6.11 Answers of German asset managers (as of September 2013) to the question what strategies they use in order to reduce the risk for the investor. *Source*: Jordan et al. (2013, p. 5)

what strategies are primarily followed nowadays and what products are used to achieve them. In general, the most important findings were[11]:

- Real capital preservation is the overarching goal of the investors.
- Diversification is *the* answer to increased risk.
- Asset allocation is the core of asset management.
- Equity is highly ranked by investors.
- Diversified strategies are in high demand.

Especially, the first target of a real and long-term capital preservation was named by 94.5 % of the asset managers as a given target by the investors. Figure 6.11 now looks at the strategies used by German asset managers to reduce the risk for the investor. In line with the list above, diversification is by far the most important strategy, followed by ongoing portfolio optimization and investments in real assets like commodities and real estate. The named strategies have to be seen as a response to the threats that German investors and asset managers see in the market environment. Figure 6.12 displays the key challenges seen mid-end of 2013. Even more important than the low interest rates environment for government bonds is the threat coming from speculative bubbles in some asset classes.

Let us now turn to the German investors and their reaction towards the crisis. Their reaction was strong, as it is evident in Fig. 6.13 which displays the EIU survey results for the question which strategy(ies) did/do German investors have in place before the global financial crisis/now to protect against tail risk events.

The most significant increase is in the use of diversification in order to prevent tail risk management followed by the increased application of managed volatility

[11]Jordan et al. (2013, p. 2).

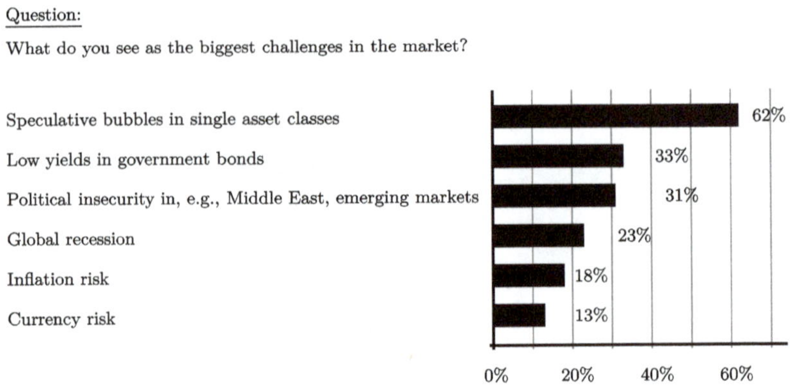

Fig. 6.12 Answers of German asset managers (as of September 2013) to the question what speculative bubbles are seen as the biggest challenge. *Source*: Jordan et al. (2013, p. 10)

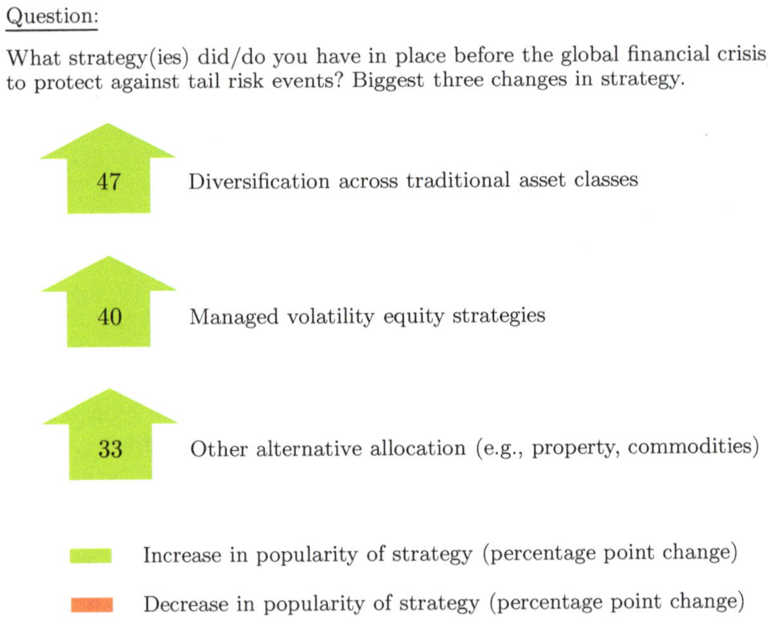

Fig. 6.13 EIU survey results for the question what strategy(ies) did/do German investors have in place before the global financial crisis/now to protect against tail risk events. Biggest three changes in strategy. *Source*: Economist Intelligence Unit (2012, p. 15)

strategies which aim to reduce portfolio volatility vs. the market. Also, the increased inclusion of alternative investments in the asset allocation was one of the top actions German investors undertook in order to prevent events like in 2008.

Let us now look at the reaction of German investors in more detail.

6.3.1 Institutional Asset Owners

German institutional asset owners largely belong to the following categories:

- Corporations
- Pension funds
- Insurance companies
- Banks
- Endowments
- Family offices

Often, these investors are advised by asset management consultants like Mercer or Watson Wyatt, which operate on a global basis, or local consultants like Feri, RMC, BFinance or SFS who primarily address the German market. This section looks at whether and how these investors (and their consultants) have changed their view on risk during the crisis compared to the time before the crisis and whether these changes were sustained after the crisis.

The business of institutional asset owners in Germany and Austria is very similar. Therefore, a distinction between these two countries is not needed.

6.3.1.1 Before the Crisis

Depending on individual characteristics like business sector, risk budget and in-house resources, institutional asset owners always had different views on risk and return. Generally speaking, before the subprime crisis, institutional investors mostly focused on risk relative to their market benchmark. Since most institutional investors (e. g., pension funds, life insurance companies) have to fulfill contractual obligations, the expected return very often provided the starting point for any strategic asset allocation.

Asset owners typically assess risk by symmetrical risk measures, for example, by volatility and tracking error relative to market benchmarks. Quite frequently, volatility as downside deviation from a portfolio's expected return was used as a measure of risk even on an absolute basis.

Furthermore, before the crisis, asset management consultants rated risk relative to benchmark by using tracking error and information ratio. Other measures that were often used included value at risk (VaR), conditional value at risk (c-VaR)[12] and maximum drawdown.

6.3.1.2 During the Crisis

During the subprime crisis, especially in 2008, investors realized that actual price changes in the capital markets were much higher than anticipated. The relative measures of risk were not considered to be sufficient anymore. In addition, the high correlation between asset classes during the downturn in 2008 led to discussions

[12]Conditional value at risk, also known as *expected shortfall*, is an aggregated risk measure to evaluate the overall risk of a portfolio. Compared to VaR, the c-VaR takes into account the shape of the loss distribution and measures the expected loss in case a loss takes place (on a certain confidence level).

about the correlation in general. Diversification as the most common method to decrease portfolio risk was not deemed reliable anymore. This was a big difference compared to the view on risk before the crisis.

As for the asset management consultants, however, significant changes of their views on risk have not been registered. It was recognizable, though, that quality assets were very much in demand. Risk avoidance was the topic of the day. As consultants took a closer look into investment strategies, people, and processes, relative risk measures became less important. Since the markets started to recover end of Q1 2009, absolute return strategies increased in importance. This is true for hedge funds, funds of hedge funds, hedge fund beta replication, and capital protection products like CPPI,[13] real estate or commodities.

6.3.1.3 The Years Directly After the Crisis

The intense discussions about investment risk during the subprime crisis led to substantial changes in policies and guidelines as well as to changes in the investors' attitude toward risk. Therefore, investor views on risk remained relatively unchanged in the years 2009–2011 even after markets recovered. On the other hand, asset owners like pension funds or life insurance companies often have to achieve return requirements. But given the low interest environment since end of 2008 which reduces the chances to earn the needed return, they also consider more risky assets for their portfolios.

In the years 2010 and 2011, relative risk measures like tracking error and volatility were still applied in individual mandates. But usually, the overall portfolio was measured by more absolute figures like VaR or c-VaR.

Asset owners, who before the subprime crisis were seeking either a constant return with the lowest possible risk or the highest possible return for a given risk budget, then tried to even more align their investments to their basic investment needs. Therefore, liability-driven investments and absolute return concepts gained more attention. Besides changing from relative to absolute risk and return measures, there was also more interest in counterparty risk and liquidity risk.

Also, the trend to assume a different view on investment risk and its measurement was well supported by a change in legislation and policies and will, presumably, be relatively stable. It became even more important, as investors were considering a wider universe of asset classes due to the low interest environment.

Among the asset management consultants, there have been no major changes to the overall risk measures and views. However, managers needed to more frequently prove their risk management expertise to the consultants and to show their ability

[13]As already explained on p. 403, CPPI stands for constant proportion portfolio insurance, a dynamic portfolio hedging strategy that allocates the assets between a risky component and a risk-free component. If fear dominates the market (a so-called *risk-off environment*), CPPI allocates a greater portion of the portfolio to risk-free assets. If investors are risk-seeking (a so-called *risk-on environment*), CPPI changes the allocation to risky assets.

to manage downside risk within the investment process.[14] Currently, there is no evidence that risk measures will change in the next 2–3 years to come.

6.3.1.4 Early 2014

Risk budgets shrank during the crisis and have not recovered much, as most investors stayed out of risky assets and did not participate in the rising asset prices since then. In order to still earn an acceptable yield, investors focused strictly on the amount of risk they incur for an anticipated return. The used risk measures were relative risk measures like tracking error, and volatility was used for institutional benchmark mandates. In absolute terms figures like VaR or c-VaR have been used.

The major change in the investors' perception of risk when entering 2014 is the present common belief that there is nothing like a risk-free rate anymore. Government bonds, especially from the eurozone and also the U.S., used to be taken as measures for risk-free rates. Currently a considerable number of participants in the capital markets regard government bonds from the southern eurozone member states as riskier then large-cap equities.

Since 2011, risk is perceived by professional investors as the situation when the long-term service of the investor's liabilities is endangered. Therefore the view on risk includes the change in liabilities as well as the likelihood of absolute losses. When entering 2014, the most prominent risk measures are relative risk measures, like tracking error, and volatility as absolute risk measure. They can still be found in individual mandates. The overall portfolio usually is measured by more absolute figures like VaR or maximum drawdown.

Given an increasing pressure from regulation, institutional investors and, therefore, consultants, are forced to differentiate more between the risk of a certain asset class and its return. Measures that asset management consultants currently take into consideration are, for example, tracking error, information ratio, VaR, conditional value at risk and maximum drawdown. The types of risk measures have not changed dramatically in the last couple of years. However, modified regulation on the institutional investors' side forces asset managers to look into the details of the specific risks of an asset class, independently from their returns.

In consequence, asset managers nowadays need to prove their risk management capabilities to the asset management consultants more frequently and show the ability to manage downside risk within the investment process. In light of the search for higher-yield asset classes given the low-yield government bond environment in early 2014, institutional investors and consultants need to more frequently check on the risk measures, in particular as regulatory reports have to be filed. Given an

---

[14] An interesting snapshot of the application of asymmetrical, i.e., downside risk measures can be found in the German *Börsenzeitung* from November 13, 2010, see Schulmerich (2010a).

increase in DC-hybrid pension plans,[15] they will have to consider shifts in the asset allocation also from a risk point of view.

Looking forward, there is definitely a trend towards measuring risk in more absolute terms. Risk management and, based on that, reporting are increasingly customized to investor needs so that the general use of risk measures, especially of absolute risk measures, will remain frequent and will most likely further increase in the future.

6.3.2 Third Party Distribution

Third party distribution to retail investors, at SSgA taken care of by the IBG group (Intermediate Business Group), is a subsegment of the asset management industry ranging from banks/global markets to retail banks. By nature, this business requires a different skill and capability set compared to institutional asset owners. Moreover, these investors operate under different legal frameworks, such as UCITS 3 (asset managers) or Basel II (banks/global markets). As a consequence, they have different views on risk and return and on whether risk should be measured in relative or absolute figures. Traditionally, asset managers tend to measure risk relative to their benchmarks, whereas family offices and wealth managers who deal with high net worth individual investors increasingly target absolute (positive) returns.

For the following, our statements are based on findings for asset managers, banks/global markets, insurance (unit-linked only), private banks, family offices and retail banks. Answers may vary depending on the subsegment or on the individual investor. Retail platforms are not included as they are not decision makers.

6.3.2.1 Before the Crisis

In the past, during bull markets (end of 1990s until 2001/2002 or from 2003 until 2007), investors were focused on relative returns rather than risk. Traditionally, this holds for those investors that are themselves subjected to benchmarking (asset managers, wealth managers, private banks). However, there have been a number of players (Pioneer Investments, PIMCO) that offered absolute return strategies before the crisis. Moreover, it should be noted that there was a large market share for guaranteed products in Germany. For banks/global markets, risk was part of their portfolio construction process, measured for example, with the VaR concept.

Clients who were themselves benchmarked (asset managers) made investment decisions by applying traditional risk measures such as tracking error, bear/bull beta and drawdown. Those who favored an absolute return approach (family offices,

[15]DC stands for *defined contribution* which refers to a pension plan where only the regular contributions are determined. The future pension outcome is not determined, but depends on the future returns of the invested pension contributions. The opposite of DC is DB which stands for *defined benefit*. In a DB pension scheme, the amount of the future pension payments is guaranteed by the pension provider who needs to make sure that the investment of the regular pension contributions yields the desired return.

wealth managers), additionally considered volatility, maximum drawdown, etc. (for example, within a CPPI framework). With the exception of banks/global markets, VaR, as a risk measure for trading book activities, was not extensively used, since the investors' investment horizon does not comply with VaR approaches.

6.3.2.2 During the Crisis

Risk avoidance was the name of the game among those investors that manage money for private investors or high net worth individuals. Rather than managing risk or applying a risk vs. return approach, risk was to be carefully avoided. This was the reason for the large outflows from mutual funds recorded by the BVI.[16]

6.3.2.3 The Years Directly After the Crisis

Despite the strong recovery of the worldwide equity markets in 2009, there were only moderate inflows into equity funds. ETFs were among the winners of the crisis. Traditional asset managers continued to follow a relative approach (using tracking error and bear/bull beta). But in other segments the picture is mixed, depending on the individual approach. Taking the lessons from the subprime crisis, an increased number of investors (wealth managers, family offices, private banks, retail banks) started to focus on absolute returns or on avoiding to lose money. Although this should have implied an absolute view of risk, an asymmetric risk and return profile became obvious. In a nutshell, investors wanted to enjoy the full upside potential while avoiding any downside risk. In order to achieve this, some were evaluating absolute return strategies and others implemented overlays via futures.

The major game after the subprime crisis was called "risk avoidance", especially as since 2010 the euro crisis took center stage. This could also be described as an absolute approach to risk. Clients tended to a more holistic view of risk (absolute and relative), if the appropriate skill set was available. A tendency towards transparency was also expected. This clearly supported the trend towards asymmetrical downside risk measures like drawdown, semi-variance, VaR, shortfall, etc.

6.3.2.4 Early 2014

When entering 2014, most of the above said still holds true. As a consequence, a number of products were launched in the market that aim to minimize the maximum drawdown in addition to downside/risk management. Also, alternative weighting schemes known as *advanced beta* or *smart beta* were designed with the objective to construct an index that is not a traditional capitalization-weighted index (as known from MPT, see Chap. 2) but composed in a way to achieve a certain target (for example, a low volatility index or investment theme indices). Typically, third party investors look at managed volatility strategies as the Isarvest survey also indicated.

More generally, two major investment streams can be observed in the German and Austria third party market nowadays: clients that aim to generate an absolute return and those that continue to focus on relative performance. The implied

[16]German association for investment funds and asset management.

asymmetric risk and return desire (for example, a beta above 1 in upward markets and a beta below 1 or even 0 in downward markets) is an issue of behavioral finance which investors increasingly take into consideration in light of the irrational market behavior since 2007.

6.4 Switzerland

6.4.1 Institutional Asset Owners

6.4.1.1 Before the Crisis

Before the crisis started in August 2007 with the quant crisis the Swiss asset owners followed market capitalization benchmark-oriented strategies for a great part of the asset allocation. Consequently, they were focused on benchmark-related risk measures, and tracking error was the main risk measure they used. They hoped for absolute returns similar to hedge funds or comparable alternatives.

Asset management consultants in Switzerland similarly concentrated on relative risk as opposed to absolute risk. They mostly relied on benchmark-related risk measures such as tracking error. For absolute risk, volatility of the portfolio was applied.

6.4.1.2 During the Crisis

During the crisis between end of 2007 and early 2009, new risk topics arose, like liquidity risk and counterparty risk. In addition, the increased correlation of asset classes was observed and discussed. Clients did not buy equities, but neither did they sell, although the market sentiment was bad.

In the view of asset management consultants, liquidity and transparency risk was the most important risk on the peak of the crisis, not only for hedge funds. To some extent, counterparty risk was also taken into account. In addition to tracking error and volatility which were applied before the crisis, VaR and c-VaR were increasingly adopted.

6.4.1.3 The Years Directly After the Crisis

Since the second half of 2009, when equity markets started to recover, investors changed their views on risk again. Asset liability management studies were called for and conducted.

Since 2010 institutional investors used additional risk measures compared to pre-crisis times. Liquidity risk was still a topic but counterparty risk was more present in discussions. Moreover, starting with the Greek crisis in spring 2010, counterparty risk had become an issue of government level discussions as well. Anyhow, downside risk was discussed more frequently, but benchmark orientation still dominated.

In the face of the Greek crisis and the developing problems in the eurozone (Ireland, Portugal), currency risk came up as a frequent topic for asset owners. There were many discussions about hedging the currency risk on different levels and for different asset classes. Country risk became a very important topic after the Greek crisis as well. Investors even more focused on passive investments and did not look for active risk.

Concerning the use of new risk measures, discussions with pension funds revealed that with respect to countries they still applied benchmark-oriented strategies. VaR was used by some of the investors, for example, health insurance companies.

For asset management consultants, however, the situation changed after the crisis. For government bond mandates, an increase of customized solutions which reduced the exposure to the biggest debtors could be observed. However, there was nothing similar to be seen in the equity field. The trend of looking at different risk measures was correlated with the uncertainty in the markets. Big pension plans definitively looked for risk reducing solutions. For instance, institutions applied customized CPPI solutions for diversification or simply looked for downside protection and absolute return via GTAA.[17]

6.4.1.4 Early 2014

The shift from relative risk to absolute risk that became apparent during the crisis did not, in the end, lead to a sustained new approach. At the end of 2013, asset allocation stayed very much the same after all. Allocations to hedge funds were reduced. However, the interest in tactical asset allocation strategies has risen a bit. Overall, there is still a strong trend to indexing and yield-seeking strategies. Currently, infrastructure as an asset class is in high demand.

The most important risk at the peak of the crisis was liquidity and transparency risk, not only for hedge funds. To some extent, counterparty risk was also taken into account. Now, in early 2014, the institutional investors are back to the typical pre-crisis risk measures: tracking error for relative return strategies and volatility for absolute return strategies. Some investors also include VaR and conditional VaR in their analyses. To reflect the economic situation in Europe, Swiss investors aim more and more for customized solutions when it comes to government-bond investing with the goal to reduce the exposure to the biggest debtors. Equity investing is still on a very low level, if at all.

A new development in the Swiss market are the discussions on *alternative sovereign* approaches. Since these concepts were never convincing, investors partly switched back to active approaches for global government bond portfolios. It is interesting to observe, how many of the *new* ideas about risk reduction strategies that were discussed in the Swiss institutional market during the crisis were actually applied. However, in the end only big investors implemented such a new approach by including, for example, hedge funds in order to hedge traditional portfolios.

[17]GTAA stands for global tactical asset allocation.

6.4.2 Third Party Distribution

6.4.2.1 Before the Crisis

Swiss third party investors always used measures like c-VaR and VaR for their large portfolios and calculated absolute returns. Since 2008 they started using these measures with the purpose to identify and manage fat tails.

6.4.2.2 During the Crisis

The application of the measures did not change during the crisis between 2007 and 2009. But analysis designs and portfolio construction concepts clearly put a stronger focus on downside risk measures.

6.4.2.3 The Years Directly After the Crisis

Since the market started to recover, risk measures such as semi-variance, VaR, c-VaR, and maximum drawdown were mainly used. Thus, there was a strong preference for instruments which were better able to measure downside risk. Assets were allocated with particular regard to the management of downside risk.

6.4.2.4 Early 2014

While benchmark-related strategies with tracking error as the risk measure dominated before the crisis, the minimization of drawdowns was given the highest priority. Since the second quarter of 2009 when the equity markets recovered *asymmetric payoffs* are in focus. Beginning 2014, this was still the key word in Switzerland's third party circles. The goal is to achieve them by smart beta investments with the down market beta being lower than one and an up market beta as high as possible. Quantitative asset management worked poorly on average, until late 2012, but 2013 signs emerged that quantitative investing is back, yielding good returns. This puts them on the radar screen for 2014 again as complements to common fundamental strategies.

6.5 United Kingdom

For U.K. investors, managed volatility strategies became increasingly important over the past years in terms of tail risk management, followed by alternative investments like commodities. This becomes evident in Fig. 6.14 which shows answers of mid-2012 to the question which strategy(ies) did/do the U.K. investors have in place before the global financial crisis/now to protect against tail risk events. Interestingly, the role of diversification diminished overall compared to pre-crisis times which is in line with the overall result in Fig. 6.8.

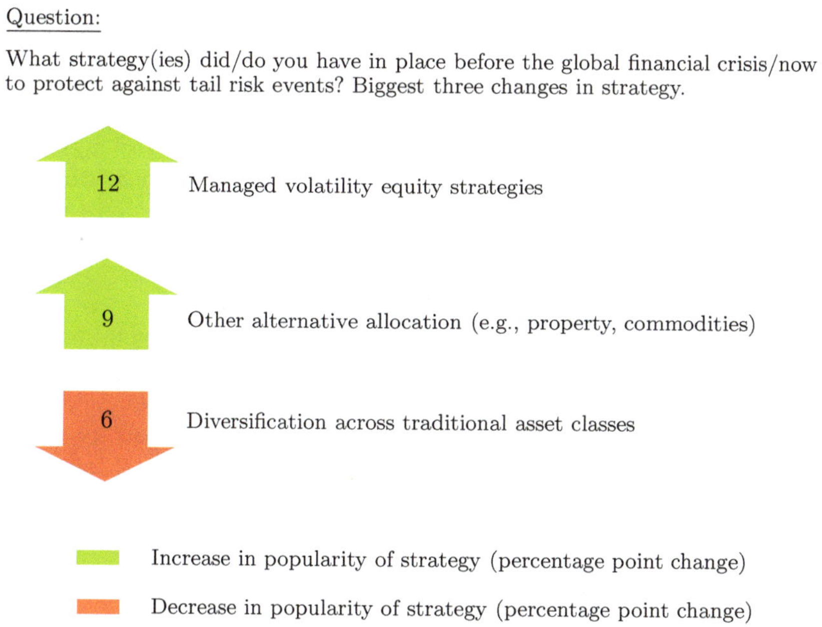

Fig. 6.14 Answers to the question what strategy(ies) did/do U.K. investors have in place before the global financial crisis/now to protect against tail risk events. Biggest three changes in strategy. *Source*: Economist Intelligence Unit (2012, p. 15)

6.5.1 Institutional Asset Owners

6.5.1.1 Before the Crisis

Before the crisis started in August 2007, tracking error was the key risk measure to assess the risk relative to benchmarks. This went hand in hand with the application of information ratio as a risk-adjusted risk measure. For absolute return mandates, Sharpe ratio was used. Some institutional investors in the U.K. also used the VaR approach. However, more sophisticated measures like shortfall risk (shortfall probability), downside risk, maximum drawdown or Treynor ratio were hardly applied at all. Interestingly, asset owners only rarely looked at volatility, no matter whether the portfolio was managed against a benchmark or not.

6.5.1.2 During the Crisis

Between end of 2007 and early 2009, risk of loss (downside risk) became much more important for investors than symmetric risk (volatility, tracking error). This was the most obvious change regarding risk measures. Asset owners also increasingly looked for absolute return products or guaranteed minimum returns and avoided benchmark-oriented products. Basically, a general flight from risk was apparent, and an increased interest in assets with low, but secure returns (for example, T-Bills etc.).

6.5.1.3 The Years Directly After the Crisis

Even after the market started to rebound mid-March 2009, institutional investors and prospects in the U.K. remained extremely cautious about investing in new funds and strategies. There was, however, a greater readiness to opt for more risky products. Regarding the risk measures which have been used after the crisis, tracking error was still a key risk measure for investors who followed benchmark-oriented strategies. But compared to pre-crisis times, many investors then also considered volatility. Other risk measures or risk-adjusted performance measures that were often used for strategy evaluation included Sharpe ratio, drawdown risk as well as maximum drawdown (absolute or relative) and VaR.

When investors made investment decisions, they also considered the correlation of their different investment strategies within the overall investor portfolio as well as the strategies' beta and VaR. Other measures like Treynor ratio, Sortino ratio, semi-variance and semi-volatility, bull/bear beta, shortfall probability (shortfall risk), expected shortfall (c-VaR), incremental value at risk, marginal value at risk, component value at risk and return on VaR (RoVaR) were not often used, if at all.

In Europe, the crisis in 2008 was followed by the Greek crisis, and further problems unfolded in the eurozone. Against this backdrop, major changes occurred among U.K. institutional investors. Compared to pre-crisis times, they especially asked for an extended measurement of risk before making investment decisions. They tried to develop skills and competencies for risk assessments, which was particularly the case for companies where only little knowledge about risk measures was available. They were much more inclined to invest in products that offered upside potential with little downside risk.

In short, institutional investors learned their lessons from the crisis. They clearly focused more on risk than before the crisis and regarded it as important as the rate of return of the financial product. Although this shift in favor of risk measurement was obvious, it was then hard to foretell whether this trend would persist over the next few years. Likewise, it was doubtful whether new risk measures like drawdown, semi-variance, VaR, shortfall, etc. would be widely used by U.K. institutional investors.

6.5.1.4 Early 2014

In early 2014, investments in both DB and DC schemes revolve around so-called matching assets and growth assets. The former are primarily beta investments like passive equity investments, passive investments in U.K. Gilts, and investments in ACWI indices.[18] The focus in terms of equities is rather on global investing (be it developed markets or emerging markets) than local U.K. investing. For these passive investments, tracking error is the key risk measure.

The latter, i.e., investments in growth assets, are linked to absolute return strategies like Libor Plus products or hedge funds. They promise returns above

[18]ACWI indices are all-country-weighted indices in equities which include developed and emerging markets.

the risk-free limit, and asymmetrical risk measures like drawdown and shortfall are applied. These strategies are often called diversified growth products with lower volatility than benchmark-oriented strategies. Diversified growth products also often include hedge funds that serve the same absolute return purpose but are more or less unconstrained. The idea of unconstrained investments or investments with less constraints is typical for this second bucket.

The combination of matching and growth assets can also be described as a core-satellite approach. The core is represented by matching assets to ensure consistent returns. The growth assets, i.e., the satellites, are represented by absolute return strategies that aim for drawdown protection.

6.5.2 Third Party Distribution

6.5.2.1 Before the Crisis

"Everyone wanted to have risky products." This was a key characteristic of the pre-crisis times, and it was true for sophisticated investors as well as for the general public. The major difference between these two groups was how they handled risk. The meanings and implications of risk differ, depending on different perspectives. Before the crisis, risk was often perceived as risk of the underlying asset or market risk. Another perception was to see risk in geographic or geopolitical terms. From the perspective of third party distribution in the U.K., these two types of risk were the main risks to be discussed. This was the active risk taken within a investor's active risk budget with volatility and tracking error being typical quantitative measures of market risk.

6.5.2.2 During the Crisis

In 2008 investors started to look at risk in a completely different way based on their crisis experience. Suddenly, formerly neglected forms of risk became prominent: counterparty risk, which was highlighted by the fall of Lehman Brothers, and tail risk, which in 2008 became increasingly important for investors.

6.5.2.3 The Years Directly After the Crisis

After 2008 new types of risk were discussed on top of the risk types mentioned above: fund risk, structure risk and liquidity risk. *What is the structure of the investment? Is it a fund or an ETF? If it is a fund, is it a mutual fund or a trust structure? What are the liquidity provisions?* These questions were raised, which not only asked about the investments of the product, but also about its actual structure. The concept of risk became more complex and included much more than simply the risk of the underlying assets or of the market.

Investors started to pay more attention on downside and shortfall risk in addition to using the traditional risk measures volatility and tracking error. However, this was mainly the case for the sophisticated investors and, then, less for the general public. Unlike the general public, the more sophisticated investors had always considered risk and were managing risk accordingly.

But after the crisis, the general public considered risk management as top priority, which raised several questions: Was the pendulum swinging too far in this direction? Would it swing back? And if so, when? There was a major concern that these investors had too much insurance or protection in their portfolio which would lead them to miss out on the upside in the market.

The euro currency had been another important aspect since early 2010. After the events in Greece and Ireland, euro sceptics in the U.K. were celebrating. They claimed that they had predicted the calamities for the euro which they regarded as a political dream that could not survive economic reality. They argued that countries without a unified economy cannot have a common currency. However, it was acknowledged that, for political reasons, the eurozone countries could not afford to let the euro die in the short or medium term and would fight very hard to keep the euro alive.

But by and large, the U.K. investors cared less about the euro currency issue. They were more concerned about the underlying country issues in Ireland and Greece and about the potential crisis candidates Portugal, Spain and Italy. This translated directly into demand for euro-denominated investment products that exclude Greece or both Greece and Ireland. Obviously, investors differentiated between the euro issue and the situation in certain troubled countries. U.K. investors did not shun the euro in 2010 and 2011, even if they do not belong to its most ardent supporters.

6.5.2.4 Early 2014

Among third party investors in the U.K., the trend seen right after the crisis continued in 2012 and 2013 such that now, in early 2014, the clear focus is on absolute return strategies. The risk appetite has increased over the past 2 years as equity markets rallied. However, this means for third party investors that they even more focus on downside risk. They may not be upset if, in an up year, returns are less than the benchmark, but they worry about negative returns even if they are (for active funds) significantly above the benchmark-returns.

This is another way of saying that third party investors in the U.K. do not, in early 2014, care about relative returns: they care about absolute returns. Therefore, asymmetrical risk measures like drawdown, semi-volatility and shortfall volatility, not symmetrical risk measures, are key risk measures. As seen in Fig. 6.14, managed volatility strategies have increased in popularity among all investor types. But in the third party segment, these strategies are less highly regarded, because they only reduce the drawdown but do not eliminate it. Since third party investors in the U.K. see the market primarily driven by behavioral aspects, they seek protection which can be offered by put options, hedge funds (or absolute return strategies in general) or structured products in case they offer a floor, i.e., a guaranteed minimum value level.

However, in order to protect from drawdown, core and simple strategies are preferred. The more complex and difficult it is to understand the product, the less success it will have with third party investors in the U.K.. For this reason, CPPI strategies are often considered as too complex, and synthetic products like synthetic

ETFs are hardly attractive for this client segment, as a synthetic product often involves a swap which can be difficult to explain. The idea of "being backed by an asset" (i.e., physically backed) as it is the case in many ETFs is key as it instills trust in the product.

6.6 Ireland

6.6.1 Institutional Asset Owners

6.6.1.1 Before the Crisis

Before the subprime crisis, risk was considered more in relation to a benchmark than in absolute terms, i.e., risk measures were relative risk measures with tracking error being the typical risk measure. This was seen relative to the achieved outperformance, i.e., institutional investors used information ratio as their key risk-adjusted relative return measure. Although absolute returns were less relevant, investors used Sharpe ratio as an absolute risk-adjusted return measure in order to compare the portfolio and the benchmark performance.

6.6.1.2 During the Crisis

As equity markets fell dramatically in 2008, downside protection came into focus. In addition, investors considered potential moves from active to passive asset management, as active management struggled to achieved a positive alpha and active management fees remained high. The focus on drawdowns implied a switch from benchmark-oriented management to absolute return portfolio management and to products which provided absolute returns and/or drawdown reduction. The traditional components of diversification were questioned as they were found to be insufficient. As a consequence, investors started to look at hedge funds which had performed well in 2008 whereas the overall confidence in hedge funds was greatly diminished.

6.6.1.3 The Years Directly After the Crisis

After the crisis, absolute return strategies and downside protection became clearly dominant. Investors more and more focused on true diversification which led to the introduction of alternative investments like hedge funds or managed volatility strategies and subsequent significant investments in these alternative strategies, especially when they offered elements of downside protection. Volatility, drawdown, VaR and correlations were key measures that investors focused on in the years right after the crisis.

6.6.1.4 Early 2014

The trend of the years in the aftermath of the crisis continued in 2012 and 2013 such that early 2014 there is still great emphasis on absolute return strategies and downside protection, especially in light of the rallying stock markets. Hedge funds are in great demand if they have proved to offer downside protection when needed, like in 2008. Therefore, downside risk measures like VaR are frequently used in early 2014 together with volatility as the standard symmetrical risk measure. Investors pay special attention to how an add-on to their asset allocation will impact the overall portfolio's risk and return profile and to the correlations of a new strategy vs. the existing components of the investor's asset allocation.

6.6.2 Third Party Distribution

6.6.2.1 Before the Crisis

In Ireland, there is no significant difference between the behavior of institutional investors and third party investors. As a result, the same investment strategies are followed. Third party investors were, before the crisis, only benchmark-relative investors using tracking error as the key risk measure.

6.6.2.2 During the Crisis

As described above, downside protection became paramount in 2008 combined with a move away from active management to drawdown reduction strategies.

6.6.2.3 The Years Directly After the Crisis

Continuing the trend, absolute return strategies and downside protection were the focus such that the launch of new products primarily took place in these areas. Like institutional investors, third party investors used volatility, drawdown, VaR and correlations as key measures while tracking error lost significance.

6.6.2.4 Early 2014

In terms of the Irish views on asset and risk measurement, 2012 and 2013 were no different from the few years before. Interestingly, Ireland became one of the most important places in Europe for hedge funds, as Irish Qualified Investors Funds (QIF) offer a perfect vehicle to implement hedge fund strategies.

6.7 Italy

6.7.1 Institutional Asset Owners

6.7.1.1 Before the Crisis

Before the crisis, many institutional investors in Italy computed their performance versus the referenced benchmark. Consequently, when the equity market grew, the institutional business in Italy tried to benefit from the trend towards passive investing (beta investing). Basically, there is a long-term trend in Italy which favors total return products as their risk and return is supposed to be not correlated to the equity market.

Investors in Italy were severely hit by the burst of the dotcom bubble. As a result, even before the subprime crisis, they did not only use symmetrical risk measures like volatility and tracking error, but they also used various other downside risk measures.

6.7.1.2 During the Crisis

Investor views on risk and applicable risk measures did not dramatically change since end of 2008. Rather, they had already changed because of the dotcom crisis in 2001. But in the aftermath of the subprime crisis, institutional investors discovered new types of risk like liquidity risk and counterparty risk for banks. Although these risks were not *new* per se, they had not played a prominent role before. After the default of Lehman Brothers and the near collapse of other financial institutions, this clearly changed. Now, these types of risk were carefully watched and considered as *serious potential risks.*

6.7.1.3 The Years Directly After the Crisis

As there was a trend toward passive beta investing in the Italian market, this risk was then actively measured. After the subprime crisis, institutional investors in Italy continued to apply all types of risk measures. They included symmetrical and asymmetrical ones, relative to the benchmark and for absolute returns.

However, the application of downside risk measures was not new and was not a consequence of the subprime crisis. Most of the Italian third party investors had already changed their way of measuring risk after the dotcom crisis. This was true for absolute and relative risk measurement. Value at risk and downside risk were good examples for this. The approach of investors in these years was to identify the risk they were not accustomed to in the past.

6.7.1.4 Early 2014

In Italy, over the years 2012 and 2013, the appetite for total return products remained high. Investors are still looking for absolute return products like risk parity strategies which have remained a key theme and, in fact, had great success during recent years. Compared to the time during and right after the subprime crisis, the perception of counterparty risk early 2014 has declined. The effects and new lessons of the crisis were very well understood by investors, since in Europe the debt crisis was particularly strong, but recent developments made counterparty risk less threatening.

There is no one specific risk measure that is generally used nowadays. Rather, many methods and measures are applied among which VaR is only one option. All of these risk measures are currently taken into account in Italy. Examples are maximum drawdown and, of course, volatility and tracking error as the typical symmetrical risk measures.

6.7.2 Third Party Distribution

6.7.2.1 Before the Crisis

In Italy, there was not much difference between the institutional investors described above and the third party business. As it was observed for the former group, many third party distributors calculated their performance versus a benchmark before the subprime crisis. When the equity market rose from 2003 until the crisis, many third party investors were focused on passive investing as well. As a result of the dotcom crisis, they not only used symmetrical risk measures like volatility and tracking error, but also other downside risk measures. Concerning risk management, there was basically no difference between asset owners and third party investors.

6.7.2.2 During the Crisis

There were no great changes in the views of third party investors on risk and applicable risk measures since end of 2008. The major shift in the risk analysis for investments came as a result of the dotcom crisis. However, new types of risk became important for third party investors as well, especially after the Lehman Brothers bankruptcy: liquidity risk and counterparty risk. Now, these types of risk were carefully watched and analyzed.

6.7.2.3 The Years Directly After the Crisis

As with institutional investors, the trend toward passive beta investing continued for Italian third party investors, and they applied all types of risk measures, symmetrical and asymmetrical. This included absolute and relative risk measures.

6.7.2.4 Early 2014

What was stated above for institutional investors in Italy also holds true for Italian third party investors in early 2014. The perception of counterparty risk is lower than in the years before and the effects and new lessons of the crisis are now very well understood given the experience of the euro crisis. Also, on the third party investor side, many methods and risk measures are now used, VaR being only one of them. Other risk measures include maximum drawdown as well as the traditional risk measures volatility and tracking error.

6.8 Nordic Region

6.8.1 Institutional Asset Owners

The Nordic region comprises Denmark, Finland, Iceland, Norway and Sweden. In all of these countries the key characteristics in the analyzed investor segments are similar. Therefore, they are grouped together.

6.8.1.1 Before the Crisis

Before 2007, investors were not really aware of certain risks that could be associated to the asset class they invested in. For example, asset-backed securities (ABS) had consistently low spreads with hardly any volatility. Volatility as the measure of market risk and tracking error as the measure of the investor's risk budget were the only applied risk measures. However, this region was generally known for having a preference for higher tracking error. In active equity management, Nordic investors focused on 130/30 strategies which came to market in 2005/2006. These products allowed for 30 % shorts in the investment, resulting in 130 % long positions.[19] The higher gross exposure of 160 % was accepted since the expected return versus the comparable index was considerably higher than in long-only products, i.e., products that did not allow short positions.

The Nordic region was generally open for new strategies, whether regional or global. Geopolitical risk was not a concern in these days. Also, *securities lending* was not seldom a part of the investor's program as it was considered *free lunch*.

6.8.1.2 During the Crisis

In 2008, the investors had to learn the hard way that the portfolio beta can wipe out big parts of your active investment. Alpha, the main concern in the years before, was not the key focus anymore. In many investment strategies, alpha was negative in 2007 and 2008. As a result, the region, especially Sweden, saw a strong move toward the separation between alpha and beta, with their *alpha portfolio* being clean of beta. Investors moved away from long-only active portfolio management.

6.8.1.3 The Years Directly After the Crisis

In the aftermath of the crisis, Nordic investors perceived risk on total portfolio level instead on asset class level. In addition to volatility and tracking error, downside risk measures like drawdown were discussed, and a considerable effort was made to analyze protection strategies for the investors' portfolio. Therefore, managing left tail risk was a key focus of Nordic investors in the years 2010 and 2011, in order to avoid drawdowns they had experienced in 2008.

6.8.1.4 Early 2014

The past few years have seen an increased interest for alternative beta solutions across the whole spectrum of institutional investors in the Nordics. Very sophisticated investors have implemented such a strategy through factor-tilted portfolios to explore the widening of factor-based allocations to other asset classes than equity. The other investors have just started the research process and are, early 2014, still in the evaluation phase. Certain investors in addition to tilting are also looking at taking equity factor exposure through traditional long positions via long-

[19] For more information about 130/30 strategies see Schulmerich (2007a).

short strategies like 130/30 or even market-neutral.[20] 130/30 long-short strategies use as benchmark-relative strategies tracking error as the main risk measure and information ratio as the risk-adjusted return measure. On the other hand, market-neutral strategies are benchmark-agnostic such that only volatility and Sharpe ratio are used as typical asymmetrical measures. Also, some investors show interest in the traditional factors like volatility, value and quality but also express interest for alternative risk premia such as FX carry and convertible arbitrage.

Some common issues faced by investors when exploring these investment themes are the identification of the appropriate strategic benchmark and the fact that pure factor exposure is not easily achieved as there is usually additional (mostly unwanted) factor exposure. The most sophisticated investors are analyzing returns of their active managers closely and differently and are looking if they can possibly replicate these through cheaper alternative beta exposure. The move towards alternative beta is also allowing these sophisticated investors to judge external managers' returns differently, i.e., away from the traditional market-cap weighted indices and more versus a factor exposure approach.

6.8.2 Third Party Distribution

6.8.2.1 Before the Crisis

Given that before the crisis returns from a quantitative investment approach were pretty consistent and that lack of performance could be explained by "usual" market behavior, risk was more considered as relative to a benchmark than viewed in absolute terms. Thus, the risk measures most frequently used before the crisis were tracking error, Sharpe ratio (of portfolio vs. benchmark) and beta.

6.8.2.2 During the Crisis

Since the former focus was on risk relative to benchmark, the logical and most common reaction of third party distributors during the crisis was a switch to passive strategies or the adoption of high tracking error strategies which were fundamentally managed. They either continued to measure risk relative to a benchmark or looked for investment opportunities with high tracking error.

6.8.2.3 The Years Directly After the Crisis

After the crisis, third party investors added on their fundamentally managed positions. They relaunched their assessment of quant models and managers and continued to look for high tracking error strategies, as the market lacked clear direction in 2010 and 2011, being primarily driven by the euro crisis. The investors remained cautious, not excluding the possibility of a double dip scenario. After they discovered that high tracking error can be coupled with lower volatility and with

[20]For more information about equity market neutral strategies see Schulmerich (2009b) and Schulmerich (2007b).

very positive effects on long-term positions, they saw a high tracking error in a different light. In these days, risk measurement was based on absolute rather than relative figures.

All eyes were then focused on the emerging markets, on emerging markets debt, and on hedge funds, as managers and selectors kept looking for strategies that could offer good returns. But investors constantly showed the capacity to become blind on past events. For example, they tended to avoid the analysis of political crises in the emerging markets and did not assess the risk that a local crisis might spread across a whole region. It seemed as if they partially forgot that after the subprime crisis economies were much more interdependent.

Investors argued that the risk of default was very low, since emerging markets countries could devaluate their currency. However, it was always the question whether to prefer to be paid little or to be paid in a currency that is worth nothing. People were running to local currency debt because they did not want to lose two percent in return by investing into ADR , GDR or hedged products.[21] As for European economies, a future flight to quality sovereign debt (of Finland, France, Germany and the Netherlands) seemed possible and, here and there, became visible.

6.8.2.4 Early 2014

Early 2014, the search for yield continues with interest in absolute return strategies which have a cash + X return target. Also, the third party investors in the Nordics still see interest in yield-seeking fixed income strategies like high yield and emerging markets debt—although to a lesser extent than in 2012 and 2013. Compliance with new and evolving regulation remains a top priority for distributors as regulators demand more transparency for the funds offerings to retail.

The use of UCITS funds to structure the risk gets more important in the Nordics which shy away from hedge funds as an asset class and specifically from funds that are not UCITS. Interestingly, the alternative beta idea which can be found everywhere among institutional investor space in the Nordics, has not reached the distributors yet, but they are looking at the developments in that field as well.

6.9 Benelux Region

The Benelux region comprises Belgium, Luxemburg and the Netherlands which show similar characteristics in terms of investor behavior. Therefore, these countries are considered together. In mid-2008, the EIU report also asked Benelux investors which strategy(ies) they did/do in have in place before the global financial crisis/now

[21] As already explained on p. 220, an ADR (American Depository Receipts) is a registered security issued by a U.S. bank representing shares of a foreign stock. ADRs trade on U.S. stock exchanges and on the OTC market. A GDR (Global Depositary Receipt) is a bank certificate issued in more than one country for shares in a foreign company. The shares are held by a foreign branch of an international bank. The shares trade as domestic shares, but are offered for sale globally through the various bank branches. GDRs and ADRs are very similar.

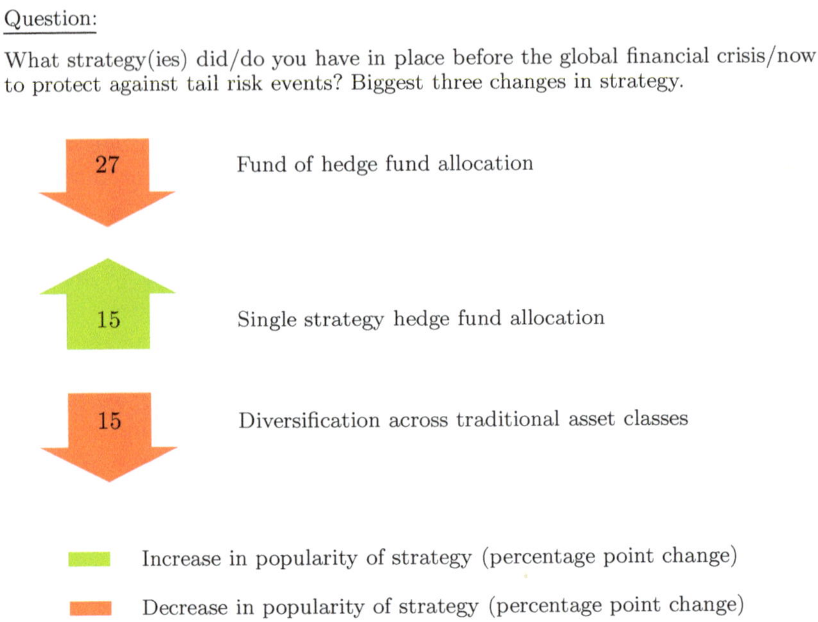

Fig. 6.15 Answers to the question what strategy(ies) did/do Benelux investors have in place before the global financial crisis/now to protect against tail risk events? Biggest three changes in strategy. *Source*: Economist Intelligence Unit (2012, p. 15)

to protect against tail risk events. Their answers show that they significantly reduced their use of fund of hedge fund allocations. They doubted the added value of funds of hedge funds, did not accept the double fee layer and considered single hedge funds much cheaper. Investments in single hedge funds increased considerably, and in early 2014, this trend was still in fact. Like in the U.K., Benelux investors increasingly questioned the effectiveness of traditional diversification. All this is reflected in Fig. 6.15.

6.9.1 Institutional Asset Owners

6.9.1.1 Before the Crisis

Before the crisis started in August 2007, risk was generally viewed relative to benchmarks and not in absolute terms. Tracking error was the most frequently used tool for risk measurement. There were sometimes discussions about calculating a risk budget or not, but as the performance of the portfolios was consistent and positive, the real risk was not identified, even if risk budgets were calculated.

6.9.1.2 During the Crisis

Between end of 2007 and early 2009, asset class diversification did not necessarily mean risk diversification. Some of the risk elements were not really considered by investors. When the equity markets started to recover in March 2009, there was not a lot of change, since the investors preferred to wait for the recovery to be sustainable. The sales and redemptions that were encountered in early 2008 were liquidity-driven.

6.9.1.3 The Years Directly After the Crisis

Risk measures relative to solvency ratios such as drawdown risk were more frequently used after the crisis. As the problems in the eurozone unfolded, credit risk and liquidity risk became more closely scrutinized. More detailed studies on risk factors and how they affect different asset classes were conducted. Drawdown, semi-variance, VaR, liability risk, credit risk, etc. were important measures that formed part of an investor's analysis toolkit. There also was a general trend to stricter monitoring, like asking for holdings reports on a monthly basis etc.

6.9.1.4 Early 2014

The pre-crisis risk appetite of institutional investors has returned such that in early 2014 in, for example, the Netherlands, the equity portions in the portfolio allocations are increasing, especially in the tactical asset allocation (TAA) schemes. However, like in almost all countries, a lot of investment decisions are driven by regulation. In 2014, the Netherlands will see a new FTK regulation,[22] and as a consequence, many institutional investors delay to make decisions. Of course, interest rate movements which will be reflected in the coverage ratio will also have a big influence. At present, in early 2014, most prominent topics are investment themes in which investors were interested also in 2013: tapering, emerging markets and the ongoing regulation of financial markets.

6.9.2 Third Party Distribution

6.9.2.1 Before the Crisis

Like in the Nordics, the quantitative approach to asset management yielded returns in the Benelux which were quite consistent before the crisis. Mediocre performances were explained by "usual" market behavior, and risk was more considered as relative to a benchmark than in absolute terms. The most popular risk measures before the crisis were tracking error, Sharpe ratio (for portfolio and benchmark), and beta (for portfolio and benchmark). Relative return thinking was predominant.

---

[22]The financial assessment framework (FTK) is the part of the Pension Act (Pensioenwet) in the Netherlands that states the statutory financial requirements for pension funds. It is built around the principles of market valuation, risk-based financial requirements and transparency.

6.9.2.2 During the Crisis

During the crisis, the situation in the Benelux and the Nordic region was very similar. While the main focus had been on risk relative to benchmark before the crisis, third party distributors now switched to passive strategies or looked for high tracking error strategies which were fundamentally managed. This means that they either considered risk relative to a benchmark, or invested in products with high tracking error. In the Benelux, investors turned to passive strategies because active behavior was not considered trustworthy anymore.

6.9.2.3 The Years Directly After the Crisis

Also in the years directly after the crisis, the situation in the Benelux and in the Nordics as described above remained similar. Every asset manager was checked again by the investors, particularly in the Benelux, and fundamental asset managers with long track records were clearly in favor.

In these years 2010 and 2011, investors tended to seek safe havens plus possible benefits from high-yielding asset classes such as hedge funds and emerging markets. Balanced funds that performed well during the crisis remained also popular. One of the trends that emerged in these years was the idea of managed volatility strategies which aim to reduce drawdown via a low-beta stock portfolio while at the same time aspiring to returns similar to the market. In 2010 and 2011, when volatility and behavioral factors had a strong impact on the markets, these benchmark-agnostic strategies (where the index is most often only used as the investment universe and not as the benchmark as such) became more important.

6.9.2.4 Early 2014

Third party investors, i.e., retail investors, show a renewed appetite for high-risk investments and are ready to invest into equity again. For 2014 and thereafter they expect a switch from fixed income investing to equity investing. All signals from retail investors confirm that they are returning to the equity markets.

Another trend already in place since the post-Lehman times is capital preservation, i.e., drawdown management. This has not changed over the past years and is not likely to change in the near future. Combining these two overarching themes in the retail market, the third party investors who collect and manage the money of retail investors show a clear focus on capital preservation and equity. Since, in the Netherlands, kickbacks to the intermediate networks are prohibited starting from 2014, an ongoing shift from active to passive funds is also expected.

6.10 France

6.10.1 Institutional Asset Owners

6.10.1.1 Before the Crisis

Before the crisis, institutional asset owners tended to look for high tracking error strategies (relative vs. a benchmark). For example, quantitatively managed active European stock portfolios were very much appreciated as core portfolio investments. As risk measures, institutional investors mostly used measures like tracking error.

6.10.1.2 During the Crisis

During the crisis, a strong tendency to reduce active risk could be noticed. This was primarily expressed by switching from active to enhanced strategies[23] or completely to passive investing.

6.10.1.3 The Years Directly After the Crisis

After the crisis, investors were still very cautious. However, they had more appetite for risk but demanded more innovative solutions (asset allocation between active and passive management, solutions to minimize drawdown). Strategies that aim to reduce portfolio volatility were also more appreciated. The main risk measure remained tracking error, but the interest in measures like drawdown or VaR grew. Passive and smart beta solutions, also known as *advanced beta*, such as non-cap-weighted benchmark approaches, including fundamental indexing[24] and low volatility strategies also seemed to look more attractive. We can expect this tendency to continue as long as volatility remains a concern for investors (probably so in the next couple of years).

6.10.1.4 Early 2014

The situation in early 2014 shows that the tracking error approach has moved into two directions:

1. Higher tracking error with the idea to look for more active and high value-adding strategies
2. A switch from tracking error to an absolute risk approach based on the concern about total risk reduction and with the goal of drawdown limitation

In addition, the investment discussions of French institutional investors have moved from *active vs. passive* to an *passive vs. advanced beta vs. absolute return* approach. Advanced beta is focusing notably on the minimum-volatility approach or sustainable income approach. The absolute return approach is focusing on a very flexible approach using risk sensibility parameters and moving away from balanced

---

[23]In equity asset management, enhanced strategies, also called *index plus strategies*, are actively managed portfolios with low alpha and tracking error targets. Often, alpha and tracking error targets of such strategies are 1 % p.a. or less. This is considerably less than active strategies in active equity portfolio management normally aim at. For the application of the index plus approach on emerging markets equity see, for example, Schulmerich (2009a).

[24]A fundamental equity index is an index that is not constructed using the firms' capitalizations but other fundamental data like sales or cash flow figures which better represent the healthiness and growth opportunities of a firm than its capitalization.

benchmarking to strategies that target an absolute return of a certain percentage figure above the cash return.

6.10.2 Third Party Distribution

6.10.2.1 Before the Crisis

Before the crisis started in August 2007, the French investors had a double view of risk versus return, i.e., they were looking for high and medium tracking error strategies (2–3 % versus benchmark for active equity portfolios). In France, a core-satellite approach was very well recognized for this type of strategy. In terms of risk measures, the French investors primarily used tracking error or volatility of the portfolio versus the benchmark.

6.10.2.2 During the Crisis

Between 2007 and early 2009, French third party investors viewed risk in a different way. They very strongly cut their investment on the equity side, sold their quantitatively managed investment strategies first, and preferred to invest in cash products and/or fixed income products such as convertible bonds, high yield and corporate bonds. Also, French private bankers who generally used high tracking error equity strategies, reduced their risk and often invested in enhanced equity strategies.

6.10.2.3 The Years Directly After the Crisis

Since the second quarter of 2009, the French third party investors applied a very specific approach to risk management. Due to the fact that the market environment was not favorable for equities, they invested very little in equity funds and preferred high tracking error or *best selection* portfolio strategies instead, in order to capture the value added of pure stock pickers. Quantitative investment strategies were usually judged as trend follower strategies despite a tracking error of 2–4 %. In 2010 and 2011, French third party investors were looking for:

- ETFs to expose their portfolio and to track the market trend with the best timing.
- High tracking error portfolios from pure French stock picking asset managers. Investors also thought that considering the situation of the financial markets, quantitative equity strategies were not appropriate due to the negative equity trends and volatile equity markets, often driven by behavioral factors.

The trend to look for new risk measures was really a new way to manage funds. But not only the view of risk measures had changed. Most of the investors knew alpha-seeking strategies very well and liked them very much, but they did rarely invest in them anymore on the equity side. Only few investors wanted to take large equity positions.

6.10.2.4 Early 2014

What was said above about institutional investors also holds true, in general, for third party investors. The two most important items amongst third party distributors are the relative ranking, i.e., being top quartile or top decile, and the absolute size of the proposed fund. Below a minimum size of €50 mn., a third party investor does not even look at a fund at all. Third party investors in France[25] use the following top three risk measures (in the order of preference):

1. Volatility, especially using the past 1 and 3 years of return data
2. Drawdown measures
3. Tracking error vs. benchmark in benchmark-related strategies

Especially when it comes to fundamental equity investing, a higher tracking error is accepted as investors are historically used to fundamental stock pickers with potentially high tracking errors. Interesting to note is also that the application of VaR as risk measure is less important as it was in the years directly after the crisis.

In France,[26] more and more investors are risk-adverse. They limit the acceptable volatility and set drawdown limits for their portfolios even if this has a consequence on returns which in some instances are disappointing. Defensive allocations are favored. Early 2014, many third party investors tend to look closely at equity investing again which, however, is rather late compared to other countries. For the months or even years to come, there is a clear trend for third party investors that limits and controls will be reinforced coupled with extended due diligence exercises. So in summary the investor behavior in early 2014 is still very cautious and risk-averse which can be seen in the use of the above-mentioned risk measures and the detailed due diligence processes.

6.11 Middle East and North Africa

6.11.1 Institutional Asset Owners

6.11.1.1 Before the Crisis

Apparently, risk was not on the radar screens of institutional investors. The majority of the investors in the Middle East measured risk in a very straightforward fashion, mostly by volatility and tracking error, with the possible exception of the most sophisticated sovereign wealth funds which might use, for example, VaR, c-VaR as well as what-if scenarios at the aggregate level of their portfolio per department and overall.

[25]The same also holds true for Monaco, Spain and Portugal.

[26]The same holds true for Spain.

6.11.1.2 During the Crisis

During the crisis, as fear and suspicion increased, confidence was much more difficult to maintain. People would go back to the drawing board and take quite radical decisions, for example, turn to a completely passive approach. Even after the crisis, the confidence of institutional investors did not rebound to the level as it was before. Investors refrained from active management during the crisis and did not opt for active quant management as they did in the time before the crisis. Previously, a good track record could be sufficient to gather assets. During the crisis, this was much more difficult.

6.11.1.3 The Years Directly After the Crisis

Since the markets recovered, in the years directly after the crisis, the attitude of the investors did not change significantly. They still felt traumatized by what had happened and were changing the way they invested.

This was particularly true for sovereign wealth funds (SWF). An SWF is a state-owned fund which invests money for the country, for example, from oil exports. SWFs generally dominate the institutional investor segment in the Middle East, and their reaction was nicely summarized by Andrew Rozanov, head of Sovereign Advisors and a managing director at State Street Global Markets in 2009.[27] This is his comparison of the investment challenges that these investors faced in mid-2009 versus the situation in 2008 and before the crisis[28]:

> *For a long time, investors have viewed portfolio diversification as the last available free lunch; incorporating a wide variety of less-than-perfect correlated asset classes can improve a portfolio's return-to-risk profile. However, the downturn has severely tested this core tenet of modern portfolio theory. Many long-established and highly respected institutional investors - sovereign wealth funds as well as pension plans, endowments and foundations - looked on in horror as their broadly diversified portfolios composed of supposedly uncorrelated asset classes suddenly registered double-digit percentage drops in value, as correlations across most investments rose in unison. Was this a statistical aberration, a so-called perfect storm, or has diversification failed? And if it has failed, what can SWFs and other large institutional investors do to protect their long-term financial wealth?*
>
> *[...]*
>
> *SWFs are likely to conduct a thorough analysis of their asset allocation, portfolio construction and risk management approaches and assumptions. A new generation of risk models will emerge that use more robust scenario analyses and stress tests for the kinds of multiple standard deviation events that have occurred with disturbing frequency over the last few years.*

Since investors understood that a simple core-satellite approach with a good selection of successful active managers was not sufficient to reach their objectives, they tried to think out of the box and became interested in new approaches. But they still remained very risk-averse and were suspicious of anything which had not been tested in a live portfolio. Examples thereof are fundamental indexing, low volatility

---

[27] State Street (2009).

[28] State Street (2009, p. 15).

approaches or equally-weighting schemes in portfolios. Concerning risk measures, they did not change their approach. Investors tried to look behind the clouds and were interested in systemic risk, particularly market systemic risk. They also had learnt about the impact of fat tails, but it remained uncertain whether they aimed to monitor this in a systematic or quantitative way going forward.

6.11.1.4 Early 2014

The situation for institutional investors in the Middle East as described above did not substantially change in 2012 and 2013. Risk management is an even more integral part of investment management than Andrew Rozanov predicted in 2009. Risk management and internal consulting departments were either established or existing ones were expanded in order to deal with all risk-related questions for whatever type of risk in the various asset classes. So, as in many other countries in EMEA, many risk professionals were hired in the years after the crisis. When it comes to decision-making, depending on the sophistication level of the institutional client, more or less all common types of risk are analyzed and return and risk-adjusted return measures are used. However, while risk management is important, the key aspect for an investment is still its return.

Benchmark-related strategies still dominate the investment range of institutional investors. However, they are very interested to learn about new investment trends, one of them being advanced beta. But today, investment in these strategies is still very limited as the investors expect to see a live track record as a proof for an asset manager's performance claims. It should also be noted that the interest in absolute return strategies is limited since the influence of the benchmark concept is strong, like before the crisis.

6.11.2 Third Party Distribution

6.11.2.1 Before the Crisis

Before the crisis, risk was not the main concern for the Middle East third party business. In fact, both end users and investment platforms that distributed investment solutions did not pay attention to risk and neither considered standard risk measures nor more sophisticated ones like downside risk measures. They rather focused on the return aspect: The higher the return of an investment strategy, the likelier they would invest in this strategy. This implies that risk-adjusted return measures were not taken into consideration by these investor groups when it came to investing. Key drivers in the decision process were the historical return and the size of the investment under consideration, for example, equity or fixed income funds. The notion of benchmarking was not a familiar one in the Middle East region before the crisis.

6.11.2.2 During the Crisis

During the subprime crisis, third party business investors suffered from lock-ups, low liquidity and, in some cases, huge losses. This triggered a sudden and substantial mindshift: They realized that neglecting the risk aspect when making investment decisions was a critical mistake. They became significantly risk-averse, and risk aspects became more and more important for the analysis of potential investments. Third party investors started to look at volatility and tracking error, and also to downside risk measures like VaR. Most third party business investors in the Middle East region stayed in cash for the most part of 2009 up to the first quarter of 2010: They simply decided that it was too risky to invest and spent most of the time trying to get out of pre-crisis investments.

6.11.2.3 The Years Directly After the Crisis

When investors and prospects made investment decisions after the crisis, they applied two main criteria: they maintained their liquidity, and asked how well the asset in question did in 2008. To a certain extent, investor meetings revolved around risk aspects. Investors were also open for absolute return investments like hedge funds and volatility reduction strategies. The events of 2008 were still in their mind and it was most likely that risk measures would be highly used in the years to come.

6.11.2.4 Early 2014

Trends observed in the couple of years after the crisis continued in the years 2012 and 2013 such that, in early 2014, all what was stated in the paragraph above still holds true: liquidity risk remains a major concern among third party investors who consider risk management aspects as vital. In addition, given the low-yielding environment in the recent years even investments in low-yielding bonds trigger the question of potential drawdowns to come. There is also a discussion about a potential switch from government bonds to equities.[29] Therefore, strategies that look at drawdown reduction, be it traditional long-only approaches like managed volatility or more sophisticated approaches from the hedge funds universe, are carefully considered.

6.12 Central Banks

Central banks view risk differently from institutional or third party investors. In the following we will take a closer look at the attitude of central banks and how this attitude has developed over the last years.

6.12.1 Before the Crisis

Prior to the crisis, central banks had adopted enhanced portfolio management techniques with the clear aim in most cases to raise the yield on their reserves without sacrificing prudence. The main reasons of this systematic enhancement of reserves, management skills and resources were the growth in reserves, the desire

[29]This discussion is often referred to as the *great rotation*.

to increase returns and the widening of the range of investments to include new and riskier asset classes.

The following statement refers to an Asian central bank: *"The central bank has added some asset classes - MBS and corporate bonds - into its investment universe, and has also upgraded its IT support system for its reserve management."*

In their search for yield, the investment policies of some central banks became adventurous in 2006. While highly rated assets remained a mainstay of central bank reserves, a growing proportion invested in low-rated papers. Central banks in developing countries displayed the tendency to diversify in the government and corporate sectors, but were less attracted to agency papers.

The majority of the central banks constructed their benchmarks in-house. A central banker in Europe remarked that in-house development allowed the creation of customized benchmarks. More than half of the central banks, especially in EM, had a benchmark for total returns on the reserve portfolio, partly for principal preservation, but also because some of them believed that this approach to benchmarking provided a means of measuring performance. Derivatives were popular with almost 70 % of the central bankers using at least one type of derivative instrument (mostly foreign currency swaps).

In 2006 and 2007, market risk was considered to become more important, and many central bankers reacted to the prospect of rising interest rates and volatility. Here are some statements of central bankers: *"Interest rate and currency volatility might become an issue given the current evolution of relative prices, and once the major economies stabilise their growth rates." "Increased volatility of foreign exchange and interest rates leads to higher market risks, and to the instability of market yields."* Some were also concerned with liquidity risk, but they did not believe in big changes in 2007. Not much thought was given to credit risk in 2006 and 2007. Market risk was taken for granted, and credit risk was underestimated. One central banker mentioned that his institution relied on rating only although this was not deemed sufficient.

Diversification was the answer to mitigate risks arising from the issuance of sovereign debt. Diversification by country of issuance and type of issues, and moving toward inflation-linked products[30] was the key strategy for many central bankers. Some of the measures were the tactical reduction of exposure to selected sovereign issuers, the ad-hoc limitation of exposure to selected issuers and the temporary suspension of new investments into issues of selected issuers.

6.12.2 During the Crisis

For most central bankers the traditional aims were safety, liquidity and return—in this order. After the crisis in 2008, they still believed that this order has remained

[30]Inflation-linked products are fixed income products whose future payoff is linked to inflation. In case of, for example, rising inflation, the payoff is adjusted upwards such that for the buyer of such products the purchasing-power-adjusted payoff does not change in the future.

unchanged, while some Asian central bankers consider safety and liquidity to be on an equal footing. European central bankers also modified their view of liquidity, but they still thought that the former priorities had not changed.

Central bankers remain wary of credit risk. These days, they are only interested in the traditional safe-haven assets of high-quality government bonds and gold. Hardly any central banker is optimistic on MBS and ABS. A minority of them still think commodities are more attractive than assets like, for example, A-rated government bonds.

In the times of crisis, there was a tendency to flee to quality by investing in top-rated government securities. Other securities and deposits appeared much riskier and less attractive. One Middle Eastern central banker said: *"12 months ago the crisis was at its height. Since then there has been a stabilisation of expectations and a reduction - it would seem to the point of overshooting - in risk premia."*

The financial crisis seriously changed the way reserve managers viewed the liquidity of markets and assets. Central bankers were shocked at how liquidity evaporated not only in markets at the lower end of the rating spectrum, but also in those for the best-rated assets, which were thought to be immune to such market dislocations. Today, half of the central bankers do not hedge any currency risk, but this may not be a direct impact from the crisis.

6.12.3 The Years Directly After the Crisis

An interesting article on risk management in the post-crisis time by John Nugée was published end of 2010.[31] He was specifically looking at official sector asset management which includes central banks and sovereign wealth funds. An excerpt of his article is the conclusion for this section:

> *Alongside the changes described in the section above, there is a growing interest among a few of the more sophisticated funds in identifying and accessing the ultimate sources of economic added value. This debate arises from the observed failure of the traditional method of risk management and risk reduction for a large portfolio, that of diversification across many asset classes.*
>
> *As has been observed before in times of extreme market stress, correlations between asset classes which seemed to be stable before the event can change dramatically in a crisis, and assets whose prices move independently in normal times all too often fall in value together in times of market turmoil. And the result - as any investor in the markets during the last few years will attest - is that diversification too often fails to provide investors and their portfolios with the expected protection against valuation losses precisely when it is needed most.*
>
> *The most dramatic example of this in the recent crisis was the breakdown in the commonly assumed relationship between debt and equity. For many investors, the general assumption has been that in normal market conditions, fixed-income securities will show some negative correlation with equity markets, allowing an investor to reap the benefit of*

[31] John Nugée was Senior Managing Director and Head of the Official Institutions Group at SSgA until his retirement end of 2013.

diversification by holding a balanced portfolio containing both asset classes. In severely stressed markets, however, this relationship does not always hold, and while a flight to quality benefitted those holding top quality sovereign debt, especially U.S. Treasuries, investors in corporate securities often found that prices of both the debt securities and the equity stock of companies were falling together as economic conditions deteriorated. This was especially the case with securities of some financial institutions, where their debt issues fell heavily in price alongside their equity, and also often started to display equity-like qualities as doubts about company solvency started to emerge.

For anyone seeking to manage risk, this poses a challenge. If orthodox or standard diversification does not deliver the risk reduction it was expected to in a crisis, then it is not just without value, but even worse than this, it is dangerous, because it can lead asset managers into a false sense of security. Moreover, it does not seem to be the case that the diversification strategies that asset managers were employing were "the right idea, but just not pursued far enough": very few sovereign funds that the author spoke to in preparing this essay think that the answer lies in further diversification.[32]

[...]

One notable result of the market crisis has been to remind sovereign asset owners of the possibility of reputational loss and damage arising from strongly negative financial performance.

[...]

This has led to two things. Firstly, funds' awareness of the desirability of protecting against extreme losses has increased. There has been a renewed interest in "tail" or "disaster" insurance, protecting a portfolio against extreme losses. The problem here for most sovereign funds is that what tail insurance is available is very richly priced, and the scale of the insurance that large sovereign funds might want to buy would risk exceeding the private sector insurance market's ability to provide it. And even if it was available, for any sovereign fund buying large-scale insurance, the transaction merely exchanges market exposure for counterparty exposure, which is not necessarily an improvement for the insurance purchaser.

Indeed, on a more fundamental level some commentators have queried even the idea of the public sector seeking large scale insurance from the private sector, given that in a financial crisis the flow of support is more usually the other way, as in a market crisis the deep pockets of the public sector are called upon to assist those in the private sector facing difficulties.

Faced with this analysis, most large sovereign funds seem to have concluded that the search for tail insurance is likely to prove unsuccessful, and that the defences against damaging reputational losses will remain the traditional ones employed by the public sector - education of their stakeholders into the limitations they face in avoiding loss, plus reliance on their longevity to repair portfolio damage ("time as the great healer").

On a secondary level, though, a number of funds have looked afresh at the various instruments they use. It is well known by official sector asset managers that it is not only the size of a loss that matters, but also how it is incurred: a \$ 1 mn. loss on a bond portfolio because yields go up is "unfortunate", a \$ 1 mn. loss on a credit portfolio because an issuer defaults is "highly damaging", and a \$ 1 mn. loss because an opaque structured product fails to live up to its promise is a disaster which risks major public relations and reputational consequences - despite the loss in all three cases being the same size in financial terms. [33]

[32]One fund manager commented to John Nugée that "*diversification was a disappointment and an illusion, and when you see how all correlations moved to 1 in the worst of the crisis, I do not think more of the same would have helped us very much.*"

[33]Nugée (2010, p. 4–5).

6.12.4 Early 2014

Today, risk measures within central banks vary. Central banks manage investment risk as well as overall liquidity risk. As far as investment risk is concerned, the majority of the official institutional investors use tracking error for benchmark-related portfolios. It is the most common measure for relative return portfolios at present.

Liquidity risk is a more complex topic. Reserve managers may use bid-offer. or bid-ask spreads as measures. Some central banks in Europe also use the TradeWeb Liquidity Index and some use duration.[34] Many try to manage liquidity by diversifying their investments; they use issue, issuer, asset class limits and keep a certain portion of reserves available within a few days.

About two thirds of central banks have not changed their exposure to dollar and euro since the end of 2011. That means a third of central banks have changed their allocation towards these traditional reserve currencies. Due to the crisis, many central banks had to "reconsider" their credit policy. Some have massively reduced their credit risk and some see sovereign risk as central to risk management. Then there is renewed attention to the yield as the boards now consider the cost of carry. Overwhelmingly, reserve managers now feel their central banks need to diversify, but progress is slow in looking at alternative instruments/markets or in diversifying geographically. As an example, roughly only one fourth of the central banks are investing or considering to invest in the Chinese renminbi.

As of early 2014, the more attractive assets for central banks are gold on the commodity side, AAA-rated government bonds, equities, AA-rated government bonds and agency papers. Still less attractive are MBS, ABS, deposits and A-rated government bonds.

Tail risk is important for central banks.[35] Each sovereign wealth fund is different, but central banks in general do not tend to invest in difficult to understand tail risk management strategies. They are more sensitive to reputational risk than other institutional investors, but more frequently also look at, for example, low-risk hedge funds to reduce portfolio drawdowns.

6.13 Summary

Based on Economist Intelligence Unit (2012), Schulmerich (2011, 2014a) and Jordan et al. (2013), this final chapter looked at the investors' risk perceptions, the changes over the past 10 years and the implications for the chosen investment strategies. However, not all investors and not all countries in EMEA reacted in the

[34]Common tools in Asia are Libor-OIS and basis swaps. A Libor-OIS is a Libor over-night index swap.

[35]The same holds true for sovereign wealth funds in general.

same way. Therefore, this chapter needed to distinguish between various investor types and countries.

The past 10 years were characterized by bubbles and crashes, both regional and global. These stressed market conditions had a significant influence on how investors deal with risk. For example, more risk measures are used nowadays compared to the pre-Lehman times, particularly the ones which look at the left tail of the return distribution. Tail risk management was a central term in the discussions between investors and asset managers especially in 2012 and, given the stellar equity market returns in 2013, which triggered fears of a bubble, also in early 2014. This fear was fed by the crisis in the emerging markets (notably the events in Turkey and Thailand brought up unpleasant memories of the start of the Asian crisis 1997) early 2014, resulting not only in negative returns for the global emerging markets indices in January but also for the global developed market indices.

Chapter 6 concludes this book which looked at the theory and practice of asset and risk management. It was especially designed as a comprehensive introduction to these fields for bachelor and master students as well as for young professionals in asset management. The book revolves around two questions:

- Why do crashes happen when in theory they should not?
- How do investors deal with such crises in terms of their risk measurement and management and what are the implications for the chosen investment strategies?

In order to answer these questions we first had to present the appropriate performance and risk measures (Chap. 1) and the key principles of modern portfolio theory (Chap. 2). At the end of Chap. 2 we saw that empirical tests often do not support the theory which is also contradicted by stock market anomalies (presented in Chap. 3). Furthermore, events like bubbles and crashes (Chap. 4) can also not be explained by standard investment theory. Behavioral finance offers a way to explain such crashes as is described in Chap. 5. Finally, this chapter presented the reaction of the investors to the market developments of the recent past that saw the most extreme movements since 1929. Thereby, Chap. 6 answered the second question and should serve the needs of senior professionals in asset management who want to get practical insights on clients' investment behavior and risk assessment over the past 10 years.

In summary, while this books aims to answer the two questions above, it also offers a thorough introduction to the theory and practice of modern portfolio management. It touches on concepts which are central to almost everyone who starts to work or already works in the asset management industry. Therefore, this book serves as a handbook for professionals in this industry on all levels. We hope, it will guide the reader well!

References

Economist Intelligence Unit (2012). Managing investments in volatile markets. *State Street Global Advisors*. http://www.ssgainsight.com/reports/Economist_Managing_Tail_Risk_Report_pr.pdf. Accessed 8 Feb 2014.

Jordan, M., Rüther, M., & Winkelhaus, N. (2013, September). *Vermögensverwaltung heute*. Isarvest & Ketchum Pleon.

Nugée, J. (2010, December 22). Current issues in official sector asset management. *State Street Global Advisors*.

Schulmerich, M. (2007a, September 10). *Neues Werkzeug fürs Portfolio - Innovative Strategie ermöglicht überdurchschnittliche Renditechancen*. Handelsblatt.

Schulmerich, M. (2007b, September 16). *Die marktunabhängigen Rendite*. Euro am Sonntag.

Schulmerich, M. (2009a, August). *Emerging markets enhanced: Die attraktive Alternative für Investitionen in Schwellenländern*. DPN.

Schulmerich, M. (2009b, March). Equity market neutral and 130/30 strategies. *State Street Global Advisors*.

Schulmerich, M. (2010a, November 13). *Asymmetrische Risikomasse finden immer häufiger Anwendung*. Börsenzeitung. https://www.boersen-zeitung.de/index.php?li=1&artid=2010220345. Accessed 9 Feb 2014.

Schulmerich, M. (2010b, November). *Active success with SSgA's managed volatility strategies*. Citywire.

Schulmerich, M. (2011). How the recent crisis has changed EMEA investors' view of risk. *State Street Global Advisors. SSgA Capital Insights*.

Schulmerich, M. (2014a). How the recent crisis has changed EMEA investors' view of risk: An update for 2014. *State Street Global Advisors. SSgA Capital Insights*.

Schulmerich, M. (2014b). The role of hedge funds in investment portfolios. *State Street Global Advisors. SSgA Capital Insights*.

State Street (2009). Sovereign wealth funds: Emerging from the financial crisis. *State Street Vision, IV*(1)

About the Authors

Dr. Marcus Schulmerich, CFA, FRM, is global portfolio strategist for actively managed quantitative equity portfolios and hedge funds in Europe, Middle East, and Africa at State Street Global Advisors (SSgA), one of the largest institutional asset managers worldwide. Before joining SSgA in 2006 he was senior product manager at PIMCO for many years and senior risk manager at Commerzbank.

Dr. Schulmerich holds a Bachelor and Master of Mathematics from the University of Mainz, where he studied mathematics, physics, and business administration. He earned an MBA degree at the M.I.T. Sloan School of Management and a doctoral degree in financial engineering at the European Business School (EBS). In 2011 he founded Pecundus (www.pecundus.com), a company that provides training services to industry professionals.

Dr. Schulmerich is a frequent conference speaker on topics of asset and risk management, behavioral finance, financial engineering, and alternative investments. He is the author of various books and lecturer at the EBS.

Yves-Michel Leporcher obtained his Bachelor of International Trade and Foreign Affairs from Toulouse University and his Master of Banking and Financial Markets from Toulouse Business School. He worked at Société Générale as middle office analyst where he dealt with a wide spectrum of financial products and trading strategies. At Banque Populaire Caisse d'Epargne, he assisted in the implementation of a secure online payment system. He also worked for State Street Global Advisors as intern in product engineering, focusing on behavioral finance.

Currently, Mr. Leporcher works for the French investment bank Crédit Agricole CIB as a computer engineer in an informatics global support team. He identifies, monitors, and manages firm-wide operational risks and takes part in the continuous improvement of the reporting processes and the front end and back end reporting software.

Ching-Hwa Eu, Ph.D., is senior quant risk manager at Deutsche Bank within the market risk management team. His work focuses on model validation in the asset classes interest rates and foreign exchange rates and includes tests of the Deutsche Bank front office pricing library, especially for stress scenarios, the implementation of appropriate tests for the model of a product. The testing also involves the

M. Schulmerich et al., *Applied Asset and Risk Management*,
Management for Professionals, DOI 10.1007/978-3-642-55444-5

implementation and development of an independent pricing library which the model validators use as a benchmark for the front office pricing tool.

Ching-Hwa Eu holds a Master of Mathematics from the Technische Universität München (University of Technology, Munich), where he studied mathematics and physics, and earned his Ph.D. in Mathematics at the Massachusetts Institute of Technology (M.I.T). He has passed all three CFA exams within 3 years and has published extensively in various mathematical journals.

Index

M. Schulmerich et al., *Applied Asset and Risk Management*,
Management for Professionals, DOI 10.1007/978-3-642-55444-5

FSC
www.fsc.org
MIX
Papier | Fördert gute Waldnutzung
FSC® C083411

Kolibri 360 GmbH
Ferdinand-Jühlke-Straße 7
99095 Erfurt, Deutschland
produktsicherheit@kolibri360.de